Statistical Methods in Education and Psychology

ALBERT K. KURTZ
SAMUEL T. MAYO

Statistical Methods in Education and Psychology

SPRINGER-VERLAG New York Heidelberg Berlin

Albert K. Kurtz, Ph.D.
1810 Ivy Lane
Winter Park, Florida 32792, USA

Samuel T. Mayo, Ph.D.
Professor, Foundations of Education
School of Education
Loyola University of Chicago
Chicago, Illinois 60611, USA

AMS Subject Classifications: 62-01

Library of Congress Cataloging in Publication Data

Kurtz, Albert Kenneth, 1904-
Statistical methods in education and psychology.

Bibliography: p.
Includes index.
1. Educational statistics. 2. Educational tests and measurements. I. Mayo, Samuel Turberville, joint author. II. Title.
LB2846.K88 370'.1'82 78-15295

Printed in the United States of America.

9 8 7 6 5 4 3 2 1

ISBN 0-387-90265-1 Springer-Verlag New York
ISBN 3-540-90265-1 Springer-Verlag Berlin Heidelberg

Preface

This book is intended for use in the elementary statistics course in Education or in Psychology. While it is primarily designed for use in the first semester of a two-semester course, it may also be used in a one-semester course. There are not five or ten competing texts; the number is much closer to fifty or a hundred. Why, then, should we write still another one?

A new statistics text for use in Education and Psychology is, to some slight extent, comparable to a new translation or edition of the Bible. Most of it has been said before—but this time with a difference.

The present writers realize that elementary statistics students know very little about the subject—even the meaning of Σ is all Greek to them. This text covers the basic course in depth, with examples using real data from the real world. It, of course, contains the usual reference tables and several new ones; it gives the appropriate formulas every time; and it accurately depicts all graphs. It is so comprehensive that if instructors can't find their own special areas of interest covered, then those interests probably don't belong in a basic text.

This book uses simple sentences written in standard English. Except for one or two sentences in this Preface, no dictionary is needed for interpretation. The present writers have taken their task seriously but themselves lightly. The book is crammed with facts and suggestions but it moves along at a breezy clip. The writers would even have referred to ecdysiasts if they had thought this would have aided in understanding any figures in the text. How to, why to, when to, and (even more important) how *not* to, are all there; poor techniques are not allowed.

In this text, the authors never utilize obfuscation in the elucidation of the untenability of doctrinaire or hypothetical tenets. They simply say that the results might well be due to chance alone.

Most of what the writers regard as basic material is in the 15 chapters; much of the other material wanted by individual instructors is in the supplements. It is expected that most instructors will wish to assign material carefully selected from several supplements; it is unlikely that many instructors will wish to assign *all* of the supplements to their classes.

The writers have spent considerable time in locating theoretical, interpretative, mechanical, and computational errors and difficulties. Then we wrote or rewrote the text in order to reduce or eliminate these troubles. The basic goal, of course, was to help the student to decide upon appropriate statistical procedures and then to use them correctly. Some specific examples are:

(a) Instead of giving a rule and then violating it by using the same data for both biserial r and point biserial r, the principle differentiating the two is clearly enunciated and then some 38 variables are individually classified. Further, 9 others that are ambiguous or equivocal are discussed, showing, in each case, under what circumstances each of the two measures (biserial r or point biserial r) is appropriate.

(b) In order to let the student know more clearly why and how various statistics are to be used, two intentionally different examples are often given instead of the usual one.

(c) Because of the large number of computational errors in published research studies, special attention has been given to checking procedures throughout the book. A new and very efficient method for checking analysis of variance computations illustrates this change.

(d) Alternate formulas have sometimes been derived in order to get the usual results with less arithmetic. Formula (10.13) for the standard error of phi is one of them. With the two tables which accompany it, the computation is now so simple that the traditional formula no longer needs to be ignored because of computational complications.

(e) Many tables are included. A number of them have been modified to make them of maximal usefulness to the student. These tables include newly derived ones, standard ones, and several designed to illustrate relationships among various statistics.

(f) Where there are several uses for or interpretations of a statistic, an effort has usually been made to present them. The outstanding example of this is the correlation coefficient. While most texts give two or three, one text gives six interpretations of the meaning of Pearson r. In Chapters 8 and 9, we present fifteen!

We have also done several things designed to make life just a little easier both for the student and for the instructor.

(a) There are many cross-references and most of them are to specific pages rather than to topics or chapters.
(b) Formulas are written so they will be easy to use—not easy to set in type. Thus, we never enclose material in parentheses which is then raised to the one-half power; we simply use a square root sign. Another example is illustrated by the derivation and use of our new formula (4.2) instead of the usual formula (4.1) for the normal probability curve.
(c) This is minor, but we even introduced a new symbol, $\mathscr{y}$, for the ordinate of the normal probability curve.
(d) A Workbook has been prepared for any instructors who may care to use it.

It is trite for authors of elementary statistics texts to deny that previous mathematics is a prerequisite. The same claim is made for this text: most of the book can be understood by an average student who has had either one year of high school algebra or none, but, of course, much of the book will be more readily comprehensible to the student with a more extensive background in mathematics.

Despite the comprehensive coverage of elementary statistics referred to earlier in this Preface, the aim throughout has been to place primary emphasis on those principles, methods, and techniques which are, and of right ought to be, most useful in research studies in Education and Psychology.

The instructor will note a greater than usual emphasis on standard errors (or other tests of significance) and on large numbers of observations. This, briefly and bluntly, is because the writers are completely fed up with conclusions which start out with, "While not significant, the results are suggestive and indicate . . ." Let's get results that *are* significant and let's interpret them properly.

The writers were aided by many different persons in the preparation of this manuscript. We wish to acknowledge our indebtedness and express our gratitude to a few of the most important of them.

First, we owe a great deal to our statistics professors who taught us much of what we know.

Next, we are grateful to our students, our colleagues, and our competitors—from all of whom we have learned much about how to present our material and, on occasion, how not to present it.

Most of all, our greatest thanks are due to:

Elaine Weimer, a former editor and statistician, who added tremendously to the clarity of the text by revising or rewriting most, we hope, of the obscure portions of it.

Melba Mayo, who provided stimulation and encouragement to both the writers, assisting us in many ways in bringing the book to fruition.

Nathan Jaspen, who provided us with valuable suggestions for improvement of earlier drafts of the manuscript.

We are, of course, grateful to the many authors and publishers who have graciously permitted us to use their copyrighted materials. More specific acknowledgements are made in the text near the cited materials.

We shall end this preface with the hope that this text will provide you with "everything you always wanted to know and teach about statistics but were afraid to ask." What more can we say?

October, 1978

ALBERT K. KURTZ
SAMUEL T. MAYO

Contents

Statistical Methods in Education and Psychology

CHAPTER 1 THE NATURE OF STATISTICAL METHODS

GENERAL INTEREST IN NUMBERS

Numbers appear in newspapers every day. Corporations assemble large quantities of data for their own use. Federal and state governments collect and summarize thousands upon thousands of pages of data. Scientists record the results of their observations. Many other individuals gather information in numerical form. These figures are not ends in themselves, but when properly analyzed and interpreted, they form sound bases for important decisions, valid conclusions, and appropriate courses of action.

The interest in figures is widespread. Statistical methods are designed to aid us in handling and getting a better understanding of figures, especially those that seem most puzzling and bewildering at first glance. Not all published figures, however, result from the proper and intelligent use of statistical methods. At times, exactly the reverse seems true. Let us read a press release prepared by the Life Insurance Association of America and published by a number of newspapers:

> **New York**—New life insurance for March was 12.8 percent more than for March of last year, ac-

> cording to a statement forwarded today by the Life Insurance Association of America to the United States Department of Commerce. For the first three months of this year the total was 2.7 percent more than for the corresponding period last year.
>
> The statement aggregates the new paid-for business, not including revivals or increases, of 39 United States companies having 80 percent of the total life insurance outstanding in all United States legal reserve companies.
>
> The new business of the 39 companies was $892,667,000 for March, against $791,695,000 for March of last year, an increase of 12.8 percent. All classes contributed to the March increase. Ordinary insurance was $644,207,000 against $565,705,000, an increase of 13.9 percent. Industrial insurance was $145,258,000 against $137,811,000, an increase of 5.4 percent. Group insurance was $103,202,000 against $88,179,000, an increase of 17 percent.
>
> The new business of the 39 companies for the first three months of this year was $2,379,682,000 against $2,317,736,000 for the corresponding period last year, an increase of 2.7 percent. New ordinary insurance was $1,759,780,000 against $1,576,805,000, an increase of 11.6 percent. Industrial insurance was $392,112,000 against $400,010,000, a decrease of 2.0 percent. Group insurance was $227,790,000 against $340,921,000, a decrease of 33.2 percent.

Now that you have read that newspaper article in its entirety, do not reread it but look away from this book and do two things: (1) quote as many of the figures as you can, and (2) state the essential sense of the article.

That press release was not prepared for life insurance people but for the general public—a public composed mostly of people without statistical training and with considerably less mathematical skill than you have. The message certainly fails to get across. The reason, of course, is that the reader is deluged with figure after figure after boring figure. In fact, we find that the last two paragraphs give 26 separate and distinct figures in 28 lines—an average of almost one figure for every single line of text! This is not the way a statistician presents statistics.

THE PURPOSES OF STATISTICAL METHODS

Historically, the need for statistical methods arose because of the inability of the human mind to grasp the full significance of a large group of figures. Hence, the simpler statistical methods were designed primarily for the purpose of describing the relevant characteristics of a large amount of data by means of only a few statistics. More complicated statistics were sometimes employed, not just to give a picture of the data, but also to identify and determine the extent of relationships that would be impossible to grasp at all from a mere examination, however minute, of the basic data. In brief, statistical methods were designed to clarify and simplify the understanding of the aggregate of figures upon which they were based.

In recent years, the use of statistical methods has expanded far beyond this aim, however worthy it may have been. Instead of descriptive statistics, increasing emphasis is being placed on statistical inference. Thus, our main concern is not with the characteristics of the sample of data that we are analyzing, but rather with the characteristics of the larger population or universe—of which the sample is only a part.

PREVIEW OF THIS TEXT

Our purpose in this text is to present most of the commonly used and better statistical methods so that you—the student—may be in a better position (1) to understand the statistics presented in textbooks and in scientific journals in the fields of Psychology and Education, (2) to evaluate the appropriateness of the statistical methods used and the conclusions drawn from them, and (3), to a certain extent, to apply statistical methods to your own data, in connection with a thesis or other research study.

To achieve these objectives, this book first presents in detail the proper statistical methods for analyzing and describing a single set of scores (such as the intelligence test scores made by 176 sixth-grade students); then takes up the problem of determining the degree of relationship, if any, between two sets of scores (such as the correlation between intelligence test scores and scores on an achievement test in geography); and, finally, shows how to determine the extent to which our various statistical methods will give us dependable and consistent results in other similar circumstances.

Most of the chapters have their own supplements. The chapters present only the currently used, practical, and scientifically sound statistical methods. The supplements contain a heterogeneous assortment of material. Crude, inaccurate, and misleading methods (such as the range and the quartile deviation when used with most educational and psychological data) are included in a supplement because some people use them, and the student may encounter references to them in the educational or psychological literature. A few simple derivations of formulas are included in the chapters. Most derivations and proofs are either given in a supplement or omitted. Some topics (such as the correlation between correlation coefficients) are mentioned only in a supplement because they are either too difficult or unnecessary in an *elementary* statistics course. Certain suggestions regarding computational details (such as exactly how to round off numbers) are put in a supplement because, to a certain extent, these are matters of opinion, and various instructors may justifiably have different but highly acceptable rules for handling such situations. Finally, the supplements contain a few tables comparing various statistical measures designed for the same general purpose. Various instructors will wish to include *some,* but different, portions of this supplementary material in their courses. It is extremely unlikely that your instructor will wish to include *all* of it.

Thus, the supplements contain a considerable amount of reference material that may be useful to students and others both during and after a course in elementary statistics. The chapters, however, contain the fundamental, most important methods and the most generally useful material. A brief summary of the contents of each of the remaining chapters follows (all technical terms are explained in detail in the appropriate chapters).

Chapter 2, Averages. If we want to know what single score is most representative of a group of scores, we need to employ some sort of average. There are a great many averages. A common one is the arithmetic mean, which is simply the total of all the scores divided by the number of them. Because it is the best one for most purposes, it is presented first. Another good measure which is discussed is the median. If the scores are arranged from high to low, the median is the one in the middle. Most of the other averages have limitations that make them considerably less useful than these two and, accordingly, they are not included in Chapter 2, but are mentioned briefly in Supplement 2.

Chapter 3, The Standard Deviation. Even though we know the arithmetic mean, we usually do not have too good an idea of the variability of the

FIGURE 1.1
Normal Probability Curve

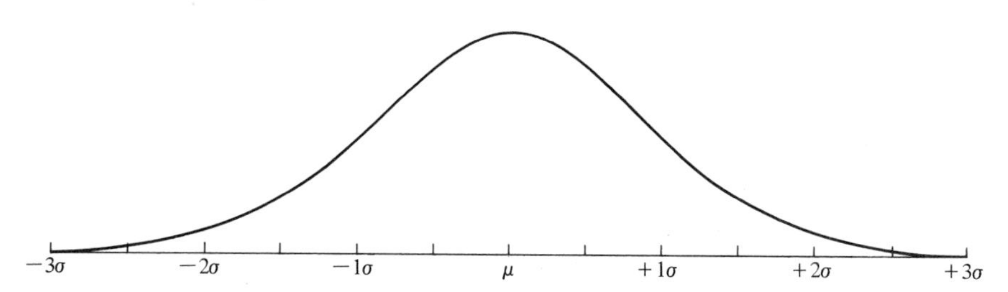

scores because they may be clustered closely about the average or they may vary widely from it. The best measure of such dispersion or variation is the standard deviation, to which this entire chapter is devoted. The next best measure of dispersion, the average deviation, is treated along with others in Supplement 3.

Chapter 4, Normal Probability Curve. The normal probability curve shown in Figure 1.1 is frequently referred to but not always accurately. Most educational and psychological measurements are distributed normally or nearly so. So are height, waist measurement, pulse rate, and errors in astronomical readings. The normal probability curve is widely used by statisticians and frequent reference is made to tables of it such as Tables D and E in the Appendix.

Chapter 5, Statistical Inference. In a research study, the group of people who are tested or measured is referred to as a sample. This sample is ordinarily drawn at random or otherwise from a much larger (maybe infinite) population. After statistics are computed from the sample, we often desire to estimate or predict corresponding values for the entire population. This process is called statistical inference. Thus, the mean or standard deviation of the larger population can be estimated or limits can be set up with a high probability (such as .99) that they will contain the corresponding mean or standard deviation of the population. Differences can be tested to determine the likelihood that they may be due to chance. Hypotheses can be tested. The types of errors that can be made in such analyses are discussed.

Chapter 6, Percentiles and Percentile Ranks. Sometimes one wishes to know the point below which a given proportion of the scores will lie. Percentiles supply the answer. For example, if 12% of a group make scores below 34, the twelfth percentile is 34.

The percentile rank of a point tells us what percentage of the cases fall below that point. These concepts find their greatest use in the interpretation of educational and psychological measurements. Two more measures of dispersion, based on percentiles, are included in Supplement 6.

Chapter 7, Skewness and Transformed Scores. Two sets of scores may have exactly the same arithmetic mean and the same standard deviation, but one set may be symmetrical, while the other set may have many of the scores concentrated a little below the mean with a few very high scores to counterbalance them. Measures of skewness tell us the direction and extent of such lopsidedness, or lack of symmetry, in the distribution of scores. Sometimes we may wish to transform such a lopsided distribution into a symmetrical one. Methods are given for doing this.

Chapter 8, Pearson Product Moment Coefficient of Correlation. Perhaps more conclusions of scientific importance are based upon the size of the coefficient of correlation than upon any other single statistic. This coefficient of correlation measures the extent of the relationship (which may be positive, zero, or negative) between two sets of scores or measures for the same individuals. It tells whether people with high scores on one test (such as arithmetic) tend to have high or low scores on another test (such as chemistry) and how close this relationship is.

Fifteen interpretations of the meaning of the coefficient of correlation are given at the end of this and the next chapter.

Chapter 9, Regression Equations. You have heard that one of the important purposes of science is to predict and control. This chapter deals with prediction. If there is a relationship between two sets of scores, then it is, of course, possible to predict from one set to the other with a certain degree of accuracy. The accuracy with which such predictions can be made depends on the size, but not the sign, of the coefficient of correlation just discussed. These predictions are made by using what is known as a regression equation. This chapter shows how to construct regression equations.

Chapter 10, More Measures of Correlation. Because the necessary figures are not always available in the proper form, it is sometimes impossible to compute the Pearson product moment coefficient of correlation. To meet this situation, various formulas have been derived to give approximations to it. Four of the best of these are explained in this chapter; three others are treated in Supplements 10 and 11.

Chapter 11, Chi Square. The measures of correlation studied in Chapters 8 and 10 indicate the degree of linear association between two variables.

However, sometimes we are not interested in the exact amount of relationship, but only in whether the variables are or are not associated in a manner consistent or inconsistent with chance (or with any theoretical expectation). Chi square provides an excellent method of determining whether our recorded figures are or are not consistent with such expectations.

Chapter 12, Nonparametric Statistics Other Than Chi Square. If our variables are measured somewhat crudely, such as by ranking, different statistical methods are often more appropriate than the more traditional ones. These other methods (called nonparametric) can be used regardless of the shape of the distribution of scores.

Chapter 13, Simple Analysis of Variance. When one wishes to compare several groups or different teaching methods in a single experiment, a procedure known as analysis of variance enables us to tell whether or not the differences that we find are of such size that they might easily have arisen solely on the basis of chance. Stated positively, we can tell whether or not the differences we obtain indicate that there are real differences among the groups or methods we are using.

Chapter 14, Standard Errors of Differences. If the same kind of statistics (such as two arithmetic means) are computed for two different groups of individuals, the two measures are apt to differ; only in rare instances will they *exactly* coincide. Are such differences real or are they just chance differences? Obviously, very small differences may well be due to chance, while very large ones almost certainly indicate real differences between the two groups. This chapter is concerned with determining such things as how large a difference must be, under any given set of circumstances, before we may conclude that it is probably a real difference and not just due to chance fluctuations.

Chapter 15, Reorientation. Because of this preview of the book and because most of the chapters end with a summary, this final chapter does not repeat what has just been presented. Instead, we disregard chapter boundaries. We show the interrelationships among nine formulas that had appeared in various parts of the text. We give brief descriptions of ten courses that might follow this one. We conclude with a discussion of ways in which the knowledge gained in this course might be assimilated and used effectively. You should then be able to decide whether you need or are interested in pursuing more advanced work in statistical methods.

Formulas and Tables Following Chapter 15, we give a comprehensive list of numbered formulas and a collection of the reference tables used in this course.

Computation and Interpretation No one has yet discovered an effective way to teach mathematics without having students solve problems. Unfortunately, that is also true to a certain extent in statistics, even though the primary aim of this book is not to develop computational skill but rather to help you to understand and interpret the statistics and statistical methods used in research studies.

ROUNDING, SIGNIFICANT FIGURES, AND DECIMALS

We have found that, unless explicit instructions are given, both students and salaried personnel will discard some digits and round the remaining figures in many different, and even fantastic, ways. In any given case, the proper procedure depends upon the characteristics of the data, the method of analysis, the use to be made of the material, and the personal preference of the investigator. However, the *precise* number of digits retained at any given stage of a statistical study is not always too important except in the interest of standardization. It is impossible to have one statistical clerk check another's work if they use different rules. It is very difficult for an instructor (or an assistant) to grade a set of 20 papers if 6 or 8 different sets of rules are used. Since there can actually be dozens of different ways to solve some statistical problems, the instructor who wants to remain calm and serene may well wish to limit this seemingly endless number of possibilities.

Hence, your instructor may give you a set of rules for determining when to drop what figures and how to round the remaining ones. There are several *acceptable* methods, but there are many more *wrong* ways of doing this. One good set of rules is given in Supplement 1. These rules are not the only good ones. They have some defects but they are easy to understand and easy to use. They give results that are practically always as accurate as the basic data justify, and they seldom involve any appreciable amount of extra work. So, some time before working out your first assignment, you should clearly understand and then follow the rules specified by your instructor—whether they are the same as or different from those in Supplement 1.

SUPPLEMENT 1

ROUNDING

Chapter 1 referred to the desirability of standardizing the procedures for rounding our computations. There the emphasis was on avoiding chaos in the grading of a set of papers in which the students used a large assortment of computing and rounding methods—some good, some fair, and, regrettably, some poor. Another consideration of somewhat greater importance is that of obtaining results of sufficient accuracy without an unduly large amount of computation time.

The Problem Let's start with our nation's bicentennial year, 1976. If we divide 19 by 76, we get .25. No problem, because the answer is .2500000000 with as many more zeros as anyone cares to write. But if we had started with the year our nation started, 1776, when we divided 17 by 76, we would get .2236842105263157894 7368. . . . Obviously, we don't want to keep all those figures, but how many should we keep?

A Method That Nearly Works People who know about some aspects of applied mathematics are apt to think of significant figures as the answer. Significant figures are all the nonzero digits of a number, plus the zeros that are included between them, plus final zeros that are used to signify accuracy. For example, 12,345, 123.45, 1.2345, 1.2300, .0012345, .00012000, and .000012345 are each correct to five significant figures. We don't know whether 1,234,500 has 5, 6, or 7 significant figures. As the song in *South Pacific* says, "you've got to be carefully taught." Therefore, if a number ends in several zeros before the decimal point, somebody has to tell you how many, if any, of the zeros to count. All intermediate zeros are counted. For example, 10305, 10.340, 10.300, 10.005, 1.0045, .0012005, and .000010000, each have 5 significant figures. Keeping a fixed number, say 3, of significant figures throughout a series of computations works fine as long as the processes involve multiplying, dividing, squaring, and extraction of roots, but it won't work when adding and subtracting are also involved. To illustrate this, let us take the numerator of a formula on page 242. It is

$$(107)(12{,}265) - (1{,}209)(1{,}044).$$

Straightforward arithmetic, with no rounding, gives

$$1{,}312{,}355 - 1{,}262{,}196 = 50{,}159.$$

Now let's perform the same computation, but let us round everything to 3 significant figures at each step. For clarification, we shall use small zeros wherever needed to fill out rounded numbers. (Thus, when 12,265 is rounded, it is printed as 12,300.) The preceding computation then becomes

$$(107)(12{,}265) - (1{,}209)(1{,}044)$$

which is rounded to (107)(12,300) − (1,210)(1,040) which equals 1,316,100 − 1,258,400. This is then rounded to

$$1{,}320{,}000 - 1{,}260{,}000 = 60{,}000.$$

Notice that although we kept 3 significant figures every step of the way, the answer contains only 1 significant figure—and even it is wrong! So much for significant figures in statistical computations. There must be a better way—and there is.

Of course, if we had kept 7 (or more) significant figures, we would have obtained the right answer. The trouble is that in most statistical computations in Education and Psychology we get more practical guidelines by using a fixed number of decimal places rather than a fixed number of significant figures in our computations.

A Practical Rule As mentioned at the end of Chapter 1, you may have been given some rules for rounding and keeping decimals. If so, use them. (Your instructor grades the papers, we don't.) In the absence of any such set of rules, here goes:

Always keep 2 decimals. This rule has several advantages:

1. It works nearly all the time with psychological and educational data.
2. It is easy to remember.
3. It is easy to use.
4. It rarely causes you to do much more arithmetic than is necessary or desirable.

There are also some disadvantages:

1. You may need more decimals in order to look something up in a table.
2. A table may give results which you should use to more than 2 decimals.
3. Sometimes you really need to use more than 2 decimals in your computations.
4. This rule may not work with data gathered in Astronomy, Economics, or other fields.

Let us consider these 4 disadvantages in more detail.

1. Suppose that you have data divided into 2 categories, and you need to know the proportion, p, in one category in order to look up some function of the normal probability curve. Then compute p to 3 decimal places, not 2, if the table allows for values of p to 3 decimal places.

2. Suppose you look something up in a table and the table gives an answer to 4 decimals. Simply use them. You won't find many functions given to 3, 5, 6, or more decimals, but if you do, use what is given. (The statistician who constructed the table may well have known how many decimals were needed.)

3. Sometimes 2 decimals are not enough. These situations will usually be obvious to you. For example, if your computations lead you to a fraction of the form $\frac{.04}{.02}$ it should be apparent that while the answer to 2 decimals *may* be 2.00, it probably is something else. In this case, the answer can be anything from a minimum of about $\frac{.03501}{.02499} = 1.40$ to a maximum of about $\frac{.04499}{.01501} = 3.00$. When you see a situation like this arising, the answer is obvious: go back and get more decimals. But don't worry much; such situations don't arise very often; that's why the 2-decimal rule is a practical one.

4. In connection with a study of the stock market, we could find that the 100 U.S. industrial corporations with the largest sales in 1975 had a median sales figure of $3,377,227,314.16. Do we need that figure to 2 decimal places? Hardly. We can't think of any circumstance in which giving it to the nearest million dollars would not be entirely adequate. But this doesn't matter. Such problems just don't arise with educational or psychological data. When you get to working with data outside your own field, different rules may apply. But we don't need to be concerned with them in this course.

To eliminate these objections, (1) keep enough decimals to be able to use all tables properly; (2) use all the decimals a table gives you; (3) use common sense in the rare situation in which 2 decimals aren't enough; and (4) if 2 decimals are too many, use your data to the nearest million or other appropriate unit. (These latter exceptions are so rare that you needn't unduly concern yourself with them.)

Details In applying the 2-decimal rule, except where tables are involved, keep 2 decimals after every operation; don't carry out computations to 3, 4, or some other number of decimals and then eventually get around to rounding. Simply do it every time.

If tables are involved, it may or may not be desirable to interpolate. If your instructor has stated a preference, follow it. Otherwise, don't interpolate; just take the nearest value the table will give you.

Round by discarding amounts of less than one half in the last place to be retained, by raising the last figure if the amount to be eliminated is more than one half, and by making the last retained digit even if the amount to be discarded is *exactly* one half. Thus, the figures in the left column, when rounded to two decimals, will become those in the right column:

12,345.6789	becomes	12,345.68
98,765.4321	becomes	98,765.43
299.9987	becomes	300.00
300.0049	becomes	300.00

$\frac{1}{8} = .12500\ldots0$	becomes	.12
$\frac{3}{8} = .37500\ldots0$	becomes	.38
$.38500\ldots0$	also becomes	.38

With one minor exception or option to be noted in Example (d) below, computations should be carried out in the order indicated, with multiplication and division coming before addition and subtraction unless otherwise indicated. For example:

(a)
$$5\left(\frac{98{,}765}{432}\right) = 5(228.62) = 1{,}143.10.$$

(b)
$$\begin{aligned} r_{\text{bis}} &= \frac{(135)(237) - (75)(237)}{(135)(.3950)\sqrt{(135)(1{,}507) - (237)^2}} \\ &= \frac{31{,}995 - 17{,}775}{(53.3250)\sqrt{203{,}445 - 56{,}169}} \\ &= \frac{14{,}220}{(53.3250)\sqrt{147{,}276}} \\ &= \frac{14{,}220}{(53.3250)(383.77)} \\ &= \frac{14{,}220}{20{,}464.54} \\ &= .69. \end{aligned}$$

In Example (b), .3950 was read from a table. Hence, it and a subsequent calculation involving it were carried to 4 instead of 2 decimals.

(c)
$$\begin{aligned} 25 + 5\left(\frac{345 - 67}{89}\right) &= 25 + 5\left(\frac{278}{89}\right) \\ &= 25 + 5(3.12) = 25 + 15.60 = 40.60. \end{aligned}$$

(d) Notice that in Example (a) we multiplied and divided one way, and got

$$5\left(\frac{98{,}765}{432}\right) = 5(228.62) = 1143.10$$

but if we had multiplied and divided in a different order, we would have obtained

$$5\left(\frac{98{,}765}{432}\right) = \frac{5(98{,}765)}{432} = \frac{493{,}825}{432} = 1143.11.$$

Similarly, in Example (c) we got

$$5\left(\frac{278}{89}\right) = 5(3.12) = 15.60$$

but we could have changed the order and obtained

$$5\left(\frac{278}{89}\right) = \frac{5(278)}{89} = \frac{1390}{89} = 15.62.$$

In all such cases, multiplying first and then dividing, even though the parentheses say not to, is permissible. This text often uses this alternate procedure; you may take your choice.

Whoops

The following was heard at the beginning of the Jet-Colt football game at the Super Bowl. A statistical comparison between the two teams appeared on the screen. Commentator Curt Gowdy said, "To give you a better idea of how the teams shape up against each other, we'll be throwing up statistics like these all during the game."

—Kermit Schafer: *Blunderful World of Bloopers*

CHAPTER 2 AVERAGES

RAW DATA

Frequently, a statistical study is to be based upon records that are given in a form such as:

Name	Test Score
Abigail Abbott	183
Benjamin Bennett	72
Carlotta Carroll	137
Dominick Dunlap	166
.	.
.	.
.	.

For most statistical purposes, the names are irrelevant and the records may be given in a more condensed form such as a single column of scores or a mere listing of them. As an illustration, Table 2.1 presents the Intelligence Quotients (IQs) of the sixth-grade students in the El Dorado School at Stockton, California as estimated from the California Short-Form Test of Mental Maturity.

TABLE 2.1
IQs of the El Dorado Sixth Grade

136, 119, 130, 109, 122, 96, 109, 116, 120, 148, 121, 107, 109, 122, 111, 111, 111, 128,
122, 96, 120, 133, 114, 90, 127, 92, 120, 101, 141, 107, 96, 116, 101, 121, 118, 97, 76,
101, 106, 123, 101, 113, 114, 97, 116, 92, 108, 122, 122, 105, 97, 88, 109, 110, 81, 98, 117,
99, 72, 67, 116, 123, 110, 109, 97, 110, 97, 106, 111, 124, 79, 98, 106, 126, 107, 114,
102, 114, 115, 129, 125, 126, 119, 111, 106, 130, 111, 126, 109, 107, 84, 116, 101, 134,
125, 102, 123, 124, 121, 130, 101, 97, 120, 135, 126, 112, 110, 98, 114, 111, 89, 122,
101, 103, 104, 111, 64, 104, 122, 116, 131, 79, 101, 105, 108, 95, 105, 128, 126, 136,
116, 123, 115, 116, 105, 111, 124, 118, 126, 120, 112, 90, 111, 116, 138, 76, 144, 126,
107, 110, 89, 95, 123, 101, 94, 92, 87, 123, 77, 112, 95, 112, 110, 106, 100, 101, 102, 75,
104, 120, 83, 106, 110, 106, 102, 127.

Such an aggregation of figures may convey some meaning to a few people, but to most people it is merely a mass of numbers. Even in their present form, these figures are not *completely* meaningless, however. It is evident that there are quite a few scores, thus indicating that the sixth grade in the El Dorado School is reasonably large. If we take the trouble to count them, we can easily find that the number of students tested is 176. Statistically stated,

$$N = 176.$$

THE MEAN COMPUTED FROM RAW DATA

Now look again at the scores in Table 2.1. Notice that the great majority of them are 3-digit numbers of 100 or more. Since these are IQs, and we know that the average IQ for the U.S. as a whole is supposed to be about 100, it is obvious, even before we do any computation, that this is a bright sixth grade. How bright? To answer this, we need a single number to represent the group. Let's face it: no one number can be a complete and accurate substitute for all 176 of the scores in Table 2.1, but we *can* get a single number that will be reasonably representative of the entire set. This will give us a good idea of the central tendency, or average, of the group of scores. How do we do this? Perhaps you already know the answer. All we do is add up all the scores and divide by the number of them. This gives us the *arithmetic* mean*. In statistical symbols, the formula is

$$\overline{X} = \frac{\Sigma X}{N}, \tag{2.1}$$

where the symbols have the following meanings: $\overline{X}$ is the symbol for the arithmetic mean; Σ is a summation sign that in statistical work is always to

---

*This word as an adjective is pronounced ăr'ĭth·mĕt'ĭk—not ȧ·rĭth'me·tik. The symbol we use for the arithmetic mean is $\overline{X}$. In other places, sometimes the symbol M is used as an alternate symbol for the mean either with or without a subscript.

be read as "the sum of" (If you happen to have heard Σ called anything else, forget it insofar as this class is concerned; the reason will be given in Chapter 3.); capital X is a symbol denoting a score (sometimes called gross score, raw score, crude score, or unadjusted score); and N, as we already know, is the number of cases. Thus, the formula says that the arithmetic mean ($\overline{X}$) is equal to the sum of the scores (ΣX) divided by the number of cases (N). It involves some routine work (and we will see later that there are easier methods of obtaining the arithmetic mean), but if we add up the scores and divide by N we obtain

$$\begin{aligned}\overline{X} &= \frac{\Sigma X}{N} \\ &= \frac{19{,}324}{176} \\ &= 109.80.\end{aligned}$$

This value of 109.80 confirms our previous observation that this was a bright sixth grade, but it also gives us a precise statement of just how far above the 100 average this sixth grade is, namely 9.80 IQ points. The arithmetic mean is not just a vague statistical abstraction. It is often a highly practical and meaningful figure that tells us at a glance how high or low the scores are. It provides us with a single number that represents the central tendency of the scores or, to repeat, it gives us just one figure which is representative of all N scores. Thus, it is a great aid in achieving the related objectives referred to in Chapter 1 (see p. 3) of describing one of the relevant characteristics of a large amount of data by means of a single figure (109.80) and of automatically clarifying and simplifying the understanding of the large number (176) of separate scores upon which it is based. We will see (in Chapter 5, p. 120) that this figure will also be valuable in making inferences concerning the mean of the larger universe or population of which our sample (of 176 cases) may be regarded as a part.

Now let us reexamine the computations that gave us this mean of 109.80. There is nothing complicated about obtaining the arithmetic mean this way (anyone who knows how to add, count, and divide can do it), but unless you have a calculator or computer available, the method is rather time consuming. Also, even one little error made any place in the long series of additions will, of course, result in a wrong answer.

Attempts to check the addition by performing the same operation a second time are unsatisfactory. First, the two totals may agree but both may be wrong because (a) the same error may be made both times or (b) compensating errors may be made. Second, the two totals may disagree, proving that at least one of the sums is wrong, although it does not follow that the other total is right. Third (yes, there *is* this possibility), both sums may agree and be right.

GROUPING OF DATA

Fortunately, there is another way to compute the arithmetic mean. It looks a bit more complicated than Formula (2.1), and it is not quite so accurate, but these are minor disadvantages when compared with the great saving of time and reduction of mistakes that result, especially when no calculators are available. It involves grouping the data into classes or *intervals* that can be handled more readily than can the raw scores.

Let us consider the number of intervals into which the data might be tabulated. Since the highest IQ in this sixth grade is 148 and the lowest is 64 (and since we have to allow for *both* of these extreme scores), the range of scores is 148 − 64 + 1, which equals 85. (When all the scores are whole numbers, the range is the same as the number of possible scores. Thus, it is one more than the difference between the highest and the lowest scores.) We *could,* then, set up a tabulation sheet with one row for each of the 85 possible scores. A portion of such a tabulation sheet is shown as Table 2.2. There are obviously too many intervals.

At the other extreme, we could use an interval width so large that all the cases would fall in one or two intervals. For instance, suppose we tabulate the data with an interval width of 50 IQ points, as shown in Table 2.3.

TABLE 2.2
Portion of Tabulation Sheet for Data from Table 2.1 with Interval Width of 1

IQ	Tallies
148	|
147	
146	
145	
144	|
143	
142	
141	|
140	
139	
138	|
137	
136	||
135	|
.	.
.	.
.	.

TABLE 2.3
Tabulation Sheet for Data from Table 2.1 with Interval Width of 50

IQ	Tallies
100–149	𝍸 III
50–99	𝍸 𝍸 𝍸 𝍸 𝍸 𝍸 𝍸 III

This grouping is obviously so coarse as to obscure most of the information provided by the original data. Hence, 85 intervals are too many; 2 intervals are too few. How many intervals should there be? The usual purpose of grouping the data is to save computational effort, but if we overdo it, as illustrated in Table 2.3, we will be unable to compute the arithmetic mean—or almost anything else.

It is clear, then, that the grouping must be neither so fine as to bring about no saving of time nor so coarse as to result in gross inaccuracy. Within these limitations, several factors must be considered in deciding upon the number of intervals and, consequently, the size of each.

First, we must consider how the records are to be handled—the purpose of the grouping. Some statistical operations, as we will see, require that we use the smallest possible interval width; while, at the other extreme, we sometimes must have only two intervals. These situations are unusual; for most statistical operations, such extremes are not desirable.

Second, so far as minimizing grouping errors is concerned, it is better to have too many intervals than too few. This point is discussed in connection with the standard deviation in Supplement 3.

Third, we must so choose the number of intervals so that the resulting interval width is a convenient number to work with. Thus, an interval width of 10 is much easier to use than one of 9, 13, or 12½.

Keeping all these factors in mind, we conclude: whenever possible, *the number of intervals should be between 15 and 20, and the interval width should be a number that is easy to work with.* Interval widths that are easy to work with are .1, .2, .5, 1, 2, 5, 10, 20, 50, 100, 200, 500, etc.; and sometimes 3 or 4. One further principle that will simplify the tabulating (and hence make it easier and more accurate) is: *the lower score limit* of the*

---

*The term score limit refers to the lowest or highest score in each interval. In the series 80–84, 85–89, 90–94, etc., the lower score limits are 80, 85, and 90; while the upper score limits are 84, 89, and 94.

interval must always be a multiple of the interval width. Thus, if the interval width is 5, the interval containing the score 72 is 70–74; and if the interval width is 2, the interval containing the score 72 is 72–73. If the scores are expressed as fractions, the rule still applies. For example, if the interval width is 5.00, the interval containing the score 72.33 is 70.00–74.99; and if the interval width is 2.00, the interval containing the score 72.38 is 72.00–73.99.

Applying the first rule to the Table 2.1 data, we find that the interval width should be 5. If we divide the range of 85 by 5, we get 17, which means that we will have about 18 intervals.* If we had tried to use an interval width of 2, we would have had too many intervals—either 43 or 44. If we had tried 3 (which is not quite so easy an interval width to work with), we would have gotten 29 or 30—still too many. Similarly, 4 would have given 22 or 23 intervals. We saw that 5 would give 17 or 18 intervals, which is fine. If we had tried 6, we would have had 15 or 16 intervals. That would be all right for the number of intervals, but 6 is an awkward number to work with. We reject 7, 8, and 9 for two reasons: awkward numbers and too few intervals. An interval width of 10 would be nice to work with, but it, along with all larger interval widths, would produce too few intervals for our tabulation. If we had been unable to get 15 to 20 intervals with an interval width that was easy to use, we would compromise and take whichever easy-to-use interval width would come closest to giving us 15 or 20 intervals. In most situations, a much easier way to determine the correct interval width is simply to look it up in Table A.

Using the correct interval width of 5, we must now arrange things so that the lower score limit of each interval will be a multiple of 5. This means that we must have intervals of 80–84, 85–89, 90–94, etc. rather than such harder-to-tabulate intervals as 81–85, 86–90, 91–95, etc. or the still worse 83–87, 88–92, 93–97, etc.

The Frequency Distribution Knowing the correct interval width and the correct lower score limit of each interval, we are now ready to prepare a *frequency distribution.* A frequency distribution consists of two or three columns, the first showing the various possible scores, the second and third showing how many of these scores there are. To make the first column, we simply list all the intervals we need from the highest (at the top) to the lowest (at the bottom, of course). Since our extreme scores were 148 and

---

*When we divide the range by a possible interval width, we sometimes get an exact whole number with no fraction remaining. In such cases, the number of intervals will be either the whole number or one more than that. Usually when we make that division we get a whole number plus a fraction. Then the number of intervals is always either one or two greater than the whole number.

64, our highest interval will be labeled 145–149 and the lowest will be 60–64. These, and all the intermediate intervals are shown in column (1) of Table 2.4.

Our next task is to make a tally mark opposite the appropriate interval for each of our N scores. Going back to Table 2.1, the first score is 136, so a tally mark is made in the space to the right of the numbers 135–139. The next tally mark, for an IQ of 119, is made to the right of the interval designated 115–119. When we come to the fifth score in any interval, we make a diagonal mark. If the diagonals are made from *upper right* to *lower left,* less confusion will later arise than if the first four marks are crossed from upper left to lower right. (In the latter case, a diagonal that is too long or too short may lead to the next diagonal being placed incorrectly.) When all 176 tally marks are made, the 176 scores should be retabulated and the two tabulations compared; if any errors are found, a third tabulation should be made very carefully.

When the tabulation is made and checked, it often facilitates later work if the number of tally marks in each interval is shown as in column (3).

TABLE 2.4
Computing the Mean, using a Frequency Distribution of IQs

			From middle		From bottom	
IQ *(1)*	Tallies *(2)*	f *(3)*	x' *(4)*	fx' *(5)*	x' *(6)*	fx' *(7)*
145–149	|	1	7	7	17	17
140–144	||	2	6	12	16	32
135–139	||||	4	5	20	15	60
130–134	卌 |	6	4	24	14	84
125–129	卌 卌 ||||	14	3	42	13	182
120–124	卌 卌 卌 卌 卌	25	2	50	12	300
115–119	卌 卌 卌 |	16	1	16	11	176
110–114	卌 卌 卌 卌 卌 ||	27	0	0	10	270
105–109	卌 卌 卌 卌 ||||	24	−1	−24	9	216
100–104	卌 卌 卌 ||||	19	−2	−38	8	152
95–99	卌 卌 卌 |	16	−3	−48	7	112
90–94	卌 |	6	−4	−24	6	36
85–89		4	−5	−20	5	20
80–84	|||	3	−6	−18	4	12
75–79	卌 |	6	−7	−42	3	18
70–74		1	−8	−8	2	2
65–69	|	1	−9	−9	1	1
60–64	|	1	−10	−10	0	0
Total		176		−70		1,690

Columns (1) and (3) are the only ones to be used in later work, and we shall refer to them, either with or without column (2), as constituting our frequency distribution.

We are now ready to compute the arithmetic mean as accurately as possible from Table 2.4 but with the smallest possible amount of work.

Let us consider some inappropriate ways first. It would be possible, for instance, to insert another column (147, 142, 137, 132, 127, . . .) showing the midpoints of each of the IQ intervals, to add each of these numbers the number of times indicated by the tallies or by the f column (thus adding one 147, two 142s, four 137s, six 132s, fourteen 127s, . . .), and to divide the total so obtained (19,362) by 176 to obtain the arithmetic mean (110.01). This method is as accurate as any other based on the tabulated data, but it is not much easier than just adding the scores in Table 2.1 and using Formula (2.1). There must be *some* advantage in tabulating the data!

Two things can be done to cut down on the amount of arithmetic in the operations just described. To reduce the size of the numbers to be worked with, we can first subtract some large number (such as 62 or 112 or some other number ending in 2 or 7) from each of the midpoints. Of course, after we obtain the mean of these residuals, we have to add the 62 or 112 to it to get the mean of the IQs. Second, since these residuals are all multiples of the interval width (85, 80, 75, etc. or 35, 30, 25, etc.), we can divide them by the interval width, 5. Again, we have to correct for this later by multiplying by the interval width. The result of these two operations (subtracting a constant and dividing by the interval width) gives us the more manageable numbers in column (4) or column (6). It should be apparent that we can get the same result just by putting a zero at the bottom or somewhere near the middle of the distribution and counting up and down from it. The longer explanation, though, shows why we are doing this and ties in with Formula (2.2).

Before further considering the nature of the formula, let us fill in the other columns of the table. We begin by placing a zero in column (4) opposite 110–114. This means that our starting place, or *arbitrary origin,* is in the middle of that interval. We then run the positive numbers up and the negative numbers down from this arbitrary origin. (If we always tabulate with the high scores near the top of the page, as we should, the high x' values will, quite logically, be high on the page also.)

Since column (3) is f, column (4) is x', and column (5) is fx', it is not too difficult to see that if we multiply each entry in column (3) by the corresponding entry in column (4), we will get the product, fx', which goes in column (5). We do this and then check our multiplication. Then we add these products. Adding the positive numbers, we get +171. Adding the negative ones, we get −241. Adding *all* these products, we get +171 +(−241) = −70. We check the addition and then record the correct total at the bottom of column (5).

THE MEAN COMPUTED FROM GROUPED DATA

Now let us see how we can take Formula (2.1) and derive the formula to use when we group our data. Obviously, we have to add, but instead of adding the X scores and getting ΣX, it might appear that we should add all N of the x' scores and get $\Sigma x'$. This would do, but the fact that the scores are grouped in intervals gives us an easier method. Instead of adding 5 + 5 + 5 + 5 for the 4 cases with x' values of 5, we simply multiply 4 by 5 and record the fx' product of 20 in the fx' column. When we add this column, it is reasonable to regard it as $\Sigma fx'$. The f is usually included in grouped data formulas but it may be omitted, provided you realize that all of the x' values are to be included as many times as they occur, and that you are not simply adding the 18 numbers in column (4).

After getting $\Sigma fx'$, we divide it by N, but, for two reasons, this does not give us the mean. First, this value, $\frac{\Sigma fx'}{N}$, was expressed in units of the interval width, which we call i; and, second, the deviations were measured from an arbitrary origin, which we call M'. Since we earlier divided by the interval width, i, we must now multiply by it. Since we earlier subtracted the arbitrary origin, M', we must now add it. Taking account of these two factors, we obtain the most useful formula for computing the arithmetic mean from grouped data:

$$\overline{X} = i\left(\frac{\Sigma fx'}{N}\right) + M'. \tag{2.2}$$

Notice that this formula is entirely general. It will give exactly the same value for $\overline{X}$ regardless of whether M' is the midpoint of an interval near the middle of the distribution, as it is in column (4), or the midpoint of the interval at the bottom of the distribution, as it is in column (6), or anywhere else. Let us compute the mean in both of these ways, first using the figures we have just worked with in columns (4) and (5).

To get a value for M', we need to know the midpoint of the interval that contains it. In this example, M' is in the IQ interval 110–114. The midpoint of any interval is halfway between the highest and the lowest scores that could possibly be tabulated in the interval. In this case, halfway between 110 and 114 is 112.00. A harder way, especially appealing to mathematicians, to find the midpoint of an interval, is to use the *exact limits*. These exact limits are always halfway between the largest score that can be tabulated in one interval and the smallest score that can be tabulated in the next higher interval. Since 110 is the smallest score in this interval, and 109 is the highest score in the next lower interval, the exact limit between these intervals is neither 109 nor 110, but 109.50. Similarly, the highest possible IQ in the 110–114 interval is 114 (since the tabulated IQs were all expressed in whole numbers, not fractions—see Table 2.1), and the smallest IQ that could be tabulated in the next higher interval is 115. Therefore, the exact upper limit of this interval (which is also the exact lower limit of the next higher interval) is 114.50. Now that we know the exact upper and lower limits of the interval, the midpoint, of course, is halfway between them. Hence, M' is $\frac{109.50 + 114.50}{2}$, or 112.00 as before.

We now have all the figures needed for Formula (2.2). Specifically, they are $i = 5$, $\Sigma fx' = -70$, $N = 176$, and $M' = 112.00$. When we substitute these in the formula, we obtain

$$\begin{aligned}\overline{X} &= i\left(\frac{\Sigma fx'}{N}\right) + M' \\ &= 5\left(\frac{-70}{176}\right) + 112.00 \\ &= -1.99 + 112.00 \\ &= 110.01.\end{aligned}$$

This value agrees rather closely with the 109.80 obtained earlier using Formula (2.1). The difference of .21 is considerably less than one IQ point. Supplement 2 (p. 42) shows that the difference between the results obtained from the two formulas can never be as great as half the interval width and it is nearly always very much less than that. Here, .21 is very much less than 2.50.

In the method just used, the arbitrary origin was fairly close to the arithmetic mean. However, the location of the arbitrary origin is just that—arbitrary. Let us see what happens if we place it at the midpoint of the lowest interval. Then we obtain the figures in columns (6) and (7) instead of those in columns (4) and (5). The arbitrary origin is now at the midpoint of the 60–64 interval, and its value is 62.00. Now we place a zero in column (6) opposite 60–64. Next, we fill in the other numbers in column (6); they are 0, 1, 2, 3, . . ., 17. As before, we multiply the f values in column (3) by the x' values, now in column (6), to obtain the fx' values in column (7). Adding column (7), we obtain $\Sigma fx' = 1{,}690$. Substituting in Formula (2.2) we obtain

$$\begin{aligned}\overline{X} &= i\left(\frac{\Sigma fx'}{N}\right) + M' \\ &= 5\left(\frac{1{,}690}{176}\right) + 62.00 \\ &= 48.01 + 62.00 \\ &= 110.01.\end{aligned}$$

This value is *exactly* the same as the one we obtained when we assumed M' was 112.00 instead of 62.00. This is not a coincidence. If we carry the computations to 30 decimal places, the two values will still be exactly the same. In other words, insofar as accuracy is concerned, one arbitrary origin is as good as another. The choice of the particular arbitrary origin to be used must be made on the basis of convenience, ease of computation, or habit rather than accuracy. (We used the alternate computing procedure described in Example (d) on page 12. If the other method had been used in both of these last two computations, the two results would still agree perfectly.)

Most statisticians place the arbitrary origin at one of two places—near the middle of the distribution or at the lowest interval. Each has advantages and disadvantages. Table 2.4 shows that when deviations are measured from the middle of the distribution, not only the deviations (x'), but also their products (fx'), are relatively small. On the other hand, when deviations are measured from the bottom, there are no negative numbers. If the computations are being made by hand, the amount of arithmetic is a little less, but errors due to signs are more likely if deviations are taken from the middle. If a calculator is available, however, the decision will nearly always be in favor of putting M' at the bottom of the distribution. (If your calculator has a memory, none of the fx' values need to be recorded; they are cumulated in the machine and only $\Sigma fx'$ is written down.) Unless your instructor prefers some other method, we will, for the sake of consistency, place M' at the midpoint of the interval having the highest frequency. If two intervals have the same highest frequency, either use the lower of these two intervals or take the one nearer the middle of the distribution. Note, however, that these conventions in no way affect the numerical value of the arithmetic mean.

Quite apart from its use in computing the arithmetic mean or other statistics, the frequency distribution gives an excellent picture of the plotted scores. This will become apparent near the end of this chapter in the section on the histogram and the frequency polygon (p. 33).

A Second Example of Computing the Mean The next illustration of the computation of the arithmetic mean is based upon highway speed data. With the cooperation of the Florida Highway Patrol, one of the writers sat in a car with a Radar Speed Meter and read the speeds of northbound and southbound cars from 3:30 to 5:23 P.M. (sunset) on a sunny spring afternoon at the side of a well-paved and fairly well-used state highway. Omitting trucks and cars whose speed might be unduly influenced by cars immediately ahead of them, speeds were obtained for 137 northbound and for 119 southbound cars. The speeds of the latter are shown in Table 2.5.

TABLE 2.5
Daytime Speeds of Southbound Cars

37, 34, 34, 40, 53, 51, 44, 43, 38, 50, 48, 35, 37, 44, 49, 41, 37, 50, 46, 35, 58, 55, 32, 51, 30, 50, 36, 40, 46, 48, 55, 53, 25, 43, 52, 50, 43, 52, 54, 50, 50, 50, 50, 53, 46, 46, 43, 53, 54, 53, 49, 48, 41, 60, 47, 43, 48, 49, 40, 48, 44, 57, 42, 38, 46, 35, 41, 44, 43, 52, 54, 52, 39, 46, 43, 44, 42, 53, 54, 50, 40, 48, 52, 59, 51, 56, 48, 39, 46, 39, 45, 39, 45, 43, 51, 38, 41, 51, 53, 53, 50, 56, 52, 58, 41, 51, 43, 62, 51, 50, 50, 46, 40, 53, 46, 57, 49, 52, 52.

Before tabulating, we need an interval width. The highest speed is 62 (3 miles per hour *under* what was then the speed limit!) and the lowest is 25. Therefore, the range is $62 - 25 + 1 = 38$. Using Table A or the two rules given on page 18, it is clear that $i = 2$ and that the lowest interval will be 24–25. This will give between 15 and 20 intervals; and the lower limit of each of them will be a multiple of 2. We now prepare Table 2.6, tabulate the data, and compute the mean.

TABLE 2.6
Computing the Mean, Using a Frequency Distribution of Car Speeds

Speed *(1)*	Tallies *(2)*	f *(3)*	x' *(4)*	fx' (5)
62–63	|	1	6	6
60–61	|	1	5	5
58–59	|||	3	4	12
56–57	||||	4	3	12
54–55	𝍸 |	6	2	12
52–53	𝍸 𝍸 𝍸 ||	17	1	17
50–51	𝍸 𝍸 𝍸 ||||	19	0	0
48–49	𝍸 𝍸 |	11	−1	−11
46–47	𝍸 𝍸	10	−2	−20
44–45	𝍸 ||	7	−3	−21
42–43	𝍸 𝍸 |	11	−4	−44
40–41	𝍸 𝍸	10	−5	−50
38–39	𝍸 ||	7	−6	−42
36–37	||||	4	−7	−28
34–35	𝍸	5	−8	−40
32–33	|	1	−9	−9
30–31	|	1	−10	−10
28–29			−11	0
26–27			−12	0
24–25	|	1	−13	−13
	Total	119		−224

The process is strictly analogous to that followed in Table 2.4 except that M' is taken from only one place (near the middle). Now let us briefly review the entire procedure:

Having determined the extreme scores, 62 and 25, and the interval width, 2, we list the possible speeds for tabulation in column (1). Going back to our basic data (Table 2.5), we tabulate the data, entering the tally marks in column (2). We check this by retabulating on another sheet of paper until we are right. We enter the corresponding frequencies in column (3) and check our work. We select the interval with the highest f and place a zero opposite it in column (4). We then fill in the other x' values. Next, we multiply each value in column (3) by the corresponding value in column (4) and record the product in column (5). Algebraically, this is simply $(f)(x') = fx'$. (In Chapter 3, p. 57, we shall learn how to use the Charlier check to check this and the next operations. Doing it over is a poor check; if the Johnny who can't read "knows" that 7 times 6 is 56, he will probably get the same wrong answer when he attempts to check his work.) We now add columns (3) and (5), checking our work by methods to be taught later, and obtain $\Sigma f = N = 119$ and $\Sigma fx' = +64 + (-288) = -224$. Noting that M' is halfway between 50 and 51, and remembering that $i = 2$, we substitute in Formula (2.2) and obtain

$$\begin{aligned}\overline{X} &= i\left(\frac{\Sigma fx'}{N}\right) + M' \\ &= 2\left(\frac{-224}{119}\right) + 50.50 \\ &= -3.76 + 50.50 \\ &= 46.74.\end{aligned}$$

A Third Example of Computing the Mean We now introduce an entirely new set of data based on what is undoubtedly the most universally acclaimed book in the field of mental measurement. Suppose we are interested in the degree to which scientific contributions in this field are made by many different authors or are made primarily by a relatively small number of prolific authors. To shed some light on this question, we looked in the Index of Names of one of Buros's Yearbooks, and we counted the number of references or citations for each person (omitting two organizations) whose last name began with Sh, which letters were selected to produce a suitable N. These results are given in Table 2.7.

If we were to follow our usual procedure for grouping data, we would find that the range is $14 - 1 + 1 = 14$, giving $i = 1$. We would then tabulate the numbers and compute $\overline{X}$. There's nothing inaccurate about this, but it's inefficient. Isn't it easier just to add the numbers? So, instead of making a

TABLE 2.7
Number of Citations to Authors

2, 1, 6, 2, 2, 2, 1, 1, 1, 1, 1, 1, 2, 1, 1, 1, 3, 1, 1, 6, 1, 1, 1, 1, 3, 1, 1, 1, 1, 3, 2, 1, 1, 1, 1,
1, 1, 1, 3, 6, 1, 9, 4, 1, 1, 1, 7, 1, 1, 1, 2, 1, 1, 1, 4, 1, 1, 2, 1, 2, 1, 1, 1, 1, 2, 1, 1, 1, 1, 3,
1, 1, 1, 2, 1, 3, 1, 1, 1, 14, 5, 2, 1, 2, 2, 1, 1, 1, 2, 1, 3, 2, 1, 2, 2, 1, 2, 1, 1, 2, 1, 2, 3, 2,
1, 1, 1, 1, 1, 2, 1, 1, 3, 1, 1, 1, 2, 1, 1, 6, 2, 1, 7, 1, 1, 1, 1, 1, 4, 2, 1, 3, 2, 1, 2, 1, 1, 1, 1,
2, 2, 1, 2, 2, 1, 1, 1, 3, 2, 1, 1, 1, 1, 3, 2, 1, 1, 1, 1, 1, 1,

Reference: Data taken from Buros, Oscar K., Editor. *The Seventh Mental Measurements Yearbook*. Highland Park, N.J.: Gryphon Press, 1972.

table similar to Tables 2.4 and 2.6, we simply add the scores, use Formula (2.1), and obtain

$$\overline{X} = \frac{\Sigma X}{N} = \frac{285}{161} = 1.77.$$

Because 103 of these are the lowest possible score, 1, the mean is bound to be low, but the few scores running up to 14 do a good job of pulling it up, and we find that the average number of citations is 1.77 per author.

Notice that both Formulas (2.1) and (2.2) are perfectly good for computing the mean. The trivial rounding errors in (2.2) are ordinarily of no consequence, so the decision between them should be based on ease of computation, convenience, previous familiarity, and personal preference. Usually, the choice is Formula (2.1) if the scores are all small positive integers or if you have a calculator. If you don't have a calculator, or even if you do, the choice may well be Formula (2.2) if the scores have any one or more of these characteristics; large scores, wide range, decimal scores, fractional scores, or some positive and some negative scores. Other considerations may also affect the choice. For example, if a graph is desired, or if you wish to see what the distribution of figures looks like, the preference is apt to be for Formula (2.2). Or if you are going to perform other calculations based on the same data, this will affect your decision.

Thus, we have two alternative formulas for the arithmetic mean, either of which may be used to compute a central figure or average which is representative of our particular sample of scores. Whether or not this arithmetic mean of the sample is also representative of some larger group will be discussed in Chapter 5.

THE MEDIAN

There are many different ways of obtaining a single number (or statistic) which can be regarded as representative of an entire set of scores. Ordinarily, the arithmetic mean is used because it gives more consistent results. To illustrate, if we gathered IQ data in many sixth-grade classes, the means of the IQs of the different classes would cluster more closely together and thus be more consistent with each other than would be the case if we used the corresponding medians. Sometimes, however, we want a different measure which may be more appropriate for some special situation. The only other measure of central tendency that is worthy of much consideration in an elementary statistics course is the *median*.

When the scores are arranged in order of size, the middle score is the median. Half of the scores are larger than the median and half of the scores are smaller than the median. If the number of scores is odd, the median will have the same value as the middle score. If the number of scores is even, the median is halfway between the two middle scores. That's all there is to computing the median if the scores are already arranged in order of size or can easily be so arranged. If desired, this can be stated in the form of a formula:

$$Mdn = \left(\frac{N+1}{2}\right) \text{th measure in order of size.} \tag{2.3}$$

Suppose, however, that we want the median of a set of tabulated data. There are usually two ways in which the median may be obtained. One is by reclassifying the data and using Formula (2.3). The other is by interpolation. The results obtained by these two procedures usually differ slightly, but the difference is unimportant and neither method has any advantage or disadvantage on the basis of accuracy. Let us take up these two methods in order.

If we already have a frequency distribution and we want to use Formula (2.3) to obtain the median, we do not need to arrange *all N* cases in order of size. For instance, look at Table 2.4 (p. 20). By simply adding up the frequencies in column (3), we see that 81 of them are included in the intervals with scores of 60–109, that 27 more are included in the 110–114 interval, and that 68 are included in the intervals with scores of 115–149. Since we want to find the $\left(\frac{176+1}{2}\right)$th or the 88½th case, it is clear that the median is somewhere in the 110–114 interval. But where? We can now go back to Table 2.1 and retabulate the 27 cases in this interval. The results are as follows:

IQ	Tallies
114	卌
113	I
112	IIII
111	卌 卌
110	卌 II

Since 81 cases are below this interval, it is clear that the 82nd through the 88th cases have scores of 110 and that the 89th through the 98th cases have scores of 111. The median, or the 88½th case, will therefore be 110.50.

Of course, we cannot use this method if we are working with someone else's grouped data and do not have the original scores, such as were given in Table 2.1. Also, although it's perfectly simple and straightforward, it often involves more work and takes more time than does Formula (2.4), which has such a formidable appearance that even we have never bothered to memorize it.

Let us use the same data of Table 2.4 and compute the median by the following formula:

$$\text{Mdn} = l + i\left[\frac{\left(\frac{N}{2}\right) - F}{f_p}\right]. \tag{2.4}$$

In this formula, l represents the exact lower limit of the interval in which the median lies, i is the interval width, N is the total number of cases, F is the sum of the frequencies of all intervals below the one containing the median, and f_p is the frequency of the interval in which the median lies.

We will now obtain numerical values for all these symbols and substitute in the formula. But before we can do this, we will have to find out in what interval the median lies. As in the previous method, we can easily find that the median is in the 110–114 interval. The first symbol, l, offers the greatest difficulty. The original scores, the IQs, were given as integers. The lowest score that could possibly occur and be tabulated in this interval is 110. The highest score that could be tabulated in the next lower interval is 109. Halfway between these two scores, 109.50, is the exact limit or the true class boundary between these two intervals. It is the exact upper limit of the 105–109 interval and the exact lower limit of the 110–114 interval. Hence, l is 109.50. The value of i is easily seen to be 5 and N is 176. In computing the median for this same distribution by the use of Formula (2.3), we have already found that 81 cases lie below this interval. Hence F is 81. Finally, f_p is 27 since there are 27 cases in this interval.

Substituting in Formula (2.4), we obtain:

$$\text{Mdn} = l + i\left[\frac{\left(\frac{N}{2}\right) - F}{f_p}\right]$$

$$= 109.50 + 5\left[\frac{\left(\frac{176}{2}\right) - 81}{27}\right]$$
$$= 109.50 + 5\left(\frac{7}{27}\right)$$
$$= 109.50 + 1.30$$
$$= 110.80.$$

Notice that the median obtained from the original (ungrouped) data by Formula (2.3) is 110.50, and the median obtained from the grouped data by Formula (2.4) is 110.80. Even though the distribution of scores within the 110–114 interval was markedly lopsided (see page 28), the difference between the two values for the median is negligible. We might add parenthetically that even this small difference is larger than is usually obtained in such circumstances.

We will now briefly illustrate the use of Formulas (2.3) and (2.4) to compute the medians for our other two sets of scores.

For southbound car speeds, we see that 57 of the 119 cars had speeds of 24 to 47 miles per hour and that 11 were in the 48–49 interval. Substituting the appropriate values in Formula (2.4), we obtain

$$\text{Mdn} = l + i\left[\frac{\left(\frac{N}{2}\right) - F}{f_p}\right]$$
$$= 47.50 + 2\left[\frac{\left(\frac{119}{2}\right) - 57}{11}\right]$$
$$= 47.50 + 2\left(\frac{2.50}{11}\right)$$
$$= 47.50 + .45$$
$$= 47.95.$$

This is 47.95 rather than 47.96 because we used the computing method described in Example (d) on page 12. Because the distribution of car speeds was nearly symmetrical, the mean and median are close together, differing by only 1.21 miles per hour.

The computation of the median of the author citations is easy. The lowest score is 1 and more than half the scores are 1. Hence, by Formula (2.3),

$$\text{Mdn} = \left(\frac{N + 1}{2}\right) \text{th measure in order of size}$$
$$= 1.00.$$

Notice that in all three cases it would have been *possible* to use either Formula (2.3) or (2.4), since we had the original scores for all three samples and we either had or could get a tabulation of the scores. We used different formulas, not because one formula was right and the other wrong, but simply because one was more appropriate in the particular situation.

On page 28 we demonstrated that if we already have a frequency distribution and if for some reason we strongly prefer to use Formula (2.3), this can be done by retabulating some of the data. Under all ordinary situations, this is not worthwhile. If the scores are tabulated, use the tabulation and compute the median by Formula (2.4). This formula *looks* bad, but it is really very easy to use. Hence, primarily from the standpoint of saving unnecessary work, we ought to use Formula (2.4) with the IQ and with the speed data, while (2.3) is appropriate for the citations simply because we never tabulated those records. Remember that either formula is permissible but use the one that is appropriate for your data—(2.3) if the records are listed in rank order, (2.4) if they are tabulated.

CHOICE OF MEAN OR MEDIAN

With certain types of data, it is impossible or inappropriate to compute the mean, yet it may be proper to compute the median. Assume that we decide to gather data on food preferences. We list 6 meat dishes and 6 vegetables. We then ask people how many of each they like, and we compute a meat-to-vegetable ratio by dividing the number of meat dishes liked by the number of vegetables liked. What would the mean of these ratios be? About .50? 1.00? 2? 20? No, it would probably be larger than any of those! If even one person likes some of the meat dishes, but can't stand any of the six vegetables, that meat/vegetable ratio will be some number divided by zero. This, of course, equals infinity. When this is averaged with all the others, no matter what they are, the mean is still infinity. To say $\overline{X} = \infty$ is not very meaningful. It would be much better to compute the median. (It might be still better to redesign the experiment using a difference score, but that's another matter.)

A situation faintly resembling this is afforded by our author citation data. Either the mean or the median can be computed easily, but the two differ very markedly; the mean being 2.43 while the median is only 1.00. In any lopsided distribution with the scores piling up at one end of the scale, the mean and median will differ. In fact, this very difference is sometimes used as a part of some formulas for measuring the extent of this lopsidedness (see the section on skewness in Chapter 7). Consequently, the question is not which is better in some general sense, but which is better for some specific purpose. For instance, if most people put from 10¢ to 50¢ in the collection plate in a small church, but one pious man always contributes $20.00, the median may be 25¢ while the mean is $1.00. If we take the worshipper's point of view, a reasonable contribution is 25¢, but if we look at the same collection data from the standpoint of the minister or the budget committee, the per capita contribution is clearly $1.00. Thus, both the mean and median are correct. They give different answers because they are different measures and are designed for different purposes.

Instances such as the meat/vegetable ratio, the author citation data, and the imaginary church collection plate deserve at least some attention because they either do or might exist. Distributions such as these may even be common in some fields of learning. But they are not common in most branches of Education or Psychology. It is desirable that we should know how to handle such situations when they do occur, but we should devote most of our attention to those statistical methods which are most appropriate for handling the kinds of data we encounter in typical rather than in unusual situations.

All this leads us to one very obvious conclusion: *Ordinarily, the arithmetic mean is the best measure of central tendency.* The few exceptions to this statement pertain to distributions departing markedly from the usual shape, and they do not usually concern us. For most psychological and educational data, we can ignore these exceptions—the conclusion holds. Why, then, do people persist in using the median and other measures of central tendency? There are quite a few reasons:

1. Sometimes people do it because of a lack of information regarding the relative merits of the two.
2. Sometimes people make errors in ordering their priorities; they devote many days or even months to gathering their data and then they save a few minutes of computational time in order to meet a deadline.
3. Sometimes people rework data published by someone else and the records may not be complete.
4. Sometimes it is impossible or ridiculous to compute the arithmetic mean (as when some of the scores are infinite or indeterminate).
5. Sometimes some other measure of central tendency forms a logical part of a series (as of percentiles) or is needed for use in some other formula (as for skewness).
6. Sometimes the interpretation or intended use of the measure of central tendency dictates that some other statistic be used.
7. With some nonparametric statistics (see Chapter 12), the median is used, especially for determining the point at which the data should be divided into two groups in order to prepare for further analyses.

Despite this long list of reasons, the number of situations in which the mean should not be used (items 4 to 7 in the preceding list) is very small. However, many other measures have been used, both correctly and incorrectly. The median is the most common and probably the best of these other measures. Several more of them are discussed in Supplement 2 at the end of this chapter.

THE HISTOGRAM AND THE FREQUENCY POLYGON

Before concluding this chapter, we should point out one very important use of the frequency distribution: it provides the basic data for drawing a graph to represent its distribution. Graphic methods often enhance the effectiveness of a presentation of data. Numbers alone, even in a frequency distribution, often fail to give an accurate picture of the characteristics of the data. A frequency polygon or a histogram will often show the relationship clearly. With the very simple frequency distribution shown in Table 2.8, the scores may be depicted in either of the two common ways shown in Figures 2.9 and 2.10. In the histogram (Figure 2.9), the area above each score indicates the frequency. Note that the tops of the rectangles run from the exact lower limit to the exact upper limit of each interval. In the frequency polygon (Figure 2.10), the frequencies are first plotted above each test score, also plotting one or more zero frequencies at each end of the distribution. Then the points are connected.

TABLE 2.8
Hypothetical Test Scores

X	f
9	10
8	0
7	20
6	30
5	10

FIGURE 2.9
Histogram for Data from Table 2.8

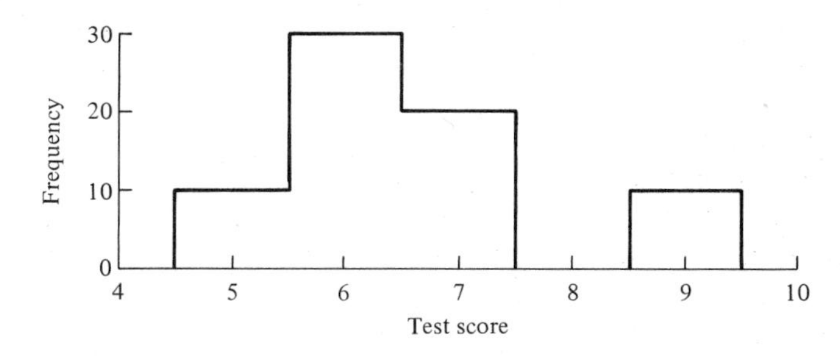

FIGURE 2.10
Frequency Polygon for Data from Table 2.8

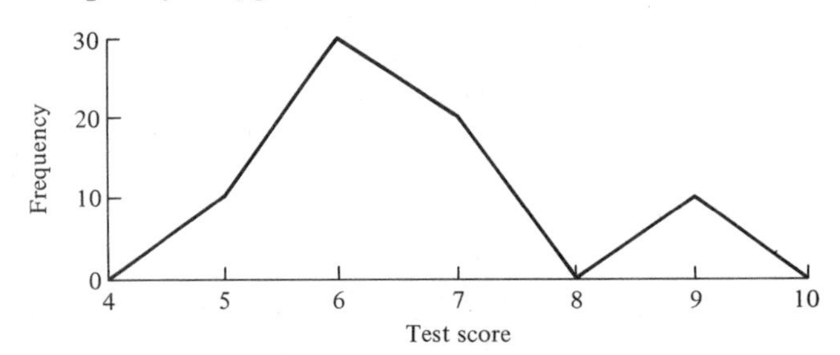

The most common error in plotting a histogram is that of not plotting the rectangles above the proper interval (such as plotting everything one-half unit to the right of where it should be). There are many different ways of making errors in plotting a frequency polygon. The most common are:

1. not labeling the test score points correctly;
2. not plotting above the midpoints of intervals;
3. not plotting a zero frequency at each end of the scale or plotting it at the wrong place (the correct place is at the midpoint of the next interval);
4. not connecting the graph of the other points to the zeros at the ends of the scale;
5. ignoring intermediate zero frequencies (as by drawing a straight line connecting the frequencies for X values of 7 and 9, thus making it appear that there is a frequency of 15 at $X = 8$); and
6. attempting to smooth the data by drawing a curve through or among the points instead of connecting successive points by straight lines.

The correct way to plot a histogram or frequency polygon is simple. Lay out the two scales to take care of all the frequencies and all the scores; plot above the midpoint of each interval; and then, *for the histogram,* draw the horizontal lines centered on the plotted points and connect them by vertical lines wherever needed or, *for the frequency polygon,* connect the points. (Actually, there are other correct ways of doing this. For instance, if we group our data, there are many correct ways to label the baseline, but there are even more wrong ways.) Also, purely for aesthetic reasons, set or adjust one of the scales so that the graph is somewhere around twice as wide as it is high; a graph that is too narrow or too flat doesn't look nice.

Figure 2.11 shows a histogram of the El Dorado IQ data, while Figure 2.12 shows a frequency polygon for the same records. Both graphs are based on the frequency distribution given in columns (1) and (3)—or (1) and (2)—of Table 2.4 (p. 20). To see how this comes about, open the book to Table 2.4 and hold it in front of a mirror, with the high IQs at your right. Compare that mirror image with Figure 2.11 and note the striking resemblance between the tabled data and the histogram. The frequency polygon is what you get if you connect the midpoints of all the horizontal lines on the histogram. Both are entirely correct ways of depicting this distribution. The choice between them is often purely a matter of preference.

An exception to this occurs when a graph is based on discontinuous data, such as our records of author citations. When intermediate scores (such as 3.62 citations) are impossible, the histogram should be used because the frequency polygon gives the impression of more continuity from one

FIGURE 2.11
Histogram of El Dorado IQ Data

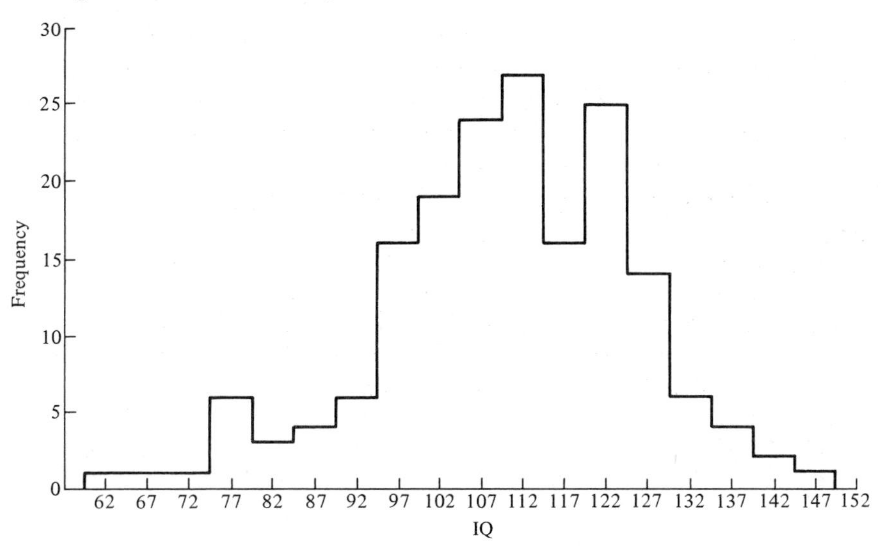

FIGURE 2.12
Frequency Polygon of El Dorado IQ Data

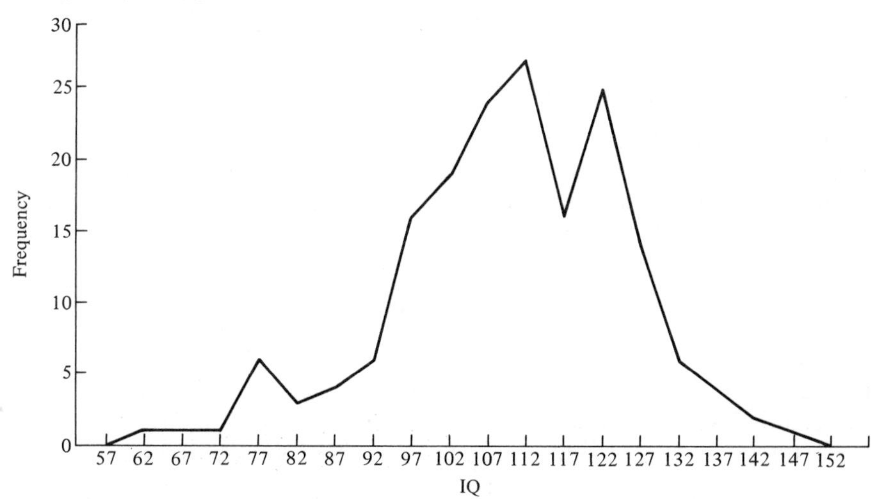

plotted point to the next. In other situations, use whichever one you happen to prefer.

There are many other types of graphs. Two more, the ogive, or cumulative frequency curve, and the bar diagram are illustrated in Supplement 2.

SUMMARY

The best measure of central tendency is usually the arithmetic mean, which is the sum of the scores divided by the number of scores. It gives the best available single value to represent an entire group of scores. The median, the middle score when the scores are arranged in order of size, is the next best measure of central tendency. Both the mean and median can be computed either from ungrouped or from grouped data. Examples are given of the use of frequency distributions in plotting histograms and frequency polygons.

SUPPLEMENT 2

OTHER AVERAGES

There are a great many averages other than the mean and the median. Most of them have limited usefulness and, as might be expected, most of them have limited use. There are, of course, circumstances for which each is appropriate but these occur infrequently with the data we obtain in Education and Psychology. Both because you may encounter references to these other averages in journal articles, and because you might possibly gather data for which one of them would be useful, here are brief descriptions of several more averages or measures of central tendency.

Geometric Mean The geometric mean is the Nth root of the product of the N scores. Its formula is

$$M_G = \sqrt[N]{({}_1X)({}_2X)({}_3X) \cdots ({}_NX)} \tag{2.5}$$

where the subscript in front of X indicates that it is the first, second, third, or the Nth (or last) score on variable X. To compute the geometric mean, we multiply all the scores together and then take the Nth root of that product. This is usually done by using logarithms, in which case the formula becomes

$$\log M_G = \frac{\Sigma(\log X)}{N}. \tag{2.6}$$

The geometric mean is useful for data in which differences between scores are regarded as being equally important if they differ by the same proportion. Thus, if we regard a difference from 6 to 8 as roughly the same as one from 30 to 40, or from 120 to 160, or from 3 to 4, then the geometric mean is appropriate. Note that M_G thus gives far less weight to differences between large scores (and, of course, far more weight to differences between small scores) than does $\overline{X}$. Also, if any score is zero, M_G is useless because it is then automatically equal to zero. Further, it won't work with a mixture of positive and negative scores. It is rarely used in Psychology or Education except to obtain the denominator of the correlation coefficient (see Chapter 8), in which case we have to take the square root of the product of two large numbers.

As just indicated, M_G is inappropriate if any one or more of the scores is negative or equal to zero. With positive scores, M_G is always intermediate in value between the highest and lowest scores. Also, M_G is always smaller than $\overline{X}$ (unless all the scores are equal, in which case all averages are equal).

Harmonic Mean Suppose we know that it takes four employees 2, 10, 20, and 50 minutes, respectively, to compose an answer to a typical letter of inquiry. Suppose, further, that a big campaign involving many more such inquiries is being considered, and the supervisor is trying to estimate the average production per employee. If $\overline{X}$ had been computed as 20.50 minutes, the supervisor might then have erroneously concluded that these employees were, on the average, turning out about 3 letters per hour; and hence the four of them would have appeared to be answering about 12 letters per hour. The absurdity of this is evident when we note that the fastest employee turns out many more than that—all alone.

For this problem, we need the harmonic mean. Its formula is:

$$M_H = \frac{N}{\Sigma\left(\frac{1}{X}\right)}. \tag{2.7}$$

Let us now solve the above problem correctly. The following tabulation will clarify the procedure:

Name	X = Time	$\frac{1}{X}$ = Reciprocal
Liv	2	.50
Al	10	.10
Yvonne	20	.05
Sam	50	.02
		$\Sigma\left(\frac{1}{X}\right) = .67$

Substituting in Formula (2.7), we get

$$M_H = \frac{N}{\Sigma\left(\frac{1}{X}\right)}$$
$$= \frac{4}{.67}$$
$$= 5.97.$$

Since the harmonic mean of the time is 5.97 minutes, it appears that each of these four employees would write about 10 letters per hour (since 60/5.97 = 10.05). If so, the four of them should turn out 4(10.05) = 40.20 letters in an hour. Let's see. Liv at 2 minutes per letter would do 30; Al at 10 minutes per letter would do 6; Yvonne would do 3; and Sam would do 1.20. When we add, we get 30 + 6 + 3 + 1.20 = 40.20, which checks, as it should. Thus, we see that for such problems as this, M_H rather than $\overline{X}$ should be used. (A time and motion study engineer might decide that the other employees should emulate the fastest, Liv—or a hard-hearted boss might fire three of the four employees—but that's another story.)

Like M_G, M_H must not be used if any of the scores are zero or negative.

Root Mean Square Suppose a city has five small square parks in one crowded neighborhood. The largest, nearly the size of a city block, is 300 feet square, another is 170 feet square, a third is 90 feet square, a fourth is 30 feet square, and the fifth is a tiny one only 10 feet on a side. What is the average size of these parks? The arithmetic mean of these five numbers is 120 feet. This is the average length of a side of these squares, but it doesn't give us much of an idea of the size of the parks. A park with this average side of 120 feet would contain $(120)^2$ = 14,400 square feet; five such parks would contain 72,000 square feet. But that's a long way from the average size, since the largest park alone is much larger than that—90,000 square feet.

Clearly we need a different measure of central tendency for this problem. Such a measure is the root mean square. Its symbol is M_R and its formula is

$$M_R = \sqrt{\frac{\Sigma X^2}{N}}. \tag{2.8}$$

With the five city parks, $\Sigma X^2 = (300)^2 + (170)^2 + (90)^2 + (30)^2 + (10)^2 =$ 128,000, and N, of course, is 5. Substituting in the formula, we get

$$M_R = \sqrt{\frac{\Sigma X^2}{N}}$$

$$= \sqrt{\frac{128{,}000}{5}}$$

$$= \sqrt{25{,}600}$$

$$= 160.00.$$

This means that if all five square parks were this same size (as indicated by $M_R = 160$), their total area would be the same as that of the original five parks. Hence, M_R is the correct average to use for this problem.

A coin collector might wish to compare the sizes of coins from different countries or of different metals. To most people, the size of a coin is indicated by its area rather than by its diameter or its volume. Because such areas are proportional to the square of the diameter, M_R could be used with diameter measurements to give the average size of copper coins, of coins from Spain, etc.

Parks and money may not sound like educational or psychological problems, and they probably aren't, but M_R has one very common use in our research. These last two illustrations, like most others, were concerned with ordinary scores denoted by X. Unlike M_G and M_H, M_R is not restricted to positive numbers; it's OK to use it—whether the numbers are positive, zero, or negative. In Chapter 3, you will learn that a measure of variability has a formula similar to that for M_R, but it contains a small x instead of a capital X. You will also learn that the formula for what is called the standard deviation of a finite population is a special case of the one just given for M_R.

Mode The mode is simply the most common score. If the scores are tabulated in a frequency distribution, it is the midpoint of the interval with the highest frequency. This may sound all right, but if someone else makes the tabulation with a smaller interval or with a larger one, the mode will be somewhere else. Even with the same size interval, there are various possible starting places, and they will give either slightly different or markedly different modes. Also, two different scores are sometimes tied for the highest frequency, resulting in what is called a bimodal distribution.

As was true of the mean and the median, the mode can be computed from ungrouped data or from grouped data. Again, these will not necessarily be equal. In fact, there may be large differences and, as just noted, there may even be two modes. For example, for our El Dorado IQ scores, there are two modes for the ungrouped data—one at 101 and the other 111 since each of these IQs, with its f of 10, has a higher frequency than any other IQ. On the other hand, the mode for the grouped data (defined as the midpoint of the interval containing the largest number of cases) is 112. If, however, we had used the same interval width, but had used different boundaries such as 97–101, 102–106, etc., the mode would have been the midpoint of the 107–111 interval, or 109. Other interval widths would give still other values for the mode. The mode may also be illustrated with the author citation data. There are more values of 1 ($f = 70$) than any other value. Therefore, the mode is equal to 1.

The mode is not frequently used, nor is it often useful or desirable. However, there may be a few instances when one actually wants the information yielded by the mode. If our variable is qualitative (that is, nonquantitative), of course we have no choice. We cannot use the mean or the median, and we would either use the mode for an average or no average at all. It should be noted that nonquantitative data would usually have unordered values of the variable. A case in point would be the number of people choosing each of several breakfast cereals. There would be no clearly correct way to arrange the cereals in rank order; the mode would be the cereal chosen by the largest number of people.

Although the mode has been used somewhat in the past, its use has declined because it fails to give unique and consistent values. We should, however, make a couple of remarks in its favor. Earlier in Chapter 2 (p. 24), in connection with getting an arbitrary origin when working with grouped data, we said to place M' at the midpoint of the interval having the highest frequency, but we didn't bother to mention we were referring to the mode. Thus, the mode is often used in computing the mean from grouped data, though few people realize this. In this section, we have been referring to what is sometimes called the crude mode. Another concept, called the theoretical mode, is used in curve fitting; it is the maximum point on a mathematically determined frequency distribution curve and it has some usefulness in advanced work in statistical theory.

Weighted Arithmetic Mean At times, we may feel that some of our observations should receive more weight than others. For instance, if we wish to give double weight to some scores, all we need to do is count them twice and divide the total by the sum of the weights instead of by N. This will give the weighted arithmetic mean. If you want a formula for the weighted mean, ${}_{\mathrm{w}}\overline{X}$, here it is:

$${}_{\mathrm{w}}\overline{X} = \frac{\Sigma WX}{\Sigma W} \tag{2.9}$$

in which each X score is multiplied by its individual weight of 2, 1, or whatever weight is decided upon, before being added. While this formula is sometimes useful, it is usually preferable to weight all scores equally.

Other Weighted Averages As is readily apparent, any average may be weighted in a manner strictly analogous to that just described for $\overline{X}$. Thus, a median may be weighted simply by listing some scores twice (to give them double weight), while others are listed only once.

Composite Mean from Means of Subgroups Suppose that we have the means of several groups, each with a different number of cases, and that we wish to use these subgroup means to compute the mean for the total (that is, for all groups combined). We should not take the average of the subgroup means, since they are based upon differing numbers of cases. Instead, we need to weight each mean according to its number of cases. A formula for doing this is:

$$\overline{X}_{\text{composite}} = \frac{N_1\overline{X}_1 + N_2\overline{X}_2 + N_3\overline{X}_3 + \cdots + N_n\overline{X}_n}{N_1 + N_2 + N_3 + \cdots + N_n}, \tag{2.10}$$

where $\overline{X}_{\text{composite}}$ is the mean of all cases in all the groups, where N_1 and $\overline{X}_1$ refer to the first group, where N_2 and $\overline{X}_2$ refer to the second group, etc. As a numerical example, suppose that a group of males and females have grade point averages as follows:

	N	$\overline{X}$
Males	21	2.21
Females	79	2.93

What is the grade point average for the combined group of males and females? Applying Formula (2.10) for $n = 2$ groups, we have

$$\begin{aligned}\overline{X}_{\text{composite}} &= \frac{N_1\overline{X}_1 + N_2\overline{X}_2}{N_1 + N_2}\\ &= \frac{21(2.21) + 79(2.93)}{21 + 79}\\ &= \frac{46.41 + 231.47}{100}\\ &= \frac{277.88}{100}\\ &= 2.78.\end{aligned}$$

If we had simply averaged the two means, we would have obtained

$$\begin{aligned}\frac{\overline{X}_1 + \overline{X}_2}{2} &= \frac{2.21 + 2.93}{2}\\ &= \frac{5.14}{2}\\ &= 2.57,\end{aligned}$$

which is different from the previous answer and obviously erroneous.

Other Means As indicated earlier, there are many other measures of central tendency. Since we feel they have very little usefulness in Education or Psychology at the present time, we will not discuss them.

PROOF THAT ERROR DUE TO GROUPING IS SMALL

The mean IQ for the 176 students in the El Dorado School was 109.80, when obtained in the traditional manner, and 110.01 (a difference of .21 IQ point), when computed from grouped data with $i = 5$.

Let us look at just one interval, the one labeled 120–124. If the 25 scores in that interval were evenly distributed, there would be 5 at 120, 5 at 121, 5 at 122, 5 at 123, and 5 at 124. The mean of these would obviously be 122, which we know is also the midpoint of that interval. No error would be made by pretending that all scores were at the midpont of that interval. That would be assuming that there were 25 scores at 122, but none at 120, 121, 123, or 124. That error of zero is, of course, the minimum. What is the maximum? If there were no scores of 120, 121, 122, or 123, and all 25 were at 124, then we would make an error of 2.00 IQ points if we regarded all 25 scores as being at the midpoint, which we do when we group our data. Similarly, if all 25 really were at 120, we again would make an error of 2.00 IQ points. If the scores were concentrated anywhere else in the interval, the error would be less.

Thus, the highest possible error of 2.00 IQ points is less than half the interval width of 5.00, as we stated near the end of our discussion of the mean of the IQ data. The maximum discrepancy is the difference between the midpoint and the highest (or lowest) score that can possibly be tabulated in that interval. This is less than the distance from the midpoint to the exact upper (or lower) limit of the interval. With this frequency distribution, we are saying that the difference of 2.00 between 122 and 124 (or 120) is less than the distance of 2.50 between 122 and 124.50 (or 119.50). We can avoid the technicalities and still be absolutely correct by stating, as we did in Chapter 2 (p. 23), that the difference can never be as great as half the interval width (here 2.50). As noted, it is usually *much* smaller than that; in fact, it is so small that statisticians just ignore it.

CUMULATIVE GRAPHS

In Chapter 2, two kinds of graphs, the histogram and frequency polygon, were considered. Each point on these graphs was based upon the frequency of cases falling within an interval.

Let us now consider two additional graphs, the cumulative frequency curve and the cumulative percentage curve. Each is based upon the cumulative frequency falling below the exact lower limit of an interval. Table 2.13 is an extension of Table 2.8 (p. 33) with a cumulative frequency column and a cumulative % column added. The appropriate cumulative frequencies and percentages are shown opposite the true class boundaries or exact limits of 4.50, 5.50, 6.50, etc.

TABLE 2.13
Cumulative Distribution of Hypothetical Test Scores

X	f	Cumulative f	Cumulative %
10.50		70	100
10	0		
9.50		70	100
9	10		
8.50		60	86
8	0		
7.50		60	86
7	20		
6.50		40	57
6	30		
5.50		10	14
5	10		
4.50		0	0

In Figure 2.14 these exact limits, of course, are between the original plotted scores. This is a line graph, and if interpolations are made, it will indicate, for any given test score, the approximate corresponding cumulative frequency below that point.

Figure 2.15 is plotted to the same scale as Figure 2.14. However, the ordinate has been converted to cumulative percentages. If interpolations are made, Figure 2.15 will indicate, for any given test score, the approximate corresponding cumulative percentage. In Chapter 6 we will find out

FIGURE 2.14
Cumulative Frequency Curve for Data of Table 2.13

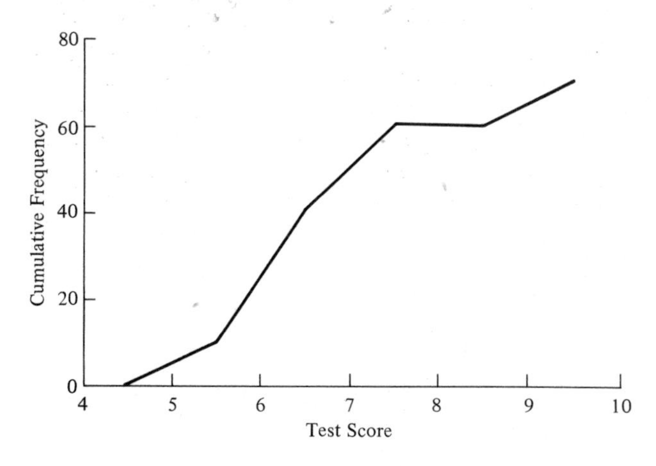

FIGURE 2.15
Cumulative Percentage Curve for Data of Table 2.13

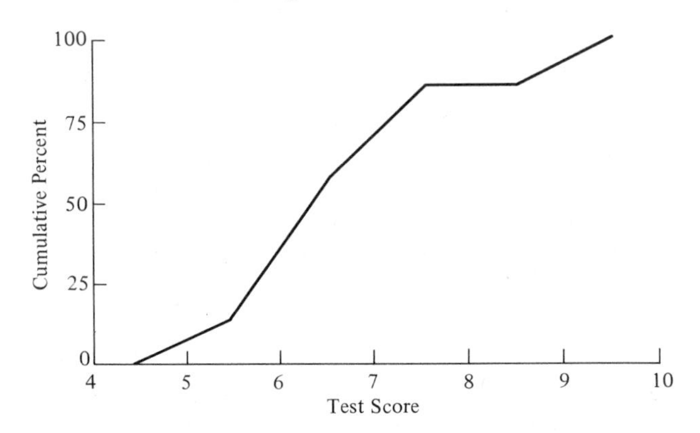

that the cumulative percentage at a given test score is the same as the percentile rank of that point. Thus, we will then see that such a curve as this can be used for graphically converting test scores to percentile ranks.

BAR DIAGRAMS

The histogram works fine for portraying data graphically, as long as the scores being tabulated are continuous. Sometimes, as in dealing with family

FIGURE 2.16
Bar Diagram: Number of Students Majoring in Various Teaching Fields in Teacher-Training Institutions

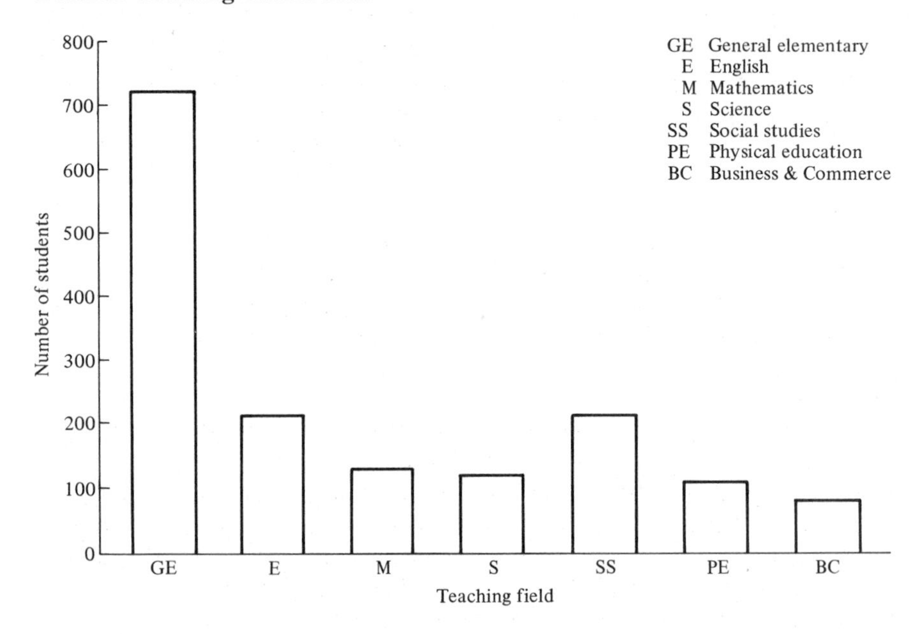

size, fractional values are impossible.* At other times, such as in classifying students by major subject, there may be no very defensible way to arrange the major subject categories. In such a case, we cannot allow adjacent sides of the rectangles to touch; there must be a gap. An example of such a case is shown in Figure 2.16.

A STATISTICS PROFESSOR used to describe himself and others in his profession this way: ''A statistician is someone who, if he puts his head in a hot oven and his feet in a bucket of ice water, would remark, 'On the average, I feel fine.''' Contributed by Janet Simons

—Reader's Digest March, 1977

*Our continuous variables are usually measured on an equal-interval (sometimes called interval) scale. Examples are length (in inches, feet, meters, etc.), time (in seconds, minutes, hours, etc.), weight (in ounces, pounds, grams, etc.), temperature (in degrees Celsius or degrees Fahrenheit), IQ, and test scores. Discrete variables are those in which there are numerical values between which there are gaps, rather than an infinite number of values, as in continuous variables. An example would be the number of children reported in a family. Categorical variables have qualitative values, such as the major fields shown in Figure 2.16.

CHAPTER 3 THE STANDARD DEVIATION

NEED FOR A MEASURE OF VARIABILITY

Sometimes two or more distributions will have the same, or essentially the same, means but still be quite different. Mexico City lies in the torrid zone, but because of its altitude of 7,556 feet, in recent years it has had an average temperature of 59.9 °F. The corresponding figure for Nashville, Tennessee, is 60.0 °F. The means are practically identical, but no one who has lived in these two cities would contend that the climates are at all similar. Why not? Because of great differences in variability. In Mexico City, the hottest month is April (with a mean of 64.0 °F), and the coldest is January (with a mean of 54.9 °F). In Nashville, the hottest month is July (with a mean of 79.5 °F), and the coldest is January (with a mean of 39.4 °F). Further, the hottest and coldest temperatures recorded during this century are 93 °F and 24 °F in Mexico City, as compared with 107 °F and −15 °F in Nashville. Truly, the means are the same, but the temperature variability in the two cities is entirely different.

Now let us consider the average intelligence test scores of third-grade pupils in two different classes. One class is the middle (or *Y*) section in a school that divides each grade into *X, Y,* and *Z* sections according to intelligence. The other class is the entire third grade in a smaller town. Clearly, the range of scores in the first class has been artificially curtailed, resulting in a more homogeneous group of students—although the means of the two classes may be close together or even identical.

As another illustration of the need for a measure of variability, let us examine the hours worked per week by two clerks. One works a 35-hour week, is seldom late or absent, and rarely works overtime. The other is supposed to work 40 hours a week but is absent so much that, in the course of a year, the number of hours actually worked averages only 35. Obviously, the means are the same, but the second clerk is far more variable in number of hours worked per week.

The mean gives us some information about the scores, but sometimes the variation among the scores is so great that the mean doesn't tell us all we would like to know about them. Sometimes the mean alone does a pretty good job of characterizing a whole set of scores; sometimes it does not; the mean alone can't do everything.* The mean is perhaps the most important statistic, but still it is only one figure and it cannot possibly tell the whole story. Clearly, we need something else. We could describe a set of scores better if we also had a measure of variability and perhaps other statistics as well. There are many statistics for measuring the variation or variability referred to in the three preceding illustrations, but the standard deviation is the best of the lot, so we'll use it.

The standard deviation, s, is related to the distances, x, of scores, X, from their mean, $\overline{X}$. Its computation involves some squaring and the extraction of a square root, but just what *is* it? It is a measure of variability or dispersion such that a distance from one standard deviation below the mean to one standard deviation above the mean includes about $\frac{2}{3}$ of the cases in most distributions of scores. That is, a range of $2s$ taken around the middle of the distribution will include about $\frac{2}{3}$ of the scores. The other third are beyond this interval—about $\frac{1}{6}$ being above it and about $\frac{1}{6}$ being below it. Further, a range twice this big, running from $\overline{X} - 2s$ to $\overline{X} + 2s$ will include about 95% of the scores. Going still further, the range from $\overline{X} - 3s$ to $\overline{X} + 3s$ takes in nearly everybody.

Now we won't always get *exactly* these results because different score distributions have different shapes, but these figures hold pretty well for most of the data we gather or use in Psychology and Education. In what is

*Occasionally, ill-informed people express skepticism about such statistics as an average of 2.8 children per family. They correctly point out that no family has exactly that many children, and then incorrectly conclude that there is something wrong with that statistic. It is indeed true that many figures do have something wrong with them. In this case, however, it isn't the figure that is wrong, it is the one who misinterprets it who is wrong.

called a normal probability curve the range from −1 to +1 standard deviation includes 68.27% of the area, that from −2 to +2 standard deviations includes 95.45%, and that from −3 to +3 standard deviations includes 99.73% of the cases.* For other distributions, the figures will differ, of course, but probably not by very much. For example, if you see the distance from one standard deviation below your mean to one standard deviation above it includes either 45% or 85% of the scores, you'd better go back and check your arithmetic—fast.

We use the standard deviation rather than some other measure primarily because it gives us more consistent results from one sample to another. This is very important since we want our results to be dependable and affected as little as possible by chance fluctuations. We do not now wish to minimize the effect of large deviations; it is often appropriate to give extra importance to them. The standard deviation does just that. You may, if you wish, regard the standard deviation as some sort of weighted average of the deviations. In any case, it is mathematically sound and will also be found useful in many ways other than its primary function of giving us a dependable measure of variability. Let us now define it exactly by giving its formula.

FORMULA FOR THE STANDARD DEVIATION

As indicated above there are many measures of variability, but because the standard deviation is usually the most appropriate, it will be considered first. Several other measures of dispersion will be considered—some in Supplement 3 and some in Supplement 6.

The standard deviation is based on squared deviations or distances from the mean. In fact, it is the square root of what is very close to the average of all such squared deviations.† In symbols,

$$s = \sqrt{\frac{\Sigma x^2}{N - 1}}, \qquad (3.1)$$

where s is the symbol for the standard deviation, x is a deviation from the mean, and N, as usual, is the number of cases. Notice the distinctions in

*These three areas were determined by using the bottom portion of Table E, doubling the entry, and then rounding it.

†This is the appropriate formula for most statistical work. Later in this chapter and in Chapter 5, we shall get into technical distinctions concerning what will then be called samples and populations. On page 59 we shall learn that formula (3.1) gives an estimate of the standard deviation in the (large) population from which our sample is drawn.

the meaning of x (a deviation from the mean), X (a score), and $\overline{X}$ (the mean). We define x by the equation $x = X - \overline{X}$.

In using Formula (3.1), after we divide Σx^2 by $(N - 1)$, we get a quotient that, especially if N is large, is very close to the average of the *squared* deviations. This actually is one measure of dispersion. It is called the *variance*, s^2, but it is given in terms of squared units. To cut it down to size and get our answer in pounds or degrees (rather than square pounds or square degrees—whatever they are), we take the square root of the variance and get the standard deviation. Because all squares are always positive, not only Σx^2, but also s^2 and s must always be positive.

What would we get if we were to attempt to shorten this process by not squaring the numbers and consequently not having to extract the square root? Nothing. Yet, that's right, nothing at all. Regardless of the variation among scores, Σx will always be zero, so $\dfrac{\Sigma x}{N - 1}$ will also equal zero. This would indeed be a very elaborate and time-consuming way to get nowhere. Since $\Sigma x = 0$, the sum of the positive deviations (above the mean) must have the same numerical value as the sum of the negative deviations (below the mean). In fact, the mean may be defined as that point in a frequency distribution from which the sum of all the deviations (both above and below the mean) is zero. We might get a measure of dispersion by just ignoring the signs of the deviations, treating them all as positive. Yes, we might, and in that case, we would get the average deviation. But that comes in Supplement 3; for the present, let us stick to the standard deviation.

Formula (3.1) is useful in illustrating the meaning of the standard deviation but, except under highly unusual circumstances, it is not very useful for actual computational purposes. In order to use Formula (3.1), it is necessary

1. to subtract the mean from each score, obtaining N deviation scores, x, each expressed to perhaps two decimal places;
2. to square each of these deviations, again keeping several decimal places;
3. to add up all these squared deviations;
4. to divide by $(N - 1)$; and
5. to extract the square root.

These processes are simple enough to describe, but they are very time-consuming and, especially in steps 2 and 3, it is extremely easy to make errors in carrying them out. It would be much simpler if we didn't have to work with decimals.

Actually, we don't have to work with decimals. We may, if we wish, work with the original scores, using an alternate formula:

$$s = \sqrt{\frac{N\Sigma X^2 - (\Sigma X)^2}{N(N-1)}} \tag{3.2}$$

in which all the symbols have been previously defined. To solve Formula (3.2) we need to compute ΣX and ΣX^2. These sums are not at all difficult to get if the numerical values of the X scores are small (say under 40 or 50) or if a calculator is available.* Note also that the value of N(N–1) can be read from Table B in the Appendix instead of computing it.

First Numerical Example with Ungrouped Data Using the data of Table 2.1, page 14, with the aid of Table Z to square the numbers for us, we get

$$\begin{aligned}\Sigma X^2 &= 136^2 + 119^2 + 130^2 + \cdots \\ &= 18{,}496 + 14{,}161 + 16{,}900 + \cdots \\ &= 2{,}161{,}528.\end{aligned}$$

From Chapter 2, we know that $\Sigma X = 136 + 119 + 130 + \cdots = 19{,}324$

Substituting these values of ΣX and of ΣX^2 in Formula (3.2), we get

$$\begin{aligned}s &= \sqrt{\frac{N\Sigma X^2 - (\Sigma X)^2}{N(N-1)}} \\ &= \sqrt{\frac{(176)(2{,}161{,}528) - (19{,}324)^2}{176(176-1)}} \\ &= \sqrt{\frac{380{,}428{,}928 - 373{,}416{,}976}{(176)(175)}} \\ &= \sqrt{\frac{7{,}011{,}952}{30{,}800}} \\ &= \sqrt{227.66} \\ &= 15.09.\end{aligned}$$

*A word of caution should be raised at this point. There is much chance for human error in the process of using a calculator or computer, probably more so than for machine error. Machine errors often are quite obvious. But consider a human error, such as multiplying 136 by 126 instead of by 136. The whole thing would be wrong and the entire set of computations would be worthless. An extremely high degree of accuracy is demanded in statistical work. *All* computations should always be checked, preferably by a different method, in a different order, and, if possible, by a different procedure. The beginner may also find it helpful to record subtotals after every 20 or so figures.

The numbers used are large, but notice that all of them are whole numbers; and there are no decimals until the last two steps of division and extraction of the square root. This means there are no decimals to get mixed up on, and there are no rounding problems until the end. And by that time they are automatic; the two-decimal rule takes care of things simply and accurately. Referring back to Chapter 2 where we calculated the mean, we can now describe the set of IQs in Table 2.1 by the following three statistics: $N = 176$; $\overline{X} = 109.80$; and $s = 15.09$.

Stated differently, these 176 students have an average IQ of 109.80 with a standard deviation of 15.09. They are, on the average, about 10 IQ points above the national average of 100, but not all of them have IQs of exactly 110 by any means. They vary and the size of the standard deviation tells us how much they vary. The bigger s is, the more the scores vary.

Second Numerical Example with Ungrouped Data Let us now compute the standard deviation for another set of data—one with a distribution quite different from those usually encountered in Education or Psychology. The data on citations to authors from Chapter 2 (p. 27) will serve this purpose admirably. Because all but one of the 161 scores are one-digit numbers, there is no difficulty in squaring and adding them. Earlier, in Chapter 2, we found that $\Sigma X = 285$. We now square all 161 scores, add the squares, and obtain $\Sigma X^2 = 939$. Substituting in Formula (3.2), we obtain

$$
\begin{aligned}
s &= \sqrt{\frac{N\Sigma X^2 - (\Sigma X)^2}{N(N-1)}} \\
&= \sqrt{\frac{(161)(939) - (285)^2}{(161)(161-1)}} \\
&= \sqrt{\frac{151{,}179 - 81{,}225}{25{,}760}} \\
&= \sqrt{\frac{69{,}954}{25{,}760}} \\
&= \sqrt{2.72} \\
&= 1.65.
\end{aligned}
$$

Even for such a lopsided distribution, the standard deviation is still a useful measure of variability. For instance, the fact that it is almost as large as the mean (a very unusual situation with raw scores) illustrates the relatively large variability of these scores. This results from most of them being the lowest possible score (1 in this case) with a few trailing off to the high value of 14. We noted in Chapter 2 that these few high scores had a marked effect on the mean. Because deviations are squared in computing s, it is altogether fitting and proper that they should have an even greater effect on the size of the standard deviation and they do.

COMPUTING THE STANDARD DEVIATION WITH GROUPED DATA

The preceding examples not only help clarify the meaning of the standard deviation but also indicate that Formula (3.2) can be used in a variety of situations. It can, but there are times when it is more convenient to tabulate the data and use a formula for grouped data. Such would certainly be true in the case of the El Dorado IQ data if no calculator were available. Let us use the tabulation of these records, as given in Table 2.4 (p. 20), for computing the standard deviation. The first three columns of Table 2.4 are, of course, copied into Table 3.1, since the basic figures are bound to be the same, no matter what statistic is being computed.

The same principles apply in selecting an arbitrary origin, whether we are computing only the mean or also the standard deviation. Hence, the next column in Table 3.1 could be either column (4) or column (6) of Table 2.4. Assuming that calculators are not available, let us take column (4) and, of course, column (5), since it is the product of columns (3) and (4). The only new column in Table 3.1 is column (6) and it involves squares of the

TABLE 3.1
Computing the Standard Deviation, using a Frequency Distribution of IQs

IQ *(1)*	Tallies *(2)*	f (3)	x' (4)	fx' (5)	$f(x')^2$ (6)
145–149	|	1	7	7	49
140–144	||	2	6	12	72
135–139	||||	4	5	20	100
130–134	𝍸 |	6	4	24	96
125–129	𝍸 𝍸 ||||	14	3	42	126
120–124	𝍸 𝍸 𝍸 𝍸 𝍸	25	2	50	100
115–119	𝍸 𝍸 𝍸 |	16	1	16	16
110–114	𝍸 𝍸 𝍸 𝍸 𝍸	27	0	0	0
105–109	𝍸 𝍸 𝍸 𝍸 ||||	24	−1	−24	24
100–104	𝍸 𝍸 𝍸 ||||	19	−2	−38	76
95–99	𝍸 𝍸 𝍸 |	16	−3	−48	144
90–94	𝍸 |	6	−4	−24	96
85–89	||||	4	−5	−20	100
80–84	|||	3	−6	−18	108
75–79	𝍸 |	6	−7	−42	294
70–74		1	−8	−8	64
65–69		1	−9	−9	81
60–64		1	−10	−10	100
Total		176		−70	1,646

deviations. It can be obtained in two ways, so we shall compute the values in column (6) by one of these methods and check them by the other. First, we note that

$$(x')(fx') = f(x')^2,$$

so we multiply the values in columns (4) and (5) together and record the products in column (6). Next we note that

$$(x')^2(f) = f(x')^2$$

and we check each value in column (6) by mentally squaring the number shown in column (4) and multiplying this square by the corresponding frequency in column (3). This should, in every case, give the same value as that previously computed and recorded in column (6). Any apparent errors should be traced down and corrected immediately. Notice that this checking process not only checks column (6) but also checks column (5), so it is not necessary to check either of these columns in any other way.* Notice also that all values in column (6) are positive, regardless of whether they are above or below the arbitrary origin.

The formula for the standard deviation computed from grouped data is

$$s = i\sqrt{\frac{N\Sigma f(x')^2 - (\Sigma fx')^2}{N(N-1)}} \tag{3.3}$$

in which all the symbols have previously been defined.

First Numerical Example with Grouped Data Next, you will substitute values for the symbols in Formula (3.3) and get a result. If it is the right answer, fine. Congratulations! If it isn't, or if you aren't sure how to proceed, go now to page 75 in Supplement 3 and do exactly what those detailed directions tell you to do. Then you will very probably get the correct answer the first time—not only to the problem you are working, but to all future problems using Formula (3.3).

Now let's utilize the appropriate values from Table 3.1 in Formula (3.3).

---

*An exception may be noted, since some people are ingenious enough to foil almost any checking system. If some row contains a frequency of 1 and the reader does not know the squares of the x' values, an error may be made and the check will not reveal it. Thus, if some reader thinks $(-9)^2$ is 89, both his original computation and his checking will agree on this erroneous value in the next to the last complete row of Table 3.1. The only cure for this is to look up or memorize the correct values for the squares with which we deal.

The only sums we need are shown at the bottom of columns (3), (5), and (6). Substituting, we readily obtain

$$\begin{aligned} s &= i\sqrt{\frac{N\Sigma f(x')^2 - (\Sigma fx')^2}{N(N-1)}} \\ &= 5\sqrt{\frac{(176)(1{,}646) - (-70)^2}{176(176-1)}} \\ &= 5\sqrt{\frac{289{,}696 - 4{,}900}{30{,}800}} \\ &= 5\sqrt{\frac{284{,}796}{30{,}800}} \\ &= 5\sqrt{9.25} \\ &= (5)(3.04) \\ &= 15.20. \end{aligned}$$

This value of 15.20 differs slightly from that of 15.09 obtained from raw scores and Formula (3.2) for two reasons: (1) there is a slight error involved in grouping the data into intervals and treating all scores within an interval as though they had the same value, and (2) there are errors due to the fact that all our computations in these problems are carried to only 2 decimals. Neither of these errors is very serious. If the number of intervals is 12 or more, the first error will average less than 1% (an error of .15 in this example), and if the number of intervals is 17 or more, this grouping error will, on the average, be less than ½ of 1%. (It is for this reason that in Chapter 2 we recommended having between 15 and 20 intervals; the grouping error is then so small that it can be ignored.) The rounding error due to retaining only a relatively small number of decimal places is illustrated in Table 3.2. The rounding error can be made as small as we wish merely by increasing the number of retained figures, but there is a real question as to just how much more work is justified in order to attain a certain degree of accuracy. The elimination of all decimals would often result in figures that were too inaccurate, but this problem certainly does not justify the retention of 6 decimals. If the original records are accurate, or nearly so, we are usually justified in keeping statistics such as $\overline{X}$ and s to one or two more significant figures than the data contained. This will generally be to about two decimal places if the raw scores are whole numbers, as they often are. Your instructor has given you a reasonably accurate method that does not involve an excessive amount of work; we have kept two decimals in this example. These rules are compromises. Important studies based on large numbers of cases often necessitate a higher degree of accuracy, while unimportant studies may justify only a small number of decimals or perhaps no statistical analysis at all.

TABLE 3.2
Effect of Number of Retained Decimal Places on the Values of $\bar{X}$ and s

Number of Decimal Places Retained in All Computations *(1)*	$\bar{X}$ by Formula (2.1) *(2)*	$\bar{X}$ by Formula (2.2) *(3)*	s by Formula (3.2) *(4)*	s by Formula (3.3) *(5)*
0	110.	110.	15.	15.
1	109.8	110.0	15.1	15.0
2	109.80	110.01	15.09	15.20
3	109.795	110.011	15.088	15.205
4	109.7955	110.0114	15.0884	15.2040
5	109.79545	110.01136	15.08843	15.20415
6	109.795455	110.011364	15.088432	15.204130
Correct values to 6 decimal places	109.795455	110.011364	15.088432	15.204131

Second Numerical Example with Grouped Data We shall now compute another standard deviation by the use of Formula (3.3), and we shall introduce another method of checking the computations. Table 2.6 gave a tabulation of the daytime speeds of southbound cars. All five of its columns are repeated in Table 3.3. Column (6), labeled $f(x')^2$, and two other columns are also included there.

If we had not already computed $\bar{X}$, we would use the data of Table 2.5 to tabulate and compute columns (1) through (5). To save time, we would omit most of the upper limits, as we have done in the new column (1). We would then tabulate the data until we were sure column (2) was right. We would then quickly fill in and check columns (3) and (4). We would then make the entries in column (5), but we would *not* check them.

Previously, we saw that column (6) could be obtained either by multiplying columns (4) and (5) or by multiplying the square of column (4) by column (3). Algebraically,

$$(x')(fx') = f(x')^2,$$
$$(x')^2(f) = f(x')^2.$$

Or symbolically, by column numbers,

$$(\text{col } 4)(\text{col } 5) = (\text{col } 6),$$
$$(\text{col } 4)^2(\text{col } 3) = (\text{col } 6).$$

We used one of these methods to check the other. This time, we are not going to check column (6) in this manner, so we'll use only one method to

TABLE 3.3
Computing the Standard Deviation, using a Frequency Distribution of Car Speeds

Speed (1)	Tallies (2)	f (3)	x' (4)	fx' (5)	$f(x')^2$ (6)	$x' + 1$ (7)	$f(x' + 1)^2$ (8)
62–63	|	1	6	6	36	7	49
60–61	|	1	5	5	25	6	36
58	|||	3	4	12	48	5	75
56	||||	4	3	12	36	4	64
54	卌 |	6	2	12	24	3	54
52	卌 卌 卌 ||	17	1	17	17	2	68
50	卌 卌 卌 ||||	19	0	0	0	1	19
48	卌 卌 |	11	−1	−1	11	0	0
46	卌 卌	10	−2	−20	40	−1	10
44	卌 ||	7	−3	−21	63	−2	28
42	卌 卌 |	11	−4	−44	176	−3	99
40	卌 卌	10	−5	−50	250	−4	160
38	卌 ||	7	−6	−42	252	−5	175
36	||||	4	−7	−28	196	−6	144
34	卌	5	−8	−40	320	−7	245
32		1	−9	−9	81	−8	64
30		1	−10	−10	100	−9	81
28			−11	0	0	−10	0
26–27			−12	0	0	−11	0
24–25	|	1	−13	−13	169	−12	144
Total		119		−224	1,844		1,515

get the entries in column (6). Thus, for speeds of 56–57 in the fourth row of the table, we may say either

$$(3)(12) = 36$$

or

$$(3)^2(4) = 36$$

And in the fifth row, we may get the answer by either

$$(2)(12) = 24$$

or

$$(2)^2(6) = 24$$

Use whichever method you prefer and compute the values in column (6), but don't bother to check them. (You should actually compute several of these values now in order to review the procedure.) Then the entries in column (6) are added to obtain $\Sigma f(x')^2 = 1{,}844$, which is recorded at the bottom of the table.

We shall now check the values of $\Sigma fx'$ and of $\Sigma f(x')^2$. The checking procedure explained earlier in the chapter is a good one, but it is not perfect. We gave an example of an error that it would miss. Also, it neither gives nor purports to give any check on the addition of the columns. As it was in the beginning, it is now and ever shall be true that no checking procedure is perfect. Nevertheless, an excellent method that will detect both types of errors just referred to, and many others as well, is called the *Charlier check*.* It may be written either in the form for use with raw scores or for use with x' scores. For the latter purpose, the Charlier check is

$$\Sigma f(x' + 1)^2 = \Sigma f(x')^2 + 2\Sigma fx' + N. \qquad (3.4)$$

To use it, we need but two more columns. Column (7) is simple enough: we just add 1 to every number in column (4). Column (8) is obtained by squaring column (7) and multiplying each square by its appropriate f from column (3). Again, just do this; don't bother to check it. We then add this column, obtaining $\Sigma f(x' + 1)^2 = 1{,}515$.

We now have all the sums we need. To apply the Charlier check, we substitute the sums from the bottom of Table 3.3 in Formula (3.4) and obtain

$$1{,}515 = 1{,}844 + (2)(-224) + 119.$$

*The Swedish statistician and astronomer, C. V. L. Charlier, devised and used this method of checking in 1906 or perhaps earlier. His name is pronounced shär·lyā′.

We now add the numbers on the right. If they add up to 1,515, we have checked not only $\Sigma f(x' + 1)^2$ but also $\Sigma f(x')^2$ and $\Sigma fx'$. (If anyone cares, we have also checked N, but if we don't know what N is, we're in sad shape.) Since the numbers do add up properly, you may place a check mark after each of them and proceed to compute the standard deviation by the appropriate formula, (3.3).

$$
\begin{aligned}
s &= i\sqrt{\frac{N\Sigma f(x')^2 - (\Sigma fx')^2}{N(N-1)}} \\
&= 2\sqrt{\frac{(119)(1{,}844) - (-224)^2}{119\,(119-1)}} \\
&= 2\sqrt{\frac{219{,}436 - 50{,}176}{14{,}042}} \\
&= 2\sqrt{\frac{169{,}260}{14{,}042}} \\
&= 2\sqrt{12.05} \\
&= (2)(3.47) \\
&= 6.94.
\end{aligned}
$$

The solution is perfectly straightforward. Simply substitute the appropriate numbers in the formula, get $N(N - 1)$ from Table B, solve, and check all the work, being sure to check the square root, not by recomputing, but by using one of the four checking methods explained in Supplement 3.

To recapitulate, we have now computed the standard deviation two times by each of the two formulas designed for computing it. When a calculator is handy, or when the scores are small, Formula (3.2) is usually the most convenient. When scores are large, or of varying sign (some + and some −), or when one wishes to see what the distribution of scores looks like, Formula (3.3) is generally preferred. We have criticized "doing it over" as an inadequate method of checking and have presented two different methods of checking the most difficult parts of the computations. This computation of the sum of the squares can be checked either by computing some terms in two different ways or, preferably, by the use of the Charlier check.

STANDARD DEVIATION OF A FINITE POPULATION

In Chapter 5, we will go into detail concerning samples, populations, and computations made from them. Before leaving the standard deviation, however, it is desirable to distinguish between the standard deviation of an entire universe or population and the standard deviation of just a portion or sample of that population.

When we have a sample of cases from an infinite population and then decide to compute a standard deviation, we usually have far more interest in the population than we do in this particular sample. Hence, it is appro-riate that in such a computation we obtain the best possible estimate of the standard deviation in the *population,* rather than in the *sample.* This is what we have done so far in this chapter. Thus, the s of 6.94 miles per hour was, technically, not the standard deviation of those 119 scores, but rather an estimate of what the standard deviation would be for a large or infinite number of *such* scores.

In some instances, we are not interested in estimating the population standard deviation; we simply want to know the standard deviation of the scores of the individuals we have measured. (In other words, we want to treat our scores as though they were the entire population.) Such a situation can justifiably arise, for example, with theoretical frequencies. When five coins are tossed, the relative frequency of getting 5 heads, 4 heads, 3 heads, 2 heads, 1 head, or no heads can be given as 1, 5, 10, 10, 5, and 1. Let us treat this not as a sample but as a finite population and compute its standard deviation.

Table 3.4 is set up much like Table 3.3. The various columns are computed in the usual manner. The Charlier check gives

$$\Sigma f(x' + 1)^2 = \Sigma f(x')^2 + 2\Sigma fx' + N$$
$$112 = 48 + (2)(16) + 32.$$

Since the check holds, we may substitute our figures in the formula for the standard deviation of a finite population, which is

$$\sigma_N = i\sqrt{\frac{N\Sigma f(x')^2 - (\Sigma fx')^2}{N^2}}. \qquad (3.5)$$

TABLE 3.4
Computing the Standard Deviation in Coin Tossing

Number of Heads *(1)*	f *(2)*	x' *(3)*	fx' *(4)*	$f(x')^2$ *(5)*	$x' + 1$ *(6)*	$f(x' + 1)^2$ *(7)*
5	1	3	3	9	4	16
4	5	2	10	20	3	45
3	10	1	10	10	2	40
2	10	0	0	0	1	10
1	5	−1	−5	5	0	0
0	1	−2	−2	4	−1	1
Total	32		+16	48		112

The only new symbol is σ_N, which is the standard deviation of a finite population.* Let us now compute σ_N for the coin tossing data

$$\begin{aligned}\sigma_N &= i\sqrt{\frac{N\Sigma f(x')^2 - (\Sigma fx')^2}{N^2}}\\ &= 1\sqrt{\frac{(32)(48) - (16)^2}{32^2}}\\ &= \sqrt{\frac{1{,}280}{1{,}024}}\\ &= \sqrt{1.25}\\ &= 1.12.\end{aligned}$$

As indicated at the beginning of this section, we usually want to compute s rather than σ_N. Some of the principal exceptions to this general rule, and hence the circumstances in which Formula (3.5), or the obvious modification of it for x or for X scores, is used, are:

1. in work with theoretical distributions,
2. in situations in which an entire population can be enumerated and measured (this may be true of the few wild whooping cranes—49 in January 1975—who now make the difference between the continuation of that species and extinction),
3. in situations in which we are interested in the standard deviation of our observations themselves without regard to any population from which they may have been drawn,
4. in cases such as preparing norms for the small group ($N = 20$ or so) who take a new objective comprehensive examination for the Ph.D. the first time it is given,
5. in curve fitting, and
6. in case σ_N is needed in other formulas, as is true in Chapter 10 and in Supplements 8, 10, and 11.

Because $s = \sigma_N\sqrt{\dfrac{N}{N-1}}$, it is obvious that when N is large, the square-root term becomes close to 1.00, so that Formulas (3.3) and (3.5) will give essentially the same answer. For instance, even with an N of only 32 in the coin tossing example just given, $\sigma_N = 1.12$, while $s = 1.14$. However, if we

---

*In Chapter 4, we shall learn that σ (the lower case Greek letter sigma) is used for the standard deviation of a (usually infinite) population. We are adding the subscript N, both to indicate that our population is finite with N cases and to indicate that this and other equivalent formulas always have denominators of N or N^2, and never $N - 1$ or $N(N - 1)$.

use similar data when only 3 coins are tossed, $\sigma_N = .87$ and $s = .93$ with 8 tosses (1; 3; 3; 1) but with 800 tosses (100; 300; 300; 100), both are .87. Since these are obviously the same theoretical distribution, we see (1) that s approaches σ_N as N increases and (2) that the standard deviations for the small and for the large number of tosses are identical, as they should be, when we (correctly) use σ_N rather than s.

STANDARD SCORES

Suppose we have two sets of scores for a group of people, and we wish in some way to obtain a combined score for each person, giving equal weight to each of the two scores.

First, let us illustrate the fact that simple averaging will not suffice. We will use body temperature and pulse rate data for a hypothetical group of nine healthy men. The body temperatures will have a mean of about 98.6 and a standard deviation of about .4, while the mean pulse rate will be around 72 with a standard deviation of about 8. The average combined score will thus be $\frac{98.6 + 72}{2} = 85.3$. Certain other combined scores are shown in Table 3.5. Three different pulse rates were selected—one $2s$ above the mean, one at the mean, and one $2s$ below the mean. Similarly, three different body temperatures were selected—one at $\overline{X} + 2s$, one at $\overline{X}$, and one at $\overline{X} - 2s$. All nine combinations of the three pulse rates with the three body temperatures are shown in Table 3.5, together with the average of each pair. Notice that the first three averages are all in the low nineties, the next three are all around 85, and the last three are all in the high seventies. There is no overlapping from one group to the next. The size of the average is deter-

TABLE 3.5
The Effect of Differences in Variability on the Average of Two Sets of Scores

Patient	Pulse Rate	Body Temperature	Average
A	88	99.4	93.7
B	88	98.6	93.3
C	88	97.8	92.9
D	72	99.4	85.7
E	72	98.6	85.3
F	72	97.8	84.9
G	56	99.4	77.7
H	56	98.6	77.3
I	56	97.8	76.9

mined almost entirely by the pulse rate; whether the body temperature is very high, medium, or very low has only a negligible effect on this average.

Why should one of these variables have so much more influence on the average than the other has? The answer is clearly and directly related to differences in the size of the standard deviations of the two distributions. Simply and solely because the s of the pulse rates (8) is 20 times as large as that of the temperatures (.4), differences in pulse rate have 20 times as much influence in determining differences in the average as do differences in temperature. To illustrate, let us examine the average scores for two pairs of people. The first pair will differ markedly in pulse rate but will be identical in temperature. The other pair will reverse this situation. In Table 3.5, B and H are a pair of persons who differ markedly (88 versus 56) in pulse rate but have the same temperature (98.6); their averages are 93.3 and 77.3, respectively. On the other hand, D and F are two people with the same pulse rate (72) but who differ markedly (99.4 versus 97.8) in body temperature; their averages are 85.7 and 84.9, respectively. Note that the former difference in the averages (93.3 − 77.3 = 16.0) is exactly 20 times as great as the latter (85.7 − 84.9 = .8).

How can we so arrange it that when we combine two sets of scores, the two sets will be equally weighted? Any method that equates the two standard deviations will accomplish this objective. One such method is the use of *standard scores*. Standard scores not only solve this weighting problem, but they also provide scores that are directly comparable, no matter what the variables may be. A standard score is simply a person's distance above (+) or below (−) the mean, expressed in units of the standard deviation. In terms of a formula,

$$z = \frac{X - \overline{X}}{s}, \tag{3.6}$$

in which all terms have been previously defined except z, the standard score, which is sometimes called the standard measure or z score.

Let us now express the data of Table 3.5 in terms of standard scores and average the standard scores. The results are shown in Table 3.6. In the first row, the standard score for A's pulse rate of 88 is determined as follows:

$$z = \frac{X - \overline{X}}{s} = \frac{88 - 72}{8} = \frac{16}{8} = +2.0.$$

The standard score for A's body temperature of 99.4 °F is similarly determined:

$$z = \frac{X - \overline{X}}{s} = \frac{99.4 - 98.6}{.4} = \frac{.8}{.4} = +2.0.$$

TABLE 3.6
The Average of Two Sets of Standard Scores

Patient	Pulse Rate	Body Temperature	Average
A	+2.0	+2.0	+2.0
B	+2.0	.0	+1.0
C	+2.0	−2.0	.0
D	.0	+2.0	+1.0
E	.0	.0	.0
F	.0	−2.0	−1.0
G	−2.0	+2.0	.0
H	−2.0	.0	−1.0
I	−2.0	−2.0	−2.0

The average of these two values is, of course, also +2.0, as shown in the last column. Notice that these averages are no longer influenced predominantly by pulse rate. Instead, both variables are now seen to be equally influential in determining variations in the averages shown in the last column.

The sizes of the two arithmetic means have no effect on the relative weight of the variables being combined. With two sets of scores, the standard deviations alone determine the relative weights that the variables receive. When scores on three or more variables are combined, the relative weights are determined both by the standard deviations and by the extent to which the variables are related to each other. But, contrary to popular opinion, *the size of the mean has no effect whatever on the weight that scores receive when they are combined.*

Suppose we wish to weight two variables unequally—say, to give three times as much weight to the first variable as to the second. All we need to do is multiply the standard scores on the first variable by 3 (or, if you prefer, divide those on the second variable by 3) and then add the two sets of scores to obtain the properly weighted combined score. Thus, a typing test might yield separate scores on speed and accuracy. After expressing both as standard scores, the scores on accuracy might be multiplied by three before combining them with the corresponding standard scores on speed. (The combined scores can then be left as is or divided by 4 to obtain a weighted average. It really doesn't matter, since the relative arrangement is unchanged by such a division. Hence, statisticians often save time—and eliminate one other possibility of error—by working with weighted totals rather than with weighted averages.)

A group of scores, X, will have a mean of $\overline{X}$ and a standard deviation of s. If we subtract $\overline{X}$ from every score, the resulting deviations, x, will have a mean of $\bar{x}$, which equals zero (as noted previously), but their standard deviation will still be s. If now we divide these x scores by s, the mean will remain zero, but the standard deviation will be $\frac{s}{s}$, which equals one. Hence, standard scores always have a mean of zero and a standard deviation of one. They often provide a convenient method for expressing a number of different scores on the same scale. For instance, if some students are given a number of different tests, direct comparisons among the tests are frequently impossible or, at best, extremely difficult to make because of differences in the means and standard deviations of the various tests. If standard scores are used, all scores become immediately comparable—not only between different people on the same test, but also between different tests of the same person, and even between one test on one person and another test on a different person. Thus, if Dorothy has a standard score of +1.8 in Economics, while Frederic has a standard score of + 1.3 in Geography, we may with complete accuracy say that Dorothy is one half a standard deviation better in Economics than Frederic is in Geography.

Standard scores have many advantages, both in comparing and in combining scores. They are easy to compute and easy to understand. They have two minor disadvantages: signs (half are positive, half negative) and decimals (usually only one decimal, but still troublesome). To overcome these disadvantages, a number of other types of standardized scores have been proposed and used to varying extents. Several of these are discussed in Supplement 3. The standard score is the most basic, and, for most purposes, standard scores will be found to be highly satisfactory.

OTHER MEASURES OF DISPERSION

There are many measures of variability in addition to the standard deviation. Perhaps the next most useful measures are the average deviation and the 10 to 90 percentile range. The former, already briefly referred to in this chapter, is treated, along with a number of other measures of dispersion, in Supplement 3. The discussion of the 10 to 90 percentile range, along with that of a few other measures of variation that are based on percentiles, is placed in Supplement 6.

SUMMARY

The best measure of dispersion is usually the standard deviation. It is ordinarily the best measure to use in determining to what extent scores cluster around or vary widely from the arithmetic mean. It may be computed conveniently either from ungrouped or from grouped data. The standard deviation is also needed for the computation of standard scores, which provide a sound method of comparing and of combining scores.

SUPPLEMENT 3
HOW TO SQUARE NUMBERS

All the various formulas for calculating the standard deviation involve dealing with squares of each score (or its derivative) and extracting a square root to obtain the final answer. This is not particularly difficult, but it is tedious and involves a lot of plain hard work! Anything we can do to reduce the amount of computational work will be to our advantage. There are many ways to simplify this labor. The choice among them will depend upon (1) your own background and temperament, and (2) what mechanical, electrical, or electronic aids you have at hand.

Let us suppose that you do not have a machine (not even an adding machine). If you are good at simple arithmetic, you can probably compute the standard deviation fairly well by a raw score longhand method, especially for a small number of cases and for scores that are all whole numbers. Grouping and computing by coded scores may be even easier.

If one computes s for X scores by longhand, then it helps to be able to square the numbers mentally in order to obtain ΣX^2. Following is a scheme for mentally squaring whole numbers from 1 to 130.

1. Memorize the squares of the first 12 numbers (1 through 12).
2. The squares of 10, 20, 30, 40, 50, 60, 70, 80, and 90 are obvious. Just square the left digit and write two zeros at the right. For example, $30^2 = 900$.
3. Numbers ending in 5 (i.e., 15, 25, 35, 45, 55, 65, 75, 85, and 95) can be squared by the "½ rule." The ½ rule is illustrated by $75^2 = (70)(80) + 25 = 5{,}600 + 25 = 5{,}625$. You simply take the product of the nearest multiples of 10 on each side and write 25 after it.
4. Numbers one smaller or one larger than a known square can be squared by the "adjacent rule." The formulas are

   $$(X + 1)^2 = X^2 + [X + (X + 1)]$$

   and

   $$(X - 1)^2 = X^2 - [X + (X - 1)].$$

   For example, if you want 31^2, you know that $30^2 = 900$, so $31^2 = 900 + (30 + 31) = 961$. Also, $29^2 = 30^2 - (30 + 29) = 900 - 59 = 841$. But, in the case of 34^2, don't use 1,225 − (35 + 34) = 1,156 because the mental subtraction is too difficult, although you may use 36^2 = 1,225 + (35 + 36) = 1,225 + 71 = 1,296.

5. The numbers from 41 to about 75 can be squared with the "near 50 rule." By this rule, the square of any number is equal to 2,500 ± 100 times the distance that number is from 50 + (never minus) the square of that distance. Thus, if we want the square of 47, which is a distance of 3 from 50, that square is 2,500 − 300 + 9 or 47^2 = 2,500 − 300 + 9 = 2,209. Similarly, 57^2 = 2,500 + 700 + 49 = 3,249. For numbers greater than 50, another way of wording this is to say that any number over 50 can be squared by adding to 2,500 as many hundreds as the difference between 50 and the number, plus the square of the difference between 50 and that number. For example, 67^2 = 2,500 + 1,700 + 289 = 4,489.

6. Numbers between 75 and 100 can be squared very easily. Start with any number X between 75 and 100. Write down the number that is twice as far from 100 as X is and put two zeros after it. To this, add the square of the difference between X and 100.

Example 1. Start with X = 97. Write down the number that is twice as far from 100 as 97 is (100 − 97 = 3; twice 3 is 6; 100 − 6 = 94) and put two zeros after it (**9400**). To this (9400) add the square (**9**) of the difference (3) between 97 and 100 (9400 + 9 = **9409**). Hence, 97^2 = **9409.** Note that all you need to record are the three numbers printed in boldface type: the number twice as far away with 00 after it (**9400**), the difference squared (**9**), and the sum of the two (**9409**).

Example 2. Start with X = 91. Write down the number that is twice as far from 100 as 91 is (100 − 91 = 9; twice 9 is 18; 100 − 18 = 82) and put two zeros after it (**8200**). To this (8200) add the square (**81**) of the difference (9) between 91 and 100 (8200 + 81 = **8281**). Hence, 91^2 = **8281**.

Example 3. Start with X = 96. From this get **9200** and **16.** Add and see that 96^2 = **9216.**

7. Numbers between 100 and about 130 (such as 112) can be squared by taking a number twice as far above 100 as the one we started with (thus getting 124), adding two zeros to it (getting 12,400), squaring the difference (or 12) between the original number and 100, and adding this square (here 144) to the number with two zeros after it (here 12,400), thus getting 12,544. For example, 109^2 = 11,800 + 81 = 11,881; 131^2 = 16,200 + 961 = 17,161.

8. The preceding rules take care of the squares of all but about a dozen of the numbers from 1 to 130. If you care to complete the set by learning them, here they are:

X	17	18	22	23	24	27	28	32	33	34	37	38
X^2	289	324	484	529	576	729	784	1024	1089	1156	1369	1444

If you see fit to memorize the squares of the numbers from 1 to 12, if you also memorize the 12 much more difficult squares just given, and if you learn the simple rules given in sections 2–7, you will be able to write down instantly the squares of 130 numbers or to enter them rapidly in your electronic calculator whenever you need them. Of course, if you don't want to do this, you can always look them up in Table Z in the Appendix at the back of the book or multiply the number by itself in order to get its square.

There is an advantage to having the square of a score in your head in calculating ΣX^2. One merely multiplies the square by the corresponding frequency to save some labor. This saves the time needed to look up the square in a table or to square the number on a calculator. This is particularly advantageous with those pocket electronic calculators that have severely limited capacity for sequential or cumulative operations.

METHODS OF EXTRACTING SQUARE ROOTS

In computing the standard deviation, there are many ways to make mistakes, but there are probably far more mistakes made in extracting square roots than in any other single operation. In view of that and in view of the emphasis on checking which seems to be present throughout this book, get ready for a surprise. In this section, we shall give you nine different methods for obtaining square roots. Pick any one you like, but don't bother to check anything yet—just plow through and get an answer. It will probably be right or somewhere near right, but don't bother about it. Your objective is to get some kind of an answer—right or wrong. Then look in the next section of this supplement and pick a method for checking it. If your answer happens to be right, you'll find it out very quickly; if it is wrong, you can probably get the right answer in less time than it would have taken you to check your work.

It is, however, necessary to get off to a good start. First, let us note that any number that consists of a string of 4s will have a square root that starts either with 2 or with 6. The reason for this is that the square root of 4 is 2, and the square root of 44 is something between 6 and 7 since 6^2 is 36 and 7^2 is 49. Further, you can look up the square root of 444 in Table Z and find it to be 21.0713. Thus, all these square roots start with either 2 or 6. Table 3.7 tells you which, and also shows you where the decimal points are. For example, the square root of 444,444, which is listed as 44′44′44 is shown to be $6xx$. This tells you that $\sqrt{44'44'44}$ starts with 6 and has two other digits before the decimal point—it is between 600 and 699. Let's now look at the numbers, N, in Table 3.7 and their square roots. The small x's in the table are merely fillers to represent numbers whose size needn't concern us now. Note that the first number, which was actually 444,444,444,444, was rewrit-

TABLE 3.7
Effect of Size of N on Decimal Point in $\sqrt{N}$

N	$\sqrt{N}$
44′44′44′44′44′44.	6*xxxxx*.
4′44′44′44′44′44.	2*xxxxx*.
44′44′44′44′44.	6*xxxx*.
4′44′44′44′44.	2*xxxx*.
44′44′44′44.	6*xxx*.
4′44′44′44.	2*xxx*.
44′44′44.	6*xx*.
4′44′44.	2*xx*.
44′44.	6*x*.
4′44.44	2*x*.
44.44	6.*xx*
4.44	2.*xx*
.44′44	.6*xxxxx*
.04′44	.2*xxxxx*
.00′44	.06*xxxxx*
.00′04′44	.02*xxxxx*
.00′00′44	.006*xxxx*
.00′00′04	.002*xxxx*

ten in groups of two digits. There were 6 such groups and there are also 6 digits in its square root. The next number also contains 6 groups and there are 6 digits in its square root. The rule is that the number of groups in N to the left of the decimal point is always the same as the number of digits to the left of the decimal point in $\sqrt{N}$. With N smaller than 1, the number of complete pairs of zeros between the decimal point and the first digit of N will be equal to the number of single zeros between the decimal point and the first digit of $\sqrt{N}$.

The idea behind the preceding tabulation is basic both to understanding where to start and to knowing where to put the decimal point when you get the answer. With the first two methods in the following list, you need not know where to start, but these methods won't work for most of the numbers whose square roots you need. With all the other methods, you *must* know whether you are starting with one or with two digits in the leftmost group; that is, are you starting with 4 or with 44?

Here are the nine methods for extracting square roots:

1. The best way to find the square root of any whole number between 1 and 999 is to look it up in Table Z. To get $\sqrt{123}$, find 123 in the N column and read the answer as 11.0905 in the $\sqrt{N}$ column.

The best way to find the square root of any 4-digit number ending in zero is to look up the first 3 digits in the N column of Table Z and read the answer from the $\sqrt{10N}$ column.

To get $\sqrt{2340}$, find 234 in the N column and read the answer as 48.3735 in the $\sqrt{10N}$ column.

2. A good way to get a good approximation to the square root of any number (either a whole number or a whole number with decimals) between 1 and 1,000,000 is to find the number closest to it in the N^2 column of Table Z and read the answer from the N column.

To get $\sqrt{123,456.7898}$, find the number closest to it, namely, 123201, in the N^2 column and read the answer as 351 in the N column.

Note A. In some instances, especially with small numbers, this method will give a result that will be too small by 1 unit in the last place. Thus, if we wish $\sqrt{71,022.33}$, we shall find that 71,022.33 in the N^2 column is closer to 70,756 than it is to 71,289, and we shall erroneously conclude that $\sqrt{71,022.33}$ is 266 instead of 267. (Since $\sqrt{71,022.33}$ is 266.500150, this isn't much of an error, but it is wrong if we want the answer to the nearest integer.)

Note B. With this method you will get a square root with no more than three digits. If this is sufficient—good. If it is not, then you should continue with some other method, such as 4. below.

3. The ideal way to get a square root is to use an electronic calculator that has a $\sqrt{\ }$ key or to use a large computer that has been programmed to give square roots. These will give square roots that will almost always be accurate to whatever number of decimal places you have specified.

4. One of us prefers and generally uses this method of division and successive approximations. This works fine with any machine that will divide, and it will quickly produce results with any predetermined standard of accuracy, up to the capacity of the machine. It is based on the fact that if A is an approximation to the square root of N, then $\frac{(A+B)}{2}$ is a better approximation where $B = \frac{N}{A}$. Let us illustrate.

Let's say we want $\sqrt{1776.}$ to six decimal places. Pointing off, we get 17'76. We know that 40^2 is 1600 and that 50^2 is 2500. So we guess that the square root is 41.

$$\frac{1776}{41} = 43.3171, \qquad \frac{41 + 43.3171}{2} = 42.15855,$$

$$\frac{1776}{42.15} = 42.135231, \qquad \frac{42.15 + 42.135231}{2} = 42.1426155,$$

$$\frac{1776}{42.142615} = 42.1426150228. \qquad \text{Correct answer is } 42.14\ 2615.$$

As another example, we want $\sqrt{(120{,}100)(68{,}446)}$ to the nearest whole number. Pointing off the product, we get 82′20′36′46′00. We know that 81 is 9^2, so we guess 90,000 as our first approximation. Then,

$$\frac{8{,}220{,}364{,}600}{90{,}000} = 91{,}337.38, \qquad \frac{90{,}000 + 91{,}337.38}{2} = 90{,}668.69,$$

$$\frac{8{,}220{,}364{,}600}{90{,}668} = 90{,}664.45, \qquad \frac{90{,}668 + 90{,}664.45}{2} = 90{,}666.225,$$

$$\frac{8{,}220{,}364{,}600}{90{,}666} = 90{,}666.45. \qquad \text{Correct answer is } 90{,}666.$$

Now let us take an example in which we start out with a really lousy guess. Suppose that we pointed off the preceding example correctly and then guessed that the square root of 82′20′36′46′00 was 66,000. How many weeks would it take us?

$$\frac{8{,}220{,}364{,}600}{66{,}000} = 124{,}551, \qquad \frac{66{,}000 + 124{,}551}{2} = 95{,}275.5,$$

$$\frac{8{,}220{,}364{,}600}{95{,}275} = 86{,}280, \qquad \frac{95{,}275 + 86{,}280}{2} = 90{,}777.5,$$

$$\frac{8{,}220{,}364{,}600}{90{,}777} = 90{,}555.59, \qquad \frac{90{,}777 + 90{,}555.59}{2} = 90{,}666.295,$$

$$\frac{8{,}220{,}364{,}600}{90{,}666} = 90{,}666.45. \qquad \text{Correct answer is } 90{,}666.$$

Note A. Nobody is likely to guess that the square root of 82′20′36′-46′00 is 66,000 or even 124,551 (which would have resulted in the same computations after the first division). But even with such horrible guesses, we got the right answer after only 3 big divisions and 3 averages. The fourth big division proved that we were right. Of course, if you can make a good guess, the answer will be obtained much more quickly.

Note B. Because the average of the two numbers will always be larger than the square root, we always discarded some of the decimals or used a number slightly smaller than the average as our next divisor. If we had made bigger deductions, we would have reached our objective sooner.

Note C. Whenever division gives you a quotient that is the same as your divisor, you have the correct square root. You might think that if you divide by 90,666 and get anything between 90,665.51 and 90,666.49 you would be right. You would, but the range is larger than that. Any quotient between 90,665.01 and 90,666.99 is all right because, when averaged with 90,666, these will still give averages closer to 90,666 than to any adjacent numbers. So, if your quotient differs from the divisor by less than 1 in the last figure you are going to use, the divisor was right.

5. What we will call the "longhand square root method" is illustrated below. It used to be taught in arithmetic classes. It is hard to explain, and it takes longer than some of the other methods, but it is accurate if you make no mistakes in arithmetic. We shall illustrate it with 1776. Write 17′76.00′00′00′ and later add more pairs of zeros if needed. Draw a line above it. Decide what number is the largest square that is smaller than the leftmost group. It is 16. Write it below 17 and write 4 above 17 and also to the left of 17. Subtract and bring down the next group (here 76). Double the part of the square root now showing above the line and write it to the left of the remainder you got after subtracting and bringing down the next group. That number is 8. Decide how many times 8 with another number after it can be subtracted from 176. The answer seems to be 2, so write 2 after the 8 and also above the group 76. Then multiply this 82 by 2, getting 164 and recording it. Subtract and bring down the next group. Double the 42 part of the square root and write it to the left of the 1200. Continue in the same way, as shown below.

```
                 4 2. 1  4  2  6
                ____________________
     4          17'76.00'00'00'00'00
                16
                ____
    82           1 76
                 1 64
                ______
   841             12 00
                    8 41
                   ______
  8424              3 59 00
                    3 36 96
                   _________
 84282                22 04 00
                      16 85 64
                     __________
842846                 5 18 36 00
                       5 05 70 76
                      ___________
842852                   12 65 24 00
```

What happens if you guess too big a number, such as 7 instead of 6? The product will be too big, so you can't subtract it, and you will almost automatically try the next smaller number. What happens if you try too small a number, such as 5 instead of 6? You'll probably notice it, but if you don't, you'll find out when you get to the next step.

We won't go into more detail. Most people will want to use easier methods, but if you like this method, try it by yourself with 1776 and then with another number for which you know the answer.

6. If you know how to use logarithms, and especially if you know how to interpolate, any table of logarithms will quickly give you square roots. Get the logarithm of N, being careful to get the characteristic correct. (If you don't know what the characteristic is, or if you don't know about linear interpolation, skip this section and get your square roots some other way. If you already know these things, this is a fairly fast and accurate method—and you don't need a calculator of any kind to use it.) Divide the logarithm of N by 2; then find the antilogarithm of this number and you have the desired $\sqrt{N}$. Without interpolation, this will give results with about three-figure accuracy. For greater accuracy, you may either interpolate or continue with method 4. With interpolation in a 5-place table of logarithms, you can get results with only small errors in the sixth digit. If this is all you need, check your result by any of the checking methods and correct the error, if any. If you need greater accuracy, about one division by method 4. will give you the accuracy you need.

7. Any slide rule can be used to give an approximation to a square root. This will be fairly accurate for the first 2 or 3 figures and some experts can come close to getting a fourth figure at the left end of the slide rule scale. This is usually not accurate enough for most statistical work, but it is a highly satisfactory way to get a first approximation before continuing with the successive approximation procedure explained in section 4. The reason for this is that if the number you start with is accurate to about 3 figures, what you get after the first division and averaging will be accurate to twice as many—or about 6 figures. After the second cycle, your square root will be accurate to about 12 figures. And there probably won't be a third cycle (except for checking purposes) because your computer won't be big enough to show you the 24 figures—and who would care anyway?

8. Nomographs are graphic computation charts based on the same principle as the slide rule. If you never heard of them, it's all right, even though one of the writers was the coauthor of a book featuring 28 nomographs.* If any line in a nomograph has a scale running from 1 to 100 on one side and from 1 to 10 on the other side of the line, the first one-digit or two-digit group of N may be found on the 1 to 100 side, and $\sqrt{N}$ may be read from the other side of the line. For example, the AB and the $\sqrt{AB}$ sides of a line in Nomograph 65 of the Handbook just cited may be so used. This gives about the same accuracy as a slide rule, so you will usually need to follow your estimate with approximations by the method described in section 4. in order to obtain the desired degree of accuracy.

9. In Table C, we present a method that was designed for use with the Friden electric calculating machine. Friden calculators are no longer

---

*Dunlap, Jack W., and Kurtz, Albert K. *Handbook of statistical nomographs, tables, and formulas*. Yonkers-on-Hudson, N.Y.: World Book Company, 1932. (NOTE: This out-of-print book is available in reprint form from Xerox University Microfilms, Ann Arbor, Michigan.)

manufactured, but Singer Business Machines has graciously granted us permission to reprint this material. Their method can be used on any machine—hand powered, electric, or electronic—that will divide and that does not have a $\sqrt{\ }$ key. (If it has a $\sqrt{\ }$ key, use it, as already suggested in method 3.)

We are not rewriting the directions. If you know how to divide on your own electronic calculator (or any other machine), the changes are obvious. Just set up the problem so that you will get plenty of decimals in your answer.

Note that while this method is extremely rapid, it will sometimes give an error in the fifth digit. So, as usual, check your answer.

Yes, there are still other methods of obtaining square roots. But these nine should give you enough choices. Methods 3, 4, and 5 will give $\sqrt{N}$ to any degree of accuracy you desire. If, however, you are used to any other method or if you prefer it for any reason, go ahead and use it. Even if it isn't accurate enough for your purpose, it will give you an excellent start, and you can then readily finish the job and check the revised answer by continuing with method 4.

HOW TO CHECK SQUARE ROOTS

At the beginning of the preceding section on extracting square roots, we stressed the importance of getting off to a good start. This is even more important in checking square roots.

Start by pointing off your number, N. Begin at the decimal point and put accent marks after every two digits in both directions. The leftmost group may contain either one or two digits; all others will contain two digits.

If N is greater than 1.00, as it usually is, the number of *groups* (counting the leftmost one) to the left of the decimal point in N is the same as the number of *digits* to the left of the decimal point in $\sqrt{N}$. Further, if there are two digits in the leftmost group, $\sqrt{N}$ will start with 4, 5, 6, 7, 8, 9, or possibly 3; while if there is only one digit in the leftmost group, $\sqrt{N}$ will start with 1, 2, or possibly 3.

If N is less than 1.00, the number of groups to the right of the decimal point that contain 00 is the same as the number of zeros between the decimal point and the first digit of $\sqrt{N}$.

Now that you are properly started with the accent marks telling you how many digits there are before the decimal point in $\sqrt{N}$ (or how many zeros there are between the decimal point and the first digit in $\sqrt{N}$), you are ready to check the value that you hope or believe is the correct square root of N.

1. No matter how you got your square root, you can check it by recomputing it by a different method selected from among the first six given in the preceding section. If both your original method and the (different) checking method are of sufficient accuracy for your problem, and if they give the same answer, there is nothing wrong with this procedure except that it may turn out to be more time-consuming than method 4 below.
2. You have a number, A, which you hope is the correct square root of N. Ignore decimals, treating A as a whole number. Square A. If A^2 is larger than N, subtract A and it should be smaller. If it is, you were right. If A^2 is smaller than N, add A and it should be larger. If it is, you were right.

   For example, with $N = 1776$, suppose we think that 42.1 is the correct square root. Then we square A, that is, 421, and get 177,241. This is smaller than 177,600. We then add 421, getting 177,662. Now this is too large, so we were right and 42.1 is correct.

   Or, with $N = 1776$, suppose we think that 43.1 is the correct square root. Square A and get 185,761. That is too large. Subtract 431, getting 185,330. That is also too large, so we are wrong. Unfortunately, we don't know how far wrong we are. If we are nearly always right, this is a good method of checking, but if we make some mistakes, method 4 below is better because it automatically enables us to correct them.
3. If you are fairly sure that one of two adjacent numbers is the correct square root of N, square the number halfway between them. If this square is too large, the larger of the two numbers is wrong. Then square a number one-half a unit smaller than the smaller number; if that square is too small, the smaller of the two numbers was correct. Similarly, if when you squared the number between the two adjacent numbers, it was too small, then the smaller number was wrong. To see if the larger number is right, square a number one-half unit larger than it. If that square is too large, then the larger of the two numbers was correct.

   For example, let's say you think $\sqrt{3'51'00'00.}$ is either 1873 or 1874. Square 1873.5 and get 3510002.25. Since this square is too large, 1874 is wrong. To find out if 1873 is right, we square 1872.5 and get 3506256.25. Since that is too small, 1873 is correct. (The decision is a close one, since the correct square root is 1873.4994, but note that this method enabled us to make the right decision.)
4. The best way to check a square root is by method 4 of the preceding section on extracting square roots. That is the best method because if you are right, you find it out (all the way to the last decimal place you used), but if you are wrong, you are automatically given a better

approximation that another cycle will usually show is correct. Just remember, if the quotient differs from your divisor by less than 1 in the last figure of the divisor, your divisor is right.

Important: Because there are so many ways to get wrong answers in computing square roots, please do not pick a method you like and then use it both to obtain and to check your square roots—unless you use method 4. It is possible to get a wrong answer by the use of any method at all; all we ask is that you not repeat yourself and get the same wrong answer again.

After you become fairly proficient in extracting square roots, method 2 for checking square roots by squaring them is very satisfactory. Its only disadvantage is that if your supposed square root is wrong, you have to start all over. That is why we feel the next paragraph shows a better way.

Regardless of your experience or proficiency, since method 4 is self-checking, we recommend it—both for extracting square roots and for checking them. You may even make several errors in copying figures or in calculations and yet method 4 will still give you the right answer and check it for you. What more do you want?

HOW TO COMPUTE AND CHECK THE STANDARD DEVIATION

To a mathematician, the correct substitution in and solution of Formula (3.3) is simple and straightforward. But most statistics students and statistical clerks are not mathematicians and, whether we like it or not, errors do occur. Let us consider the most common types of errors—*and then avoid them!* The first term under the radical is $N\Sigma f(x')^2$. We know how to multiply N by anything else, but just exactly what does $\Sigma f(x')^2$ mean? It means that each x' value is first squared and then multiplied by the corresponding frequency to form the $f(x')^2$ product. This is the method we used in checking all the values in column (6) of Table 3.1. Then all these products are added. The resulting total may be written in a variety of ways. Thus,

$$\Sigma f(x')^2 = \Sigma f(x'^2) = \Sigma(fx'^2) = (\Sigma fx'^2) = \Sigma fx'^2.$$

Study these expressions until it is clear that each one means that an x' is squared, then multipled by its frequency and that these products are then added. In Table 3.1, each of these five expressions would have the same value, 1,646. There should now be no difficulty in determining N and multiplying by it to obtain the first term in the numerator under the radical sign of Formula (3.3). The second term, $(\Sigma fx')^2$, may look somewhat similar to $\Sigma f(x')^2$, but it is entirely different. Notice that the fx' products are obtained, then they are added, and that only after this sum is obtained is anything squared. No individual x' values are squared. As a matter of fact, the material within the parentheses is exactly the same as we used in obtaining the mean by Formula (2.2). After squaring this sum, the resulting value (always positive) is subtracted, as indicated, from the previously obtained value of $N\Sigma f(x')^2$. This gives the numerator. The denominator may be obtained either by multiplying N by $(N - 1)$ or by looking up this product in Table B. Next, divide the numerator by the denominator and record the quotient, paying special attention to the correct location of the decimal point. (With the usual 15 to 20 intervals, this quotient will practically always be between 1 and 99 and if it is, there will be only *one* digit to the left of the decimal point.) We are then ready to extract the square root.

Square root—ah, there's the place to make errors! The most common error is that of not pointing off properly in the beginning and thus starting to extract the square root of a number $1/10$ or 10 times as great as that desired. (This gives a sequence of numbers that does not even faintly resemble the correct one, one set being $\sqrt{10}$, or 3.1623, times the other.) A related error is the exasperating one of obtaining the digits of the square root correctly and then putting the decimal point in the wrong place. Another error is that of having an error of 1 in the last retained digit—usually caused by getting tired and not bothering to determine whether the next digit would be more or less than 5. Then there is that very large category that can be described only as miscellaneous. Clearly, square root is taught in junior high school, but it is either not learned or most effectively forgotten. Your instructor will either teach you one good method or will refer you to one of the 9 methods of extracting square root that were given earlier in this Supplement. We care little what method is used—there are many good methods—provided the answer is checked. (Four methods of checking the square root are also given in this Supplement.) To return to Formula (3.3), after the square root is extracted, checked, and pointed off (probably with one digit to the left of the decimal point), we multiply by i, the interval width. This gives the correct value of the standard deviation.

SHEPPARD'S CORRECTION FOR COARSENESS OF GROUPING

Whenever we take our scores, group them in a frequency distribution, and then compute with the x' instead of X scores, we introduce a bit of inaccuracy. With the IQs in Table 2.1, we found that the mean, calculated from X scores did not agree exactly with $\overline{X}$ calculated from x' scores, but the difference of .21 IQ point was negligible. In this chapter, we found that we got two slightly different values of s. Again, we regarded the difference of only .11 IQ point as negligible.

The chance errors that affect both $\overline{X}$ and s are usually negligible but we encounter another problem leading to systematic rather than chance errors in computing s from grouped data. We can easily illustrate this. Let us consider a relatively wide interval located a standard deviation or so above the mean. With most distributions, we would expect more cases at the low end of the interval (nearer to $\overline{X}$) than at the high end. In an interval below the mean, we would similarly expect more cases at the high end of the interval (nearer to $\overline{X}$) than at the low end. Notice that these tendencies cancel out insofar as the mean is concerned, but that on both sides of the mean the deviations tend to be overestimated for the standard deviation. As a result, our computed s will systematically tend to be larger than it ought to be.

This error due to grouping is small when we have a reasonably large number of intervals. That is one reason why on page 18 we said that the number of intervals should be between 15 and 20. What happens when, for any reason, the number of intervals is not in that range? If the number is larger, there is no problem. If it is smaller, it is appropriate to make an adjustment in the size of the standard deviation in order to compensate for this unintentional and unwanted overestimation in its size. This is done by means of Sheppard's correction.

The easiest way to apply Sheppard's correction is simply to subtract $\frac{1}{12}$ before extracting the square root. If that sounds very simple, it is. Here are Formulas (3.3) and (3.5) rewritten to incorporate this correction with the subscript in front of the symbol referring to the correction:

$$ {}_{c}s = i\sqrt{\frac{N\Sigma f(x')^2 - (\Sigma fx')^2}{N(N-1)} - .0833}, \qquad (3.7)$$

$$ {}_{c}\sigma_N = i\sqrt{\frac{N\Sigma f(x')^2 - (\Sigma fx')^2}{N^2} - .0833}. \qquad (3.8)$$

Let us try to illustrate the use of Formula (3.7) with the daytime speeds of southbound cars. When we used Formula (3.3) we obtained

$$\begin{aligned} s &= 2\sqrt{12.05} \\ &= (2)(3.47) \\ &= 6.94. \end{aligned}$$

If we now use Formula (3.7) we will obtain

$$\begin{aligned} {}_{c}s &= 2\sqrt{12.05 - .0833} \\ &= 2\sqrt{11.9667} \\ &= (2)(3.46) \\ &= 6.92. \end{aligned}$$

The illustration is a flop. There were so many intervals that the correction became unimportant. That will be true, we hope, with nearly all of our other illustrations. But, solely to illustrate Sheppard's correction, let us retabulate those auto speeds into ten mile per hour intervals: 20–29, 30–39, 40–49, 50–59, and 60–69. Then Formula (3.3) will give us

$$\begin{aligned} s &= 10\sqrt{.5919} \\ &= (10)(.7694) \\ &= 7.694, \end{aligned}$$

while the use of Formula (3.7) will give

$$\begin{aligned} {}_c s &= 10\sqrt{.5919 - .0833} \\ &= 10\sqrt{.5086} \\ &= (10)(.7132) \\ &= 7.132. \end{aligned}$$

For the data tabulated in intervals of 10 mph, we see that the usual formula (3.3) gave 7.69, which grossly overestimated the correct value of 6.92 or 6.94, and that Sheppard's correction lowered the 7.69 to 7.13. This failed to hit the correct value on the nose but its miss of .21 is far closer than the .77 discrepancy given by Formula (3.3).

Other textbooks sometimes point out that Sheppard's correction becomes unimportant when the number of intervals reaches 10, 12, 15, or 18. With this we can wholeheartedly agree; the 20 intervals we used for these car speeds made the correction completely unnecessary and accounted for the ''flop'' referred to a couple of paragraphs ago. Notice also that the rule we gave of trying to get between 15 and 20 intervals takes care of the situation very nicely.

One final comment: If you see a formula that purports to be Sheppard's correction but which contains $\frac{i^2}{12}$ or $\frac{N^2}{12}$ in it, don't panic. It's just a more complicated way of getting exactly the same answer we get with Formula (3.7) or (3.8).

SOME OTHER MEASURES OF DISPERSION

There are at least eight measures of dispersion. For most situations, the best of these is the standard deviation, which was the subject of Chapter 3. Four other measures of variability will be discussed in this supplement. Two of them, the variance and the probable error, are minor variations of the standard deviation. Then two more, the average deviation from the median and the average deviation from the mean, will be shown. Three others will be treated in Supplement 6, since they are closely related and are based upon percentiles, the subject of Chapter 6.

Variance The variance is the square of the standard deviation. Naturally, if the records are expressed as inches, the variance will be in terms of square inches. Okay, but if the scores are IQ points, the variance will be in terms of squared IQ points—whatever they are. Hence, the variance is not ordinarily used directly as a measure of dispersion, but it is used in other ways; for example, it is basic to the analysis of variance in Chapter 13.

Probable Error At one time the probable error was widely used. It still is used to a limited extent in some other scientific fields, but its present use is extremely limited in Psychology and Education. This is as it should be. Its formula is

$$\text{PE} = .6745s. \tag{3.9}$$

It came into existence because one-half of the cases in a normal distribution are included between a point about two-thirds of a standard deviation below the mean and another point—about two-thirds of a standard deviation above the mean.* There will probably be no need for you to use it, but you should know what it is if you see references to it in the literature you read.

Average Deviation from the Median The average deviation as its name clearly indicates, is simply the arithmetic mean of the absolute values of the deviations from a measure of central tendency. (The expression *absolute value* means the size of a value or quantity, taken without regard to its sign.) Although the average deviation could be computed from any measure of central tendency, in practice, the deviations are nearly always measured from the median or from the mean. Because the average deviation is a minimum (and hence a unique value) when measured from the median, *AD from Mdn* is probably slightly preferable to *AD from* $\overline{X}$, so we are presenting it first. The formula is

$$AD\ from\ Mdn = \frac{\Sigma|X - Mdn|}{N}, \tag{3.10}$$

where the symbol $|\ \ |$ means "the absolute value of." This means that we completely ignore signs and add all the deviations together. Thus, if some of the deviations are +4, −9, +2, +5, −1, and −3, we simply add 4, 9, 2, 5, 1, and 3, getting 24.

We have said that the standard deviation is usually the best measure of variability. Why, then, should we use the average deviation? Let's be frank about it. One of the chief reasons that people want to avoid using the standard deviation is related to laziness. This difficulty has been greatly exaggerated although it is true that under most circumstances other mea-

---

*More precisely, .67448975s.

sures of dispersion can be computed in less time than it takes to compute the standard deviation. (Whether this really is a good way to save time is doubtful, however.) Another reason for wanting another measure of dispersion is a desire to give only a relatively small amount of importance to extreme deviations. The average deviation takes care of this objection quite well; it does not ignore extreme deviations, but it does attach less weight to them than does the standard deviation. If you feel this way, then you should probably prefer the median to the mean for the same reason. Hence you should probably use the preceding formula, rather than the one that will follow for the average deviation from the mean.

Average Deviation from the Mean By implication, *AD from* $\overline{X}$ was defined in the preceding section of this supplement. Its formula is

$$AD \textit{ from } \overline{X} = \frac{\Sigma|x|}{N}, \tag{3.11}$$

in which all the terms have previously been defined. The best way to get $\Sigma|x|$ is: (1) add all the positive deviations and record their sum; (2) add all the negative deviations and record their sum; (3) compare the two sums, ignoring signs, and, if they ever differ by as much as $\frac{N}{2}$, recompute until you find the error; and finally (4) ignore the signs and add the two (nearly equal) sums together. Then when you divide by N, you have computed AD from $\overline{X}$.

In a normal distribution, both AD from the median and AD from the mean are equal to $.7979s$. For other types of distributions, this will not be true, AD varying from being as large as s to being only a small fraction of s in size.

TYPES OF STANDARD SCORES

Once you have the basic type of standard score (z) it is easy to convert zs to many new standard score scales. This can be done with the appropriate conversion formula from the general formula

$$\text{New standard score} = (S')(z) + \overline{X}', \tag{3.12}$$

where S' is the standard deviation of the new standard score scale, and $\overline{X}'$ is the new mean. For example, to convert to Z scores we would use the formula

$$Z = 10z + 50.$$

For college board scores, the formula would be

$$\text{CEEB} = 100z + 500.$$

TABLE 3.8
Means and Standard Deviations of Some Well-Known Standard Score Scales

Standard Score	Mean	Standard Deviation
T Scores*	50	10
Z Scores	50	10
CEEB Scores	500	100
AGCT Scores	100	20
Deviation IQs	100	15
Wechsler Subscales	10	3
ITED Scores	15	5
Stanines	5	2

*These *T* scores result from an area transformation; *Z* scores do not. See pages 189–191.

Table 3.8 shows the means and standard deviations for several common standard score scales. In Supplement 7, we shall learn how to normalize distributions. In some of the following cases, the distribution is normalized; in others it is not.

—Sarasota (Florida) Herald Tribune

CHAPTER 4 NORMAL PROBABILITY CURVE

By far the best known distribution curve in statistics is the normal probability curve.* Unlike some curves with two peaks, this one has only one high point. One reason it is so well known is that it has so many uses. Since the normal probability curve is widely used, it is important for us to know about some of its uses.

THE NATURE OF THE NORMAL PROBABILITY CURVE

The normal curve serves as a model for many sets of numerical data which we encounter in practice. Being able to justify the use of the normal curve means that we can make much better descriptions and interpretations of our data than otherwise. We should hasten to add, however, that the normal curve is a mathematical abstraction. It does not exactly describe nature, but in many cases it gives us the best explanation we can get of our data.

*Some of the many other names that it has been called are Gaussian error curve, theoretical frequency curve, bell-shaped curve, curve of the normal law of error, normal distribution curve, normal error curve, and, most briefly, the normal curve.

The normal distribution has been very useful in educational and psychological measurement in developing measuring scales which describe the status of a person or object or stimulus on that scale relative to other persons, objects, or stimuli. Among the most familiar examples of such scales would be the three-digit scores reported for an applicant on the Scholastic Aptitude Test or the Graduate Record Examination.

The normal curve has been used as a basis for the mathematical derivation of many formulas in statistics. In some formulas and methodologies the normal distribution of data is assumed. Therefore, in order to know if we are meeting the assumptions, we need to know what the normal distribution is. These are not the only uses of the normal curve, but they are three of the most general uses. There is a multitude of specific uses in many disciplines.

We said that the normal curve makes a good model for many sets of data. But how do we know when it would be likely to provide a good model? The answer is to be found from a consideration of or an examination of the kind of data we have under study. We know that certain kinds of data consistently yield what are called *symmetrical distributions* (where the mode, median, and mean are close to each other). Examples of such data would be:

1. scores on many educational and psychological tests, assuming a fairly large number of test items (say several dozen) and at least a fairly large sample of persons taking the test (say 100 or more);
2. measurements on some physical variables (depending in part on heredity) such as height, weight, etc.;
3. many measurements in which chance is involved (These are most easily illustrated by games of chance such as the number of times each possible point comes up when two dice are rolled, or any score which is obtained solely by flipping coins.);
4. measurements which depend upon many causes (An example of this would be test scores again, because there are so many variables which affect test performance. Another example would be a distribution of mental ability for a freshman class in a college which drew graduates from many different high schools.);
5. scores that, as we know from past experience, usually yield a normal distribution (This is not a good enough reason to jump to the conclusion that the distribution is normal. We should go to example 6 to be sure.);
6. data which we have tested for normality (This refers to an empirical test, which means that we rigorously examine how well our distribution of scores can be fitted by a normal probability curve. There are several ways to do this but we do not need to go into them at this time.).

We have looked at the applications of the normal curve. But what about its theoretical nature? There are two ways we can approach the theoretical aspects of the normal curve. We can (1) study the mathematical equation for the curve itself, plotting the actual curve from ordinates, or we can (2) study the relationships among various distances along the base line (horizontal scale) of the curve to the area under the curve between the mean and these distances. We will make both kinds of studies in this chapter.

As you may remember, we saw a picture of the normal probability curve in the preview of chapters on page 5. You have probably already noticed both from the tally marks in Table 2.4 on page 20 and from Figure 2.12 on page 35 that our distribution of IQs conforms closely to this theoretical curve.

Most test scores and many other mental measurements are distributed either normally or nearly so. If a trait such as intelligence is normally distributed, but all we have is a sample such as 176 IQs, the normal probability curve enables us to make inferences and estimates that are no longer specific to our one limited sample of 176 cases.

Thus, with our known $\overline{X} = 110.01$ and $s = 15.20$ we can by means of Table E make estimates of how many students have IQs above or below any given score or even between any two scores we care to choose. These estimates are not obtained by counting (which, of course, is quite accurate for this specific sixth grade class), but by, in effect, smoothing our obtained distribution so that our results will be applicable to other classes. This smoothing is done by fitting a normal probability curve to the scores shown in Table 2.4. We don't have to fit a curve to the data to do this. Later in this chapter, we shall see that by merely looking up one or two numbers in Table E we can make such statements as:

1. 90% of all such students have IQs between 85 and 135;
2. 15% of all such students have IQs below 94; or even
3. only 1 student in 1000 would be expected to have an IQ above 157.

This last is a remarkable statement since *nobody* in Table 2.4 scored that high. What basis could we possibly have for making such a wild statement? Answer: The assumption that the IQs are distributed normally; and in this case, at least, this is a reasonable assumption.

THE ORDINATES OF THE NORMAL PROBABILITY CURVE

We have said that the normal probability curve is symmetrical, and its height continues to decrease, though not uniformly, as we deviate in either direction from the maximum height, which is at the mean. The decline is at first very gradual, but it soon increases so that at one standard deviation above (or below) the mean, the height of the normal probability curve is

FIGURE 4.1
The Normal Probability Curve

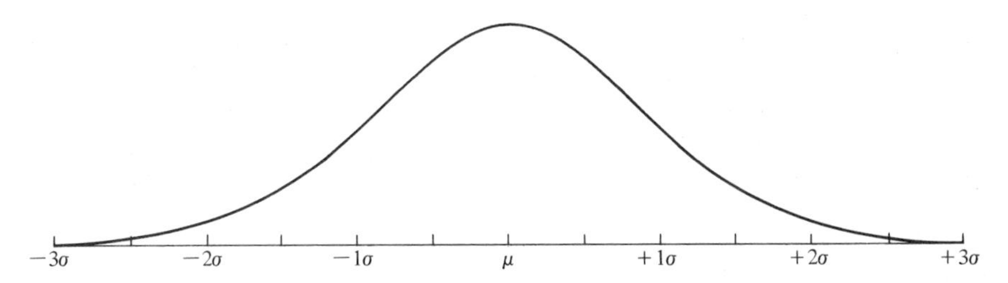

changing at its most rapid rate.* The descent continues, and at two standard deviations above the mean its height is less than one-seventh of the maximum. By the time we reach three standard deviations, the height is only 1% of the original value. The curve never reaches the base line, but it gets extremely close to it. Of course, the most precise way to define this curve is to give its equation. This we shall now do, but we shall then give illustrations that, together with Figure 4.1, should clarify its meaning.

The equation of this remarkable curve is

$$\mathscr{y} = \frac{N}{\sigma\sqrt{2\pi}} e^{-x^2/2\sigma^2}. \tag{4.1}$$

The symbol $\mathscr{y}$ stands for the ordinate,† or height, of the normal probability curve at the point that divides the area into two parts—one containing the proportion p and the other the proportion q of the N cases. Because this is a theoretical curve, it is appropriate to use σ, the symbol for the standard deviation in a large theoretical population, rather than s, the symbol for the standard deviation in our sample. The symbols N, x, and π have their usual meanings: the number of cases, a deviation from the arithmetic mean, and 3.14159265, respectively. The symbol e is widely used in mathematics and is the base of the system of natural logarithms. For our own purposes, it is simply a constant whose value is 2.71828183.

Since the normal probability curve is defined by an equation, it is, of course, a mathematical abstraction which happens to occur frequently in our own and many other fields. Because Equation (4.1) is hard to visualize,

---

*This point is frequently called the "point of inflection." On a ski slope it would represent the steepest part of the slope. One can easily look at a normal curve and estimate the position of this point; then it is easy to estimate the length of the standard deviation by dropping a perpendicular from the point to the base line and measuring the distance to the mean.

†Note that we are using $\mathscr{y}$ as the ordinate of the normal probability curve in order to distinguish it from the usual printed y, which is often used as the ordinate of *any* curve or as a deviation from the mean of the second variable.

even for a mathematician, one of the writers simplified it so that we could get a better idea of it and actually see how it works. We shall now present this new equation and illustrate it by computing a few values to reveal the shape of the curve.

Let us now simplify Equation (4.1) a bit in order to make it easier to compute a few values of y at various distances from the mean. We can get rid of the N by working with relative frequencies, which is another way of saying that we are treating N as 100% of our cases, or $N = 1$. Similarly, we can get rid of σ by working with standard scores in which $z = \frac{x}{\sigma}$. These two changes, together with a little algebra and arithmetic as shown in Supplement 4, give the following greatly simplified formula:

$$y = (.3989)(.6065)^{z^2}. \qquad (4.2)$$

Equations (4.1) and (4.2) don't look much alike, but, except for the factor of N, they yield exactly the same answers. Equation (4.2) is obviously easier to experiment with, and if we multiply the answers obtained from it by N, we shall get the same values as would have been obtained through Equation (4.1). We shall now substitute a few values of z in Equation (4.2) and solve for y.

Let us first find the height of the maximum ordinate, which is at the mean. Here the deviation, z, will be zero; hence the exponent will also be zero. Since any finite number (except zero itself) raised to the zero power is equal to 1, it follows that $(.6065)^0$ will equal 1 and y will equal .3989. Hence, we may say:

Whenever $z = 0$, y will equal .3989.

Let us see what the situation is a standard deviation away from the mean. If $z = +1$, the exponent will be 1. If $z = -1$, the exponent will also be 1. Since both positive and negative values of z produce identical values of the exponent, it is obvious that the curve is symmetrical. That is, it doesn't matter whether we go 1σ above the mean (where $z = +1$) or 1σ below the mean (where $z = -1$); the height of the ordinate, y, of the normal probability curve will be the same in either case. When $z = 1$, the exponent, z^2, will equal 1, so y will equal (.3989)(.6065). This multiplies out to .2419, but if we had carried more decimals, we would have obtained the correct value of .2420. Hence:

Whenever $z = \pm 1$, y will equal .2420.

If we go two standard deviations away from the mean, z will equal +2 or −2 and the exponent in Formula (4.2) will equal 4. This means that .6065 should be raised to the fourth power, yielding .1353. Again multiplying by the constant, .3989, we obtain .0540.

Whenever $z = \pm 2$, y will equal .0540.

Additional arithmetic is involved in the computation at $z = +3$ and the resulting value of the ordinate is very small. To raise .6065 to the 9th power, first square .6065 on your calculator, square that answer getting .1353 as we just did for the fourth power, then square the .1353 getting the eighth power, and finally multiply this by the .6065 getting only .0111 as the ninth power. As usual, we multiply this by our constant, .3989, and obtain only .0044 as the ordinate at $+3\sigma$. The normal probability curve has not reached the base line, but it is certainly getting close.

Whenever $z = \pm 3$, y will equal .0044.

We have now computed 4 of the 350 values that appear in the top part of Table D. These are definitely among the easiest ones to solve. For example, to obtain the ordinate at 1.53σ, we would first have to raise .6065 to the 2.3409th power. Want to try it? It can be done all right, either slowly by using logarithms or rapidly by using a hand calculator with a y^x button, but why bother? Statisticians before us have computed all these values several times; all we have to do is look them up in Table D whenever we need any of them.

To find out just what some of these numbers mean, let us construct a normal probability curve that is a mile high. This is going to be a bit awkward to handle but perhaps we can anchor it at the bottom and rest it against the 1,454 foot Sears Roebuck Building or the 984 foot Eiffel Tower on a calm day. We can let the standard deviation be one-half mile or any other convenient distance—it doesn't matter. How high will the curve be at various distances away from the mean? The answers are shown in column (3) of Table 4.2. The first few of these values are also shown in comparison with other objects in Figure 4.3.

The height at the mean is one mile, the distance that Roger Bannister ran in 3:59.4 when, in 1954, he was the first man to run a mile in less than 4 minutes.

One standard deviation away from the mean, the curve is making its steepest descent (see Figure 4.3), but it is still 3,202 feet above the base line (see Table 4.2). This is about the height of Stromboli Volcano (3,040 feet) on the island of the same name in the Mediterranean Sea.

As the curve continues its rapid descent, at 1.2σ from the mean we come to a practically unknown waterfall that just happens to be the highest in the world. It is El Salto Angel (Angel Falls) in Venezuela and is 2,648 feet high.

A little further away, 1.6σ from the mean, we see the loftiest building in the world, the Sears Tower, 1,454 feet tall. At 1.8σ, we find the 984 foot Eiffel Tower that, built for the Exposition of 1889, has remained standing for about 90 years.

TABLE 4.2
Heights of Equidistant Ordinates of Hypothetical Mile-high Normal Probability Curve

Distance from Mean (in σ units) *(1)*	Height of Ordinate, y, from Table D *(2)*	Height of Ordinate of Mile-High Normal Probability Curve *(3)*	Approximate Fraction of Height at Mean in Terms of % *(4)*
0	.3989	5280 feet	100%
1.0σ	.2420	3202 feet	61%
2.0σ	.053,99	715 feet	14%
3.0σ	.004,432	59 feet	1%
4.0σ	.000,133,8	21¼ inches	3% of 1%
5.0σ	.000,001,487	¼ inch	4% of 1% of 1%
6.0σ	.000,000,006,076	.001 inch	2% of 1% of 1% of 1%
7.0σ	.000,000,000,009,135	microscopic	23% of 1% of 1% of 1% of 1% of 1%

At $+2\sigma$, the curve is 715 feet high. This is a little lower than either the Golden Gate Bridge tower (746 feet) or the Prudential Tower (750 feet) in Boston. It is also about the same height as the Oroville and Hoover Dams, which, at 770 and 726 feet, are the highest dam structures in the United States.

At a deviation of 2.2σ we come to the largest of the pyramids, the Pyramid of Cheops, built well over 4,000 years ago. Its original height is estimated at 482 feet, and it is now 450 feet high. At 2.6σ we come to two famous tourist attractions, the Leaning Tower of Pisa (179 feet tall and leaning 16 feet out of the perpendicular) and Niagara Falls (167 feet).

FIGURE 4.3
Hypothetical Mile-high Normal Probability Curve Plotted in Relation to Natural and Man-made Wonders

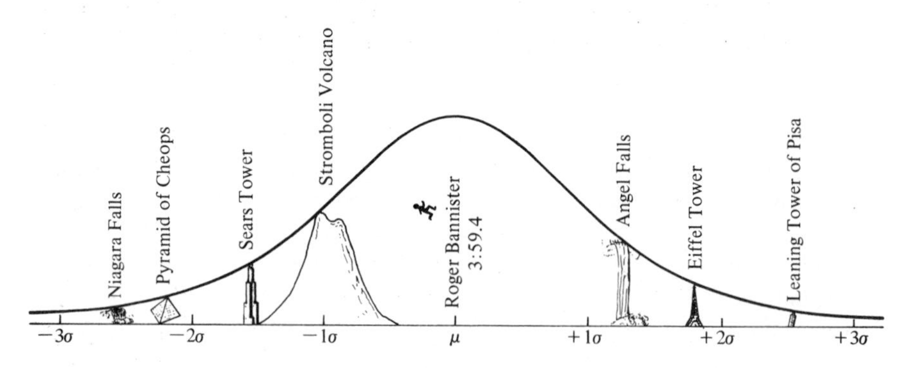

At 3σ from the mean, the height of 59 feet is about the same as that of many five- or six-story buildings. From here, the curve drops to $21\frac{1}{4}$ inches at 4σ and to only $\frac{1}{4}$ inch at 5σ. By 6σ it is even less than a thousandth of an inch. At 7σ the only way to describe the height is to call it microscopic; its height is only $\frac{1}{2000}$ of the diameter of a human red blood corpuscle; and this is under one forty-billionth of the height of the mean. The last column of Table 4.2 describes this as 23% of 1% of 1% of 1% of 1% of 1%.

Table 4.2 stops at 7σ and Figure 4.3 stops sooner than that—not because the curve stops, but because the values are so infinitesimal. The lower portion of Table D shows that at around $\frac{x}{\sigma} = 4.5$, each time we go out another half a standard deviation, the curve drops to one-tenth its previous height. This relative change becomes even more rapid as the distance from the mean increases; by 9σ, another half a standard deviation brings about a drop to only one percent of the previous ordinate.

By now, we should be well aware of the nature and shape of the normal probability curve. We should realize that it is infinite in extent, never quite reaching the base line, but coming almost inconceivably close to it. Despite this, we should also realize that, for practical purposes, we can ordinarily neglect the portions of the curve more than about 3σ away from the mean, since the height of the curve is so small from there on out.

Table D is useful in several ways. (1) It, along with Table 4.2 and Figure 4.3, which were derived from it, provides the basis for an understanding of the shape of the normal probability curve. (2) A second use is to provide values of y that are needed in order to solve various formulas we shall encounter later, such as Formula (6.2), for the standard error of a percentile. (3) It is very useful in plotting normal probability curves. We have already seen such an instance of its use in the computation of the values shown in Table 4.2 and their plotting in Figure 4.3. (4) Closely related to the plotting of curves is the problem of fitting a normal probability curve to a given set of data. We do this in Supplement 4.

BINOMIAL COEFFICIENTS AND THE NORMAL PROBABILITY CURVE

Let us now see what kind of data will produce a normal probability curve. As noted in the footnote on page 82, this curve is sometimes called the normal error curve because many independent errors will affect scores in such a manner as to produce a normal probability curve. The simplest illustration of this is given by the binomial expansion. If we square $(p + q)$ we obtain $(p^2 + 2pq + q^2)$, the coefficients of the successive terms being 1, 2, and 1. If $p = q$, the numerical values of the successive terms will also be proportional to 1, 2, and 1. If we raise $(p + q)$ to the third power, we obtain $(p^3 + 3p^2q + 3pq^2 + q^3)$, the coefficients of the terms being 1, 3, 3, and 1.

If we raise $(p + q)$ to the fourth power, the coefficients are 1, 4, 6, 4, 1. Such coefficients are known as the binomial coefficients. They can be obtained in a very interesting way, shown in Figure 4.4 and known as Pascal's triangle. In this triangle, any number is equal to the sum of the two numbers just above it, one to the right and one to the left. Calling the top row the 0th row, the 1st row (1 1) gives the coefficients of $(p + q)^1$, the 2nd row (1 2 1) gives the coefficients of $(p + q)^2$, etc. Thus, from the 10th row we see that the coefficients of $(p + q)^{10}$ are (1 10 45 120 etc.), and hence the expansion of $(p + q)^{10}$ is $(p^{10} + 10p^9q + 45p^8q^2 + 120p^7q^3$ + etc.). Figure 4.4 gives these coefficients through the 15th row. (The number of each row is given by the second figure in the row.) That is usually far enough, but if anyone should want the coefficients of $(p + q)^{16}$, they could easily be obtained by adding another row to the table, each entry being the sum of the two entries above it (one slightly to the left, the other slightly to the right). These values would be (1 16 120 560 1820 4368 8008 . . .). Successive rows may then be added until the computer gets tired.

FIGURE 4.4
Pascal's Triangle

1
1 1
1 2 1
1 3 3 1
1 4 6 4 1
1 5 10 10 5 1
1 6 15 20 15 6 1
1 7 21 35 35 21 7 1
1 8 28 56 70 56 28 8 1
1 9 36 84 126 126 84 36 9 1
1 10 45 120 210 252 210 120 45 10 1
1 11 55 165 330 462 462 330 165 55 11 1
1 12 66 220 495 792 924 792 495 220 66 12 1
1 13 78 286 715 1287 1716 1716 1287 715 286 78 13 1
1 14 91 364 1001 2002 3003 3432 3003 2002 1001 364 91 14 1
1 15 105 455 1365 3003 5005 6435 6435 5005 3003 1365 455 105 15 1

These coefficients have an interesting property. If they are regarded as ordinates, successive values may be connected to form a frequency polygon. Then the frequency polygon of the 0th row will be a triangle; so will that of the 2nd; but as we plot later and later rows of Pascal's triangle, these frequency polygons come to look more and more like the normal probability curve. In fact, in the limiting case in which the number of the row approaches infinity, the frequency polygon becomes the normal probability curve.

Fortunately, we don't have to get very close to infinity in order to see this. In Figure 4.5 we have plotted frequency polygons resulting from the use of the exponents 2, 4, 6, 8, 10, 12, 20, and ∞. This last one, the normal curve, has been plotted to the same scale. All the individual distributions have been so plotted as to have the means* centered, to have the same standard deviation after plotting, and to cover the same area on the page. In other words, we have used standard scores with a constant area, thus making all of the distributions equal with respect to μ, σ, and N.

FIGURE 4.5
Frequency Polygons for Binomial Expansion

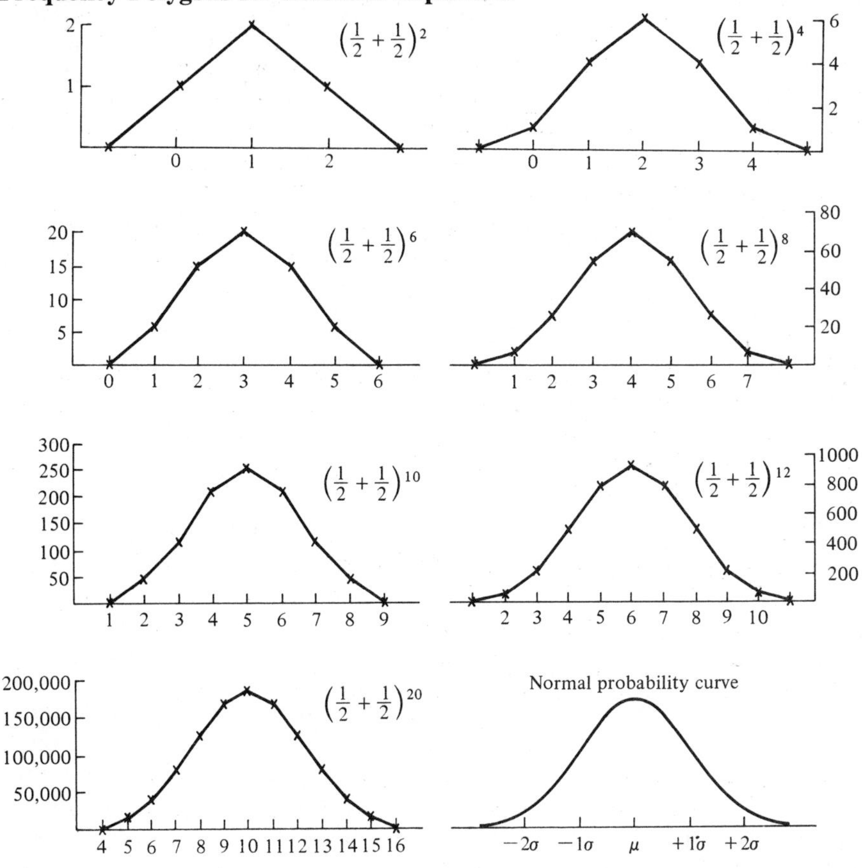

The drawing of $(p + q)^2$ in the upper left corner of Figure 4.5 is a triangle. Whether a person thinks a triangle looks something like a normal probability curve depends on how good an imagination he has. We cannot see too much similarity.

*The Greek letter μ is used instead of $\overline{X}$ to signify a theoretical or population mean. This symbol μ is used in the next sentence and its location was shown in Figure 4.1 on page 85.

The drawing of $(p + q)^4$ in the upper right corner begins to look a little like the normal probability curve. The next ones of $(p + q)$ to the 6th, 8th, 10th, and 12th powers bear more and more resemblance to that curve.

Skipping to what would be the 20th row of Pascal's triangle, note the striking resemblance it bears to the normal probability curve drawn in the lower right corner. Notice also that since the standard deviation of the distribution of $(p + q)^{20}$ was 2.24, only the frequencies for scores from 4 through 16 are shown on the graph. Even though this expansion shows frequencies of 1, 20, 190, and 1,140 that we did not plot at either end, these values are so small when compared to the maximum ordinate of 184,756 or with the total frequency of 1,048,576 that it is impossible to plot them on this scale as different from zero. In fact, the last plotted values (of 4845) at 4 and 16 are only $1/38$ as high as the maximum ordinate; the ignored values are far smaller than this.

The plots of these binomial coefficients show that nearly all of the frequencies (and hence nearly all of the areas) are included within the range $\mu \pm 3\sigma$. When the exponent of $(p + q)$ gets large (10 or more) there are some frequencies outside this range, but they constitute a very small proportion of the total and may ordinarily be neglected—just as we ordinarily may neglect both extremes of the normal probability curve even though we know that it extends infinitely far beyond $\mu \pm 3\sigma$ in both directions.

Thus, we see that the distribution of the binomial coefficients is very similar to that of the normal probability curve. In fact, one way of deriving Equation (4.1) for the latter is by means of these coefficients.

APPLICATIONS OF THE BINOMIAL COEFFICIENTS

Gamblers have made considerable (and profitable) use of the binomial coefficients and of the normal probability curve. In fact, questions raised by various gamblers provided the stimulus for much of the early work on the theory of probability. The next few pages will illustrate the extent of the applicability of the binomial coefficients to a wide variety of different situations.

Roulette A roulette wheel has 18 red and 18 black compartments, plus positions 0 and sometimes 00, which are added to provide a profit to the management. If we ignore 0 and 00, it is obvious that the chance of the ivory ball coming to rest in a red compartment is one-half, assuming that the wheel is unbiased. This assumption is frequently false, although it is rumored that there are some honest roulette wheels in Monte Carlo and in Nevada. Suppose a gambler, down to his last $1300, superstitiously decides to bet it all on the black, $100 at a time, on 13 consecutive turns of the roulette wheel. Ignoring the 0 and 00, the chances of 0 wins, 1 win, 2 wins, . . . , 13 wins are given by the coefficients in the 13th row of Pascal's triangle. The gambler has only 1 chance in 8192 of doubling his money, but he also has only 1 chance in 8192 of losing everything. There are 13 chances in 8192 of his winning 12 times, losing once, and ending up with $2400. Combining the first four terms, there are 1 + 13 + 78 + 286 or 378 chances in 8192 that he will wind up with $2,000 or more. Lest this appear to be an extremely easy way to acquire $2,000, it must be pointed out that there is exactly the same probability $\frac{378}{8192}$ that he will lose at least $700 and wind up with $600 or less. Of course, in real life, the odds cannot be even, because the management needs some money and sometimes takes a large cut to pay for overhead, taxes, and miscellaneous expenses, and to provide a profit.

True–False Tests Students sometimes toss coins to decide how to answer true–false questions; at other times, they guess. There is a belief that these guesses are as random as coin tosses. If this were true, a student who had answered all the other questions and then decided to guess at the remaining 7, would have the chances indicated in row 7 of Figure 4.4. He would have 1 chance in 128 of missing all of them, 7 chances of getting one right, 21 chances of getting two right, etc. (There are some exceptions, but most students actually do better than this. Although they may think they know nothing, they usually have *some* information on the subject, and they improve their scores more often than they reduce them by "guessing.")

Sexes in Families For small- and medium-sized families, Pascal's triangle shows approximately the number of boys and girls to be expected. Thus, the 4th row shows that if 4 children are born in each of a number of families, in about $^1/_{16}$ of the families, all the children will be girls; in about $^4/_{16}$, there will be 1 boy and 3 girls; about 6 times in 16, there will be the same number of each; etc. Why did we say "about" and restrict these statements to relatively small families? Because more boys than girls are born, despite the fact that there are about six million more adult women than men in the United States.

Compounding of Small Differences Ever since the U.S. Public Health Service started gathering such data in 1915, the proportion of male births has been between .51 and .515 instead of .50. In recent years this ratio has stayed fairly close to .5133. This certainly isn't very far away from one-half, but in a family of 13 children, the chance of all 13 of them being born boys is exactly *twice* as great as of all 13 being girls. Of course, neither of these probabilities is very great—we would expect about 7 boys—but the probability that all 13 births will be boys is 172 in one million (or .000172) while the chance of all 13 being girls is .000086, or only one-half as great. The first probability was obtained by raising .5133 to the thirteenth power, since the chance of a birth being a boy is .5133, of two successive births being boys is (.5133)(.5133), etc. The second probability is, of course, $(.4867)^{13}$. Families in which 13 or more children have been born alive are very rare now in the United States, but among every eleven or twelve thousand such families, we can expect about 2 in which the first 13 births were all boys and about 1 in which they were all girls.

Dice This leads directly into the question of the computation of other probabilities and of the shape of the resulting distributions. Earlier in this chapter, we pointed out that if $p = q$, the numerical values of the successive terms in the binomial expansion will be proportional to the binomial coefficients. When the two probabilities are unequal (that is, when p and q are not each equal to .50) their values must be considered and the resulting frequency distribution will not be a symmetrical one. That is, when we raise $(p + q)$ to the fourth power, we must consider the five terms $p^4 + 4p^3q + 6p^2q^2 + 4pq^3 + q^4$, rather than merely the coefficients 1, 4, 6, 4, and 1.

Let us set up a game in which we have a markedly better than 50–50 chance of winning. Dice provide a simple way of doing this. Since a die has 6 faces, let us say that we will win if it comes up showing a 1, 2, 3, 4, or 5, and our opponent will win if the 6 is on top. If we roll 1 die, our probability of winning, p, is $5/6$ and our probability of losing, q, is $1/6$. If we roll 2 dice and continue with these rules, the probabilities for various numbers of wins are given by the terms of the binomial expansion, $p^2 + 2pq + q^2$. These probabilities are $\left(\frac{5}{6}\right)^2 + 2\left(\frac{5}{6}\right)\left(\frac{1}{6}\right) + \left(\frac{1}{6}\right)^2$ or $\frac{25}{36} + \frac{10}{36} + \frac{1}{36}$. That is, the probability of our winning both times is $\frac{25}{36}$, that of winning once and losing once is $\frac{10}{36}$, and that of losing both times is $\frac{1}{36}$. These probabilities add up to 1.00. They always do.

If we play the above dice game with 4 dice, the probabilities are given by $(p + q)^4$, which was written out two paragraphs ago. The values of the terms are

$$\left(\frac{5}{6}\right)^4 + 4\left(\frac{5}{6}\right)^3\left(\frac{1}{6}\right) + 6\left(\frac{5}{6}\right)^2\left(\frac{1}{6}\right)^2 + 4\left(\frac{5}{6}\right)\left(\frac{1}{6}\right)^3 + \left(\frac{1}{6}\right)^4,$$

which equals $\frac{625}{1296} + \frac{500}{1296} + \frac{150}{1296} + \frac{20}{1296} + \frac{1}{1296}$.

With 6 dice, the probabilities resulting from the expansion of $\left(\frac{5}{6}+\frac{1}{6}\right)^6$ are:

$$\left(\frac{5}{6}\right)^6 + 6\left(\frac{5}{6}\right)^5\left(\frac{1}{6}\right) + 15\left(\frac{5}{6}\right)^4\left(\frac{1}{6}\right)^2 + 20\left(\frac{5}{6}\right)^3\left(\frac{1}{6}\right)^3 + 15\left(\frac{5}{6}\right)^2\left(\frac{1}{6}\right)^4$$

$$+ 6\left(\frac{5}{6}\right)\left(\frac{1}{6}\right)^5 + \left(\frac{1}{6}\right)^6$$

which equals

$$\frac{15{,}625}{46{,}656} + \frac{18{,}750}{46{,}656} + \frac{9{,}375}{46{,}656} + \frac{2{,}500}{46{,}656} + \frac{375}{46{,}656} + \frac{30}{46{,}656} + \frac{1}{46{,}656}$$

For the first time, the probability of all of the dice winning for us is lower than that of some other combination—in this case, 5 wins and 1 loss.

Trend to Normality All these probabilities, together with corresponding ones for 8, 10, 12, and 20 dice, are shown in Figure 4.6. Notice how, as the number of dice increases, the successive drawings in the graph become less and less lopsided and come to look more and more like the normal probability curve drawn in the lower right corner.*

You may be bothered by the fact that we did not plot very many of the frequencies for these curves. The reason is simply that the missing ones are too small to see on the graph. For instance, it is perfectly possible that we should lose on all 20 dice, but for our opponent to win by having 6s showing on all 20 dice is an event that can happen only once in over three quadrillion times. For our opponent to win on even half the dice is highly unlikely, even though there are nearly two trillion ways in which he can do this. The explanation is that there are so many, many more ways in which he can fail to do it. The probability of exactly 10 of the dice showing 6s (he wins) and exactly 10 of them showing any other number (we win) is

$$\frac{1{,}804{,}257{,}812{,}500}{3{,}656{,}158{,}440{,}062{,}976}$$

which is almost equal to .0005, or roughly 1 chance in 2,000. This obviously cannot be plotted as perceptibly different from zero on the scale we are using.

Thus, we see that the normal probability curve is approached as a limit by the successive terms of the binomial expansion. This is true regardless of whether the two probabilities are equal or unequal.

---

*Instead of the colloquial word "lopsided" as used here for effect, we later will use the term "skewed" for a curve that departs from symmetry. Chapter 7 is devoted to the study of skewness.

FIGURE 4.6
Frequency Polygons for Selected Dice Games

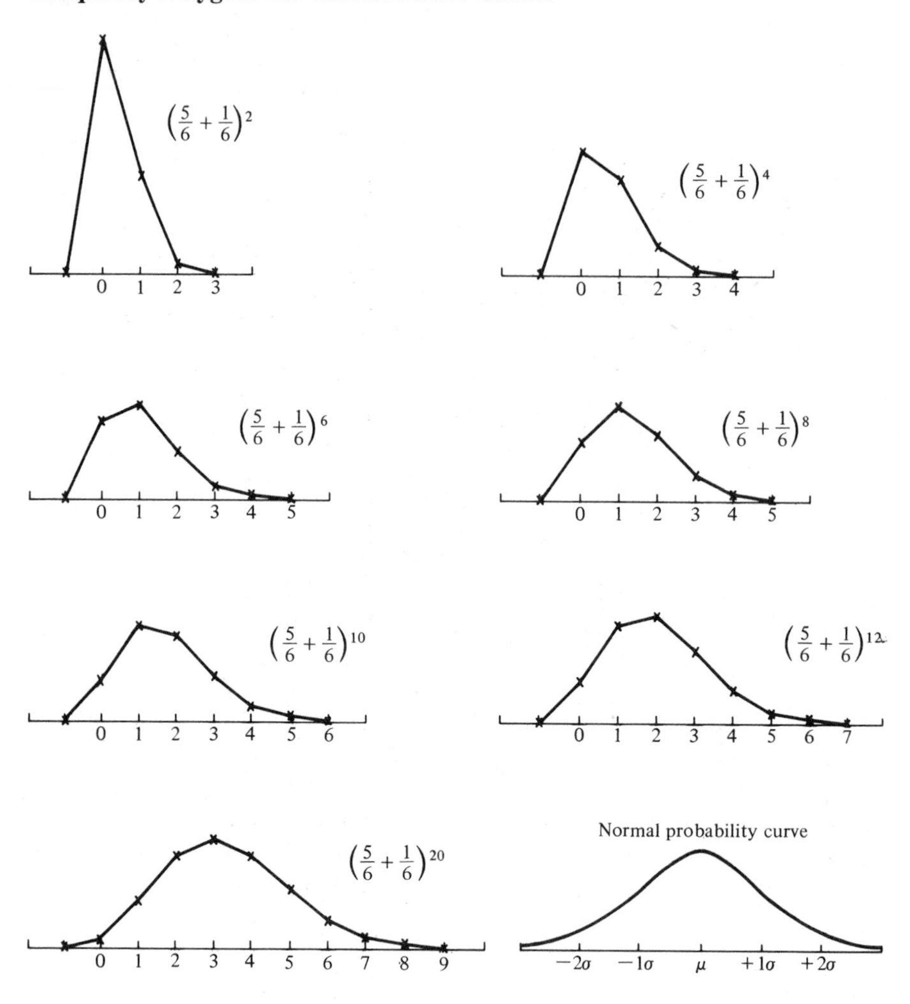

Many people are surprised by the fact that even when p and q depart markedly from equality the resulting distributions are, nevertheless, essentially symmetrical. Take a look at the symmetrical distributions in Figure 4.5 which were based on $p = q = 1/2$, then look at those in Figure 4.6, based on $p = 5/6$ and $q = 1/6$. Notice the marked skewness for the top two distributions but then note how quickly, as the number of dice increases, the distributions become almost exactly symmetrical. The graph for 20 dice looks very much like the normal probability curve—no matter whether $p = q$ or whether p is 5 times as great as q.

Missing Trumps If you don't play bridge, don't read this section.

Many people believe that the missing trumps in bridge are distributed according to the binomial coefficients. This is essentially right when our side has nearly all of the trumps, but not if several are missing. Table 4.7 shows the situation. If we look at Pascal's triangle (Figure 4.4) and express each row as a proportion (= dividing the entries by their sum), this will be evident. For example, if 4 trumps are missing, we would get:

Row 4 of Table 4.7	.0478	.2487	.4070	.2487	.0478
$\frac{1}{16}$ of Pascal's row 4	.0625	.2500	.3750	.2500	.0625

But if we go to row 6, the discrepancies are greater:

Row 6 of Table 4.7	.0075	.0727	.2422	.3553	.2422	.0727	.0075
$\frac{1}{64}$ of Pascal's row 6	.0156	.0938	.2344	.3125	.2344	.0938	.0156

Synopsis These examples have illustrated the close relationship between the normal probability curve and the binomial expansion, both when $p = q$ and when they differ markedly. We have seen the relations between the binomial coefficients and true–false tests, sex in families, and other concepts. Our hurried glance at gambling games illustrates that some of them have close relations to the binomial coefficients but that other factors are also involved. Because we assumed that such games were operated honestly, the old adage still holds and we must tell it like it is: Don't gamble with strangers.

THE AREA UNDER THE NORMAL PROBABILITY CURVE

So far, we have dealt with ordinates rather than with areas of the normal probability curve. In much of our discussion of the binomial coefficients and of the binomial expansion, we were in an intermediate position: we were actually dealing with ordinates, but we were often treating them as though they were areas. In this section, we shall deal directly with the areas between or beyond various ordinates of the normal probability curve. These areas will be expressed as proportions of the total area. Thus, we shall find out such facts as these: the proportion of the area under the normal curve (and above the base line) between an ordinate at the mean

TABLE 4.7
Probability of Missing Trumps Being Held by One Opponent, West, in Playing Bridge

Number of Trumps Held by Opponents	Number in West Hand								
	0	1	2	3	4	5	6	7	8
1	.5000	.5000							
2	.2400	.5200	.2400						
3	.1100	.3900	.3900	.1100					
4	.0478	.2487	.4070	.2487	.0478				
5	.0196	.1413	.3391	.3391	.1413	.0196			
6	.0075	.0727	.2422	.3553	.2422	.0727	.0075		
7	.0026	.0339	.1526	.3109	.3109	.1526	.0339	.0026	
8	.0008	.0143	.0857	.2356	.3272	.2356	.0857	.0143	.0008
9 or more	You need a new partner—or he does.								

and another ordinate one standard deviation above the mean is .3413. Or, the area more than 1σ above μ is .5000 − .3413, or .1587 of the total area area. Or, the point below which .0250 of the area lies is 1.96σ below the mean. Or, the central portion of the curve, including 95% of the cases, with 5% lying beyond, runs from 1.96σ below the mean to 1.96σ above the mean.

The place to which we go for information such as that contained in the preceding four sentences is to a table showing the area cut off by ordinates at various distances from the mean. Table E is such a table. It shows the area between an ordinate (the mean ordinate) erected at the middle of the curve and another ordinate erected at any distance from the mean. The latter distances are measured in standard scores, using an interval of $.01\sigma$. This is usually accurate enough for all ordinary uses without the necessity for interpolation. Hence, unless your instructor tells you to interpolate for certain problems demanding an unusually high degree of accuracy, it is not necessary to interpolate in either Table D or Table E.

Table E consists of two parts, the top part being all that is ordinarily required. Although the bottom section gives a large number of decimals, it covers only a few values of $\frac{x}{\sigma}$. In using the top section, the standard score is found at the side and top (or bottom). Thus, .92 is located by finding .9 at either the left or right side and .02 at either the top or bottom of the table (note that $.9 + .02 = .92$). At the intersection of the .9 row and the .02 column, we find the entry .3212. This means that the area under the normal probability curve between an ordinate at the mean and another ordinate $.92\sigma$ away from the mean is .3212, or about 32% of the total area under the entire curve from $-\infty$ to $+\infty$. This is fundamentally a very simple concept, but one that must be *thoroughly* understood, since it is essential to all use of Table E. Let us review it: To find the area between the mean and a deviation of $.92\sigma$, find the .9 row and the .02 column. The number, .3212, at the intersection of the .9 row and the .02 column is the required area. (Is that perfectly clear—or perhaps even boring? If not, reread this paragraph until it *is* clear.)

When we look up one $\frac{x}{\sigma}$ value (such as .92) we get one area (such as .3212). By adding this area to or subtracting it from .5000 and by doubling some of these values, we can make eight entirely separate and meaningful statements about eight different areas. Let us assume that we have a normal distribution and then show, both graphically and verbally, what these eight meaningful statements are. The statements and graphs are shown in Figure 4.8.

FIGURE 4.8
Areas under the Normal Probability Curve Corresponding to a Given Standard Score

(a)

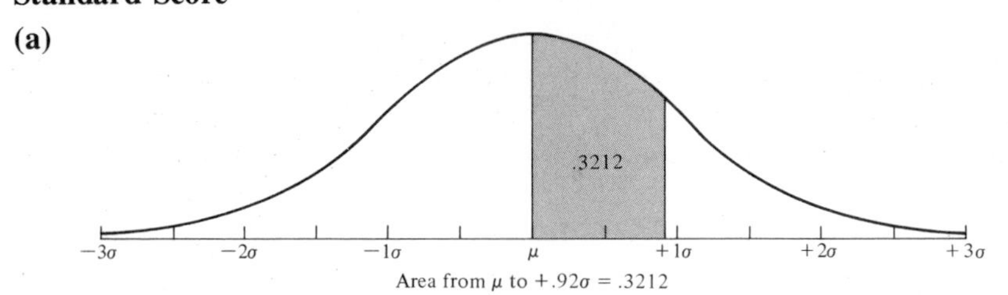

Area from μ to $+.92\sigma = .3212$

(b)

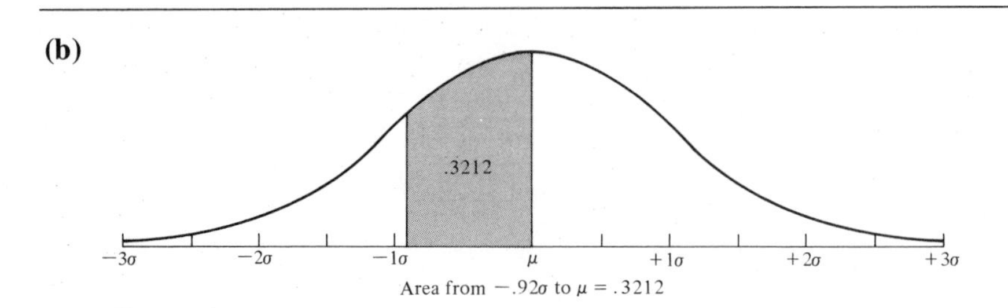

Area from $-.92\sigma$ to $\mu = .3212$

(c)

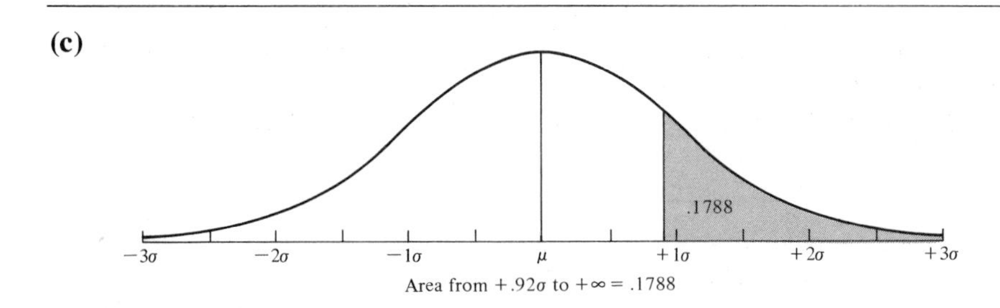

Area from $+.92\sigma$ to $+\infty = .1788$

(d)

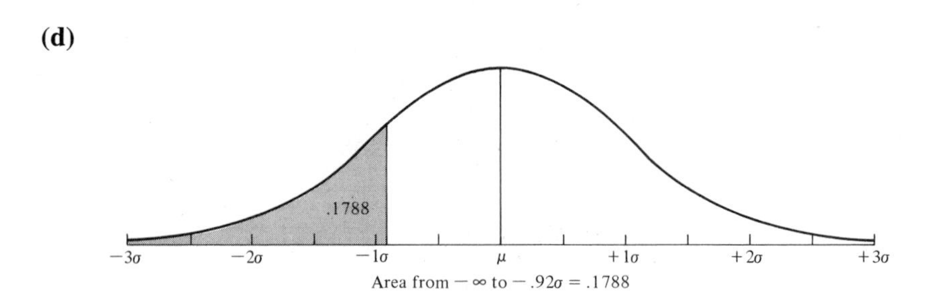

Area from $-\infty$ to $-.92\sigma = .1788$

(e)

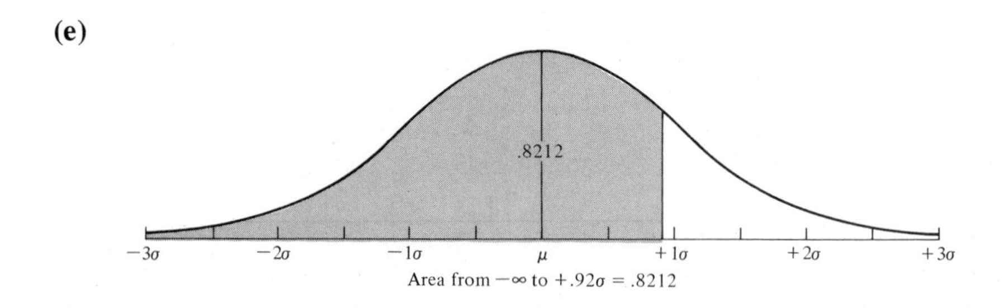

Area from $-\infty$ to $+.92\sigma = .8212$

(a) The area under the curve from the mean ordinate to the ordinate at the given positive $\frac{x}{\sigma}$ distance is shown by the shaded area in the curve at the left. The probability that a person selected at random will be between μ and a standard score of .92 is .3212. There are 3,212 chances in 10,000 that any randomly selected score will be in the shaded area.

(b) Since the normal probability curve is symmetrical, the area between an ordinate at $-.92\sigma$ and the mean is the same as that given in the preceding paragraph for the area from the mean to $+.92\sigma$. This shaded area is also .3212. Thus, $\frac{3{,}212}{10{,}000}$ of the scores are between $-.92\sigma$ and μ. The probability that a person will score below the mean but higher than a standard score of $-.92$ is .3212.

(c) Since the normal probability curve is symmetrical, half the area is above the mean. Table E shows that .3212 of the area is between the mean and $+ .92\sigma$. Therefore, $.5000 - .3212 = .1788$ of the area is above $+.92\sigma$. The chance that a score will deviate more than $.92\sigma$ and in the positive direction is .1788. Only 1,788 times in 10,000 could we expect a deviation this large or larger in this direction.

(d) Because of the symmetry of the normal probability curve, the area below $-.92\sigma$ must also equal $.5000 - .3212 = .1788$. A difference as great as $.92\sigma$ or more and in the negative direction could occur by chance only 1,788 times in 10,000. The probability that a person will make a score more than $.92\sigma$ below the mean is .1788. The number of standard scores between $-\infty$ and $-.92\sigma$ is $\frac{1{,}788}{10{,}000}$ of the total number of cases, N. If the passing score is set $.92\sigma$ below μ, 17.88% of the applicants will fail the test.

(e) Half the area is below the mean, and Table E shows that .3212 of the area is between μ and $.92\sigma$. Hence the total area below $.92\sigma$ is $.5000 + .3212 = .8212$. In a group of 10,000 people, we expect 8,212 of them to score below $+ .92\sigma$. In screening students for admission to a law school, if we demand a standard score of at least $+ .92$, we will reject 82.12% of the applicants.

FIGURE 4.8 *(continued)*

(f)

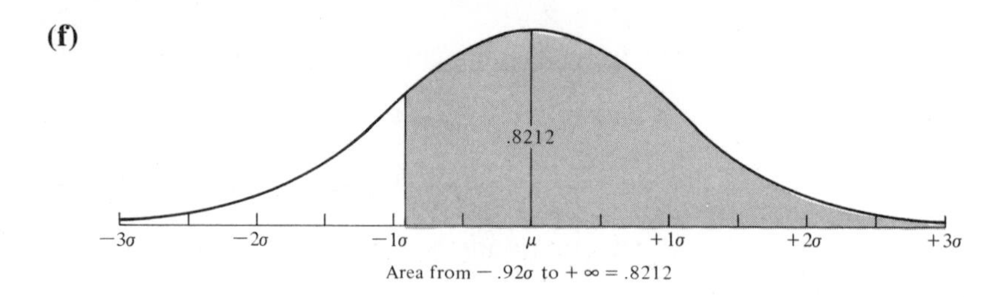

(g)

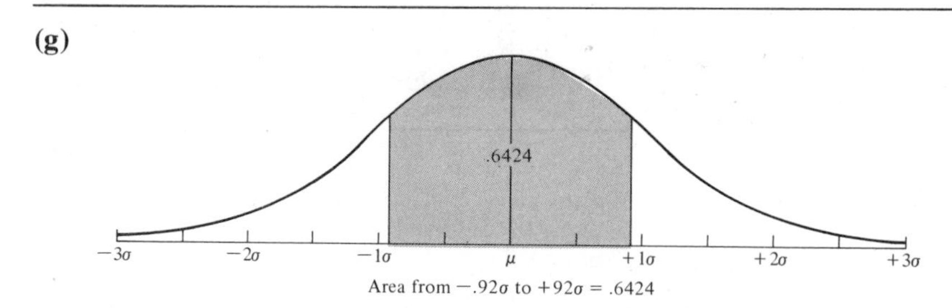

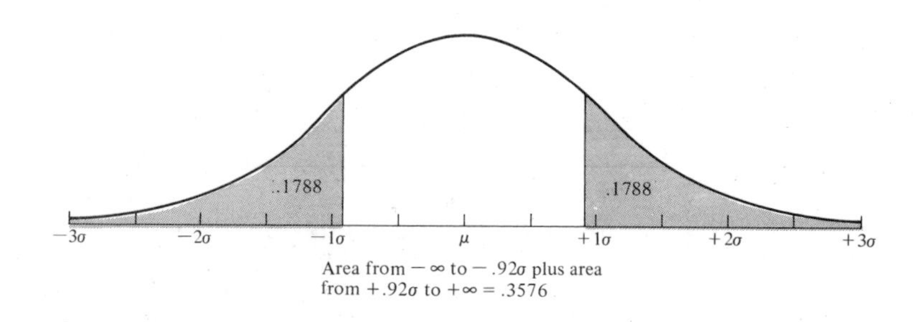

Notice that in each of the eight preceding examples, we took one distance ($.92\sigma$) and read one answer (.3212). Now we shall review, using the same procedures with eight different readings from Table E.

Be sure you see why each of these problems corresponds to the preceding interpretation bearing the same letter.

(a) What proportion of the cases lie between the mean and $+ 1.00\sigma$?
Answer: .3413

(b) What is the probability that a given score will be below the mean but no more than 2.00σ below the mean? *Answer:* .4772

(c) How often will a score be expected to be more than 3.00σ above the mean? *Answer:* 13 times in 10,000

(d) Agnes dislikes Anatomy, which may account for the fact that she is $.85\sigma$ below the class mean. Assuming that the scores are distributed normally, what proportion of scores will be lower than hers?
Answer: .1977

(e) Mabel's standard score in Music is +.44. What percent of the students make lower scores than hers, if the scores are normally distributed?
Answer: 67%

(f) The area between $-.92\sigma$ and μ is given as .3212; the area above the mean is .5000. Hence, the area above a standard score of $-.92$ is .3212 + .5000 = .8212. The probability that a person will do better than a standard score of $-.92$ is .8212. There are 8,212 chances in 10,000 that a score selected at random will be better than $-.92\sigma$. If the passing score is set at $-.92\ \sigma$, then 82.12% of the applicants will pass.

(g) Sometimes we want to know what proportion of the cases are within a certain distance of the mean. Since Table E shows that there are .3212 of the cases between the mean and either $+.92\sigma$ or $-.92\sigma$, there must be twice that many, or .6424, between $-.92\sigma$ and $+.92\sigma$. Hence, .6424 of the cases are within $.92\sigma$ of the mean. And 64.24% of the area under the normal probability curve is within $.92\sigma$ of the mean. The range within which the middle 64.24% of the cases lie is from $-.92\sigma$ to $+.92\sigma$.

(h) This is the only one of our eight examples in which the area we want is divided into two segments. To determine the proportion of cases that are beyond certain limits, we can either subtract the area given in the table (.3212) from .5000 (getting .1788) and double the value so obtained (getting .3576), or we can double the area given in the table (getting .6424) and subtract it from 1.0000 (again getting .3576). The chances are 3,576 out of 10,000 that a case selected at random will lie outside the limits of $\mu \pm .92\sigma$. And 35.76% of the area is more than $.92\sigma$ away from the mean. A deviation as great or greater than $.92\sigma$ in either direction will occur by chance 3,576 times in 10,000. The probability that any case will lie beyond $.92\sigma$ in either direction from the mean is $\dfrac{3{,}576}{10{,}000}$.

(f) Becky is $.68\sigma$ below the class average in Biology, but is showing signs of improvement. Assuming normality, what proportion of the students are now achieving higher scores? *Answer:* .7517

(g) What percentage of the cases in a normal distribution are within 2.58σ of the mean? *Answer:* 99%

These limits of -2.58σ and $+2.58\sigma$ include .99 or 99% of the cases in the distribution. On page 128 of Chapter 5 on Statistical Inference we will learn that, reasonably enough, these limits are called the .99 confidence limits.

(h) What proportion of the area under a normal probability curve is more than 1.96σ away from the mean? *Answer:* .0500

Here .05, or 5%, of the cases in the distribution are more than 1.96σ away from the mean (counting both tails of the curve). Page 139 will tell you that results that are more than 1.96σ away from the mean are said to be significant at the .05 level.

FIGURE 4.9
Areas under the Normal Probability Curve Between Two Standard Scores

(a)

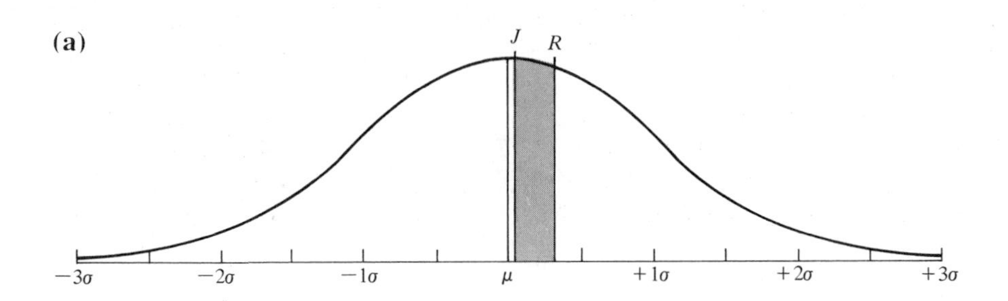

(b)

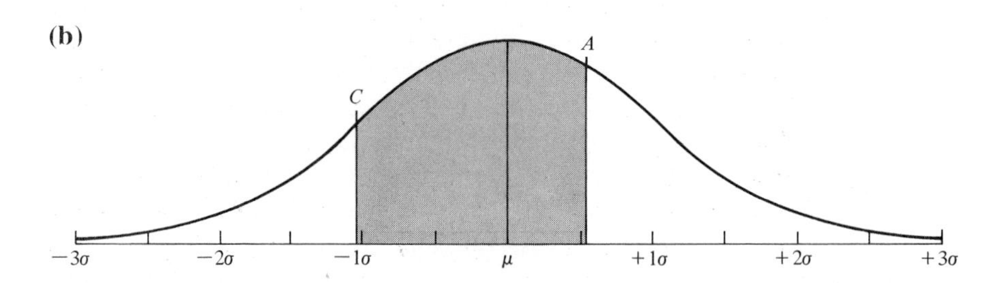

(c)

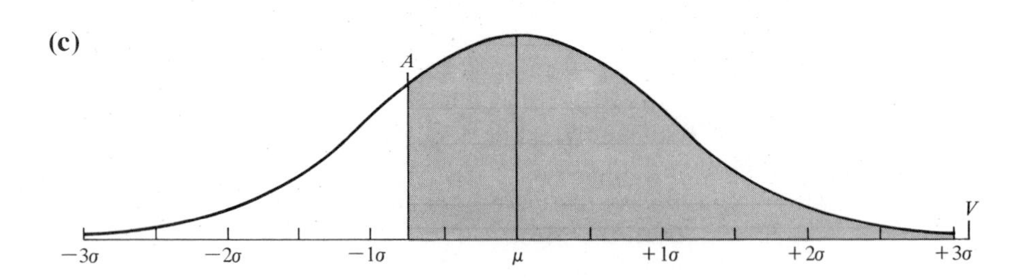

(a) On a test of Roman History given one July during summer session, Romeo's score was $.33\sigma$ above the mean while Juliet's score was $+.05\sigma$. Although both scores are above average, these two people are close together in History, as they were in other ways. Table E shows that .1293 of the area is between μ and Romeo's score of $+.33\sigma$. It also shows .0199 of the area to be between μ and Juliet's score of $+.05\sigma$. Simple subtraction shows us that $.1293 - .0199 = .1094$ of the area is located between these two standard scores. Thus 10.94% of the Roman History students are expected to score higher than Juliet and less than Romeo.

(b) On the trait, consideration for other people, Antony's rating was $.52\sigma$ above the mean of ancient military leaders, while Cleopatra's rating was exactly twice that far below the mean. Table E shows that .1985 of the leaders were rated between the mean and Antony (or $.5000 + .1985 = .6985$ of them were judged as showing less consideration for other people than Antony did). Cleopatra's standard score of -1.04 leads to an area of .3508 between her rating and the mean (or $.5000 - .3508 = .1492$ of such leaders were rated as showing less consideration than she did). Diagram (b) shows that $.1985 + .3508 = .5493$ of the leaders fall between Cleopatra's coldbloodedness and Antony's relative softheartedness. (As a check we may obtain this same answer, though with more work, by noting that $.6985 - .1492 = .5493$.)

(c) On a test of social intelligence, Venus scores 3.09 standard deviations above the mean, while Adonis is $.76\sigma$ below the mean. How much of the area of the normal probability curve is bounded by their scores? From Table E we obtain the answer: $.4990 + .2764 = .7754$. Notice that Venus' score is so outstanding that it makes little difference whether we ask what proportion are between Venus and Adonis (.7754), or what proportion are superior to Adonis (.7764).

FIGURE 4.9 (*continued*)

(d)

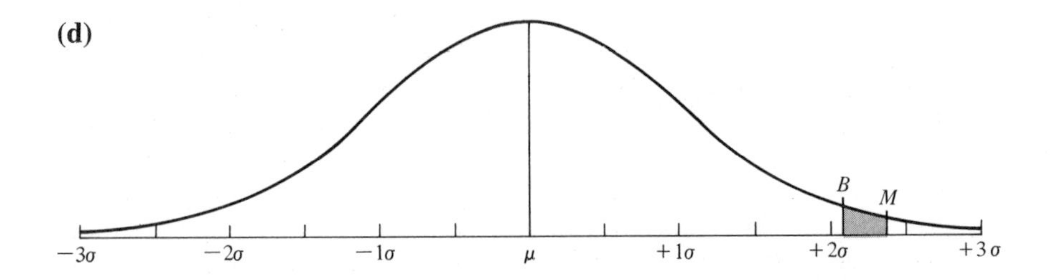

(e)

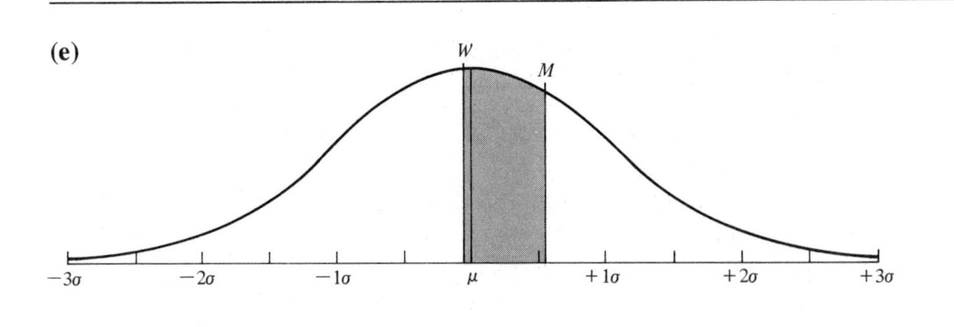

Table E may also be used in reverse. If we know the proportion of cases between the mean and any other point, we can find out how many standard deviations that point is above or below the mean. Thus, 40% of the cases next above the mean are included between the mean and $+ 1.28\sigma$. (Actually, .4000 is not given in the body of the table, but .3997 and .4015 are given. Choosing the closer of the two, we see that .3997, and hence, .4000, and 40% also, corresponds to a standard score of + 1.28.) We can also use Table E if the figures needed to enter the table are not directly given but may be inferred. For example, if Julius Caesar's physical strength was greater than that of 99% of the men of his time, then .9900 − .5000 = .4900 of the men were between the mean and his strength. The closest entry to .4900 in the body of the table is .4901, which corresponds to 2.33. Hence, Caesar's standard score in physical strength would have been + 2.33.

(d) Mac and Beth correctly spell 87 and 84 words, respectively, on a 95-item test with a mean of 62.04 and a standard deviation of 10.71. By substituting in Formula (3.6) we readily find that their standard scores are + 2.33 and + 2.05. Table E shows that .4901 of the scores lie between the mean and $+2.33\sigma$, while .4798 lie between the mean and $+2.05\sigma$. Hence, 1.03% of the scores will be expected to lie between Mac's and Beth's scores. Let us compare Mac and Beth with Romeo and Juliet. In each case, the difference in terms of standard scores is the same: $2.33 - 2.05 = .33 - .05 = .28$. The corresponding areas are quite unequal, however.

	Standard Score	Proportion Below This Score	Difference
Mac	+2.33	.9901	.0103
Beth	+2.05	.9798	
Romeo	+ .33	.6293	.1094
Juliet	+ .05	.5199	

The normal probability curve is much higher near the mean than it is a couple of standard deviations away and hence the central area from $+.05\sigma$ to $+.33\sigma$ contains over ten times the area that is included between $+2.05\sigma$ and $+2.33\sigma$. In each example, the standard scores show the true abilities of the people and show us that Romeo is just as much better than Juliet in Roman History as Mac is better than Beth in Spelling—the difference in standard scores being .28 in each case.

(e) On a test of self-sufficiency, Mary scores $.60\sigma$ above the mean. She is the wife of Windsor, who scores one-tenth that far below the mean. What proportion of the group made scores between those of Mary and Windsor? Table E shows .2257 between the mean and Mary's score of $+.60\sigma$ and an additional .0239 below the mean but above Windsor's standard score of $-.06$. Hence, .2496 (or about one-fourth) of the population made scores which were lower than Mary's but higher than Windsor's.

Again, if Prince Hamlet's melancholy was such that he could have been regarded as happier than only one man in a thousand, then there were $.5000 - .0010 = .4990$ of the men who were between his degree of happiness and the mean. Table E lists the value .4990 three times. Taking the middle one of these three values, we would assign the melancholy Dane a happiness score of -3.09σ, the minus sign, of course, indicating **un**happiness.

Another type of problem involves determining the proportion of cases between two standard scores, making the usual assumption of a normal distribution. If both standard scores are on the same side of the mean, it will be necessary for us to obtain two areas from Table E and subtract the smaller from the larger. If one standard score is above the mean (positive) while the other is below the mean (negative), the two obtained areas are added together. The five examples in Figure 4.9 illustrate this.

SUMMARY

The usual equation for the normal probability curve, (4.1), is given, followed by our greatly simplified version, (4.2). A few ordinates are computed and a complete table of ordinates, Table D, is presented. To illustrate their meaning, a mile-high curve is assumed and the ordinates (in feet, inches, or fractions at various distances) are given and are made more meaningful by comparing them with various tourist attractions.

Some of the curve's uses are mentioned. An example is given to show how a normal probability curve may be fitted to a given set of data.

Binomial coefficients are introduced by giving Pascal's triangle (Figure 4.4) and then showing (Figures 4.5 and 4.6) that the frequency polygons formed by plotting successive rows of Pascal's triangle (or by plotting successive powers of the binomial expansion) give polygons that fairly rapidly approximate the shape of the normal probability curve.

Several other uses and meanings of the binomial coefficients are given. Some situations in which they are inappropriate are also pointed out.

Table E gives the area under the normal probability curve between the mean and any given deviation. By such simple arithmetical processes as subtracting these values from .5000 or 1.0000, or by doubling them, eight meaningful statements can be made on the basis of just one value read from the table. These are illustrated by eight paragraphs, each with its own graph (Figure 4.8), and are followed by eight practical questions reviewing the same principles.

Five more questions illustrate the use of Table E in determining the area between *two* points on the normal probability curve (Figure 4.9). These complete the presentation of the direct use of this curve.

SUPPLEMENT 4

SIMPLIFYING THE EQUATION FOR THE NORMAL CURVE

The well-known equation for the normal curve was given in this chapter as Equation (4.1),

$$y = \frac{N}{\sigma\sqrt{2\pi}} e^{-x^2/2\sigma^2}$$

The equation can be simplified by making two changes: (1) get rid of N so that we can work with relative frequencies by letting $N = 1$; and (2) get rid of σ so that we can work with standard scores by letting $\frac{x}{\sigma} = z$ and $\sigma = 1$.

Applying these, we have

$$\begin{aligned}
y &= \frac{N}{\sigma\sqrt{2\pi}} e^{-x^2/2\sigma^2} \\
&= \frac{1}{1\sqrt{2\pi}} e^{-z^2/2} \\
&= \left(\frac{1}{\sqrt{(2)(3.1416)}}\right)\left(\frac{1}{\sqrt{2.7183}}\right)^{z^2} \\
&= \left(\frac{1}{\sqrt{6.2832}}\right)\left(\frac{1}{\sqrt{2.7183}}\right)^{z^2} \\
&= \left(\frac{1}{2.5066}\right)\left(\frac{1}{1.6487}\right)^{z^2} \\
&= (.3989)(.6065)^{z^2}
\end{aligned}$$

This last equation is the simplified one, (4.2).

FITTING A NORMAL PROBABILITY CURVE TO ANY FREQUENCY DISTRIBUTION

We often have a distribution of scores similar to some of those given in Chapter 2. Sometimes the distribution (as the IQs in Table 2.4 on page 20) looks much like the normal distribution; at other times (as the citations in Table 2.7 on page 27) it obviously doesn't; but sometimes it is hard to tell. We may want to know, for example, whether college grades really are distributed about the way the normal probability curve would require. In such cases we may wish to try to fit this curve to our data. Of course, it is more difficult to fit some curves than others. It is very easy to fit a normal probability curve to any given set of data. There are several ways to do this; we shall use the method of moments. In one sense, all normal probability curves are alike; any one can be transformed into any other simply by altering the vertical and horizontal scales. In another sense, most normal probability curves are different; they differ in the location of the origin and in the horizontal and vertical scales.

If we were to ask for a normal probability curve with a given N, mean, and standard deviation, we would completely specify the curve. One and only one normal probability curve exists for any given set of values of N, μ, and

σ. In the method of moments, we construct a normal probability curve that has exactly the same N, μ, and σ as the corresponding figures for our data. These will be N, $\overline{X}$, and either s or σ_N, depending on the purpose we have in mind. If, as is often the case, we wish to fit the curve to the sample data we actually have, rather than to our estimate of the population distribution, we should use σ_N. We can now fit a normal probability curve to a set of grades. Before doing so, we should first try to set straight one of the most common misconceptions concerning the normal probability curve: the notion that "the curve" forces teachers to fail their students. Actually, that curve never forced any teacher to fail anybody, yet hundreds (yes, thousands) of teachers have hidden behind it, answering a student's complaint with some such nonsense as, "I'm sorry you failed, but I grade 'according to the curve' and I have to flunk a certain percentage out of each class." A teacher might draw a line 3σ below the mean of a normal probability curve (note that the area is far less than 1%) and then decide that she will grade "according to the curve" and never fail anybody. Another teacher may draw a line through the mean, note that half the cases are below it, and say that she, too, is grading "according to the curve," while flunking half the class. Both statements are obviously misleading—to put it mildly. A correct statement is simply that each teacher should courageously shoulder his own responsibility for assigning grades, rather than attempting to maintain the pretense that he is a helpless robot enslaved by a Frankenstein which is known as the normal probability curve.

With this preliminary, let us look at the grades assigned during an academic year by the faculty of the Psychology Department at the University of Florida. In recent years, the number of A and B grades has increased markedly while the number of low and failing grades has dwindled. To get a distribution of all five grades rather than just A, B, and maybe C, we went back a few years to get the grades shown in Table 4.10. The figures given in the table are based on course registrations, rather than on credit hours, thus counting each student only once (unless he registered for more than one Psychology course). For some purposes, it would be better to weight courses according to the credit they carry but since almost all of these were 3-hour courses, the two graphs would have almost exactly the same shape. A few students (27 incomplete and 8 absent from final examination) who did not complete their courses are omitted. The grades for all others are shown in Table 4.10.

Before we do any fitting we must compute the mean and standard deviation. The latter should be computed by Formula (3.5) which gives σ_N, the standard deviation of a finite population. Thus, our basic statistics are $N = 1304$; $\overline{X} = 2.46$; and $\sigma_N = 1.02$.

TABLE 4.10
Psychology Grades at the University of Florida

Grade *(1)*	Fall *(2)*	Spring *(3)*	Academic Year *(4)*
A	100	126	226
B	184	207	391
C	245	254	499
D	60	75	135
E	27	26	53
Total	616	688	1,304

Now let us fit a normal probability curve to the grades just tabulated in Table 4.10 by following these simple and straightforward steps.

1. Write the midpoints of the successive intervals.
2. Determine the value of each of these midpoints as a standard score.
3. Look up the value of y in Table D for each of these standard scores.
4. Evaluate the constant $\frac{Ni}{\sigma_N}$.
5. Multiply .3989 and each value of y by this constant.
6. Plot the points obtained in step 5.
7. Draw a curve through these plotted points.

The totals shown in column (4) of Table 4.10 do not look very much like figures we associate with a normal distribution. Nevertheless, let us do the best job we can of fitting a normal probability curve to the data, following the seven steps just listed.

1. On a suitable sheet of graph paper (10 small squares to the inch in this case), copy the midpoints of the successive intervals. We need to substitute some numbers for the letters A, B, C, D, and E. We shall use the common equivalents of 4 for A down to 0 for E, though any other set of equidistant numbers would give us exactly the same normal probability curve. To facilitate plotting the ends of the curve, we shall extend the scale a couple of units in each direction and record the numbers from 6 to −2 in the X column of Figure 4.11.

2. Determine the standard score corresponding to each of these X scores. This can be done by substituting σ_N for s in Formula (3.6) so that it reads

$$z = \frac{X - \overline{X}}{\sigma_N}$$

FIGURE 4.11
Normal Probability Curve Fitted to Psychology Grades

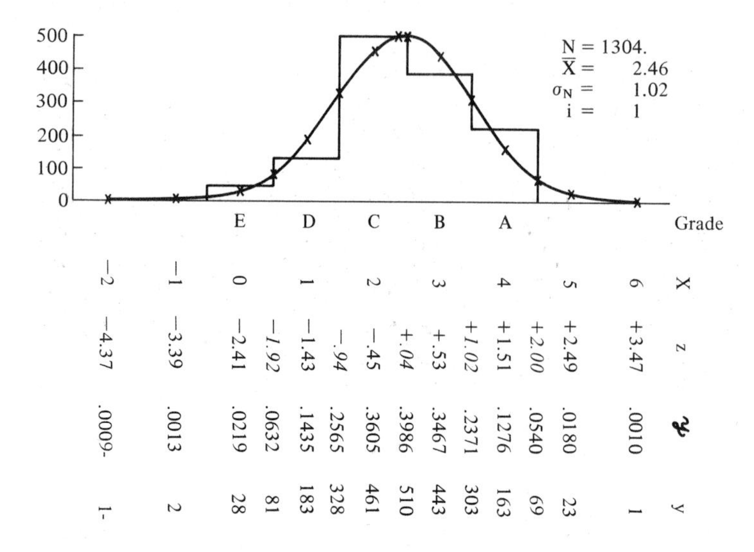

X	z	$\mathscr{y}$	y
6	+3.47	.0010	1
5	+2.49	.0180	23
	+2.00	.0540	69
4	+1.51	.1276	163
	+1.02	.2371	303
3	+.53	.3467	443
	+.04	.3986	510
2	−.45	.3605	461
	−.94	.2565	328
1	−1.43	.1435	183
	−1.92	.0632	81
0	−2.41	.0219	28
−1	−3.39	.0013	2
−2	−4.37	.0009-	1-

Using our known values of $\overline{X}$ and σ_N, we find that the standard score corresponding to an X of 6 is

$$\begin{aligned} z &= \frac{6 - 2.46}{1.02} \\ &= 3.4706 \\ &= 3.47 \end{aligned}$$

To obtain the other z scores, we should keep four decimals and successively subtract $\frac{i}{\sigma_N}$, which, in this example is $\frac{1}{1.02}$ or .9804 from this 3.4706. This gives the values shown in regular type in the z column of Figure 4.11, the last value being −4.3726. Direct calculation gives

$$\begin{aligned} z &= \frac{(-2) - (2.46)}{1.02} \\ &= -4.3725 \end{aligned}$$

Since the difference between these two determinations is only .0001, the agreement is eminently satisfactory. Recomputing all the values from −4.3725 on up gives the same values, to two decimals, as those already recorded, thus checking all nine values of z.

3. Look up these values of z in Table D and record the corresponding values of $\mathscr{y}$ in the next column of Figure 4.11. One of the values, the last one, is beyond the range of the main table, so we simply record it as .0009−, knowing that the value we get in the next column will be too small to plot as measurably different from zero anyway.

4. We evaluate the constant $\frac{Ni}{\sigma_N}$, which turns out to be $\frac{(1{,}304)(1)}{1.02} =$ 1,278.43. Our check on this, multiplying the .9804 used in step 2 by 1,304 gives 1,278.44, the negligible difference of 1 being due to rounding errors. Recomputation shows that 1,278.43 is the correct value, so we shall use it. If the difference should ever be as great as 2 in the last retained figure, the cause of the error should be found and the correct value used.

5. We multiply the mean ordinate, .3989, and all the other $\mathscr{y}$ ordinates by 1,278.43 and record the resulting values in the y column of Figure 4.11. About this time, we should realize that, through oversight, we have caused ourselves a little extra work. It is evident from looking at the successive y ordinates, and it will be quite obvious when they are plotted, that the values are too far apart. We don't have enough values to enable us to plot a really smooth curve. We especially need another point about halfway between A and B and between C and D. We could compute these two values, but it is about as easy to go back to step 2, divide our constant, .9804, by two and compute all the other values that seem useful to us. This we have done and these values are entered in italics in the z column of Figure 4.11. The resulting values of $\mathscr{y}$ and y are then recorded. After we have fitted a few curves, we will realize that when the points are about a half σ_N apart (as they now will be in Figure 4.11) it is easy to plot an accurate curve through them. Only an expert can do an accurate job when the points are a standard deviation or so apart (as they originally were in Figure 4.11).

6. Plot the values of y just obtained by placing a small x either above the midpoints of the intervals or on the boundaries between them, as the case may be. The mean is 2.46 with a mean ordinate of 510 and the boundary between B and C grades is 2.50, also having an ordinate of 510. These points almost coincide.

7. Finally, we draw a smooth curve through the plotted points. The normal probability curve gives a surprisingly good fit to the original grade distribution, which is shown as a histogram drawn to the same scale. The histogram and the normal probability curve have the same mean (2.46), the same standard deviation (1.02), and are based on the same number of cases (1304). Comparing the curve and the histogram, we see that the fit is far better than we would probably have expected by looking at Table 4.10.

We will give a slightly different method of fitting a normal probability curve in Chapter 11.

Weather forecaster, on phone, to wife: "I'll try to remember to pick up a loaf of bread, dear, but there's a 38-percent probability I'll forget."

—L. Herman in *The Wall Street Journal*

CHAPTER 5 STATISTICAL INFERENCE

DEPENDABILITY OF FIGURES

This chapter is devoted to various aspects of statistical inference, as are parts of nearly all of the following chapters, especially Chapter 14. We wish to know how dependable the figures are that we have computed so far. When we see a result computed to several decimal places, can we trust all of them? We have computed quite a few statistics so far. Are some of them unrealistic? If so, at least we have company, for unrealistic figures exist in other places. Here we are asking two questions: (1) Are the figures we have computed accurate measures of the specific data they are supposed to represent? (2) To what extent do the figures we have calculated apply to other samples we might have measured? Let us now examine the first of these questions.

In official world's records, automobile speeds are given to 6 figures, or to 3 decimals. The highest speed yet recorded by a wheeled vehicle was achieved by Gary Gabelich on October 23, 1970, in a car powered by a rocket engine. On his first mile run, he was timed at 5.829 seconds; his second run, in the other direction, took 5.739 seconds. Thus, the average for the two runs was 5.784 seconds. Dividing 3,600 seconds by 5.784 seconds gives 622.407 mph, a world's record.*

This was a fine record, but it is practically a certainty that 622.407 mph was not his average speed. Why? Because the times are given to 4 significant figures and neither of them can be expected to be exact to 5, 6, or more digits. That is, the 5.784 does not mean exactly 5.784000, but it covers the entire range from 5.783500 to 5.784500. These correspond to speeds of anything from 622.460 down to 622.353 mph and Gabelich's actual speed could very well be any one of these 108 different values. Stated differently, even if we assume that each time was accurate to the last of the 4 digits to which it was recorded, the resulting speed may well be in error by anything from zero to .054, which means that it is probably 622.4 but that it might be 622.5 mph. In other words, it is for sure accurate to only 3 of the 6 figures to which it is given; timing to the nearest thousandth of a second cannot give accuracy to anything like the nearest thousandth of a mile. Finally, times that are accurate to 4 significant figures cannot produce speeds that are any more accurate than to about 4 significant figures. In this case, they aren't even that good.

Our own car speeds, in Table 2.5, somewhat slower than Gabelich's, probably don't average out to the stated 46.74 either, although that is the value of $\overline{X}$. An error of 1 mph in any one of them would change the mean by about .01. Several such errors are not merely possible—they are probable.

There is still another source of error—and a more important one—that we must consider. We shall now turn to the second question we raised in the opening paragraph of this chapter—the extent to which our computed figures may be applicable to other groups. To illustrate this problem, suppose we select another group of 176 students similar to those in the El Dorado School and administer the same California Short-Form Test of Mental Maturity to them. Would their mean also be 110.01? Suppose we select still another such group. And another. And another. Would all these groups have arithmetic means of 110.01? Of course not.

Clearly, we would not be justified in placing all our faith in the particular figure we compute from our first sample. We certainly cannot rely completely on it. To what extent *can* we rely on such a figure? This is a far more complicated question than it appears to be and a large part of this chapter will be devoted to answering it. To do this we will have to do some speculating, we will need to predict what would happen if we knew some

*Guiness Book of World Records, Revised Ed., 1975.

things we shall never know, and we will have to investigate some aspects of a topic called *statistical inference*. Only then can we answer the question of how well we can rely on the statistics we have computed.

SPECULATION

Let us now make a few assumptions and see what we can deduce. Suppose we knew the shape of the frequency distribution of the extremely large population or universe of which our sample of 119 car speeds or 176 IQs is a part. There might then be a way to determine the most probable mean and standard deviation of such samples. As a matter of fact, there is. From the mean of the frequency distribution of the entire population, we can easily estimate the mean of a large number of random samples taken from this large population. It happens that the two means are the same. We can also estimate the standard deviation within each of the samples. It turns out to be just a bit smaller than that of the population. From figures for the population, we can determine, for instance, whether our sample mean is close to or far from the true mean of the population and hence how far that true mean deviates from our own sample mean.

This is nice. It is mathematically sound. It is of theoretical interest. It will help in understanding some of the concepts of sampling. It is even useful in hypothesis testing, as we will see later in this chapter. But from a hard-boiled practical point of view, it's looking at things backwards; it isn't directly useful for our present problem, and it may even be misleading. The difficulty is that we need more information than is at all likely to be available. We need to know the size of the samples, *N*, and the mean and standard deviation of the population distribution. We know *N*, but in nearly all practical situations, nobody (with the possible exception of God) knows the other two values. After all, if we knew them, there would be no problem. Why select a sample, gather the data, compute the mean and standard deviation for the sample, and then try to estimate corresponding figures for the population if you already know what they are?

But before dismissing this as fantasy, let us daydream just a little bit more. If we selected a large number of groups at random from the same population, we have already agreed that they would not all have identical means. That is, they would vary and, if we wanted to do so, we could plot a frequency distribution of the means. The mean of these means, as you might guess, would tend to be very close to the population mean. But the means don't vary nearly as much as the individual scores do, so the standard deviation of this distribution of means would be much smaller than the standard deviation of the individual scores.

An interesting thing about the shape of the distribution of the means is that it tends to be normal whether or not the original scores are normally distributed. Consequently, if we had this distribution of the means, we could then tell not only what particular sample mean is most probable (it happens to be the true mean of the population), but we could also state how probable any other means are, deviating from it by any specified amount. Or, if we preferred, we could state the upper and lower limits that would include any given proportion (say .95) of such means. But, again, this is *if* we knew the true values for the population. We don't. So let's find out how to solve the general problem with what we do know. The preceding discussion has purposely been somewhat nontechnical. We shall now be more precise and, in order to do so, we shall have to define a number of terms. Let us begin with the fundamental definitions pertaining to the distinction between the figures we get from our sample and from the population.

SAMPLES AND POPULATIONS

In much statistical work, we gather data or obtain scores for the individuals in a group, called a *sample,* which it is hoped will provide us with information about a larger (often infinite) group, called a *population.* The figures we compute from the data we gather are called *sample statistics* or *statistics.* The corresponding figures for the entire population are called *population parameters* or *parameters.* (Note that the **s**s go together in **s**ample and **s**tatistic, while the **p**s go together in **p**opulation and **p**arameter.) Statistics, always based on a finite amount of data, are usually symbolized by ordinary (roman) letters, while parameters, typically characterizing very large or infinite populations, usually have Greek letters as symbols. The following diagram may help clarify these relationships:

	Scope of Data	Character-istics	Type of Symbol	Examples
Sample	Finite	**S**tatistics	Usually roman	$N\ \overline{X}\ s\ s^2$
Population	Large or infinite	**P**arameters	Usually Greek	$\infty\ \mu\ \sigma\ \sigma^2$

In the study of statistical methods, one type of sample is more commonly referred to and is more basic than any other. A *random sample* is one that is so drawn that every individual in the population has an equal and independent chance of being selected. The first part (equal) of this definition is clear; the second part (independent) means that the selection of any one case must have no influence, favorable or unfavorable, on the selection of any other case for possible inclusion in our sample. Thus, we *cannot* get a random sample of the population of 12,000 students attending a university by taking 100 attending a fraternity party, by picking the next 100 walking by the Education Building, or even by taking any 100 of the 238 enrolled in the elementary Psychology course, since in each case every student in the university does not have an equal and independent chance of being selected. We *can* get a random sample of this population by borrowing the 12,000 student cards from the registrar, mixing them thoroughly, and then, without looking, drawing any 100 cards. There are, of course, simpler ways involving less disruption of the university records. If the students are numbered, and if a so-called table of random numbers is available, it is fairly simple to start anywhere, taking every student number that appears until the desired 100 have been selected. Or coins may be tossed. Or cards may be drawn. But we rarely select samples in Education or Psychology in this manner. Further, we too often assume that our sample is typical not just of the 12,000 students of Fourfront University but of all students in coeducational colleges, of all American students, of Americans, or even of people. Why do we try to generalize so much? Simply because practically nobody is interested in the sample as such. A teacher we know was fond of making a statement slightly stronger than, "We don't give a darn about the sample." What we want to know is something about the population from which the sample came. Now, most unfortunately, the only practicable way most such generalizations can be made is from a random sample to the population. Hence, we sometimes tend, inaccurately to be sure, to assume that our samples are random, even though we know that this is not strictly true—or not true at all. We are not always, or even often, this irresponsible. Frequently, we can, with considerable justification, generalize from our sample to a population or populations from which such a sample might reasonably have been randomly drawn. We must, however, exercise discretion and avoid anything even faintly resembling a generalization from the students in a class in educational administration to all students or to all people.

SAMPLING DISTRIBUTIONS AND STANDARD ERROR OF THE MEAN

Let us again assume, as we did earlier, that we have a population of scores. From this population, assume that we draw a large number of random samples, each composed of N cases. We now compute the same statistic, say $\overline{X}$, for each sample. If we plot these means, they will tend to be symmetrically and normally distributed. This frequency distribution is called the *sampling distribution* of the mean. Similarly, if any other statistic, such as s, is computed for each of the infinite number of random samples that can be drawn from the parent population of scores, then the distribution of these standard deviations is called the sampling distribution of the standard deviation.

The sampling distributions of many, but not all, statistics are either normal or nearly so. When the sampling distributions are essentially normal, as they are for the three statistics ($\overline{X}$, *Mdn,* and s) we have so far considered, the mean and standard deviation of the sampling distribution may be estimated fairly readily, either from the parameters of the parent population or from the statistics of our sample.

Confining ourselves to samples of size N, where N is of any size we wish, let us assume that we have drawn all possible samples from our population. If the population is infinite, there will be an infinite number of such samples. Let us now compute the mean, $\overline{X}$, of each of these samples. We already know that the population mean, μ, must by definition be the mean of all the X scores in the population. Now it so happens that the population mean, μ, is also the mean of all the possible sample means, $\overline{X}$, that may be obtained by drawing samples from that population. Thus, our best estimate of the mean of the sampling distribution of the mean, $\overline{X}$, is

$$\tilde{\mu} = \overline{X}; \tag{5.1}$$

in which the tilde, or wavy line, over the μ indicates that we are predicting it.

With the standard deviation of these sample means, the situation is different. The standard deviation of the sampling distribution of the mean is much smaller than either the standard deviation, s, of the sample or the standard deviation, σ, of the population, both of which tend to be of about the same size. Our best estimate of the standard deviation of the sampling distribution of the mean, also called the *standard error of the mean,* is simply the standard deviation divided by the square root of N. In symbols, this is

$$\tilde{\sigma}_{\bar{X}} = s_{\bar{X}} = \frac{s}{\sqrt{N}}, \tag{5.2}$$

where the tilde, as before, signifies an estimate or prediction. This formula is often written with only the last two terms,

$$s_{\bar{X}} = \frac{s}{\sqrt{N}},$$

and is referred to as the standard error of the mean.

We shall give precise interpretations of a common application of this formula in the next section of this chapter, but in order to clarify the meaning of $s_{\bar{X}}$, we shall now compute it and give an approximate interpretation of its meaning for the El Dorado IQ data. This will be followed by a correct application of $s_{\bar{X}}$ in the estimation of a population mean.

We already know that $\overline{X} = 110.01$, $s = 15.20$, and $N = 176$. Hence,

$$\begin{aligned} s_{\bar{X}} &= \frac{s}{\sqrt{N}} \\ &= \frac{15.20}{\sqrt{176}} \\ &= \frac{15.20}{13.27} \\ &= 1.15. \end{aligned}$$

Table E shows that 2 times .3413 or .6826 of the cases lie between points one standard deviation on each side of the mean. This is true both for frequency distributions and for standard errors of various statistics such as the mean. Hence, if our population mean, μ, had been 110.01, about .6826 of the time we would expect our obtained mean, $\overline{X}$, to lie within the limits bounded by one standard deviation on each side of it or, in this case, within the limits of 110.01 ± 1.15, or 108.86 and 111.16. Roughly speaking, we can reverse this and say that since our sample mean, $\overline{X}$, is 110.01, about two-thirds of the time we would expect the population mean, μ, to be between 108.86 and 111.16.

Such formulas as these relating to the mean are very useful in theoretical work and they, or simple modifications of them, can be used in testing hypotheses about known or assumed population parameters. Formula (5.2) is especially useful to us in the t test, which we shall now discuss.

THE *t* TEST FOR MEANS

Let us again assume that we have a sample randomly selected from a population. Now, any time we have a random sample, there are some inferences that we can make concerning the parameters of the population from which it was drawn. This is important. This is interesting. This is useful. This is the most common of all situations in which we estimate parameters.

In our discussion of the El Dorado IQs in Chapter 2, we stated that this is a bright sixth grade. Can we substantiate this statement statistically? In a sense, we have already done this, for we have computed the sample mean, $\overline{X}$, and found it to be 110.01. This is clearly larger than what we may wish to regard as the population mean, μ, of 100. But is the difference truly significant or is it just a chance fluctuation? Assuming that our El Dorado sample might reasonably have been drawn at random from some larger population, the *t test* will answer this question for us and will require no more information than is contained in our sample. Let us see what this t test is. First, we need to define t.

$$t = \frac{\overline{X} - \mu}{\frac{s}{\sqrt{N}}} = \frac{(\overline{X} - \mu)\sqrt{N}}{s}. \tag{5.3}$$

All the symbols are familiar. The numerator of the expression between the two equal signs gives us a figure indicating a deviation from the population mean. The denominator was just given in Formula (5.2) as the standard error of the mean. This expression for t is thus similar to a standard score and, like z, gives us a figure indicating a deviation from the mean of a distribution. But whereas z is interpreted in terms of the normal probability curve, t is interpreted in terms of the t distribution. When compared with the normal probability curve, the t distribution has more area under its tails, especially when N is small. When N becomes large, the t distribution rapidly approaches the normal probability curve.

Before going further we need to define one term, *degrees of freedom* (nearly always written as *df*). This is the number of the observations or deviations that are free to vary when we are computing s. Since Σx must be zero, only $N - 1$ of the observations are free to vary, the last one necessarily being of whatever size is needed to make $\Sigma x = 0$. In problems involving a single computed mean, *df* is always $N - 1$.

Actually there are many t distributions—one for every value of *df*. It would not be feasible to show them all, or even a few of them, in detail. It is, however, feasible to pick out and show the relevant portions of a large number of the t distributions for the few significance levels in which we are interested.

The portions of the distributions we are particularly interested in are those areas distant from the mean which contain only a small proportion of the cases, say .05, or .01, or .001. Then we can identify those scores which have only a small chance (50, 10, or 1 in 1000) of occurring by chance in each distribution. The .05, .01, and .001 levels are *levels of significance*. A significance level of .001, for example, means that if there really were no difference between $\overline{X}$ and μ and chance alone were operating, a difference of this size would occur .001, or 1/10 of 1%, of the time. (English translation: Such a difference is probably not due to chance.) Table G shows the t values at the three commonly used significance levels; let us now examine the t table and note some of the characteristics of the t distribution:

The values of t required for significance at any level are very large when *df* is ridiculously small, as the first few rows of the table show.

The value of t needed for any specified significance level decreases as df increases, rapidly at first and then more slowly.

As df increases, each column of t values approaches and eventually reaches the figure in the ∞ row of the t table. These figures are identical to those obtainable from Table E of the normal probability curve. Thus, Table E shows us that .05, or 5%, of the area of the normal probability curve (.025 at each end) lies beyond an $\frac{x}{\sigma}$ distance of 1.96, while Table G shows that as df increases, the t value required for significance at the .05 level drops closer and closer to 1.96 and finally reaches 1.96 somewhere between 400 and 500 df.

Thus, we see that when df is infinite, the t distribution is normal; when df is large it is nearly normal; and only when df becomes very small does it depart much from the normal probability curve. Then the departure is greatest as we get further from the mean: The t distribution is slightly higher at the mean, lower a little distance from the mean, and then higher in the two tails than is the normal distribution. That is why when $df = 10$, for example, it is necessary to go out to t values of -4.59 and $+4.59$ in order to include all but .001 of the area, whereas a distance of only ± 3.29 would have done it if the curve had been exactly normal. This difference will appear relatively much greater if, instead of comparing the distances needed to include .999 of the area, we compare the areas beyond 4.59. As just indicated, for $df = 10$, the area in the t distribution beyond $\pm 4.59s$ is .001. In the normal probability curve, however, the area beyond $\pm 4.59\sigma$ is only .000004432. The t area of .001 is 226 times as large as this. (Both areas are, however, so small that we would ordinarily regard them as negligible.)

Let us now return to our El Dorado IQ data and use Table G to determine whether or not our observed difference between means, $\overline{X} - \mu$, is of such size that it can reasonably be regarded as a chance fluctuation from zero. We shall use the t table to test whether our $\overline{X}$ of 110.01 is significantly different from the assumed population mean, μ, of 100. From Chapter 3, page 54, we find that $s = 15.20$. Also, from Chapter 2, page 23, we know that $N = 176$, so $df = 176 - 1 = 175$. Substituting in Formula (5.3), we obtain

$$t = \frac{(\overline{X} - \mu)\sqrt{N}}{s}$$

$$= \frac{(110.01 - 100)\sqrt{176}}{15.20}$$

$$= \frac{(10.01)(13.27)}{15.20}$$

$$= \frac{132.83}{15.20}$$
$$= 8.74.$$

The t table does not have a row for $df = 175$, but 175 is clearly between 150 and 200. Since the t values for these two rows are similar, and since 8.74 is far beyond the highest value in either row, our $\overline{X}$ of 110.01 is significantly different from the population mean of 100 at the .001 level. If chance alone were operating, a sample mean (110.01) would differ this much (10.01) from the population mean (100) less often than once in a thousand times. This means that although this difference, or any other one, *may* be due to chance, it is extremely unlikely that such is the case. In statistical language, we may state that the difference is significant at (or beyond) the .001 level, or that the difference is very highly significant. This is sometimes written as $P < .001$ to indicate that the probability is less than .001 of obtaining a t this large if chance alone had accounted for all deviations from the parameter μ.

LEVELS OF SIGNIFICANCE

We have just concluded that the difference between the El Dorado sample mean IQ and the assumed population parameter is very highly significant. How big does a difference have to be in order to be called significant? This is a little easier to answer than "How high does a tree have to be in order to be called tall?" or "How heavy does a person have to be in order to be called overweight?" For such questions, there are no absolute standards. There is nothing sacred about any particular level of significance—various levels such as .0027 have been used in the past—but at the present time most statisticians and research workers use one or more of the three levels shown in Table G. These levels of significance are often designated by the Greek letter α. If a person decides to use the .01 level alone ($\alpha = .01$), he will then treat all values of t smaller than those in the .01 column as not significant (often abbreviated as *NS* or *N.S.*) while he will regard all values of t equal to or greater than those in the .01 column as significant.

A more flexible and less arbitrary procedure is to place all values of t in one or the other of four, rather than two, classes depending on which of the three levels of significance they reach or fail to reach. We can illustrate this in Table H.

Column (1) gives the precise designations for the four intervals. Column (2) is correct only on the assumption (now fairly common in psychological literature) that the investigator will claim as great significance as possible for his results. Column (3) shows a currently growing trend in the labeling of differences in tables; thus, the figure -3.18^{**} means that the difference is 3.18 in the negative direction and that this difference is significant at the .01 but not at the .001 level.

It is because our El Dorado data gave a t of 8.74, which yielded a P of less than .001, that we said the difference of 10.01 IQ points was "very highly significant."

As another illustration in comparing a sample mean with a theoretical parameter, assume that we have $t = 2.34$ and $N = 238$. Then df will be 237. Our t of 2.34 is clearly greater than the 1.97 in the .05 column, but less than the 2.60 or 2.59 in the .01 column. Since P is less than .05 but greater than .01, we may write $.05 > P > .01$ and state that the difference is barely significant.*

For all values of df of 20 or more, t has to be about four-thirds as large to be significant at the .01 level as at the .05 level. Further, by an interesting coincidence, this same relation holds between the next two levels also: to be significant at the .001 level, t must be four-thirds as large as would be necessary to be significant at the .01 level. (The actual figures, of course, are not exactly four-thirds, but they are very close to it. They vary only from 1.31 to 1.36 for the first comparison and from 1.28 to 1.35 for the second.)

As we go from the large toward the small values of df, both on an absolute and on a relative basis, the values of t needed for significance change relatively little at the .05 level (becoming 5% larger than 1.96 only when df is as small as 27), while they change somewhat more at the .01 level (becoming 5% larger than 2.58 around $df = 40$) and still more rapidly at the .001 level (becoming 5% larger than 3.29 when df is about 60). Of course, no one of the changes cited is very large—each one is 5%—so when df is 50 or more, the t distribution is very much like the normal distribution shown in the last row of the t table.

THE *z* TEST FOR MEANS

Now that we have learned how to use the t test, let us consider a similar kind of test for testing hypotheses about means. It is called the z test, and the z here has the same meaning as the z we studied in Chapter 4. A z score tells us how many standard deviations we are away from the mean (either above or below). We compute z in a manner similar to that we used for t, except that (1) we use Formula (5.4) for the standard error of the mean; (2)

*Notice that a research worker who consistently used .01 as the significance level would state that this difference is not significant, while one who used the somewhat lax standard of .05 would state that the same difference is significant. Either investigator can—and will—be wrong on some statements. The two types of errors they can—and do—make are discussed on pages 134–137.

we use Formula (5.5) for our test statistic z; and (3) we would refer our obtained z to a table for normal curve areas.

The formula for standard error of the mean for z is based upon the assumption that you either know definitely what the population standard deviation is or you feel so sure about an assumed value that you use it as if you did know for sure. The formula is similar to Formula (5.2) except that it uses σ in place of s. The formula is

$$\sigma_{\bar{X}} = \frac{\sigma}{\sqrt{N}}. \tag{5.4}$$

The formula for z is

$$z = \frac{\bar{X} - \mu}{\dfrac{\sigma}{\sqrt{N}}}. \tag{5.5}$$

We do not really have to go to a table for normal curve areas, such as Table E. Instead we can use the bottom row of Table G for an infinite sample size. This gives us exactly the same values of z for the .05, .01, and the .001 significance levels which we would have obtained by going to Table E. These zs for the three significance levels above are as follows:

Significance Level	Value of t for Infinite Degrees of Freedom
.05	1.96
.01	2.58
.001	3.29

At this point you may be wondering why we have two different methods which seem to have the same aim, that of testing hypotheses about an observed mean. You may also be wondering how we go about choosing between z and t. We have also said that if we know σ, the population standard deviation, we can use z. When we do not know σ, we must use s as an estimate of σ and must choose the t test for an observed mean.

Some textbooks make a distinction between "large samples" (so-called) and "small samples" (so-called). They recommend that z be used for a *large sample* (defined as larger than $N = 30$) and that t be used for a *small sample* (defined as smaller than $N = 30$). This is an arbitrary rule and you can satisfy yourself that it is indeed arbitrary by noting that the values of t change gradually around $N = 30$ in Table G, rather than suddenly. Those who use z for "large samples" must resort to t at some point in order to prevent their results from being biased due to the effect of small samples. We recommend that you always use t, so that you will always be protected against bias.

POINT AND INTERVAL ESTIMATES

When we considered the calculation of the mean in Chapter 2 and when we calculated the value for the El Dorado IQ data, we were making what is known as a *point estimate* of μ. The sample mean, $\overline{X}$, is our point estimate of μ.

However, at other times, we may wish to make a second kind of estimate, an *interval estimate,* of an interval along the scale of our variable within which μ could be expected to fall some given proportion of the time. The proportion is actually a probability and is usually called that.

Let us approach interval estimation by considering in turn several different assumed values of μ and asking how consistent the hypothesis is that a certain assumed μ is the true population mean.

In using the t test with the El Dorado mean IQ data, we were attempting to determine whether or not our sample mean was consistent with an assumed parameter located at a specific point (in this case, 100). We found that it was not. But we found more than that. Our t of 8.74 was far beyond any value in the table for our df of 175. We may then reasonably ask, is our sample mean of 110.01 consistent with the hypothesis that the corresponding population parameter is 105? Or 110? Or 115? Or 120? Or 90? Let us compute t for the parameter of 105:

$$
\begin{aligned}
t &= \frac{(\overline{X} - \mu)\sqrt{N}}{s} \\
&= \frac{(110.01 - 105)\sqrt{176}}{15.20} \\
&= \frac{(5.01)(13.27)}{15.20} \\
&= \frac{66.48}{15.20} \\
&= 4.37.
\end{aligned}
$$

Again, P is less than .001 and the difference is very highly significant. But if we assume that $\mu = 110$, our difference of $110.01 - 110$ yields a t of only .01, giving a P far greater than .05. Of course, the difference is not shown to be significant.

If t had happened to be 3.35, Table G shows that for $df = 175$, the difference would be the smallest possible one that could be significant at the .001 level. By trial and error, perhaps aided by interpolation, we could find out what value of μ would yield a t of 3.35. Since negative differences are just as important as positive ones, we could also find out what value of μ

would give $t = -3.35$. A somewhat easier way would be to substitute in Equation (5.3). A still easier method to obtain these confidence limits is to use Equation (5.6):

$$\mu = \overline{X} \pm \frac{ts}{\sqrt{N}}. \tag{5.6}$$

All these methods will give the same answer. Let us use the easiest procedure of substituting in Equation (5.6).

$$\begin{aligned}
\mu &= \overline{X} \pm \frac{ts}{\sqrt{N}} \\
&= 110.01 \pm \frac{(3.35)(15.20)}{\sqrt{176}} \\
&= 110.01 \pm \frac{50.92}{13.27} \\
&= 110.01 \pm 3.84 \\
&= 106.17 \text{ or } 113.85.
\end{aligned}$$

This means that if μ really is either 106.17 or 113.85, a difference as great as 3.84 IQ points (which would land us right on our sample mean) would happen, on the average, exactly once per 1000 samples drawn. All values of μ between these limits will, of course, be closer to our sample mean; and all values of μ outside these limits will be further away from it. The values 106.17 and 113.85 are *confidence limits,* which are the end points for what we call the *confidence interval,* which includes all values between the two limits. Thus, the .999 confidence limits for the population mean IQ, in the population of which our El Dorado scores are a random sample, are 106.17 and 113.85.

If we want confidence limits for the mean at the .99 or at the .95 level, all we need to do is look in the proper *df* row of Table G for the appropriate value of t (in this case, 2.60 or 1.98) and use it instead of 3.35 in Equation (5.6). If we do this for the .99 confidence limits, we obtain

$$\begin{aligned}
\mu &= \overline{X} \pm \frac{ts}{\sqrt{N}} \\
&= 110.01 \pm \frac{(2.60)(15.20)}{\sqrt{176}} \\
&= 110.01 \pm \frac{39.52}{13.27} \\
&= 110.01 \pm 2.98 \\
&= 107.03 \text{ or } 112.99.
\end{aligned}$$

Similarly, the .95 confidence limits for the mean are

$$\begin{aligned}\mu &= \overline{X} \pm \frac{ts}{\sqrt{N}} \\ &= 110.01 \pm \frac{(1.98)(15.20)}{\sqrt{176}} \\ &= 110.01 \pm \frac{30.10}{13.27} \\ &= 110.01 \pm 2.27 \\ &= 107.74 \text{ or } 112.28.\end{aligned}$$

Thus, you may say, with about 999 chances in 1000 of being right, that the population mean IQ is between 106 and 114. Or, with approximately 99 chances in 100 of being right, you may state that it is between 107 and 113. Finally, with roughly 95 chances in a 100 of being correct, you may give the limits as 108 and 112. Which of these statements you use depends upon your preference; we recommend that you choose the first one because IQs of 106 and 114 are not far apart. The choice is always a compromise between our desire to be sure of ourselves and our desire to give limits that are narrow enough to be meaningful. (A statement that μ almost certainly lies between 50 and 170, though true, is so vague as to be pointless. At the other extreme, a statement that has 80 chances in 100 of being correct is fine for gambling, but not good enough for scientific decisions.) Note that the more sure we wish to be of what we say, the wider the confidence interval must be. If we want a narrow interval we can get it, but only at the price of being less sure of ourselves.

Such confidence limits or confidence interval estimates of parameters can be very useful. If we pick the value of t from the column corresponding to the level of significance (such as $P = .01$) at which we wish to operate, Equation (5.6) gives us limits including all values of the parameter, μ, from which our sample could be regarded as a random sample with the corresponding (such as .99) confidence limits. Our sample, then, might reasonably be considered as one drawn at random from a population with a parameter anywhere between these limits. Further, our sample differs significantly from all populations with parameters outside these limits.

STATISTICAL INFERENCE

In establishing confidence limits for a parameter, as we just did for the mean IQ, we are engaging in statistical inference, the process of estimating the parameters of a population from our computed sample statistics. Because, as mentioned earlier, we are rarely or never interested merely in the sample but rather in the population of which it is a part, statistical inference is extremely important. Probably more inferences are made with respect to means than are made concerning any other parameters. You have already seen how to use the t table to determine whether or not a sample mean differs significantly from an assumed parameter. You have also used the t table and Formula (5.6) to establish confidence limits for the mean. Later (in Chapter 14) you will learn how to determine whether two sample means are significantly different from each other, or whether they may be regarded as coming from the same population. But let us now turn from the use of the t table to some other aspects of statistical inference.

SAMPLING DISTRIBUTION AND STANDARD ERROR OF THE MEDIAN

We will assume, as before, that we have drawn all possible samples of size N from our population. However, this time N must be 100 cases or larger. We will further assume that we have computed the median of each of these samples. Now we will make another assumption that was not necessary in dealing with the mean—the assumption that the scores in the population from which our samples are drawn are distributed normally or nearly so. This assumption will not ordinarily handicap us; as we stated on page 5, it holds for most of the measurements with which we shall be working. We now plot a distribution of the sample medians and thus obtain the sampling distribution of the median. This sampling distribution will be of the familiar normal shape.

Now the *population median,* which we shall designate by the symbol Mdn_{pop}, is by definition the median of all the X scores in the population. Using the tilde, as before, to indicate a predicted value, our estimate of the population median turns out to be the sample median. In symbols, this is

$$\widetilde{Mdn}_{\text{pop}} = Mdn. \tag{5.7}$$

But we know that the sample medians will vary about the (unknown) population median. How widely? The standard error of the median gives the answer. Its value is

$$\tilde{\sigma}_{Mdn} = s_{Mdn} = 1.2533\left(\frac{s}{\sqrt{N}}\right). \tag{5.8}$$

We cannot use the t test to establish confidence limits for the median as we did for the mean. We can, however, assume any value we wish for the population median and then find out whether our obtained Mdn may reasonably be regarded as a chance deviation from that assumed value. Let us do this with the El Dorado IQ data. From Chapters 2 (p. 30) and 3 (p. 54), we obtain the necessary basic data: $N = 176$; $Mdn = 110.80$; and $s = 15.20$. Substituting in Formula (5.8), we obtain

$$\begin{aligned} s_{Mdn} &= 1.2533 \left(\frac{s}{\sqrt{N}}\right) \\ &= 1.2533 \left(\frac{15.20}{\sqrt{176}}\right) \\ &= \frac{19.0502}{13.27} \\ &= 1.44. \end{aligned}$$

The interpretation is, of course, the same as in the case of the standard error of the mean, discussed earlier on page 119. About two-thirds of such medians (each based on a different set of 176 cases selected at random from the same infinite universe) will be within 1.44 IQ points of the true median; .99 of them will be within 2.60 times 1.44, or 3.74, IQ points of the true median of the infinite universe.

These limits are somewhat wider than were those for the mean. In fact, as a comparison of Formulas (5.2) and (5.8) shows, they are each 1.2533 times as great. Thus, we see that successive medians are distributed more widely about the true median than are the successive means around the true mean. In statistical terminology, the median has a larger standard error than does the mean.

Can our sample median of 110.80 reasonably be regarded as a chance fluctuation from the assumed population median (and mean) of 100? To answer this, we have to obtain what is really a standard score corresponding to the deviation of our sample median from a fixed point, the population median. To get this standard score, we may simply divide the difference, 10.80, by the standard error of the median, 1.44, getting 7.50. Or, we may be more formal, starting with Formula (3.6), then rewriting it by noting that in the sampling distribution of the median, z may be regarded as a score, Mdn, deviating from the mean of such scores, Mdn_{pop}, and divided by their standard deviation, s_{Mdn}. Thus, we get

$$\begin{aligned} z &= \frac{X - \overline{X}}{s} \\ &= \frac{Mdn - Mdn_{\text{pop}}}{s_{Mdn}} \\ &= \frac{110.80 - 100}{1.44} \\ &= \frac{10.80}{1.44} \\ &= 7.50. \end{aligned}$$

In either case, we get the same result—a standard score or normal curve deviate of 7.50. We have assumed that the medians are normally distributed. Since standard scores in normal distributions are interpreted in Table E, we can use that table to determine the significance of the deviation of our median. Much less than .001 of the area of the curve is contained in both tails beyond $z = 7.50$; hence, P is much less than .001 and this difference is very highly significant.

Now let us try it for an assumed population median of 115. In this case, $z = \frac{(110.80 - 115)}{1.44} = -2.92$. Table E gives an area of .4982, which means the tail contains an area of .0018, or both tails contain an area of .0036. Our result is thus significant at the .01 but not at the .001 level. Notice that our $\bar{X}$ of 110.01 was significantly different from μ of 115 at the .001 level but that our Mdn of 110.80 was significantly different from Mdn_{pop} of 115 at only the .01 level. The reason for this is partly because the sample median is a little larger than the sample mean, but primarily because the standard error of the median is considerably larger, in fact, 25% larger, as a comparison of Formulas (5.2) and (5.8) will show, than that of the mean. This difference in the standard errors means that, if we are sampling from a normal distribution, as we often are, and if we are interested in estimating the population mean or median with a fixed degree of precision, we shall need about 57% more cases if we use the median than we shall need if we use the mean. This point is discussed at greater length in Supplement 5.

SAMPLING DISTRIBUTION AND STANDARD ERROR OF THE STANDARD DEVIATION

From a mathematical point of view, it would be simpler if we discussed the distribution of the variance, s^2, rather than of the standard deviation. From the practical standpoint, however, it is more meaningful to discuss s, so we shall do so. The best estimate of the population variance, σ^2, is the sample variance, s^2. For standard deviations, it is not *exactly* true, but to a close approximation we may still make the corresponding statement that

$$\tilde{\sigma} = s, \tag{5.9}$$

where the tilde again indicates a predicted value.

How widely will our sample standard deviations vary about the population parameter? The formula is quite satisfactory when N is over 100, gives reasonably good results when N is between 30 and 100, but is appreciably in error when N is below 30. Subject to this restriction, the standard error of the standard deviation is

$$\tilde{\sigma}_s = s_s = \frac{s}{\sqrt{2N}}. \tag{5.10}$$

Because $N = 176$, we have well over 100 cases and it is entirely appropriate to use this formula for the determination of the standard error of the El Dorado standard deviation of 15.20. Thus, we readily obtain

$$\begin{aligned} s_s &= \frac{s}{\sqrt{2N}} \\ &= \frac{15.20}{\sqrt{(2)(176)}} \\ &= \frac{15.20}{18.76} \\ &= .81. \end{aligned}$$

Does our computed sample standard deviation differ significantly from an assumed population parameter of 17? We may solve this by determining z.

$$\begin{aligned} z &= \frac{X - \overline{X}}{s} \\ &= \frac{s - \sigma}{s_s} \\ &= \frac{15.20 - 17}{.81} \\ &= -\frac{1.80}{.81} \\ &= -2.22. \end{aligned}$$

Using Table E or the last line of the t table, we see that this value of z is significant at the .05 but not at the .01 level. Since $.05 > P > .01$, according to the rule we gave in Table H, this difference is a barely significant one. Our sample, therefore, is probably not drawn from a population with a standard deviation of 17.

Does our sample standard deviation differ from a population value of 16? This time $z = \dfrac{(15.20 - 16)}{.81} = -.99$. Table G shows that $P > .05$ and Table E shows that the area beyond .99 in both directions is (2)(.1611) or .3222. Our obtained s of 15.20 has certainly not been shown to differ significantly from a σ of 16. Some people would say that the difference is not significant. This statement is technically true if no misunderstanding results. It is definitely incorrect to conclude that since the difference is not at all significant, we have shown that there is no difference between them. We can never prove that a difference is zero. To illustrate, notice that if we compare our s of 15.20 with an assumed σ of either 14 or 15, we shall in each case also get a small z yielding $P > .05$. Certainly, we cannot simultaneously prove that there is no difference between our sample s and each of three different population standard deviations. We shall return to this point in later chapters because there is considerable confusion with respect to it. The correct point of view is simply that *we can sometimes demonstrate or prove (with a high degree of probability such as .99 or even .999) that a difference does exist; we can never prove that there is no difference* either between two statistics or between a statistic and a fixed point such as an assumed parameter.

HYPOTHESIS TESTING

Much research starts out with the idea that, according to some theory, certain results should appear. The research study is carried out in order to check up on the theory. If the results turn out to be as predicted, or not significantly different, they are obviously consistent with what the hypothesis called for. So we retain the hypothesis and say that the results substantiate (not prove) it. If the results are significantly different from those expected by the hypothesis, we reject the hypothesis and state that our results disprove it.

Notice that a hypothesis need be nothing more than an assumption or a guess. It may or may not be true. It must, however, be of such a nature that it can be tested. Thus, the statement that blue is a better color than red is not testable; the statement that more 9-year-olds prefer blue than red is testable; so is the statement that 9-year-olds will divide evenly on their preferences for blue and red. This last statement is made in a form that is very common in statistical theory. It is called a *null hypothesis* (or H_0) because it specifies that there is no difference. Null hypotheses are often easier to test, statistically, than are others. Thus, to test the hypothesis that there is no difference in the red versus blue preference of 9-year-olds, we draw a random sample of such children, get their preferences, and analyze the results. The null hypothesis calls for no difference in the color preferences. If the sample proportion differs significantly from the parameter of .50 assumed under the null hypothesis, we reject the hypothesis and state that such children prefer blue or red, as the case may be. If the results do not differ significantly from .50, we must, temporarily, at least, retain the hypothesis that there is no difference.

When we state a hypothesis, and especially when we formulate a null hypothesis, we do this not because we believe it to be true (we often don't), but because it provides us with a firm base for carrying out a definite procedure for making decisions. More often than not, the person who formulates a null hypothesis does so because of a feeling (based on a theory or for some other reason) that the difference will *not* be zero. That is, the scientist states the null hypothesis and carries out an experiment for the purpose of rejecting that hypothesis.

Type I and Type II Errors In hypothesis testing, one of four circumstances will occur. In two of these, we will be right; in the other two, we will be wrong. We know this, even though in any one specific experiment, we can never be *sure* whether our conclusion is correct or not. Let us first consider the errors that we may make. They are almost universally designated as Type I or Type II:

A. *The Type I error* arises when the hypothesis is true, but our experiment gives results that are not consistent with it. We erroneously reject the true hypothesis and conclude that our results disprove it. This is quite obviously bad. How often will it happen? Oddly enough, this can be decided in advance. If a scientist conducts a number of different experiments and decides to work at the .01 level of significance, this research worker will regard a difference as conflicting with the hypothesis only if it is of such a nature that it could arise no more than .01 of the time if it were a chance fluctuation (due to random sampling) from the hypothesized value. Thus, automatically and inevitably, of all the times that the various hypotheses are true, .01 of these times the research worker will obtain results so rare that they could happen only .01 of the time. Hence, in the long run, the scientist will be wrong in .01 of these instances. If another person decides to use the .05 level, and if the hypotheses are all true, the proportion of errors will, on the average, be .05. More generally, if α (the first letter, alpha, of the Greek alphabet) is the specified significance level, and if the hypotheses are all true, the proportion of Type I errors is α. This situation is shown diagrammatically in the cell labeled A of Table 5.1. Notice that, to facilitate memory and understanding, this table has been so arranged that in the *first* of the four cells, the *first* roman numeral (**I**), the *first* letter of our Latin alphabet (**A**), and the *first* letter of the Greek alphabet ($\boldsymbol{\alpha}$) all go together.

B. The *Type II error* arises when the hypothesis is false but our experiment fails to show this and we erroneously retain the false hypothesis. This, too, is bad. How often will it happen? Unfortunately, we don't

TABLE 5.1
The Nature of Type I and Type II Errors

	Hypothesis Really Is True	Hypothesis Really Is False
Our results are inconsistent with the hypothesis, we *reject* it; we disprove it.	**A** Type **I** error $\boldsymbol{\alpha}$	C Correct conclusion
Our results are consistent with the hypothesis; we fail to reject it; we *retain* it.	D Correct conclusion	**B** Type **II** error

know, and there is no way to find out. We do know, however, that, in all ordinary circumstances, the proportion of Type II errors gets larger (1) if α is made smaller or (2) if N is made smaller. To illustrate, if we have been using the .01 level of significance and change to the .001 level, then a difference will have to be larger in order for us to regard it as significant. If the hypothesis is true, our Type I errors are thus reduced. But if the hypothesis is false, when we insist on a larger difference in order to reject the hypothesis, we are retaining more of these false hypotheses and making more Type II errors. On the other hand, if we change our significance level from .01 to the more lax standard of .05, we shall regard more of both kinds of hypotheses as false. If the hypothesis really is true, we increase the proportion of Type I errors from .01 to .05. If the hypothesis really is false, we shall (correctly) reject it more often, thus reducing the proportion of Type II errors.

As far as N is concerned, a decrease in N requires that a difference be larger in order to be regarded as significant. Hence, if the hypothesis is false, it will be rejected less often, thus increasing the proportion of Type II errors.

C. The hypothesis is false and our experiment gives results that are not consistent with the hypothesis. Hence we reject the hypothesis and conclude that our results disprove it. Our conclusion is correct, but only rarely are we in a position to know this. (If we already knew that the hypothesis was false, why would we conduct the experiment to find out whether or not it was false?)

D. The hypothesis is true and our experiment gives results that are consistent with it. In this case, we conclude that we have no basis for rejecting it, so we retain it. This conclusion is correct, but in any specific experiment we are not at all likely to know this.

Choosing Between Type I and Type II Errors "Out of the frying pan into the fire." A nasty choice. And a nasty choice we have to make with respect to Type I and Type II errors. In order to reduce Type I errors, whenever we make our significance level more stringent (by making α smaller, such as going from .01 to .001), we succeed, but we increase the number of Type II errors. Whenever we use a more lax significance level in order to reduce Type II errors, we succeed, but we increase the number of Type I errors.

If we use a stringent level of significance, such as $\alpha = .001$, we will seldom make unjustified claims, but we will often fail to detect the existence of real differences. If the null hypothesis is true (no difference), we will be doing fine, but if the null hypothesis is false (a real difference), we'll often fail to find this out. To go to a ridiculous extreme, if we used $\alpha = .000001$, we would practically never make the error of claiming that a true hypothesis was false, but we would also fail to detect the falsity of a large proportion of the false hypotheses.

If we use a very lax significance level such as .05, we will correctly classify practically all true differences, but we will also claim that many chance fluctuations in the data represent real differences. Here, if H_0 really is false (a real difference), we'll probably know about it, but if H_0 really is true (no difference), we'll often be going off half-cocked and claiming real differences that just don't exist. Again, to take an extreme position, if we used a .25 level of significance, we would claim that we had disproved large numbers of null hypotheses. With respect to the false ones, this is fine, but it is horrible to claim mistakenly that we have disproved a fourth of the ones that really are true.

How do we get out of this mess? First of all, we don't let $\alpha = .000001$ or .25. We use one of the intermediate values such as .01. That reduces the Type I errors to .01 of the number of experiments. Thus, if over the years, we test 200 hypotheses that are, in fact, true (plus others that aren't), we will be wrong in claiming the falsity of only about two of them. But what do we do about the Type II errors? We increase N. It's as simple as that. We may not know the proportion of Type II errors but we do know that we can reduce this proportion by increasing N. For this reason, we would recommend that the .01 level be used and that the data gathering process be extended, if possible, until N exceeds 100. This will not only cut down on both the Type I and the Type II errors, but on many others as well.

Regardless of the size of N, the problem of the size of α remains. As we implied in the preceding paragraph, we prefer .01. Excellent reasons can, however, be given for either the .05 or the .001 levels. If your instructor prefers one of these levels, we feel that neither you nor we should argue the point. All three of these levels can be well defended; more extreme ones probably cannot.

One-tailed Versus Two-tailed Tests Suppose it is known that in a number of schools apparently comparable to the El Dorado school, the median IQ is very close to 107.40. Suppose, further, that the El Dorado principal introduced some new teaching methods, designed to increase achievement test scores and perhaps also intelligence test scores. The principal now gets the data we used earlier in this chapter. Using 107.40 as the assumed population median, simple arithmetic gives $z = \frac{(110.80 - 107.40)}{1.44} = 2.36.$

Should the principal call a meeting of the teachers and tell them the news? What news? First of all, the 107.40 is an accepted value, and the 110.80 is the correct median for this sixth grade, so the difference of +3.40 may be regarded as arithmetically accurate. But how sure can the principal be that it is not just a chance difference? Table E shows that the area above $+2.36\sigma$ is .0091. The area below -2.36σ is also .0091. Hence, the area beyond 2.36σ in both directions is .0182. Is this differerence significant at the .01 level? There was a time when different statisticians would give different answers to this question. In fact, there still is, but the majority favoring what we shall call Interpretation B seems to be increasing. Here are the two ways of looking at the situation:

Interpretation A: One-tailed Test Since the new teaching methods were designed to increase intelligence scores or, at worst, leave them unchanged, the true difference should be positive or, at worst, zero. Negative differences are not expected and, if one should occur, it should be attributed to chance. Hence, our question should be worded as, "Is this *gain* of 3.40 IQ points significant?" The answer would then be that a gain this large or larger would occur only .0091 of the time, so $P < .01$ and the difference is significant at the .01 level.

Interpretation B: Two-tailed Test Regardless of the intent of the new teaching methods, they may turn out to be either better or worse than the old ones. Hence, whether or not negative differences are expected, they very well may occur and they may or may not be due to chance. The hypothesis that we are testing is the null hypothesis that the difference is zero and that chance alone is operating. If we get a difference that is too large to make the null hypothesis appear reasonable, we will reject it. This holds regardless of the direction of the difference. Hence, our question should be worded as "Is this *difference* of 3.40 IQ points significant?" The answer would then be that a difference this large or larger would occur $(2)(.0091) = .0182$ of the time, so $.05 > P > .01$ and the difference is significant at the .05 but not at the .01 level.

Further Comparison of One-tailed and Two-tailed Tests If we are using .01 as our standard of significance, Interpretations A and B will here lead us to opposite conclusions. This is not always, or even often, the case, but it is important that we know how to take care of such situations. We should point out that in experiments conducted with no prior hypotheses concerning the outcome, everyone agrees that Interpretation B is correct. It is only where a directional difference is hypothesized that Interpretation A is sometimes advocated. Let us see under what circumstances this makes a difference.

Table 5.2 shows that there is disagreement in a number of places at both ends of the curve. Close inspection shows, however, that there is complete agreement in several places, including the entire area from $z = -1.95$ to $z = +1.64$. Since these two limits include 92.50% of the entire area under a normal probability curve, the extent of disagreement is not nearly so great as might at first appear. There is no way to tell exactly how often these two methods of interpretation would lead to different conclusions. For instance, if the directions of the differences were predicted with uncanny accuracy,

TABLE 5.2
The Significance of Differences Under One-Tailed Versus Two-Tailed Tests When a Positive Difference Is Predicted

z	One-Tailed	Two-Tailed
$-\infty$ to -3.29	NS	.001
-3.28 to -2.58	NS	.01
-2.57 to -1.96	NS	.05
-1.95 to $+1.64$	NS	NS
$+1.65$ to $+1.95$	.05	NS
$+1.96$ to $+2.32$	.05	.05
$+2.33$ to $+2.57$	.01	.05
$+2.58$ to $+3.08$	.01	.01
$+3.09$ to $+3.28$	.001	.01
$+3.29$ to $+\infty$	.001	.001

Note: NS means not significant, even at the .05 level.

and if N was large enough that z in each case was greater in absolute value than 3.29, there would be no disagreement at all. At the other extreme, if the predictions of the direction of the differences turned out to be wrong as often as right and if again N was large enough so that z in each case was greater in absolute value than 3.29, disagreements would occur 50% of the time. But if there were no differences, the null hypothesis being consistently correct, the area of disagreement would be as shown in Figure 5.3, a total of only .0555, or less than 6% of the entire area.

In most real-life situations, the extent of disagreement will be between the zero for perfect prediction and the .0555 for the null hypothesis holding. It will be even lower if we confine our statements to one level of significance, instead of giving complete coverage to all three of the more or less standard levels. But, nevertheless, a decision must be made. As hinted at in earlier examples in this chapter, we feel that Interpretation B is to be preferred.

FIGURE 5.3
One-Tailed and Two-Tailed Disagreements

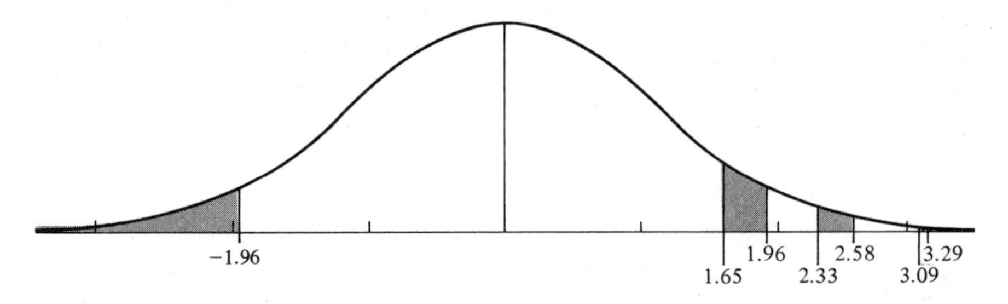

Hence, unless in certain circumstances your instructor gives you reasons for doing otherwise, you should routinely use two-tailed tests. Table G of the t table is so set up as to make this easy to do.

SUMMARY

Successive samples differ from each other, even though they are drawn randomly from the same population. We rarely know the population parameters. We use our sample statistics to estimate them, to establish confidence limits for them, or to test hypotheses about them. For these purposes, we need to know the sampling distribution that tells us how the statistic is distributed when many random samples are taken from the same population. We also need to know the standard error of the statistic, which is the standard deviation of its sampling distribution. We are then in a position to test hypotheses concerning the size of the corresponding population parameter. The t test is widely used in connection with means. The z test, based on the normal probability distribution, is used in connection with many other statistics. We can reduce Type I errors by using a relatively stringent level of significance such as .01 or .001. We can reduce Type II errors by using a larger N. In the great majority of situations, two-tailed tests are to be used rather than one-tailed tests.

SUPPLEMENT 5

STABILITY OF THE MEDIAN

Suppose, as we did for the arithmetic mean, that we selected a number of similar groups such as the 176 sixth-grade El Dorado School pupils. How widely would the successive medians vary around the true median of the infinite population from which we assume that these samples were drawn at random? The answer is given by a formula that is strictly analogous to Formula (5.2). Formula (5.8) for the standard error of the median was given as

$$s_{Mdn} = 1.2533\left(\frac{s}{\sqrt{N}}\right).$$

Substituting the known values of s and N, using the El Dorado IQ data, we obtained

$$\begin{aligned} s_{Mdn} &= 1.2533\left(\frac{15.20}{\sqrt{176}}\right) \\ &= \frac{19.0502}{13.27} \\ &= 1.44. \end{aligned}$$

It may at first appear that the standard error of any median is 25% larger than the standard error of the corresponding mean. This is absolutely correct, but it does not tell quite all of the story. An important question is: How can we determine the median with the same degree of accuracy as the mean? If we want the two standard errors to be the same, then the value obtained from Formula (5.2) must be identical with that obtained from Formula (5.8). The two values of s, being both based on the same distribution will be identical, but it will be necessary to distinguish between the two Ns. Let us temporarily use the symbol N_2 for the N in Formula (5.2) and N_8 for the N in Formula (5.8). We may now equate the two formulas:

$$s_{\bar{X}} = s_{Mdn},$$

$$\frac{s}{\sqrt{N_2}} = 1.2533 \left(\frac{s}{\sqrt{N_8}}\right).$$

Multiplying both sides of both equations by the two denominators, we obtain

$$s\sqrt{N_8} = 1.2533\, s\sqrt{N_2}$$

from which

$$\sqrt{N_8} = 1.2533\sqrt{N_2}.$$

Squaring both sides,

$$N_8 = 1.5708\, N_2.$$

Thus, we see that the same degree of accuracy may be obtained either by computing the mean or by gathering 57% more data and computing the median. Rare, indeed, is the situation in which it could possibly be easier to collect 57% more records and compute a median than simply to compute the mean for the original data. Let us consider a specific instance. In the El Dorado School, 176 pupils were tested, and the mean IQ was found to be 110.01 with a standard error of 1.15. Suppose, however, that someone likes medians and doesn't like means, but wishes to determine the median with equal precision. Substituting in the immediately preceding formula shows that when $N_2 = 176$, N_8 will be 276—exactly 100 more. Hence, all that this person will need to do is to locate 100 more students similar to these (though sometimes this is impossible), test them, score the tests, check the scoring, record the scores, either tabulate them or arrange them in order with the original 176, and compute the median. We conclude: *It would be just as accurate and much less work to compute the mean.*

The preceding illustration was specific to the situation in which $N_2 = 176$. The fact that essentially the same relationship holds for numbers of cases varying from very small on up is illustrated in Table 5.4.

TABLE 5.4
The Number of Cases Needed to Determine Means and Medians with Equal Accuracy

N_2 for Mean	N_8 for Median
20	31
40	63
60	94
80	126
100	157
200	314
500	785
Other Numbers	$1.57N_2$

Our other examples may also be used to illustrate this same principle. The standard error of the median of the car speeds is:

$$\begin{aligned} s_{Mdn} &= 1.2533\left(\frac{s}{\sqrt{N}}\right) \\ &= 1.2533\left(\frac{6.94}{\sqrt{119}}\right) \\ &= \frac{8.70}{10.91} \\ &= .80. \end{aligned}$$

This, of course, means that the true median is likely to be between $Mdn \pm s_{Mdn}$, which is 47.96 ± .80, or 47.16 and 48.76. Even though these limits are not very great, they are, of course, 1.2533 times as wide as those for the mean.

If we wish to cut this range down to the same size as that for the mean, we shall need to have

$$\begin{aligned} N_8 &= 1.5708\ N_2 \\ &= (1.5708)(119) \\ &= 187. \end{aligned}$$

Thus, to determine the median speed as accurately as the corresponding mean car speed would require 187 instead of 119 cases. Obviously, any real or imagined time saved by computing the median instead of the mean is completely negligible compared to the time and cost of gathering enough more data to get an equally accurate estimate of the true mean or median in the population from which our sample is drawn.

Perhaps the best way to conclude this section on the stability of the median is to call your attention to a highway sign.

STANDARD ERROR OF A PROPORTION

In cases where events can be classified into one of two categories or characteristics, we are often interested in the proportion of occurrence of one specified characteristic. Examples might be the number of boys or girls in a set of pupils or the number of persons answering Yes or No to an opinion question. To calculate the proportion of persons having a designated characteristic is a simple matter at the descriptive level. However, sometimes we wish to go beyond the descriptive level to infer population characteristics.

If we consider the sample proportion p as an estimate of the population proportion, then we need an estimate of the limits of error in this estimation. The standard error of a proportion is given by the formula

$$s_p = \sqrt{\frac{pq}{N}}, \tag{5.11}$$

where $q = 1 - p$. The formula is illustrated here by a numerical example from *Methods of Educational Research* by Max D. Engelhart (Rand McNally, 1972), p. 263. For a sample of 92 seventh-grade pupils, the proportion of pupils reading below grade level is .36. Then,

$$s_p = \sqrt{\frac{(.36)(.64)}{92}}$$
$$= \sqrt{\frac{.2304}{92}}$$
$$= \sqrt{.002504}$$
$$= .05.$$

To estimate confidence limits, we use the same procedure as for the mean. The .95 confidence interval is .36 ± (1.96)(.05), which round off to .26 and .46. The limits for the .99 confidence interval are .36 ± (2.58)(.05), which round off to .23 and .49.

The number of cases in the smaller group must be at least 5, preferably larger, in order to justify the use of this formula for s_p. If the inference drawn from the sample is to apply, not to all seventh-grade pupils, but strictly to the population of 368 seventh-grade pupils in the school system from which the 92 pupils were drawn at random, then s_p must be multiplied by

$$\sqrt{\frac{M - N}{M - 1}} \quad \text{or} \quad \sqrt{\frac{368 - 92}{368 - 1}} = .87,$$

where M is the size of the finite population. This is the *finite population correction*. For these records the corrected s_p = (.87)(.05) = .04. The .95 and .99 confidence intervals would become .28 to .44 and .26 to .46, respectively.

What's up, Doc?

NEWSCASTER: "Dr. John Emory of this city was killed in a hunting accident on Friday. Dr. Emory, a prominent and respected citizen, left a wife and twelve children. Police say that Emory was mistaken for a rabbit."

—Kermit Schafer: *Blunderful World of Bloopers*

CHAPTER 6 PERCENTILES AND PERCENTILE RANKS

A *percentile* is a point below which a given percentage of the cases occur. For instance, 67 percent of the cases are below the sixty-seventh percentile, which is written as P_{67}. The middle of a distribution, or the point below which 50 percent of the cases lie, is, of course, the fiftieth percentile. In Chapter 2 we learned another name for this value—the median. The median is probably computed, used, and discussed more than any other percentile.

There are three common uses for percentiles:

1. *A small number of percentiles may be used to give a fairly accurate picture of the shape of a frequency distribution.* Thus, if we know P_{10}, P_{25}, P_{50}, P_{75}, and P_{90}, we have the basic data needed for a rather complete understanding of the shape of the distribution.
2. *Percentiles are often used in interpreting test scores or other measures.* Thus, Robin may make a score of 134 on a standardized test in spelling. Material supplied by the test publisher enables the teacher to find out that such a score is at the fifty-fourth percentile—or that Robin scored higher than 54% of the children on whom this test was standardized. Caution should be exercised in such a use of percentiles, since standard scores (see p. 61) are often superior to percentiles for such purposes.

3. *Percentiles are used in computing other statistics.* We have already seen a simple example of this, namely, the median is the fiftieth percentile. We shall later see that some other attributes of a frequency distribution, such as dispersion (scores bunched together or spread out), skewness (lopsided or symmetrical frequency distribution), and kurtosis (peaked or flat distribution), may be measured by statistics that are combinations of two or more percentiles. From a theoretical point of view, their use in other formulas may be the most important use of percentiles, although their use in interpreting scores is probably their most common function.

Because percentiles have been so very widely used, if we are going to understand the educational and psychological literature (especially that dealing with mental tests), we had better have a thorough understanding of them. The concept of the percentile is fundamentally simple and easy to grasp. Two terms are to be distinguished: percentile and percentile rank. Going back to section 2 above, 54% of the children made scores below 134. Hence, the fifty-fourth percentile for that spelling test is 134. Further, the score of 134 gives Robin a *percentile rank* of 54. Notice that these are but two different ways of expressing the same information: the relation between a test score and the percent of people who make scores below it. We will define and give some of the characteristics of percentiles and of percentile ranks in the next two paragraphs.

PERCENTILES

A percentile is a point, measured along the scale of the plotted variable, below which a given percentage of the measures occur. The percentile is named for the percentage of cases below it. Thus, 54 percent of the observations are below the fifty-fourth percentile, P_{54}. Percentiles have subscripts running from 1 to 99. Each percentile is equal to a raw score whose value can be determined by Formula (6.1), which will be explained shortly. It is perfectly possible to calculate the values of all 99 percentiles, but ordinarily we don't do that. Usually there are a relatively small number of percentiles, such as P_{25}, P_{50}, or P_{90}, whose values we seek. We compute a percentile when we want to know what score corresponds to some predetermined point. Thus, when we start with a fixed proportion or percent of the scores, such as 90%, we then compute the corresponding percentile, in this case P_{90}, and our formula gives us its value in raw score units. This value is, of course, the point below which our proportion, here 90%, of the scores lie.

PERCENTILE RANKS

A percentile rank is the number used to designate any one of the 101 successive intervals in a frequency distribution, the middle 99 of which each contain 1% of the cases. (The 99 percentiles will be near the middle of these 99 intervals.) The highest and lowest percentile ranks each include only one-half of 1% of the cases. Thus, in theory, in a continuous distribution with a very large number of different scores, one-half of 1% of the scores would receive a percentile rank *(PR)* of 0; 1% would receive *PR*'s of 1, 2, 3, . . . , 99; and one-half of 1% would receive the highest *PR* of 100. In practical applications, the *PR* for any score shows, to the nearest whole number, what percent of the group made lower scores. Thus, if someone makes a score of 134 and we find that the percentile rank for this score is 54, then we know that between 53½% and 54½% of the group made lower scores. (In all such comparisons, the people scoring exactly at 134 are regarded as split 50–50, half above and half below 134.) It is possible that one or more adjacent scores, such as 135, may also have *PR*'s of 54. This illustrates the fact that each *PR* is an interval or range rather than a point. We can, but ordinarily do not, compute one or two percentile ranks; more commonly, we start with a frequency distribution and compute the *PR*s corresponding to every possible score (or score group).

Despite their differences, percentiles and percentile ranks are obviously closely related. In many situations, they simply indicate two alternate ways of expressing essentially the same idea. Thus, if 54% of the students have spelling scores below 134, then $P_{54} = 134$; and corresponding to a score of 134 we find that $PR = 54$. These two differences in orientation lead to two different computational methods.

COMPUTATION OF PERCENTILES

In Chapter 2 we saw that there were two ways to compute the median. Since the median is a percentile, you might expect (or hope) that the two formulas for the median in Chapter 2 might somehow be applicable here. Oddly enough, these are not the same two ways that we use in computing percentiles and percentile ranks. It is true that it is possible to generalize Formula (2.3) but, except in the case of the median, it becomes very awkward to use. Formula (2.4), however, can readily be generalized into a formula that is quite useful for the calculation of predetermined percentiles; such as the ninety-third percentile, the seventy-fifth percentile, or the twentieth percentile. One other method of computation is extremely simple and straightforward when one wishes to compute percentile ranks for all the possible scores on a test. It is explained and illustrated in Supplement 6.

First, let us take another look at Formula (2.4):

$$Mdn = l + i \left[\frac{\left(\frac{N}{2}\right) - F}{f_p} \right].$$

Most of the terms do not look as though they pertain specifically to the median. This is correct. The formula for the computation of any percentile looks very similar to this one, and the meanings of the terms are also either similar or identical. The formula is

$$P_p = l + i\left(\frac{pN - F}{f_p}\right). \tag{6.1}$$

The pth percentile is designated by the symbol P_p, the subscript p corresponding to the percentage of cases below the given percentile. Here l represents the exact lower limit of the interval in which the percentile lies, i is the interval width, p is the proportion of cases below the given percentile, N is the total number of cases, F is the sum of the frequencies of all intervals below this interval, and f_p is the frequency of the interval in which the percentile lies. There is a slight difference in meaning between p as an entire symbol and p as a subscript in P_p: as a symbol it refers to a proportion, but as a subscript to a percentage. Thus, in computing the thirty-eighth percentile, p will have the value .38, but the percentile is ordinarily written P_{38} rather than $P_{.38}$. In passing, it should be noted that if we let $p = .50$, Formula (6.1) becomes identical with Formula (2.4), as it should, since P_{50} is the median.

As a first example, let us compute P_{60} for a distribution for which we have already computed the median—the data of Table 2.4 (p. 20). First it is necessary to find out which interval contains P_{60}. The number of cases lying below any percentile is pN, which in this case is (.60)(176), or 105.60. Thus 105.60 cases lie below the sixtieth percentile. Counting from the bottom in Table 2.4, we find 81 cases with scores of 60–109 inclusive, and 108 with scores of 60–114. Hence P_{60} must be in the interval that is labeled 110–114. The median was in this same interval, so many of the figures are the same, as a comparison of the following with the computations in Chapter 2 will show.

First, let us note again that the true class boundary between the 105–109 interval and the 110–114 interval is halfway between the highest score (109) that could be tabulated in the lower interval and the lowest score (110) that could be tabulated in the upper interval. Hence, $l = 109.50$. Notice that this 109.50 is both the exact upper limit of the 105–109 interval and the exact lower limit of the 110–114 interval. Except for pN, all the other values in Formula (6.1) are the same as those used earlier in computing Mdn. Let us substitute.

$$P_p = l + i\left(\frac{pN - F}{f_p}\right),$$

$$P_{60} = 109.50 + 5\left(\frac{(.60)(176) - 81}{27}\right)$$

$$= 109.50 + 5\left(\frac{105.60 - 81}{27}\right)$$

$$= 109.50 + 5\left(\frac{24.60}{27}\right)$$
$$= 109.50 + 4.56$$
$$= 114.06.$$

As a second example, let us compute P_{10} from the same table. The value of pN is $(.10)(176) = 17.60$. There are 16 cases in the intervals from 60–89 inclusive, and 22 in the range from 60–94. Hence P_{10} is in the 90–94 interval, whose exact lower limit is 89.50. The simple computation follows:

$$P_p = l + i\left(\frac{pN - F}{f_p}\right),$$
$$P_{10} = 89.50 + 5\left(\frac{(.10)(176) - 16}{6}\right)$$
$$= 89.50 + 5\left(\frac{17.60 - 16}{6}\right)$$
$$= 89.50 + 5\left(\frac{1.60}{6}\right)$$
$$= 89.50 + \frac{8.00}{6}$$
$$= 89.50 + 1.33$$
$$= 90.83.$$

To illustrate the computation for a different set of data, we shall compute P_{25} for the car speeds shown in Table 2.5 (p. 24). Using the tabulation in Table 2.6 (p. 25), we see that 29 of the 119 cars had speeds of 24–41 and that 11 more were clocked at 42 or 43. Substituting in Formula (6.1), we obtain

$$P_p = l + i\left(\frac{pN - F}{f_p}\right),$$
$$P_{25} = 41.50 + 2\left(\frac{(.25)(119) - 29}{11}\right)$$
$$= 41.50 + 2\left(\frac{29.75 - 29}{11}\right)$$
$$= 41.50 + 2\left(\frac{.75}{11}\right)$$
$$= 41.50 + \frac{1.50}{11}$$
$$= 41.50 + .14$$
$$= 41.64.$$

In the immediately preceding examples of the computation of percentiles, we were working with grouped data. What happens when, for any reason, we do not wish to group our data? Nothing much. We just tabulate the raw scores without grouping, note that i then equals 1, and go right ahead with Formula (6.1).

PERCENTILES AND PERCENTILE RANKS COMPARED

Figure 6.1 provides what is essentially a magnified section of a part of a frequency distribution. It shows the similarities and differences in general orientation, method, and result that lead to a set of percentile ranks and of percentiles for the same data.

Column (1) shows the successive scores, labelled X. Column (2) gives the frequency of each and this is also plotted in the histogram further to the right. Column (3) gives the cumulative frequency to the *lower limit* of each interval, while column (4), also called F, shows it to the *middle* of each interval. When column (4) is divided by N and then the quotient is multiplied by 100, we get column (5). It shows the percentage of the 300 cases that are below the midpoint of each of the intervals. That's all. The percentile rank of each interval in column (1) is the percentage (to the nearest whole number) given in column (5). Thus, looking at the bottom row, we see that the percentile rank for a score of 129 is 50. This is indicated in column (6) by the equation $PR_{129} = 50$. Notice that the subscripts of PR increase regularly as we go up this column, each being identical with the X value in column (1), but that the percentile ranks do not. For instance, 3 scores have percentile ranks of 54 and there is no score that has a percentile rank of 56.

We shall now compute a percentile for every test score shown in the Figure. The first two columns, (1) and (2), of Figure 6.1 give a small part of the hypothetical frequency distribution of 300 scores. From the first F column, column (3), of Figure 6.1, we see that 149 cases are included up to 128.5 (the true class boundary or exact limit between scores of 128 and 129) and 151 are included when we go up to 129.5. If it is not already obvious that $Mdn = 129$, either Formula (2.4) or (2.5) or (6.1) will show that such is the case. The little marks to the right of the histogram depict graphically what Formula (6.1) assumes: that the frequency within an interval is distributed evenly throughout that interval. Thus, of the 2 cases with scores of 129, one is regarded as occupying the area from the lowest little mark (at 128.5) to the next (at 129), the other taking up the space from 129 to 129.5. The next interval contains 3 cases, so the little marks must be closer together. With this histogram and the little marks showing the individual boundaries between successive cases, the process of obtaining percentiles is simple indeed. To determine P_p, just find the score below which pN of the cases lie. Thus, if we wish P_{51}, we first compute $pN = (.51)(300) = 153$ and then find the point below which 153 cases lie. The F column tells us that 151 lie below 129.50. The second little mark above this will indicate 153. This mark is two-thirds of the way from 129.50 to 130.50, which is $129.50 + .67 = 130.17$.

FIGURE 6.1
Relation of Frequency Distribution and Histogram to Percentiles and Percentile Ranks

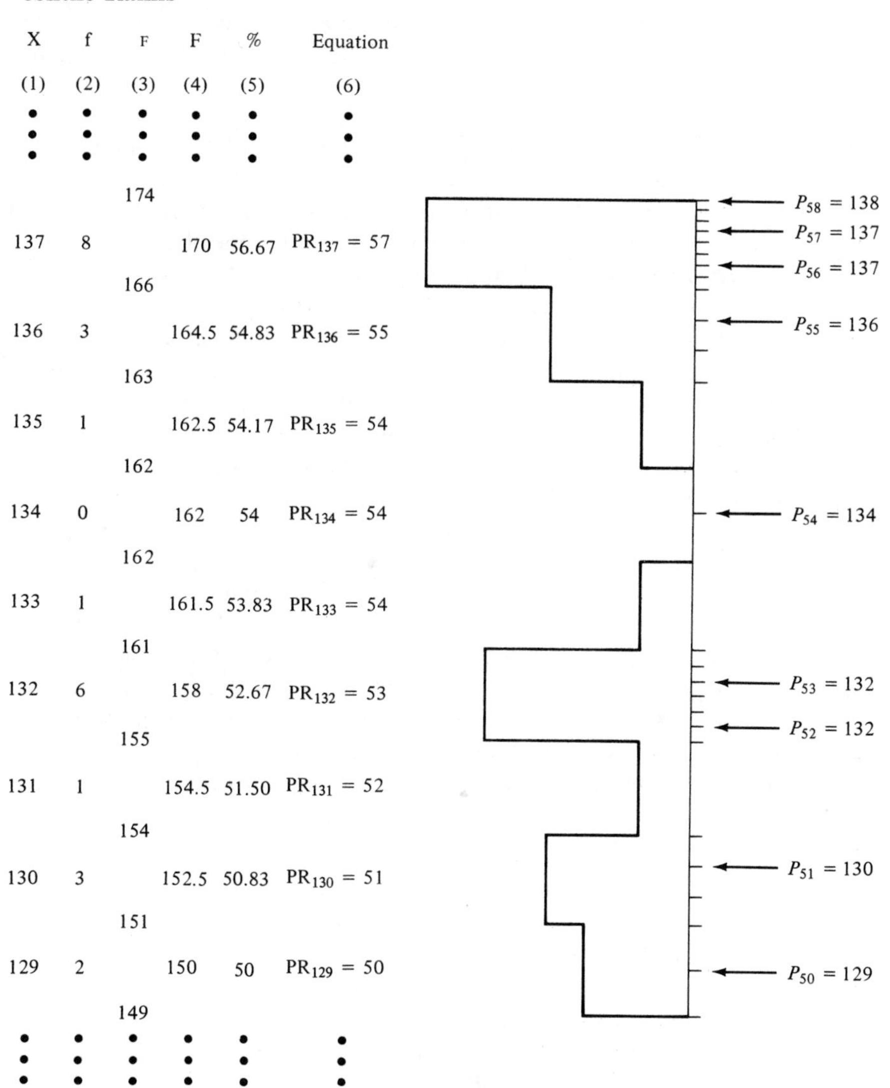

X (1)	f (2)	F (3)	F (4)	% (5)	Equation (6)
⋮	⋮	⋮	⋮	⋮	⋮
		174			
137	8		170	56.67	$PR_{137} = 57$
		166			
136	3		164.5	54.83	$PR_{136} = 55$
		163			
135	1		162.5	54.17	$PR_{135} = 54$
		162			
134	0		162	54	$PR_{134} = 54$
		162			
133	1		161.5	53.83	$PR_{133} = 54$
		161			
132	6		158	52.67	$PR_{132} = 53$
		155			
131	1		154.5	51.50	$PR_{131} = 52$
		154			
130	3		152.5	50.83	$PR_{130} = 51$
		151			
129	2		150	50	$PR_{129} = 50$
		149			
⋮	⋮	⋮	⋮	⋮	⋮

To determine P_{52}, we find the point below which $pN = (.52)(300) = 156$ cases lie. From the F column we see that we must go to the next little mark above 131.50. Since there are 6 cases in this interval, the one next to the bottom will be located at $131.50 + 1/6 = 131.67$. Notice that the values of successive percentiles are equidistant in terms of the little marks but that they are not usually equidistant in terms of X scores. Note that when an interval has a relatively large *f*, several percentiles may, after rounding to the nearest digit, have the same value. Thus, P_{52} and P_{53} are each equal to 132. At $X = 134$, a question arises. Note that any point between 133.5 and 134.5 would satisfy our definition of P_{54}, but in such cases, by custom, we assign to the percentile the score at the middle of the empty interval or intervals.

It is now very easy to compare and contrast percentiles and percentile ranks. There are 99 percentiles, P_1 to P_{99}, each being equal to some raw score. There are as many percentile ranks, *PR*, as there are intervals in the distribution, each percentile rank being equal to a number between 0 and 100. While each percentile refers to a point, each percentile rank covers an entire interval. When we start with a given proportion or percentage, we compute a percentile whose value is some raw score; when we start with a given raw score or interval, we compute a percentile rank whose value is between 0 and 100. In the rare instances in which either *P* or *PR* turns out to be an integer with no fraction to be rounded, an interesting reciprocal relationship exists. This is illustrated by the lowest interval in our example, where we find that $P_{50} = 129$ and $PR_{129} = 50$.

DECILES

Just as the 99 percentiles divide a frequency distribution into 100 equal parts, so the 9 deciles divide it into 10 equal parts. A *decile* is defined as one of the 9 points, measured along the scale of the plotted variable, which divide the frequency distribution into 10 parts or intervals, each containing exactly one-tenth of the cases. Deciles are named for the proportions of cases below them. Thus, 30 percent of the observations are below the third decile, D_3. There is nothing new about deciles. Not only are they strictly analogous to percentiles, all deciles *are* percentiles, thus:

$$\begin{aligned}
D_9 &= P_{90},\\
D_8 &= P_{80},\\
D_7 &= P_{70},\\
D_6 &= P_{60},\\
D_5 &= P_{50} = Mdn,\\
D_4 &= P_{40},\\
D_3 &= P_{30},\\
D_2 &= P_{20},\\
D_1 &= P_{10}.
\end{aligned}$$

Other than the median, the two most commonly used deciles are D_9 and D_1. They are used in formulas for dispersion, skewness, and kurtosis, which will be presented in the supplement to this chapter and in the next chapter. However, they are usually designated as P_{90} and P_{10}.

One common misuse of deciles should be mentioned. After giving a test to a small number of people, the test author often wishes to present norms or standards so that the test scores can be interpreted by those who use the test. Realizing that not enough people have yet taken the test to justify the computation of percentiles, the author erroneously concludes that deciles should be computed. Because all deciles are percentiles this is clearly not acceptable. A much better procedure would be to gather more data and compute percentiles. In many instances, a still better procedure would be to gather more data and compute standard scores.

Other uses of deciles need not be discussed. They are obviously the same as those of percentiles.

QUARTILES

Quartiles are very similar to percentiles and deciles except that they divide a distribution into 4 parts instead of into 100 or 10. A quartile is defined as one of the 3 points, measured along the scale of the plotted variable, which divide the frequency distribution into 4 parts, each containing exactly one-fourth of the cases. There are only 3 quartiles, each of which is also a percentile:

$$
\begin{aligned}
Q_3 &= P_{75}, \\
Q_2 &= P_{50} = D_5 = Mdn, \\
Q_1 &= P_{25}.
\end{aligned}
$$

The term Q_2 occurs so rarely that one almost forgets it exists. This value is nearly always designated by the symbol *Mdn* rather than by the other equivalent symbols listed above. Q_3 is often called the upper quartile and Q_1 the lower quartile.

Q_3 and Q_1 probably find their greatest use in the formula for one of the measures of dispersion discussed in Supplement 6, but they are also commonly used (along with the median) in describing frequency distributions. No new formulas are needed for computing deciles or quartiles; just use Formula (6.1). In fact, Q_1 was computed earlier in this chapter, where it was called P_{25}.

Beginners (and also other people who really ought to know better) sometimes mistakenly refer to a person as "in the third quartile," or they say that a score is "in the top quartile of the distribution." They should say, "in the next to the top (or next to the bottom?) quarter," "in the highest fourth of the distribution," etc. Quartiles, deciles, and percentiles are points—not ranges or areas.

STANDARD ERROR OF A PERCENTILE

Percentiles, deciles, and quartiles have standard errors, just as other statistics do. The size of the standard error depends on the percentile. It is smallest for percentiles near the middle of the distribution and gets much larger for the very high and very low percentiles. In fact the standard error of P_3 (or of P_{97}) is twice as large as the standard error of P_{50} (the median) for the same distribution. The formula is

$$s_{P_p} = \frac{s}{y}\sqrt{\frac{pq}{N}}. \quad (6.2)$$

In Formula (6.2), s, as usual, represents the standard deviation of the frequency distribution. The symbol y was defined in Chapter 4 as an ordinate of the normal probability curve. The value of y does not have to be computed; it can be looked up in Table P. This is the fundamental formula for the standard error of any percentile, but by rearranging the formula and using a different table, we can markedly reduce the amount of work required in obtaining such standard errors. Let us rewrite Formula (6.2) as

$$s_{P_p} = \left(\frac{\sqrt{pq}}{y}\right)\left(\frac{s}{\sqrt{N}}\right)$$

If we define F_3 as $\frac{\sqrt{pq}}{y}$, the formula becomes

$$s_{P_p} = F_3\left(\frac{s}{\sqrt{N}}\right), \quad (6.3)$$

in which the values of F_3 are obtained from Table Q. Neither Formula (6.2) nor (6.3) is applicable when p or q is more extreme than about .07 or .93, unless N is very large.

Let us use Formula (6.3) to compute the standard errors of the percentiles we computed for the IQs. Earlier in this chapter, we found that P_{60} was 114.06, and P_{10} was 90.83. For P_{60}, we note that p and q are .60 and .40. Looking up $p = .60$ in Table Q, we see that $F_3 = 1.2680$. Since we already know that $s = 15.20$ and $N = 176$, simple substitution in Formula (6.3) gives us

$$
\begin{aligned}
s_{P_p} &= F_3\left(\frac{s}{\sqrt{N}}\right),\\
s_{P_{60}} &= (1.2680)\left(\frac{15.20}{\sqrt{176}}\right)\\
&= (1.2680)\left(\frac{15.20}{13.27}\right)\\
&= (1.2680)(1.15)\\
&= 1.46.
\end{aligned}
$$

Similarly, for P_{10} we see that $p = .10$ and $q = .90$. Looking up .90 in Table Q, we find that the function F_3 has the value 1.7094. The value of $\frac{s}{\sqrt{N}}$ will be the same for all percentiles in the same distribution. Hence,

$$s_{P_p} = F_3\left(\frac{s}{\sqrt{N}}\right),$$
$$s_{P_{10}} = (1.7094)(1.15)$$
$$= 1.97.$$

Let us check these computations by doing them a different way. (It is always desirable to avoid recomputing—and thereby possibly making the same error twice—in checking calculations.) Using the value of $\mathscr{y}$ from Table P in Formula (6.2), we obtain

$$s_{P_p} = \frac{s}{\mathscr{y}}\sqrt{\frac{pq}{N}},$$
$$s_{P_{60}} = \frac{15.20}{.3863}\sqrt{\frac{(.60)(.40)}{176}}$$
$$= 39.35\sqrt{\frac{.24}{176}}$$
$$= 39.35\sqrt{.001364}$$
$$= (39.35)(.0369)$$
$$= 1.45.$$

Similarly, we can check the standard error of P_{10}.

$$s_{P_p} = \frac{s}{\mathscr{y}}\sqrt{\frac{pq}{N}},$$
$$s_{P_{10}} = \frac{15.20}{.1755}\sqrt{\frac{(.10)(.90)}{176}}$$
$$= 86.61\sqrt{\frac{.09}{176}}$$
$$= 86.61\sqrt{.000511}$$
$$= (86.61)(.0226)$$
$$= 1.96.$$

Several comments are in order concerning these different methods of computing the standard errors of percentiles. First, the figures check essentially, illustrating that either method may be used. Second, there are minor discrepancies in the last retained figure. These discrepancies are due to rounding errors; if Tables Q and P were given to many more decimal places, and if subsequent computations were also carried out to many more decimal places, the two results could be made to agree to any predetermined standard of accuracy. Third, it was necessary to depart from our usual rule of keeping only two decimal places. If we had not done so, we would have extracted the square root of .00 and both standard errors would erroneously have appeared to be zero. Finally, the method using Table Q requires less time and is much easier than the method in which Table P was used. In fact, if the standard errors of a number of percentiles for the same distribution are needed, each one after the first is computed by one multiplication—that's all—in contrast to two multiplications, two divisions, and one square root for the traditional method.

SUMMARY

Aside from their use in describing a frequency distribution, percentiles serve two distinct functions. First, they are used as an aid in interpreting such measures as test scores. Second, they appear as parts of other statistical formulas.

Deciles, quartiles, and the median are all percentiles. Most percentiles are about as easy to compute as the median is. If an individual wants percentile ranks for *all* the scores in a distribution, the method shown in Supplement 6 is easy, rapid, and accurate. With the aid of Table Q, the computation of the standard error of any percentile is extremely simple.

SUPPLEMENT 6

METHOD OF OBTAINING PERCENTILE RANKS FOR GROUPED DATA

Percentile ranks can be obtained easily from a frequency distribution by using either a desk calculator or a pocket calculator. Table 6.2 shows all the steps.

Columns (1) and (2) are already familiar. Note that column (3) gives the cumulative frequency to the TOP of each interval. However, we need to obtain the percentile rank of the midpoint of each interval (since that point would be representative of all scores tallied in the interval). Hence, we need the cumulative frequency to the MIDPOINT of each interval, as shown in column (4). There are three easy ways to obtain the figures in column (4). Let us first compute them by going down one row, thus getting the cumulative frequency to the BOTTOM of this interval from this next lower row and adding ½ of f to it. For example, for the interval 130–139, we

TABLE 6.2
Obtaining Percentile Ranks from Grouped Data

(1) *Interval*	(2) *f*	(3) *Cumulative Frequency to TOP of Interval*	(4) *Cumulative Frequency to MIDDLE of Interval*	(5) *Percentile Rank*
300–309	1	140	139.5	100
290–299	0	139	139.0	99
280–289	0	139	139.0	99
270	1	139	138.5	99
260	4	138	136.0	97
250	4	134	132.0	94
240	8	130	126.0	90
230	5	122	119.5	85
220	10	117	112.0	80
210	6	107	104.0	74
200	8	101	97.0	69
190	2	93	92.0	66
180	2	91	90.0	64
170	6	89	86.0	61
160	4	83	81.0	58
150	7	79	75.5	54
140	14	72	65.0	46
130	12	58	52.0	37
120	8	46	42.0	30
110	11	38	32.5	23
100	8	27	23.0	16
90	12	19	13.0	9
80	3	7	5.5	4
70–79	2	4	3.0	2
60–69	1	2	1.5	1
50–59	1	1	0.5	0
	140			

$$\frac{100}{N} = \frac{100}{140} = .71428571$$

take 46 (which is the cumulative frequency to the TOP of the 120–129 interval and also the cumulative frequency to the BOTTOM of the 130–139 interval) and add ½ of 12 to it, getting $46 + \frac{12}{2} = 52$ as the cumulative frequency for the MIDDLE of the 130–139 interval. After getting all these entries for column (4), we check them by, for each row, adding the entry in column (3) to the entry for the next lower row in column (3) and getting double the entry in column (4). Thus, in checking the 130–139 row, we add 58 to 46 and get double 52.0 as $58 + 46 = 104 = 2(52.0)$. The third way to obtain the figures in column (4), which may be used as an alternative way of checking, is to subtract ½ of f from the entry in column (3). Thus for this same 130–139 interval, 58 − (½) (12) = 52.0.

After computing and checking every entry in column (4), we, in effect, divide each of them by N, multiply by 100, and record the answer to the nearest integer in column (5). These are the percentile ranks we seek. In the actual computation, we first get $100/N$, record it to six or eight decimals, use it as a constant in our calculator, and then multiply every entry in column (4) by this constant, recording the product to the nearest integer in column (5). Thus we divide 100 by 140, getting .714286 or .71428571. Using this as a constant, starting from the top we multiply by 139.5, getting 99.64 which is recorded as 100 in column (5). Continuing through 130–139, we multiply by 52.0, getting 37.14 which is recorded as 37. The best way to check column (5) is to successively divide each figure in column (4) by $\frac{N}{100}$ which, in this case, is 1.40. Thus, for the row 130–139, we divide 52.0 by 1.40 and again get 37.14 which, as before, rounds to 37.

When all these figures have been checked, we may use the table to determine the percentile rank for any score. For example, any score from 210 through 219 will have a percentile rank of 74. A score of 220 will have a percentile rank of 80. If you are bothered by this change of zero in the percentile rank as the scores increase from 210 to 219 followed by a change of 6 when the scores increase to 220, let us first assure you that a change in percentile rank from 74 to 80 is not terribly great. If, however, it still bothers you, the remedy is simple: cut the interval from 10 to 5 and the sudden changes in percentile ranks will be about half as great.

If the conversions from raw scores to percentile ranks are to be made by several people, you might want to retype the table showing only the first and last columns. Then proofread it to be sure that, after checking all your work, things haven't been spoiled by a typographical error.

MEASURES OF VARIABILITY BASED UPON PERCENTILES

There are many measures of variability involving percentiles. The best of the common ones is D, the *10 to 90 percentile range*; the most widely used is Q, the quartile deviation. These are the only ones we shall discuss.

The 10 to 90 percentile range is exactly what its name says it is. It is defined by the equation

$$D = P_{90} - P_{10}. \tag{6.4}$$

We cannot improve on a description of it given by the statistician who derived it. Kelley* showed that "for a normal distribution the interpercentile range having the minimal error is that between the 6.917 and the 93.083 percentiles. A range but slightly different from this and having nearly as great reliability is that between the 10th and 90th percentiles. This distance is called D and is given as the most serviceable measure of dispersion based upon percentiles."

The standard error of D is

$$s_D = .8892\left(\frac{D}{\sqrt{N}}\right), \tag{6.5}$$

and it is interpreted in a manner strictly analogous to s_s in Formula (5.10).

The *quartile deviation* is also known as the semi-interquartile range. The interquartile range is $Q_3 - Q_1$ (or $P_{75} - P_{25}$). The *semi-interquartile range*, more commonly called the quartile deviation, thus is

$$Q = \frac{P_{75} - P_{25}}{2}. \tag{6.6}$$

Because half the scores are between P_{25} and P_{75}, in a symmetrical distribution Q tells us how far above the median and how far below the median we have to go to include half of the cases.

The standard error of Q is

$$s_Q = 1.1664\left(\frac{Q}{\sqrt{N}}\right). \tag{6.7}$$

The computation of D, Q, and their standard errors is very simple, so we'll review it very hurriedly.

We have already computed P_{10} for the IQs in Table 2.4 on page 20. Its value is given on page 149 as 90.83. Similarly, we compute P_{90} and find it to be 127.86. Then we solve for D:

$$\begin{aligned} D &= P_{90} - P_{10} \\ &= 127.86 - 90.83 \\ &= 37.03. \end{aligned}$$

---

*Kelley, Truman L. *Statistical method*. New York: Macmillan, 1924.

Its standard error is

$$
\begin{aligned}
s_D &= .8892\left(\frac{D}{\sqrt{N}}\right)\\
&= .8892\left(\frac{37.03}{\sqrt{176}}\right)\\
&= \frac{32.9271}{13.2665}\\
&= 2.48.
\end{aligned}
$$

Similarly, for Q we need P_{75} and P_{25}. Their values, computed by Formula (6.1), turn out to be 121.10 and 101.08, respectively. Hence

$$
\begin{aligned}
Q &= \frac{P_{75} - P_{25}}{2}\\
&= \frac{121.10 - 101.08}{2}\\
&= \frac{20.02}{2}\\
&= 10.01.
\end{aligned}
$$

The standard error of Q is

$$
\begin{aligned}
s_Q &= 1.1664\left(\frac{Q}{\sqrt{N}}\right)\\
&= 1.1664\left(\frac{10.01}{\sqrt{176}}\right)\\
&= \frac{11.6757}{13.2665}\\
&= .88.
\end{aligned}
$$

THE RANGE

The *range* is sometimes used for an ungrouped set of raw scores. It is equal to the difference between the highest score and the lowest score, plus one unit of measurement on the particular scale. For example, if whole number scores are being used, 1 would be added to the difference. If the scores were reported to the nearest tenth of a point, then .1 would be added to the difference. Consider the following example

Highest score	= 99
Lowest score	= 21
Difference	= 78
Plus 1	+1
Range	= 79

TABLE 6.3
A Comparison of Various Measures of Variability

(1)	Standard error in terms of *s* (2)	Ratio of the measure to *s* (3)	Standard error in terms of the measure (4)	Relative size of constant (5)	Relative number of cases required (6)
s	$.7071 \frac{s}{\sqrt{N}}$	1.0000	$.7071 \frac{s}{\sqrt{N}}$	1.0000	100
AD	$.6028 \frac{s}{\sqrt{N}}$	.7979	$.7555 \frac{AD}{\sqrt{N}}$	1.0684	114
D	$2.2792 \frac{s}{\sqrt{N}}$	2.5631	$.8892 \frac{D}{\sqrt{N}}$	1.2575	158
Q	$.7867 \frac{s}{\sqrt{N}}$	.6745	$1.1664 \frac{Q}{\sqrt{N}}$	1.6496	272
Approx. Range: .01–.99 ($N = 50$)	$5.2529 \frac{s}{\sqrt{N}}$	4.6527	$1.1290 \frac{\text{Range}}{\sqrt{N}}$	1.5967	255
.005–.995 ($N = 100$)	$6.8810 \frac{s}{\sqrt{N}}$	5.1517	$1,3357 \frac{\text{Range}}{\sqrt{N}}$	1.8890	357
.002–.998 ($N = 250$)	$9.9557 \frac{s}{\sqrt{N}}$	5.7563	$1.7295 \frac{\text{Range}}{\sqrt{N}}$	2.4459	598

but

$$
\begin{array}{lcr}
\text{Highest score} & = & 18.7 \\
\text{Lowest score} & = & 9.1 \\
\hline
\text{Difference} & = & 9.6 \\
\text{Plus } .1 & & +.1 \\
\hline
\text{Range} & = & 9.7
\end{array}
$$

The range is the easiest measure of dispersion to compute but is less useful than the standard deviation or semi-interquartile range. It is not indicative of variability in the middle region. Furthermore, as Table 6.3 shows, it lacks the stability of the other measures in successive sampling.

RELATIVE VALUE OF MEASURES OF VARIABILITY

Beginning on page 140, there was an extensive discussion of the stability of the median in which it was demonstrated that the mean was markedly superior to the median. Because s_{Mdn} turned out to be 1.2533 times as great as $s_{\bar{X}}$, we found that to attain the accuracy and stability obtainable from a mean based on 100 cases we would need an N of 157 if we preferred to use the median.

A similar analysis can, of course, be made for measures of variability. This we have done in Table 6.3. Column (1) lists the measures of variability we are considering. Column (2) gives the corresponding standard errors. Column (3) gives the ratio of the measure to the standard deviation. Thus, in a normal distribution, since the 10th percentile is located at $-1.28s$, the 10–90 percentile range will be $2.56s$. Hence 2.56 or the more exact ratio, 2.5631, is listed in column (3). We now divide the constants in column (2) by the divisors in column (3) and get the constants in column (4). Note that these are now the constants for the standard errors in terms of their own magnitudes. Thus, the standard error of AD is given in a formula utilizing the AD we must have computed rather than the more theoretical formula in terms of s which was given in column (2).

The constants in column (4) have an interesting property already illustrated by the constant 1.2533 in the formula for s_{Mdn}. They show the relative sizes of the standard errors of these various measures of variation. Thus, D is a large measure over 2½ times as large as s, but the .8892 in column (4) allows for this. (If we had defined D_2 as (.10D or 8D, the D in column (4) would have been replaced by D_2, but the constant multiplier would still be .8892.)

We next divide all these constants by .7071 to facilitate comparison with the standard deviation; and these quotients constitute column (5). We then square these ratios, multiply by 100, and record the result in column (6), labeled "Relative number of cases required." Column (5) makes crystal-clear our reason for preferring s to any other measure of dispersion, but note that AD is a close second.

How much better is s than the rest of the measures of variability? Column (6) gives the answer. If we want a measure of dispersion but don't want to use s, there's no need to use s if we want to use AD and are willing to gather data on 114 subjects for every 100 we would have used with s. We could also attain this same degree of precision by testing 158 and computing D or by testing 272 and computing Q. Sure, it's easier to compute D or Q than to compute s, but if you have to collect data on 158 or 272 people instead of 100, is it worth it?

What about the range? The formula for the standard error of the range is different for every range. If $N = 50$, the midpoints of the lowest and highest scores would be the 1st and 99th percentiles, but if $N = 250$, they would be more extreme and less stable. It's hard to say just how large N should be if anyone insists on using the range instead of s, but such a person would have to collect, interpret, and analyze records on around 250 to 600 cases to get as dependable a measure of variation as s would yield with a mere 100 cases.

There are situations in which s is inapplicable or its use is inadvisable, but they are rare. Ordinarily s will do a good job for you. If, however, you disagree, just gather 14% more data and compute AD. If that won't work, then by all means get 58% more data and use D, but don't use Q or the range if you care at all about what your results may mean.

A PROFESSOR of statistics, about to retire, bought himself a tiny island—then christened it "Percent Isle." —Contributed by Richard L. Barclay

—Reader's Digest January, 1976

CHAPTER 7 SKEWNESS AND TRANSFORMED SCORES

SKEWNESS

We have already noticed that many frequency distributions are symmetrical, or nearly so, while some (such as the two shown in Figure 7.1) are markedly lopsided or skewed. *Skewness* is defined as the extent to which a frequency distribution departs from a symmetrical shape.

Thus, the distribution of the author citation data on page 27 was markedly skewed—more than half of the scores were 1, the others trailing off to a maximum of 14. Instead of our saying "markedly skewed," it would be far better if we could use a number to show the exact degree of lopsidedness and a + or − sign to show its direction. This we shall do. Our world often emphasizes symmetry. If there is a departure from symmetry, it is important that we recognize it. Sometimes we may wish to change the situation, sometimes not, but first we must find out if there are any such discrepancies. Formulas for skewness enable us to do this. Thus, it is possible to measure skewness, and the skewness of any distribution can be expressed in quantitative terms. If we had considerable confidence in our data and felt that computation time need not be kept low, then the best measure of skewness would be one of the functions of β_1 that are given as formulas for skewness in Supplement 7.

In practical situations, however, time is important—sometimes very important. Further, the mere fact that a distribution is markedly skewed sometimes leads us to question the accuracy of some of the extreme scores. In such circumstances, it is imperative that our measure of skewness be not unduly influenced by the very scores that are most likely to be in error.

These considerations lead us to adopt a measure of skewness such as

$$Sk = \frac{\dfrac{P_{90} + P_{10}}{2} - Mdn}{D}, \tag{7.1}$$

where D is the 10–90 percentile range. So far as the writers know, this formula may never before have been published, but it is only a minor modification of one derived by Kelley.*

Note the simplicity of the formula. If the distribution is symmetrical, the average of the 10th and 90th percentiles will equal the median, and the value of Sk will be zero. If P_{90} is further above the median than P_{10} is below the median, the skewness is positive. Such is the case in the distribution plotted in the right half of Figure 7.1. Look at that distribution of scores. Note that when Sk is **positive,** the scores deviate further from the mean or

FIGURE 7.1
Two Histograms Illustrating Negative Skewness and Positive Skewness

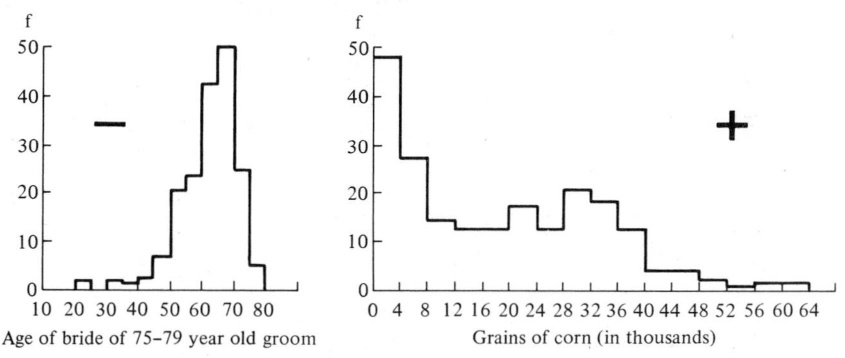

*Truman L. Kelley derived a formula and first published it in the June, 1921, issue of the *Quarterly Journal of the American Statistical Association.* We have reversed the sign, to agree with most other measures of skewness, and divided by D in order to get a pure number. If weights were expressed in ounces, Kelley's original formula would give a value for the skewness 16 times as great as if the same weights were expressed in pounds. Our modification, Formula (7.1), will give the same value of Sk regardless of whether a distribution of weights is expressed in pounds, ounces, grams, drams, stones, carats, piculs, or candareens.

median in the **positive** direction than they do in the negative direction. Also, when we plot the distribution of scores, if the skewness is positive, there is room for us to write a + sign at the side of the curve where the **positive** deviations are. Negative skewness is, of course, the exact opposite of positive skewness. Figure 7.1 depicts these relations.

In using Formula (7.1) we need to compute only three percentiles (P_{10}, *Mdn,* and P_{90}) before substituting in it. The formula for the standard error of *Sk* is even simpler. It is

$$s_{Sk} = \frac{.5185}{\sqrt{N}}. \qquad (7.2)$$

Let us compute *Sk* for the data of Table 2.4. In Chapters 2 and 6 we computed the following statistics for these IQs:

$$\begin{aligned} Mdn &= 110.80, \\ P_{10} &= 90.83, \\ P_{90} &= 127.86. \end{aligned}$$

Substituting these values in Formula (7.1), we obtain

$$\begin{aligned} Sk &= \frac{\dfrac{P_{90} + P_{10}}{2} - Mdn}{D} \\ &= \frac{\dfrac{127.86 + 90.83}{2} - 110.80}{127.86 - 90.83} \\ &= \frac{109.34 - 110.80}{37.03} \\ &= \frac{-1.46}{37.03} \\ &= -.0394. \end{aligned}$$

We now wish to use s_{Sk} to determine whether this IQ distribution is skewed significantly, or whether it may well be merely a chance fluctuation from a skewness of zero. Here, the corresponding standard error is

$$\begin{aligned} s_{Sk} &= \frac{.5185}{\sqrt{N}} \\ &= \frac{.5185}{\sqrt{176}} \\ &= \frac{.5185}{13.27} \\ &= .0391. \end{aligned}$$

The value of Sk is thus just about equal (in absolute value) to its standard error. Notice that for the sake of accuracy we carried the two final divisions to four decimals instead of to the usual two. Dividing the skewness, $-.0394$, by its standard error, .0391, we obtain -1.01 as the value of z. Since this is smaller than 1.96, it is obviously not significant. Hence, although there actually are a few more extreme negative than positive scores, the obtained skewness of only $-.04$ is not significantly different from zero. So we conclude that there is no evidence to indicate that IQs are not distributed symmetrically. Note carefully that conclusion. Do not make the erroneous inference that we have thereby proved that the scores *are* distributed symmetrically.

Suppose we watched two employed typists for an hour and found that Ann typed 17 letters while Bob typed 15. Realizing all the complexities and distractions involved in typing, we would not be willing to conclude that Ann is faster than Bob. She may be; she may not. We don't yet have enough evidence to reach that conclusion, but this does *not* justify us in concluding that the two typists are equally fast. Similarly, in the preceding paragraph, *the fact that the skewness is not significantly different from zero does not justify us in concluding that the skewness is zero.*

The histogram in the right half of Figure 7.1 is based on guesses of how many grains of corn a goose would eat in 30 days. The distribution is obviously skewed, but let us obtain a numerical value for Sk. From the data corresponding to the histogram at the right of Figure 7.1, we computed $P_{90} = 37{,}966$, $P_{10} = 1{,}722$, and $Mdn = 16{,}769$.

To obtain the value of the skewness, we now substitute in Formula (7.1):

$$Sk = \frac{\dfrac{P_{90} + P_{10}}{2} - Mdn}{D}$$

$$= \frac{\dfrac{37{,}966 + 1{,}722}{2} - 16{,}769}{37{,}966 - 1{,}722}$$

$$= \frac{19{,}844 - 16{,}769}{36{,}244}$$

$$= \frac{3{,}075}{36{,}244}$$

$$= +.0848.$$

The standard error of this value is

$$s_{Sk} = \frac{.5185}{\sqrt{N}} = \frac{.5185}{\sqrt{211}} = \frac{.5185}{14.53} = .0357.$$

This time the value of the skewness is more than twice its standard error; $z = +\frac{.0848}{.0357} = +2.38$. We see from the unnumbered table on page 126 that $.05 > P > .01$; thus the obtained value of *Sk* is barely significant. Hence, we infer that the true distribution of all such guesses is also skewed in the same direction as indicated by this sample of 211 guesses.

You may want to know why we deviated from our rule of keeping only two decimals. In our first example on page 167, it would have made no particular difference, since if we had kept two decimals when we divided $-.04$ by $.04$, we would have obtained -1.00 instead of -1.01. In our second example, however, if we had divided $+.08$ by $.04$, we would have obtained $+2.00$. This is quite different from $+2.38$ and is too large a mistake to make simply as a result of not carrying enough decimals in division. The possible errors are even greater than this. The quotient $+\frac{.08}{.04}$ might mean anything from about $+\frac{.07500001}{.04499999}$ to $+\frac{.08499999}{.03500001}$. These quotients are $+1.67$ and $+2.43$. They correspond to probabilities of about .0950 and .0150, respectively. Since one of these is over six times as great as the other, it is clear that when we divide we need to keep more decimals so that we can get the right answer, which in this case is $+2.38$.

Use of the Skewness Measure The chief use of *Sk* is, naturally, to measure the extent to which a frequency distribution or curve departs from symmetry.

What do we do if a distribution is skewed? There are several answers:

1. We compute *Sk* and its standard error. If the skewness is *not* significantly different from zero, we ignore it.
2. We compute *Sk* and s_{Sk}. If the skewness *is* significantly different from zero, we think about it, look at the distribution of scores again, decide that the distribution is not *very* skewed, but say or do little or nothing beyond reporting *Sk* and s_{Sk}. This sounds like a very casual procedure.

It is, but in many situations it is good enough, especially if either (a) Sk is between $-.05$ and $+.05$ or (b) $\frac{Sk}{s_{Sk}}$ is between -2 and $+2$.

3. We transform our scores by using some function of the scores which will give a distribution with little or no skewness. Thus, if the distribution of X scores is positively skewed, such functions as $X_2 = \sqrt{X}$ or $X_3 = \log X$ will increase the spacing among the low scores and reduce the spacing among the high scores, thereby reducing the value of Sk. These and other methods of transforming scores are described later in this chapter.
4. We alter the shape of the distribution in an arbitrary manner, forcing it to become normal, thus making $Sk = .00$. This process, called *normalizing,* is explained later in the chapter. It has been widely used in determining "T scores" and "CEEB Scores," to interpret standardized tests, and to determine Army General Classification Test Scores during World War II.

Interpretation of the Skewness Measure The maximum value that Sk can have is $+.50$ and the minimum is $-.50$. Symmetry or lack of skewness is denoted by .00. If Sk is positive and if $\frac{Sk}{s_{Sk}}$ is $+1.96$ or greater, we may conclude that the lopsidedness evident in our sample of cases is not just accidental, but that the trait we are measuring really is positively skewed. That is, if we were to select some other group similar to the one we did use, we would again find that the deviations would go further away from the median (and also from the mean) in the positive, rather than in the negative, direction.

If Sk is negative and if $\frac{Sk}{s_{Sk}}$ is negative but equal to or greater than 1.96 in absolute value (that is, between -1.96 and $-\infty$), we may similarly conclude that the trait really is skewed, but in the negative direction.

Whenever $\frac{Sk}{s_{Sk}}$ is between -1.95 and $+1.95$, we have failed to prove that the skewness exists. Note two things: (1) This holds regardless of the value of Sk. (2) This does not mean that we have proved that the curve is symmetrical.

What are some of the situations that lead to skewness? There are at least five. First, there may be a natural restriction at the low end of the scale, resulting in positive skewness. For example, number of offspring, number of leaves on clover, commissions earned by life insurance agents, minimum speed attained by cars at stop signs, time required by 10-year-old children to run 40 yards, or the population of cities.

Second, there may be a natural restriction at the high end of the scale, resulting in negative skewness. For example, the number of fruit flies that can live in a 100 cm^3 flask.

Third, there may be an artificial restriction at the low end of the scale, resulting in positive skewness. An example is the distribution of scores made by students on an extremely difficult test.

Fourth, there may be an artificial restriction at the high end of the scale, resulting in negative skewness. This is illustrated by scores of students on a test that is too easy, or by the merit ratings supervisors give to their employees.

Fifth, the skewness may be a function of the measuring system. One example is a record of the time required to complete a task when we are really interested in speed. Another is body weight. Sometimes, as in the case of body weights or volumes, the skewed distribution may be regarded as more fundamental than the symmetrical distribution that may result when the scores are transformed by extracting cube roots.

Suppose we find that $\frac{Sk}{s_{Sk}}$ is beyond -1.96 (and hence significantly skewed in the negative direction), and we also find that the value of *Sk* is, say, $-.20$. What does this mean? Of course, it means that the distribution is more skewed than it would have been if *Sk* had been $-.19$ and less skewed than it would have been if *Sk* had been $-.21$, but such statements are not very meaningful. Eventually, each statistician, through his experience with skewed curves, develops his own understanding of the meaning of various amounts of skewness. In order to shorten (or nearly eliminate) the time required for the accumulation of so much experience, Figure 7.2 has been prepared.

Looking down the center column of Figure 7.2, we see, in order, distributions with skewness of $+.40$, $+.20$, $.00$, $-.20$, and $-.40$. These curves vary widely in skewness, but all have the same *N*, nearly the same *Mdn*, nearly the same *D*, and the same average degree of kurtosis (to be discussed on page 177). Thus, because other characteristics are held constant, you can get a very clear idea of the appearance and meaning not only of the five plotted values of *Sk*, but also, by the use of interpolation, of all intermediate values as well. Notice that the curve in the center of Figure 7.2 looks very much like the normal probability curve. That is exactly what it is, as nearly as we were able to construct it, subject to the limitations that we imposed with respect to the size of *N, Mdn, D, Q, Sk,* and *Ku*.

Is it possible to make similar comparisons within each of the other two columns of Figure 7.2? Yes, it is. In the first column, all the curves should have relatively higher peaks or longer tails than those in the middle column. (We shall learn later that they are leptokurtic rather than mesokurtic.) Nevertheless, they all have the same amount of kurtosis; and the differences in skewness (varying from +.40 to −.40 again), as we go down the first column, are clearly evident.

In the third column, the curves are flatter in the middle or even bimodal, and they have shorter tails. (We shall find out that such curves are called platykurtic instead of mesokurtic.) Again, the differences in skewness from top to bottom of the column are clearly evident. Notice the rectangular distribution in the middle of this column, all frequencies (except at the very ends) being identical.

The best notion of skewness is provided by the middle column. It will pay the reader to spend five or ten minutes studying Figure 7.2 now, especially the center column. Note the changes in the shape of the curves and the relation of these changes to the five little pointers on the base line of each curve. The three triangles pointing up show P_{10}, Mdn, and P_{90}, while the triangles pointing down refer to Q_1 and Q_3. Note how their spacing varies—and especially how the distances from P_{10} to Mdn and from Mdn to P_{90} change—as the skewness changes. Take that five or ten minutes and study Figure 7.2—*right now*.

Stability of the Skewness Measure We have already given the very simple formula, (7.2), for s_{Sk}. Its use and interpretations are similar to those of other standard error formulas.

As we mentioned earlier, there are many formulas for skewness other than Formula (7.1). For most purposes, Formulas (7.1) and (7.2) for Sk and s_{Sk} will be quite adequate. As already implied, perhaps this is not the very best measure of skewness in the world, but only rarely can we justify spending the much greater computational time required by the theoretically superior formulas in Supplement 7 based on β_1. Perhaps one other justification for presenting the more difficult, though elegant, Formulas (7.19) and (7.21) is that β_1 is a built-in feature of some "package" computer programs. Nevertheless, whenever there is considerable doubt concerning the accuracy of some of the extreme scores (as there was in the goose guessing contest), our Formula (7.1) will be the most appropriate.

KURTOSIS

The relative degree of flatness or peakedness of a curve, as compared with a normal probability curve with the same variability, is known as *kurtosis*. This term is widely misunderstood, largely because it is often inaccurately or inadequately explained in books that refer to it. We shall not discuss it at great length, but we shall attempt to clear up the most common misconception.

FIGURE 7.2
Skewness and Kurtosis

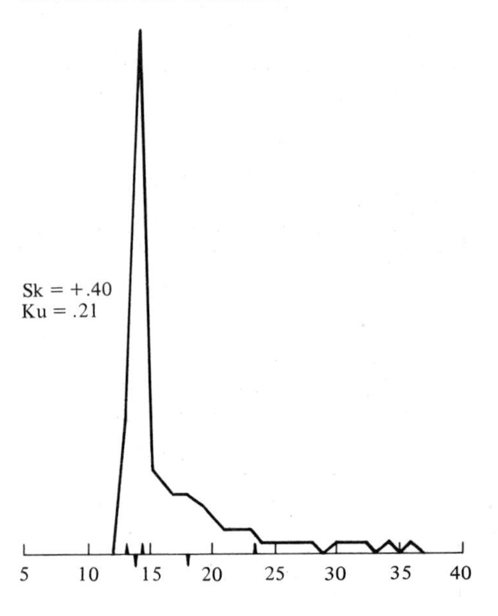

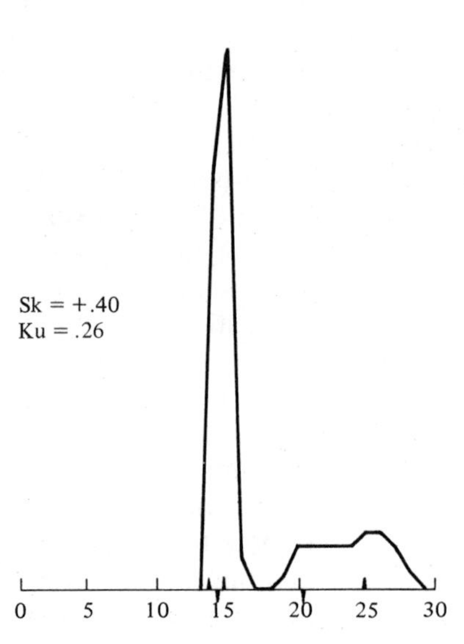

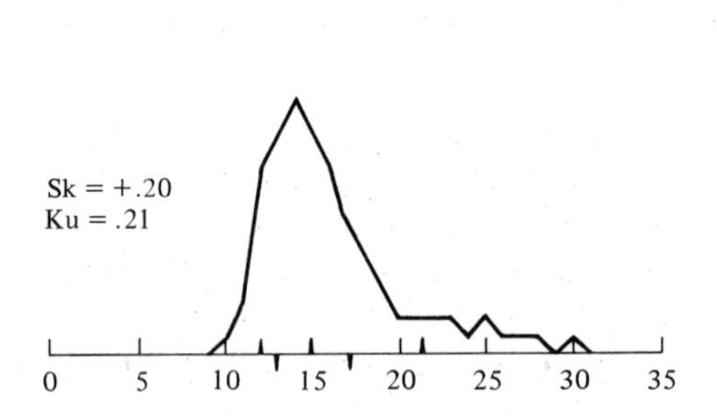

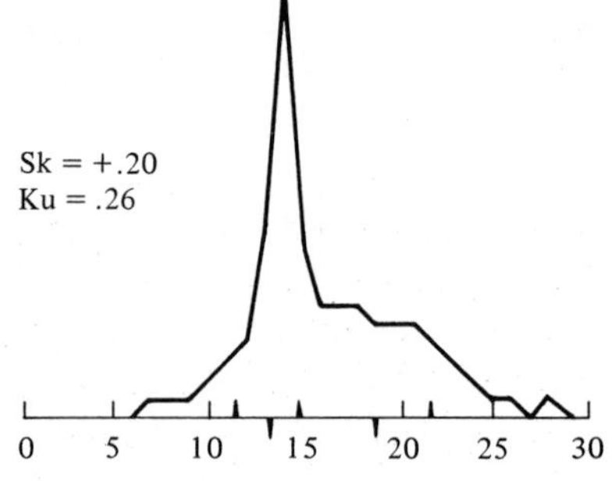

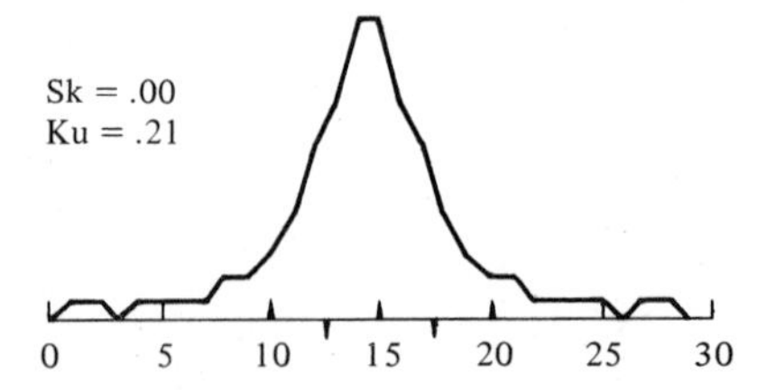

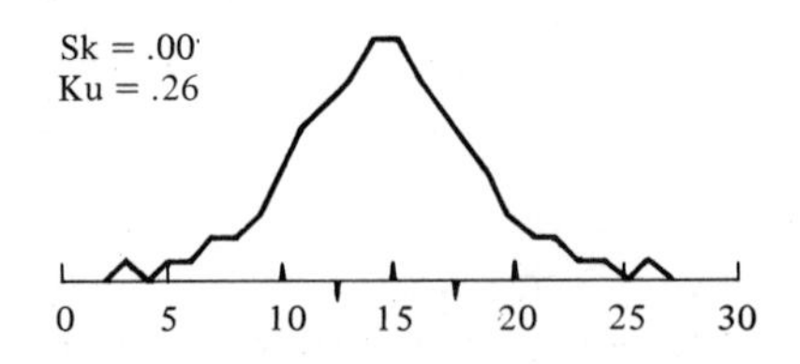

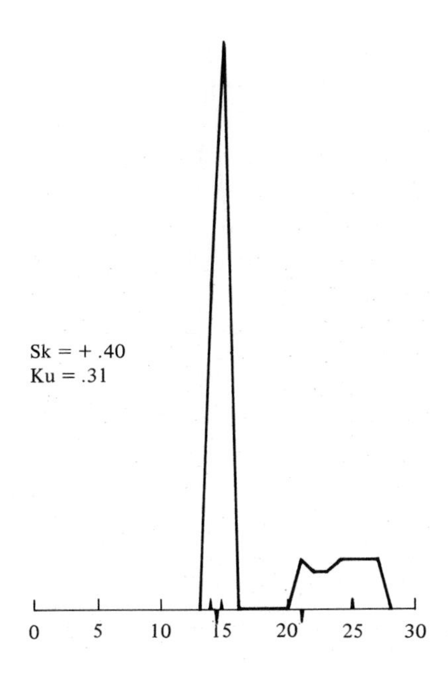
Sk = + .40
Ku = .31
0
5
10
15
20
25
30

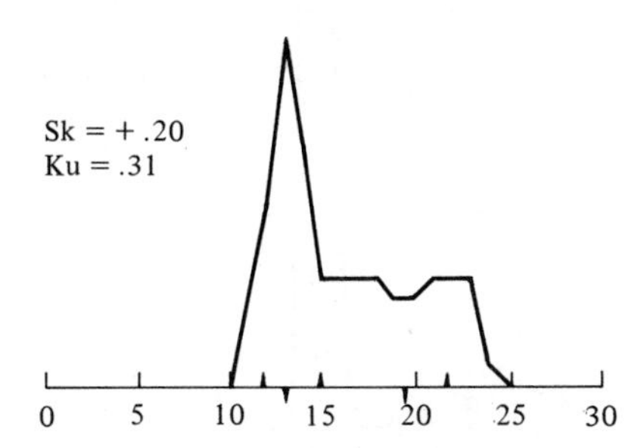
Sk = + .20
Ku = .31
0
5
10
15
20
25
30

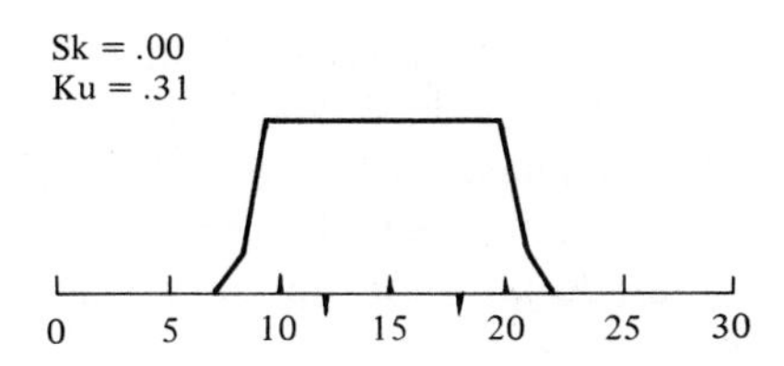
Sk = .00
Ku = .31
0
5
10
15
20
25
30

FIGURE 7.2 (*continued*)

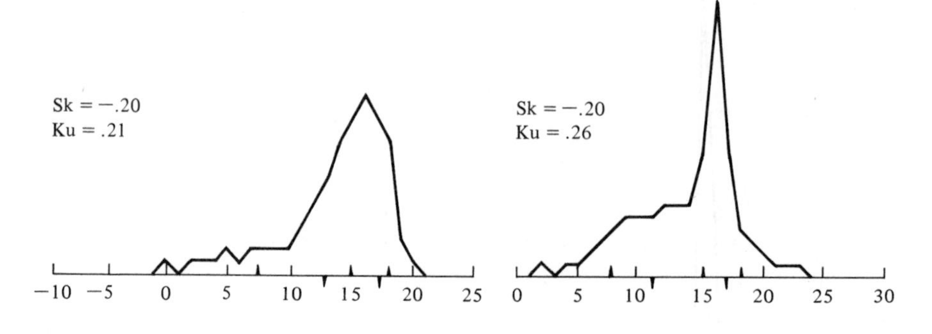

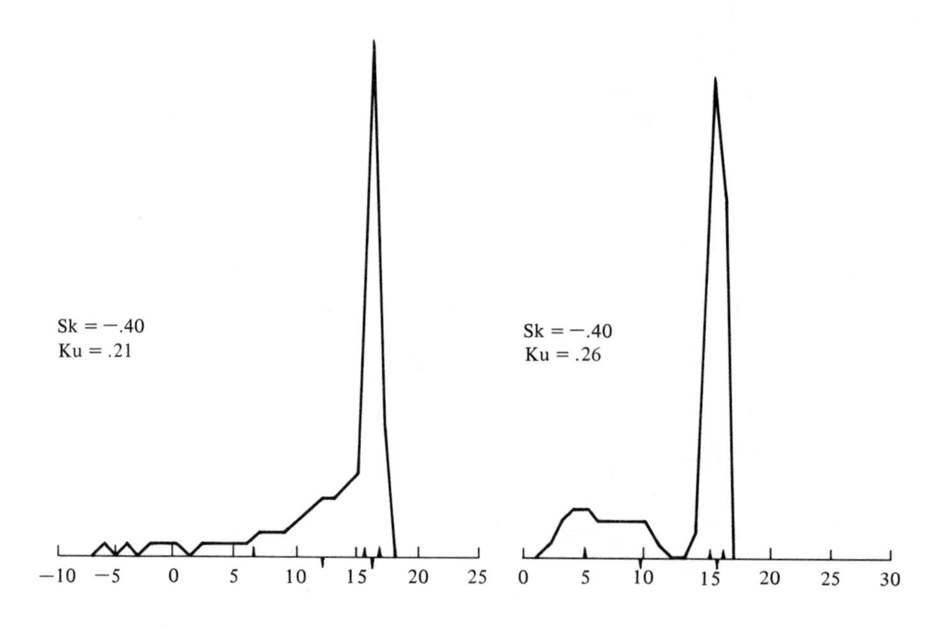

All three curves in Figure 7.3 have *the same* kurtosis. All three of them also have the same mean, the same median, the same standard deviation, the same average deviation, and the same skewness. These curves look different only because they have been drawn to different scales. To repeat, all three curves depict exactly the same data; they do not differ in number of cases, central tendency, dispersion, skewness, or kurtosis.

What, then, is kurtosis? It is the extent to which a curve is more or less peaked than a normal probability curve with the same standard deviation. How can a curve be more peaked than a normal probability curve and still

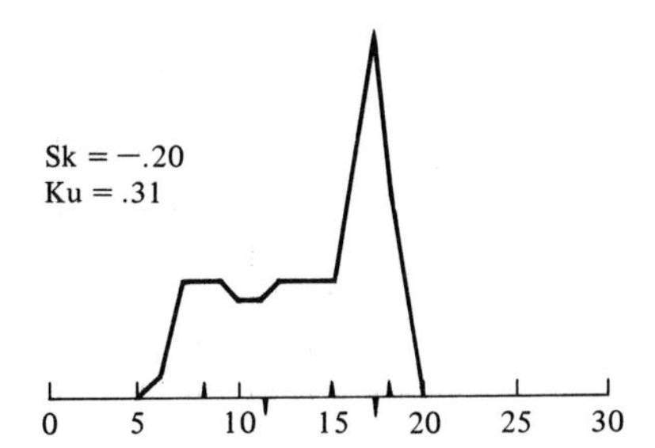

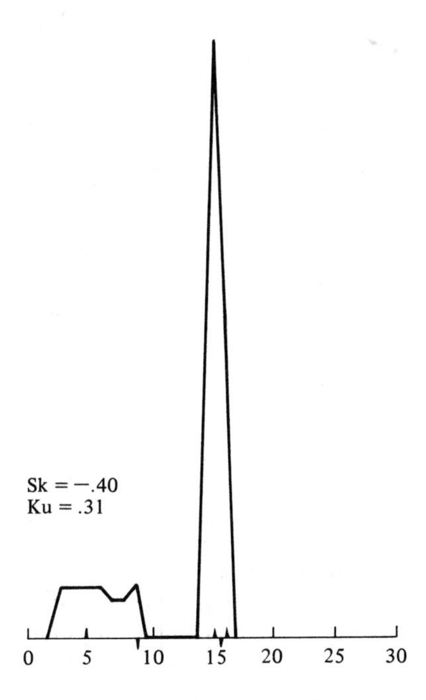

have the same standard deviation? By having more cases at the mean and at both ends of the curve, and by having correspondingly fewer cases about a standard deviation or so on each side of the mean. When a curve has this shape, it is called *leptokurtic*. (In order for anyone to have **leapt** over an obstacle he must have jumped either a high or a long distance. The **lept**okurtic curve can be remembered because it is higher in the middle and has longer tails than the corresponding normal curve.)

A curve may also be less peaked than the normal probability curve. Such a curve will have shorter tails and a more nearly uniform distribution of

FIGURE 7.3
Number of Heads Expected if 10 Coins Are Tossed 1,024 Times

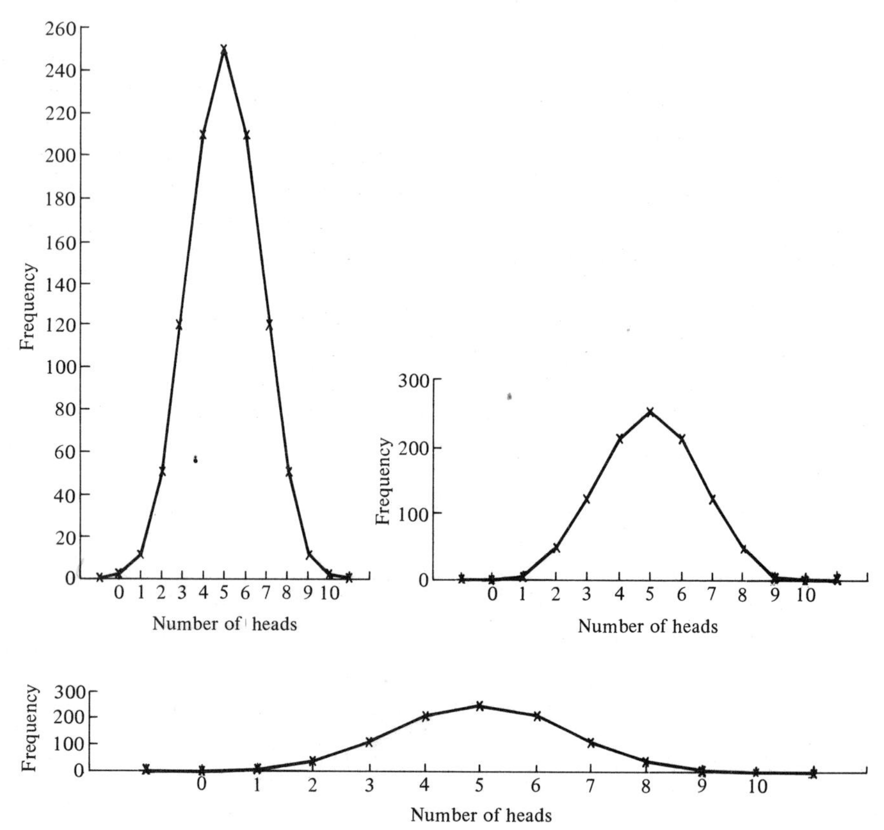

scores throughout its length. An example is a rectangular distribution, with no cases at all below or above two limits, but with exactly the same number of cases at every score between these limits. These relatively **flat** distributions are called **plat**ykurtic.

Finally, if the curve is neither peaked nor flat-topped, but is the same as the normal curve in kurtosis, this **m**edium kurtosis is described as **m**esokurtic.

In order to make these differences perfectly clear, Figure 7.4 has been prepared showing a leptokurtic, a platykurtic, and mesokurtic curve, all plotted on the same set of axes. The mesokurtic curve is the same one that was shown in Figure 7.3; the others were constructed to be as nearly like it as possible in all respects except kurtosis. In fact, the three curves differ by less than one-tenth of 1% in D and are *identical* in N, $\overline{X}$, Mdn, s, and Sk. They differ greatly, however, in kurtosis. Let us discuss them in alphabetical order, which, fortunately, is also in order of size of Ku and also in order of their heights from top to bottom at their centers or means in Figure 7.4.

FIGURE 7.4
Three Frequency Polygons on Same Axis Illustrating Leptokurtic, Mesokurtic, and Platykurtic Distributions

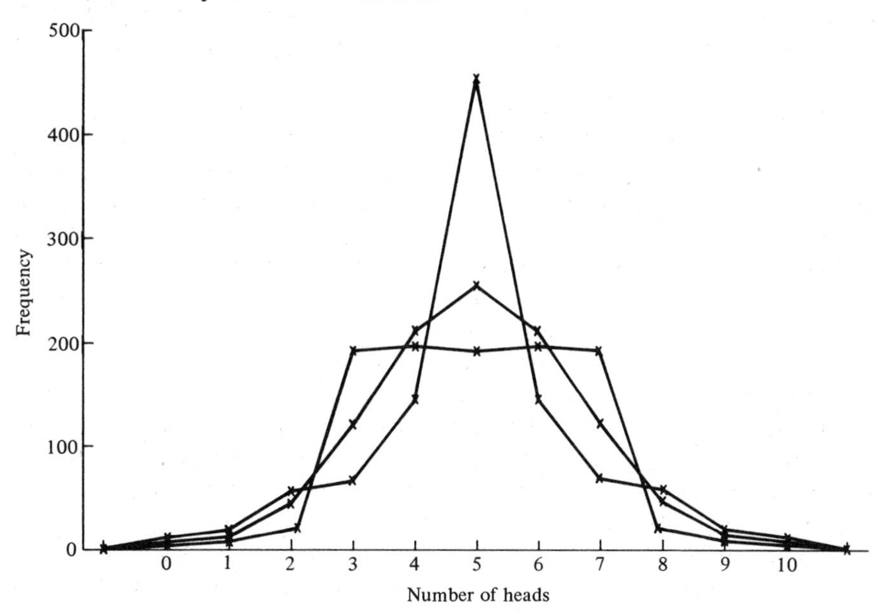

The leptokurtic curve has a kurtosis of only .17. (How this was computed will be explained on page 178.) It rises to a peak of 452; that is, almost half of the total of 1,024 cases in its frequency distribution are concentrated in one interval. Despite this, this leptokurtic curve has higher frequencies than either of the others at both ends, and its tails are longer.

The middle curve is almost exactly mesokurtic, since its kurtosis of .26 (or, more precisely, .2648) differs only negligibly from the theoretical value of .2632. It is a graph of the coefficients of the successive terms in the expansion of $(p + q)^{10}$. As such, it gives a close approximation to the normal probability curve, which, by definition, is mesokurtic. This curve remains in the middle, between the other two curves, throughout most of its length.

The platykurtic curve has a kurtosis of .31. It obviously fits its description of being flat-topped, the five central frequencies being 190, 196, 194, 196, and 190. It also has the shortest tails of the three curves and the lowest peak (if we may call it that). If the kurtosis were still larger, the curve would

Further insight into kurtosis is provided by Figure 7.2. Looking across the center row, we see a leptokurtic, a mesokurtic, and a platykurtic curve.
show an indentation instead of a peak in the center, the tails would be shorter, and there would be two peaks or modes. Such curves are rare.

These three curves are symmetrical—just as those in Figure 7.4 are. In fact, looking across any row we see (in alphabetical order) a leptokurtic, a mesokurtic, and a platykurtic curve. The second and fourth rows show what happens when these three degrees of kurtosis are combined with a moderate amount of skewness, while the top and bottom rows show three degrees of kurtosis for very skewed curves.

Since those distributions of scores that show any appreciable departure from mesokurtosis are often skewed, as well, you should note in Figure 7.2 what the distributions look like when both skewness and kurtosis vary. Later, when a distribution is described in terms of *Sk* and *Ku,* a glance at Figure 7.2 will give you a very good idea of the shape of the new distribution. Nearly all distributions of psychological and educational scores will fall within the limits shown in Figure 7.2, most of them not far from the curve with $Sk = 0$ and $Ku = .26$, which is shown in the center.*

Computation of the Kurtosis Measure Essentially the same reasons we gave for preferring a simple formula for skewness apply to kurtosis.† Hence, we recommend

$$Ku = \frac{Q}{D}. \tag{7.3}$$

This formula, like Formula (7.1) for skewness, is a pure number. That is, it is independent of the unit of measurement and will, for example, give the same value, regardless of whether heights are measured in feet, inches, millimeters, puntos, or pies.

To compute *Ku* for the data of Table 2.4, we note that in Supplement 6, the values of Q and D have already been computed as 10.01 and 37.03, respectively. Hence,

$$\begin{aligned} Ku &= \frac{Q}{D} \\ &= \frac{10.01}{37.03} \\ &= .2703. \end{aligned}$$

The numerical value of *Ku* must always be between .00 and .50—though most values will fall well within the limits of .21 and .31, which are shown in Figure 7.2.

---

*We would like to take time out to settle a purely personal matter: Kurtosis was not named for or by one of the writers. The term kurtosis was first used by Karl Pearson in 1906. It is derived from a Greek word meaning curvature or convexity.

†For the rare study in which the most efficient measures of skewness and kurtosis are desired, see Supplement 7.

Stability of Kurtosis The standard error of Ku is given by the very simple formula

$$s_{Ku} = \frac{.2778}{\sqrt{N}}. \tag{7.4}$$

So, for the El Dorado IQ data, we have

$$\begin{aligned} s_{Ku} &= \frac{.2778}{\sqrt{176}} \\ &= \frac{.2778}{13.27} \\ &= .0209. \end{aligned}$$

To determine whether the El Dorado kurtosis is significantly different from .2632, signifying mesokurtosis, we simply take its deviation from the fixed value, .2632, and divide the difference by its standard error. Thus,

$$\frac{.2703 - .2632}{.0209} = +.34.$$

Since this quotient is very much smaller in absolute value than 1.96, corresponding to the .05 level of significance, we conclude that there is no evidence to disprove our natural assumption that the curve was mesokurtic.

TRANSFORMED SCORES

There are a very great many ways in which one set of scores may be transformed into a second set of scores, the two sets differing in some respects, but remaining similar or the same in others. Some changes are so simple that we scarcely think of them as transformations at all. One such is changing a set of X scores to x scores by subtracting $\overline{X}$ from each X. Another is transforming X scores into x' scores. Still another, almost as simple, is the conversion of X scores to z scores, this transformation involving both subtraction and division. In these three cases, the shape of the distribution remains unchanged; if it was negatively skewed and leptokurtic, it still is.

In the following paragraphs, we shall discuss some transformations involving greater changes than the simple ones just referred to, but, first, let us ask why we want to transform scores anyway. There are many reasons.

1. There is sometimes a reason to believe that some particular function of our variable gives a better representation of the underlying trait which we are seeking to measure.
2. Closely related is the situation existing when our scores are expressed one way, but not the way we are interested in having them. For

instance, we may see how long it takes an auto to travel a measured mile, but it is nearly always more meaningful to transmute this elapsed time into miles per hour.

3. Some statistical formulas assume that the distribution of our scores is a normal one or some reasonably close approximation to it. If we are not to misuse such a formula, we must either abandon it or transform our scores so as to give a distribution that is approximately normal.
4. A relatively minor reason for using transformations is that some people just naturally seem to prefer symmetrical distributions to skewed ones.
5. A much better reason for using transformed scores is that the latter will sometimes show higher correlations with other variables. This is very important in any problem in which we are interested in making predictions, a matter to which all of Chapter 9 is devoted.

Assume, then, that one (or more) of these reasons applies and that we wish to transform our scores. How can we do this? The following paragraphs will give the answers.

(1) *Percentile.* A transformation with which we are already familiar is that of converting a set of X scores into percentile ranks. Here, the distribution is changed from normal or whatever other shape it may have been to an approximately rectangular shape. This is a good example of a typical transformation. Except for ties on one scale or the other, all the scores remain in exactly the same order; but the relative spacing among them is changed, and the shape of the two frequency distributions is different.

(2) *Square root.* This last statement leads directly into the principal reason for transforming scores. We can illustrate this best by an artificial example. For this purpose suppose that we are given the following scores, arranged in order of size:

0, 2, 4, 6, 8, 9, 11, 13, 15, 17, 19, 21, 23, 25,
27, 30, 32, 34, 36, 38, 41, 44, 47, 50, 54, 58,
62, 66, 70, 75, 80, 85, 91, 98, 110, 132.

When we plot these, we get the positively skewed distribution shown in the left side of Table 7.5. In one sense, the scores are all right and the distribution is all right—after all, it is simply a frequency distribution of the given scores. But it is *so* lopsided! If we could change it some way or other so that it would be more nearly normal—or even just symmetrical—it would not only look nicer, but it would also be more like the many other distributions we encounter. As such, it would be more meaningful, it would be easier to interpret the scores, and the scores could then be analyzed by various statistical methods which assume that the scores are normally distributed. We need a transformation that will squeeze the high scores

TABLE 7.5
Two Tabulations of Hypothetical Scores

A Distribution of Scores		B Distribution of Square Roots of Scores	
X	f	$\sqrt{X}$	f
130–139	I	11.00–11.99	I
120–129		10.00–10.99	I
110–119	I	9.00–9.99	III
100		8	IIII
90	II	7	IIII
80	II	6	~~IIII~~
70	II	5	~~IIII~~
60	II	4	IIII
50	III	3	IIII
40	III	2.00–2.99	III
30	~~IIII~~	1.00–1.99	I
20–29	IIII	.00– .99	I
10–19	~~IIII~~		
0– 9	~~IIII~~ I		

closer together and spread the low ones relatively farther apart. Although many procedures will do this, let us first try one of the simplest and take the square roots. Doing this, our previous distribution becomes

0.00, 1.41, 2.00, 2.45, 2.83, 3.00, 3.32, 3.61, 3.87
4.12, 4.36, 4.58, 4.80, 5.00, 5.20, 5.48, 5.66, 5.83
6.00, 6.16, 6.40, 6.63, 6.86, 7.07, 7.35, 7.62, 7.87,
8.12, 8.37, 8.66, 8.94, 9.22, 9.54, 9.90, 10.49, 11.49.

When we plot these scores we get the beautifully symmetrical distribution plotted in the right-hand side of Table 7.5. Sure, this is a tricky example made to look nice by the use of artificial data, but it does show exactly what type of distribution (see the left-hand side of Table 7.5) can become nearly normal (see the right-hand side of the table) by the use of the square root transformation.

(3) *Logarithmic*. Another transformation that is sometimes useful for positively skewed distributions is the logarithmic one. In the example we just used, if we took log X, we would overdo it and have a distribution skewed in the negative direction. This transformation cannot be used if there are any negative numbers. Even a zero score offers difficulties, since its lograithm is $-\infty$. It is, of course, possible to eliminate this difficulty by taking, for example, $\log(X + 10)$, but there is ordinarily no logical reason for using 10 rather than 3 or 100 as the constant to be added.

Distributions of numbers with dollar signs in front of them will often become nearly symmetrical when a square root or a logarithmic transformation is used. The difference in selling ability of two people who sell \$500,000 and \$600,000 of life insurance in their first year is small while there is a vast difference between the \$100,000 and the \$200,000 person. The dollar differences are identical but the square root and the logarithmic transformations give differences that are intuitively much more reasonable. They should be considered for such financial data as income, living expenses, earnings based solely on commissions, and stock market prices.

(4) *Cube root.* Distributions of heights are often close to normal, while those of body weights often show positive skewness. Since most people will just barely float in fresh water, it is obvious that body weight is very closely related to volume. Volume is proportional to the product of length, width, and thickness. Among people of the same shape, then, body weight ought to be proportional to the cube of any one of these dimensions—and it is. Consequently, if we take the cube root of body weight, we ought to get a new set of numbers that will conform much more closely to the normal form. Hence, the cube root is a useful transformation when we are dealing with weights or volumes.

(5) *Reciprocal.* The transformations just discussed in sections (2), (3), and (4) above all decrease the amount of skewness and are appropriate only for positively skewed distributions. The use of $\frac{1}{X}$ as a variable has varying effects. Often, but not always, it will reverse the sign of Sk, but it will increase its absolute value about as often as it decreases it. It is particularly appropriate when we have time scores and wish to transmute them to speed scores. Two illustrations of this are included in this text. On page 316 we find that this transformation will not take care of the troublesome problem of nonswimmers. On page 383 we show that, although response times on an assembly test are not normally distributed, the corresponding speed scores are entirely consistent with the hypothesis that the speed with which people complete the task is normally distributed. This transformation has the added characteristic (usually an advantage) that very high time scores or even total failure to complete the task can be included, since such scores usually transmute to zero or to one of the two lowest intervals. The reciprocal transformation is widely used by experimental psychologists for transforming resistance in ohms into conductance in mhos, where mho = $\frac{1}{\text{ohm}}$.

(6) *Miscellaneous.* Statisticians also use many other types of transformations. Several of the most common of these are trigonometric functions,

one of the most useful being the inverse sine (or arcsin) transformation. Thus, if p is a proportion, we find that angle whose sine is equal to p. Sometimes we first extract the square root of p, thus obtaining arcsin $\sqrt{p}$. The student who finds that the transformations given here are not satisfactory for his data should consult the Fisher and Yates Statistical Tables.* These authors discuss and provide tables for some of those transformations we have already discussed, as well as for four other transformations (called probit, angular, logit, and loglog).

(7) *Inverse transformations.* In sections (2), (3), and (4), transformations were given that would reduce or eliminate positive skewness. But what do we do about negative skewness? Perhaps one or more of the transformations discussed in sections (5) or (6) will be satisfactory. But there is a more obvious method: simply reverse the process. Thus, if taking the square roots of a set of scores reduces skewness, squaring a number of scores will increase the skewness—thus altering a negatively skewed distribution so that its skewness becomes less negative, zero, or perhaps even positive. The logarithmic or cube root transformations may also be reversed by taking antilogarithms or cubes, respectively. If we try to reverse the reciprocal transformation, we simply get the reciprocal transformation again. The miscellaneous transformations can be reversed, of course, but such transformations have so far had little applicability.

That takes care of all the inverse transformations, except the inverse of the percentile transformation discussed in section (1). The percentile transformation transforms distributions of any shape into the rectangular form. What happens if we try to reverse this process? Almost anything, but one possibility is of particular interest. We can transmute any distribution of percentile ranks into a very close approximation to the normal probability curve. We can even go farther, as is noted in the supplement to this chapter, and transform any distribution at all into the normal shape. This process, however, should not be used unless we have a very large amount of data, probably some hundreds or thousands of cases.

THE DESCRIPTION OF FREQUENCY DISTRIBUTIONS

It is interesting that some of the common ways of describing distributions of scores are based upon the sums of successive powers of those scores or of their deviations from the mean.

---

*Fisher, Ronald A., and Yates, Frank. *Statistical tables for biological, agricultural and medical research,* Sixth Ed., Darien, Conn.: Hafner, 1970.

The zero power: Early in algebra, students are taught that if we raise any number (except 0 or ∞) to the zero power, we get 1. Thus, $X^0 = 1$. Suppose we have N such X^0 scores and we add them up. We would be adding one N times and we would get N. Thus, $\Sigma X^0 = N$.

The first power: If we raise all our numbers to the first power (that is, leave them alone) and divide by the number of cases, we get the arithmetic mean. Thus,

$$\frac{\Sigma X^1}{N} = \overline{X}. \tag{7.5}$$

The second power: If we express our scores as deviations from the mean and then square them, we are well on our way toward computing the standard deviation. Thus,

$$\sqrt{\frac{\Sigma x^2}{N}} = \sigma_N \tag{7.6}$$

The third power: The third power comes up with respect to skewness, the main subject of this chapter. Although Formula (7.1) makes no reference to the third power, many other skewness formulas do. Hence, if we express our scores as deviations from the mean, cube them, and then average these cubes, we get the numerator of one of the basic skewness formulas discussed in Supplement 7. The denominator is an appropriate power, the third, of the standard deviation. Thus,

$$\frac{\dfrac{\Sigma x^3}{N}}{s^3} = \sqrt{\beta_1}. \tag{7.7}$$

The fourth power: In measuring kurtosis, the fourth powers of scores are usually involved, even though they do not appear in Formula (7.3). Hence, if we express our scores as deviations from the mean, raise them to the fourth power, and then average these powers, we get the numerator of one of the basic kurtosis formulas discussed in Supplement 7. The denominator is an appropriate power, the fourth, of the standard deviation. Thus,

$$\frac{\dfrac{\Sigma x^4}{N}}{s^4} = \beta_2. \tag{7.8}$$

The fifth power: So far as we know, no one has seriously proposed using the fifth, sixth, or higher powers of the scores or of their deviations as the basis for any other statistics useful in describing a frequency curve. It could be done, but it does not seem worth the trouble. In fact, we regard the formulas at the end of the two preceding paragraphs as far too cumbersome for ordinary use and recommend Formula (7.1) for skewness and (7.3) for kurtosis, not because they are better, but because they are so much easier to compute than β_1 and β_2. Anyway, by the time we have given N, $\overline{X}$, s, Sk (or β_1), and Ku (or β_2), we have given an extremely accurate description of the frequency distribution. In fact, most statisticians are quite willing to settle for only the first four of these five measures.

SUMMARY

Most frequency distributions of educational or psychological data have a shape very similar to that of the normal probability curve discussed in Chapter 4. Some, however, are lopsided or skewed. This skewness can be measured by Formula (7.1) or by other formulas given in Supplement 7. There are many causes, both natural and artificial, for skewness.

Kurtosis refers to the extent to which a curve is more peaked (leptokurtic), as peaked (mesokurtic), or more flat-topped (platykurtic) than a normal probability curve with the same degree of variability. Measures of kurtosis are not widely used; hence, a simple formula was given for its measurement.

Several graphs were used in order to clarify the meaning of skewness and kurtosis, and to distinguish kurtosis from variability.

When, for any reason, it becomes desirable to change the skewness and kurtosis of a frequency distribution, thus transforming it into one with a different shape, this can be done by using one of the large number of transformations explained or referred to in the text.

This chapter completes the discussion of the statistics used to describe a frequency distribution. The next chapter will consider the relations of two variables to each other.

SUPPLEMENT 7

ADDITIONAL MEASURES OF SKEWNESS AND KURTOSIS

In many—perhaps most—cases one might not go to the trouble of assessing skewness by means of a formula. If one uses a formula, it would probably be Formula (7.1) or one similar to it which is based upon percentiles. Formula (7.1) is a time-saver. For one who has more time and wants to investigate skewness in some depth, the following formulas are given.

The desirable characteristics of a measure of skewness are: (1) it should be a pure number, thereby independent of the units in which our variable is measured; and (2) it should be zero when the distribution is symmetrical.

Because there are at least a dozen skewness formulas in use, we shall use subscripts to identify the ones we are publishing. To avoid confusion, we shall use the same subscripts as were used by Dunlap and Kurtz.* Our next formula, their Sk_6, is

$$Sk_6 = \frac{Q_1 + Q_3 - 2Mdn}{Q_3 - Q_1}. \tag{7.9}$$

This formula seems intuitively appealing to us, since it is based upon the fact that Q_1 and Q_3 are equally far from the median in a symmetrical distribution but become unequally distant from the median whenever there is skewness.

A similar formula is

$$Sk_7 = \frac{Q_1 + Q_3 - 2Mdn}{Q}. \tag{7.10}$$

It is very similar to Sk_6, the only difference being that each value of Sk_7 is exactly twice the corresponding value of Sk_6.

A formula in terms of mean, median, and standard deviation, and which is appropriate for slightly skewed distributions, is given by

$$Sk_5 = \frac{3(\overline{X} - Mdn)}{s}. \tag{7.11}$$

A very simple formula in terms of mean, mode, and standard deviation is given by

$$Sk_4 = \frac{\overline{X} - Mo}{s}. \tag{7.12}$$

Its standard error is given by

$$s_{Sk_4} = \sqrt{\frac{3}{2N}}. \tag{7.13}$$

---

*Dunlap, Jack W., and Kurtz, Albert K. *Handbook of statistical nomographs, tables, and formulas,* Yonkers-on-Hudson, N.Y.: World Book Company, 1932.

The other type of skewness formula is based upon *moments about the mean*. A moment about the mean is given by the general formula

$$\mu_i = \frac{\Sigma x^i}{N}. \tag{7.14}$$

Thus, the first through fourth moments about the mean are

$$\mu_1 = \frac{\Sigma x}{N} = 0 \qquad \text{(always zero)}, \tag{7.15}$$

$$\mu_2 = \frac{\Sigma x^2}{N} = s^2 \qquad \text{(variance)},$$

$$\mu_3 = \frac{\Sigma x^3}{N}, \tag{7.17}$$

$$\mu_4 = \frac{\Sigma x^4}{N}. \tag{7.18}$$

Several formulas for skewness and kurtosis have been based upon the moments. Two intermediate values often used in such formulas are called β_1 and β_2 and are defined as

$$\beta_1 = \frac{\mu_3^2}{\mu_2^3} \tag{7.19}$$

and

$$\beta_2 = \frac{\mu_4}{\mu_2^2}. \tag{7.20}$$

The simplest formula for skewness, using the Dunlap and Kurtz subscript, is

$$Sk_1 = \sqrt{\beta_1}. \tag{7.21}$$

In a normal curve, $\sqrt{\beta_1} = 0$. Its standard error is

$$s_{Sk_1} = \sqrt{\frac{6}{N}}. \tag{7.22}$$

Another formula for skewness, which is based upon the third moment and the standard deviation, is

$$Sk_2 = \frac{\sqrt[3]{\mu_3}}{s}. \tag{7.23}$$

Still another formula for skewness, which is based upon β_1 and β_2, is defined as

$$Sk_3 = \frac{\sqrt{\beta_1}(\beta_2 + 3)}{2(5\beta_2 - 6\beta_1 - 9)}. \tag{7.24}$$

Its standard error is

$$s_{Sk_3} = \sqrt{\frac{3}{2N}}. \tag{7.25}$$

Two formulas for measuring kurtosis are based upon β_2, which is a function of the second and fourth moments. One of these was given previously, as Formula (7.20). Its standard error is

$$s_{\beta_2} = \sqrt{\frac{24}{N}}. \tag{7.26}$$

And since in a normal distribution $\beta_2 = 3$, the other measure of kurtosis can be given as

$$Ku_1 = \beta_2 - 3. \tag{7.27}$$

In a normal distribution $Ku_1 = 0$. Its standard error is

$$s_{Ku_1} = s_{\beta_2} = \sqrt{\frac{24}{N}}. \tag{7.28}$$

Area Transformation to Normalize Scores Sometimes it is an advantage to transform a set of raw scores to standard scores. If we should transform raw scores to z scores by Formula (3.6), we would have numbers with decimals, and about half of them would be negative. However, there is a way to rid ourselves of these disadvantages by transforming our data into a close approximation to a normal distribution which will have a mean of 50 and a standard deviation of 10. We call this process an *area transformation*. The systematic method for converting an entire set of grouped raw scores to T scores is shown in Table 7.6. The figures from Table 6.2 on page 157 are used, and the cumulative frequency to the midpoint of each interval is copied from column (4) in Table 6.2 to column (4) in Table 7.6. The values in column (5) are obtained by dividing each figure in column (4) by 140, the total number of cases. Next, from every proportion greater than .5000, the constant .5000 is subtracted, while those proportions below .5000 are subtracted from .5000. In each case we then have the proportion

TABLE 7.6
Conversion of Grouped Raw Scores to *T* Scores

Intervals (1)	Midpoint (2)	*f* (3)	Cumulative Frequency to Midpoint (4)	Proportion below Midpoint (5)	*z* for Proportion (6)	*T* Score (7)
300–309	304.5	1	139.5	.9964	2.69	77
290–299	294.5		139.0	.9929	2.45	74
280–289	284.5		139.0	.9929	2.45	74
270	274.5	1	138.5	.9893	2.30	73
260	264.5	4	136.0	.9714	1.90	69
250	254.5	4	132.0	.9429	1.58	66
240	244.5	8	126.0	.9000	1.28	63
230	234.5	5	119.5	.8536	1.05+	61
220	224.5	10	112.0	.8000	0.84	58
210	214.5	6	104.0	.7429	0.65+	57
200	204.5	8	97.0	.6929	0.50	55
190	194.5	2	92.0	.6571	0.40	54
180	184.5	2	90.0	.6429	0.37	54
170	174.5	6	86.0	.6143	0.29	53
160	164.5	4	81.0	.5786	0.20	52
150	154.5	7	75.5	.5393	0.10	51
140	144.5	14	65.0	.4643	−0.09	49
130	134.5	12	52.0	.3714	−0.33	47
120	124.5	8	42.0	.3000	−0.52	45
110	114.5	11	32.5	.2321	−0.73	43
100	104.5	8	23.0	.1643	−0.98	40
90	94.5	12	13.0	.0929	−1.32	37
80	84.5	3	5.5	.0393	−1.76	32
70–79	74.5	2	3.0	.0214	−2.03	30
60–69	64.5	1	1.5	.0107	−2.30	27
50–59	54.5	1	0.5	.0036	−2.69	23

between the mean and the given point. These proportions are then found in the body of Table E and the corresponding value of $\frac{x}{\sigma}$ (which is z) is then read from the side and top of the table. For example, the value of .9964 for the top interval in Table 7.6 is transformed to .9964 − .5000 = .4964. This exact value of .4964 is found in Table E, and the corresponding value of z is 2.69. For the proportion .9714, the exact value .4714 is not in Table E, but the closest value is .4713 and it yields a z of 1.90.

FIGURE 7.7
Frequency Polygon of Table 7.6 Raw Scores

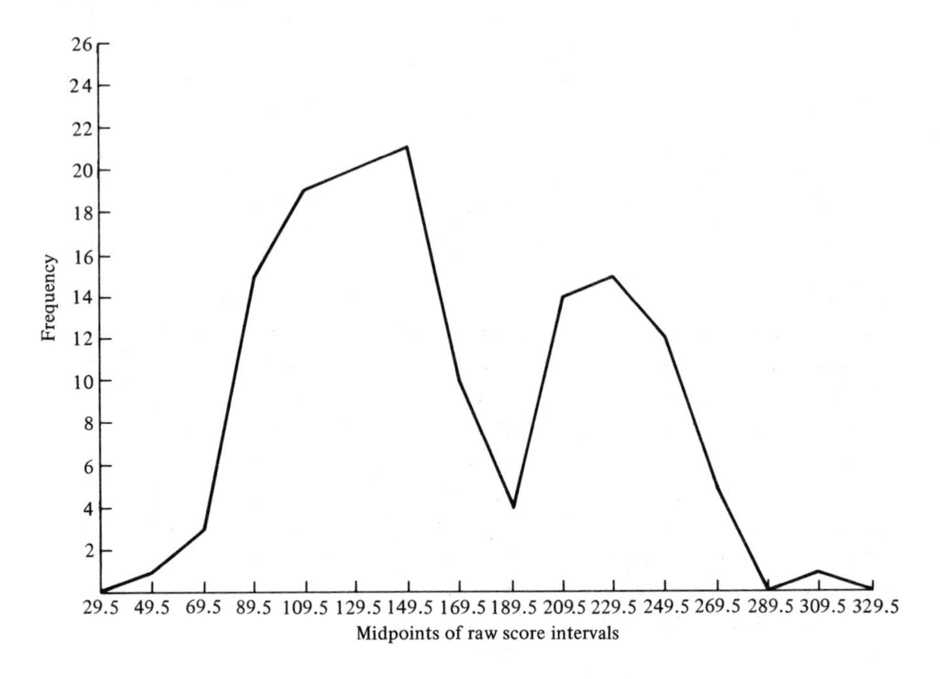

The transformation formula we shall use is

$$T = 10z + 50. \tag{7.29}$$

The T scores in the last column can be obtained mentally (that is, without using pencil and paper or a calculator) by the following process. The z values are multiplied by 10 by moving the decimal point one place to the right. Then the result is rounded (mentally again) to the nearest whole number. Thus the z's of 2.69, 2.45, −0.33, and −2.03 are mentally transformed into whole numbers 27, 24, −3, and −20, respectively. These resulting whole numbers are added to the constant of 50, yielding T scores of 77, 74, 47, and 30, respectively. The process is quite easy to carry out.

The accompanying frequency polygons, Figures 7.7 and 7.8, show what our distribution looks like before and after normalizing. By glancing at the two graphs, one can readily see that the effect has been to squeeze the data into

FIGURE 7.8
Frequency Polygon of Table 7.6 *T* Scores

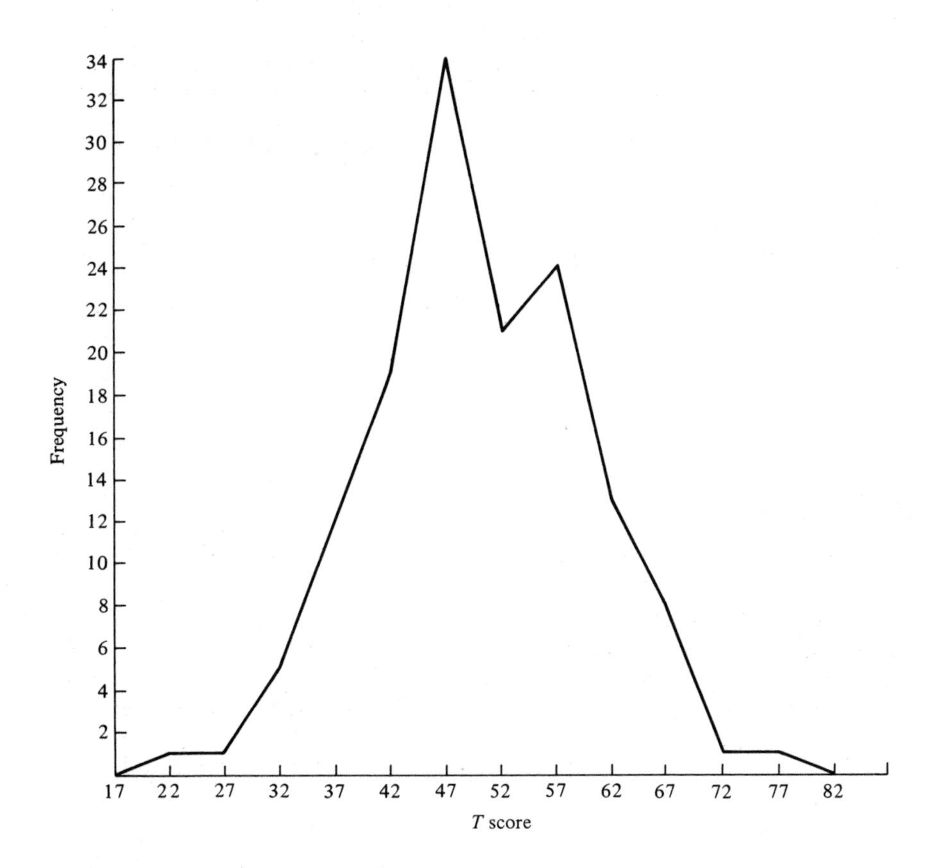

a more nearly peaked and symmetrical shape which rather closely approximates that of the normal probability curve.

However, the distribution of T scores should be distinguished from the distribution of Z scores. Both have a mean of 50 and a standard deviation of 10, but the distribution of T scores is nearly normal while the distribution of Z scores remains exactly the same as that of the original X scores, however skewed or distorted that distribution may have been.

For still more transformations of test scores, see page 81.

CHAPTER 8 PEARSON PRODUCT MOMENT COEFFICIENT OF CORRELATION

In the first seven chapters of this text, we have dealt with situations in which each one of the N people in our study had a score, X, on some test or other measurement. It is now time for us to expand our view and consider the perfectly realistic situation in which each individual may have scores on two (or more) different variables. Suppose that we have the height and weight of each one of a group of 12-year-old girls. We can, of course, compute the mean and standard deviation of height; and we can do the same for weight. But a more important question may well be: Is there a correlation, or relationship, between height and weight? Are the tall girls apt to weigh more or less than the short girls? Do the heavy girls tend to be above or below the average in height?

When two such sets of measurements are associated so that the measurements in one set are related to those in the other set, we say that the two sets of measurements are correlated. Thus, there is a (positive) correlation between the height of these 12-year-old girls and their weight. There is also a (negative) correlation between the score made on an intelligence test and the length of time required to solve certain puzzles and mazes.

Notice that the correlation does not need to be perfect. While most of the tall girls weigh more than the average, a few very slender girls may be both tall and below the average weight for girls of their own age group. If there are only a few such exceptions, the degree of correlation is obviously higher than if there are many of them. Obviously higher, yes, but just what do we mean by that? Clearly we need a more precise statement of the extent of the relationship—preferably in quantitative terms. Statisticians have derived more than 20 different formulas for measuring the degree of relationship between the pairs of measurements people have on two variables such as height and weight. But we don't need to consider all 20 of the formulas every time we want a correlation coefficient. For most purposes, a particular one of them is outstandingly superior to all the others. Hence unless we have *very* good reasons for preferring some other measure in a particular situation, we should use the *Pearson product moment coefficient of correlation*. Its name is long, but its symbol is short—merely r or r with two appropriate subscripts, such as r_{12} or r_{xy}.

DEFINITION OF PEARSON r

The Pearson product moment coefficient of correlation is a pure number, between the limiting values of $+1.00$ and -1.00, which expresses the degree of relationship between two variables. It may be considered as a measure of the amount of change in one variable that is associated, on the average, with a change of one unit in the other variable, when both variables are expressed in standard scores. For example, if the correlation between intelligence and average high school grade is .60, and if Richard is 2.00 standard deviations below the mean in intelligence, we may expect his high school average to be $(.60)(2.00) = 1.20$ standard deviations below the mean. In one sense, that's pretty good for him, since we predict that he will be only .60 as far below the mean in high school grades as he is in intelligence. Or if James has a standard score of $+2.22$ in intelligence, then his high school average, expressed in standard scores, should be about $(.60)(+2.22)\ s = +1.33s$. In each case, the actual high school record may vary somewhat from the predicted, as we shall see in Chapter 9.

When people who make high scores on one variable tend to make high scores on the other variable, and when people who make low scores on one variable tend to make low scores on the other variable also, Pearson r is *positive*. When there is only a chance relationship and people with low scores on one variable are just as likely as are people with high scores to make a high score on the other variable, the correlation is *zero*. When people who are high on one variable are likely to be low on the other (and when people who are low on either are apt to be high on the other), r is *negative*.

Uses of the Pearson Product Moment Coefficient of Correlation There is only one main use for r: to state in quantitative terms the extent of the relationship existing between two variables. If we could be satisfied with a statement that (a) there is a relationship or (b) there isn't, we wouldn't need Pearson r. We need Pearson r because we wish to quantify our statements. This makes comparisons possible. For instance, proficiency in receiving international Morse code may be predicted by means of the Army General Classification Test to the extent of +.24, a rather low correlation. For the same group of 221 soldiers, on the other hand, the Speed of Response Test correlates +.55 with proficiency. This last correlation is much higher than the first and is, in fact, about as high a correlation as is ordinarily obtained in predicting an industrial or military criterion of success. Higher correlations, often between .50 and .70, are found between intelligence tests and college grades.

Besides providing a direct measure of the degree of association existing between two variables, Pearson r occurs in equations which are used in predicting one score (such as success on the job) from another (such as an aptitude test score). These regression equations were alluded to in defining r and they form the subject of the next chapter.

There are a great many uses for correlation coefficients in mental measurement and psychological theory. They provide evidence concerning the interrelationships among test scores and other variables. They are often crucial in determining whether certain data are consistent with a hypothesis or whether they refute it.

When several variables are correlated with each other, the many correlations among them can often be explained in terms of a smaller number of underlying factors. These are found by a technical process known as factor analysis. Although many more factors have been located, it has been possible to define 25 in the aptitude-achievement area, each of which has three paper-and-pencil tests that are relatively pure measures of it. Correlation coefficients are the basic data that form the foundation upon which all such studies are based. Further, Pearson r's are virtually indispensable in evaluating the relative merits of various aptitude and achievement tests, both in vocational guidance and in vocational selection.

Illustrations of Pearson r's of Various Sizes Some idea of the meaning of r is, of course, contained in the definitions already given. Further meanings of r will be discussed in the last section of this chapter, entitled "Interpretation of Pearson r." One of the best ways to get to know how to understand and interpret correlation coefficients of various sizes is through actual contact with such correlations. That experience will be provided, both by means of a number of scatter diagrams showing r's ranging from +1.00 through .00 to −1.00; and then by citing a number of correlations that have been computed by earlier investigators in the fields of Education and Psychology. But first we need a brief review of plotting in rectangular coordinates, so that the basic nature of the scatter diagram will be clear.

TABLE 8.1
Measurements of Imaginary Psychologists

Name	Weight in Pounds	Height in Inches
Albert Adams	193	70¼
Bruce Bennett	190	74
Chet Coffman	182	71½
Dael Dunlap	163	68¼
Edward Engelhart	164	69½
Floyd Flanagan	218	73
George Glass	168	69¼
Harold Harris	172	70½
Irving Iverson	131	67¾
Jesus Jaspen	140	66¼
Kenneth Kuder	181	69¾
Leonard Lorge	204	72
Max Mayo	147	69¼
Nathan Norman	163	69¾
.	.	.
.	.	.
.	.	.

PLOTTING A SCATTER DIAGRAM

Assume that we were given the data shown in Table 8.1. If we were going to tabulate weight in order to compute $\overline{X}$ and s, we would first select a suitable interval; then we would tabulate the various weights. That is what we do now, except that we must also take height into account; that is, we must work with two variables simultaneously. The weights run from 131 to 218 pounds and the heights from 66¼ to 74 inches. Hence, suitable intervals are 5 pounds and ½ inch. Using any suitable cross-section paper, we draw a scatter diagram, as shown in Figure 8.2, and plot the first set of scores for Albert Adams. Notice that the plotting is done with a single tally mark. Since Albert Adams weighs 193 pounds and is 70¼ inches tall, that tally mark will have to be in the 190–194 column and in the 70–70¼ row. Is it clear that the single tally mark in Figure 8.2 represents someone whose weight is in the low 190s, and whose height is in the 70–70¼ inch interval? In order to plot a scatter diagram we must have *two* different scores (in this instance weight and height) for *each* person (such as those named in Table 8.1). On the other hand, each *single* tally mark in a scatter diagram (such as Figure 8.2 or 8.3) always corresponds to *two* scores for *the same* person.

Let us now tabulate the remaining scores. Bruce Bennett's weight of 190 pounds puts him in the same column as our first tally mark, but he is much taller than Albert Adams and his tally mark belongs in the very top row. This is shown in Figure 8.3. The other cases are tallied in the same way, in each case placing the tally mark in the little cell formed by the intersection

FIGURE 8.2
Scatter Diagram for Plotting Weight and Height

Height in inches \ Weight in pounds	130–134	135–139	140–144	145–149	150–154	155–159	160–164	165–169	170–174	175–179	180–184	185–189	190–194	195–199	200–204	205–209	210–214	215–219
$74-74\frac{1}{4}$																		
$73\frac{1}{2}-73\frac{3}{4}$																		
$73-73\frac{1}{4}$																		
$72\frac{1}{2}-72\frac{3}{4}$																		
$72-72\frac{1}{4}$																		
$71\frac{1}{2}-71\frac{3}{4}$																		
$71-71\frac{1}{4}$																		
$70\frac{1}{2}-70\frac{3}{4}$													/					
$70-70\frac{1}{4}$																		
$69\frac{1}{2}-69\frac{3}{4}$																		
$69-69\frac{1}{4}$																		
$68\frac{1}{2}-68\frac{3}{4}$																		
$68-68\frac{1}{4}$																		
$67\frac{1}{2}-67\frac{3}{4}$																		
$67-67\frac{1}{4}$																		
$66\frac{1}{2}-66\frac{3}{4}$																		
$66-66\frac{1}{4}$																		

of the appropriate weight column and height row. Notice that all the tally marks in any one cell need not correspond to identical scores, though all such tally marks must have scores falling in the same intervals on both variables. Thus Edward Engelhart and Nathan Norman differ both in weight and in height, but the differences are slight, and since both sets of scores fall in the same intervals, both of these scores are plotted in the same cell in Figure 8.3. When the first 14 scores (all that are listed in Table 8.1) are plotted, the scatter diagram first shown as Figure 8.2 becomes Figure 8.3.

That shows the construction of a scatter diagram and the plotting of the first few tally marks. The correlation appears to be positive and fairly high, but we had better withold judgment, because a correlation based on only 14 cases is highly unreliable. When more tally marks have been plotted, we can then compute r. But these 14 cases were imaginary, as Table 8.1 indicated. Let us shift to reality and look at some scatter diagrams of real data.

FIGURE 8.3
Scatter Diagram after Weight and Height have been plotted for the first 14 Cases

Weight in pounds (columns); Height in inches (rows)

Height in inches	130–134	135–139	140–144	145–149	150–154	155–159	160–164	165–169	170–174	175–179	180–184	185–189	190–194	195–199	200–204	205–209	210–214	215–219
$74-74\frac{1}{4}$													/					
$73\frac{1}{2}-73\frac{3}{4}$																		
$73-73\frac{1}{4}$																		/
$72\frac{1}{2}-72\frac{3}{4}$																		
$72-72\frac{1}{4}$															/			
$71\frac{1}{2}-71\frac{3}{4}$											/							
$71-71\frac{1}{4}$																		
$70\frac{1}{2}-70\frac{3}{4}$									/									
$70-70\frac{1}{4}$													/					
$69\frac{1}{2}-69\frac{3}{4}$							//				/							
$69-69\frac{1}{4}$				/				/										
$68\frac{1}{2}-68\frac{3}{4}$																		
$68-68\frac{1}{4}$							/											
$67\frac{1}{2}-67\frac{3}{4}$	/																	
$67-67\frac{1}{4}$																		
$66\frac{1}{2}-66\frac{3}{4}$																		
$66-66\frac{1}{4}$			/															

ILLUSTRATIONS OF PEARSON *r*'s OF VARIOUS SIZES

Figure 8.4 shows a set of 11 scatter diagrams of real data, carefully selected to cover the range of possible values from +1.00 through .00 to −1.00. We tried to make them equidistant, but in order to get data meeting our other requirements, some minor departures from the 1.00, .80, .60, . . . sequence resulted. Each of these scatter diagrams contains at least 200 tally marks or cases. This means that the obtained Pearson *r*'s are relatively stable and that if we were to get data from other groups, selected in essentially the same way that these were selected, the correlations would not change much. It is practically certain that these 11 correlation coefficients would still be in exactly the same order. Some of them might change their values by .05 or so, but there would be very few, if any, changes of more than .10 in either direction.

The intervals and other essential features of the scatter diagrams are those used by the original experimenters. In some cases, other statisticians might

Explanation of Figure 8.4

Dear Student:
First of all, it is not true that we do not know which way is up. Nor is it true that we don't know an X axis from a Y axis.

Yes, some of the type looks sideways or upside down. It *is* sideways or upside down. Believe it or not, there is a reason for this. The reason is simple: when r has a high positive value, the scatter diagram looks somewhat like

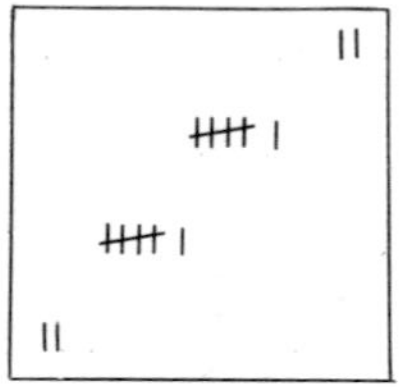

When r is about zero, it looks about like

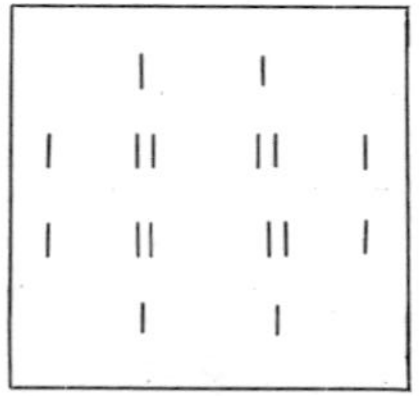

And when r has a high negative value, it resembles

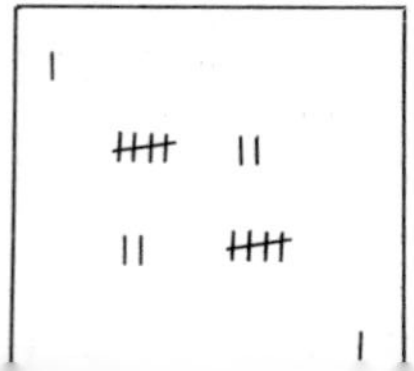

But these three diagrams are synthetic or even phony. We decided that it would be a good idea to show you some *real* data based on *real* studies made by *real* people. We wanted to show you a set of scatter diagrams from research that happened to have correlations of exactly +1.00, +.80, +.60, etc. through .00 and ending with −.80 and −1.00. That wasn't possible so we did the best job we could of approximating it and we also produced some more real data to fill in the gaps. To maintain realism, we printed the exact scatter diagrams the various authors printed in their books or articles. OK so far?

Now comes the problem. We wanted to let you see how the pattern formed by the various tally marks changes as r decreases. We decided that the best way to do this was to take the diagrams as printed and rotate them so that the high values of both variables would *always* be in the upper right corner of the page as you looked at it. In other words, the sole purpose of Figure 8.4 is to let you see exactly what the successive correlations look like and to show you clearly how the diagrams change their appearance as the degree of correlation changes. So please look at the diagrams on the next eleven pages and pay attention to the changes as you go, by steps of about .20, from r = +1.00 to r = −1.00.

Finally, if any of the scatter diagrams intrigue you and you wish to examine them in detail, turn the book around all you want to and satisfy your curiosity.

FIGURE 8.4
Pearson r's Illustrated by Eleven Scatter Diagrams Ranging from $r = +1.00$ for (A) to $r = -1.00$ for (K).

(A)

$r = +1.00$
$N =$ 234 tosses of 25 coins
$X =$ Number of heads on 25 coins
$Y =$ Number of heads on the same toss of the same 25 coins

Remarks: Any variable correlated with itself will give $r = +1.00$. Except for minor rounding errors, such variables as height in inches and height in centimeters will correlate $+1.00$. Similarly, the number of items right on a test will correlate $+1.00$ (or nearly so) with the percent of items right on the same test.

Y \ X	6	7	8	9	10	11	12	13	14	15	16	17	18	Total
18													I	1
17												卌 II		7
16											卌 卌 卌 III			18
15										卌 卌 卌 卌				20
14									卌 卌 卌 卌 卌 卌 IIII					34
13								卌 卌 卌 卌 卌 卌 II						32
12							卌 卌 卌 卌 卌 卌 卌 卌 IIII							44
11						卌 卌 卌 卌 IIII								24
10					卌 卌 卌 卌 卌 IIII									29
9				卌 卌 卌 II										17
8			卌 I											6
7		I												1
6	I													1
Total	1	1	6	17	29	24	44	32	34	20	18	7	1	234

Length in mm. (Central Values). $\longrightarrow +x$

Breadth in mm. (Central Values). $+y \longleftarrow$

Breadth \ Length	17·0	16·5	16·0	15·5	15·0	14·5	14·0	13·5	13·0	12·5	12·0	11·5	11·0	10·5	10·0	9·5	Totals
9·125	—	2	—	—	3	—	—	—	—	—	—	—	—	—	—	—	5
8·875	4	8	17	19	—	—	—	—	—	—	—	—	—	—	—	—	48
8·625	2	23	101	156	93	23	2	—	—	—	—	—	—	—	—	—	400
8·375	—	18	105	494	574	227	56	9	—	—	—	—	—	—	—	—	1483
8·125	—	4	44	375	956	913	362	73	12	3	—	—	—	—	—	—	2742
7·875	—	—	7	81	385	871	794	330	89	19	3	—	—	—	—	—	2579
7·625	—	—	1	4	65	236	469	361	175	55	27	4	—	—	—	—	1397
7·375	—	—	—	—	6	23	91	137	124	78	37	22	11	—	1	—	530
7·125	—	—	—	—	—	1	13	18	28	35	25	32	11	6	1	—	170
6·875	—	—	—	—	—	—	—	1	9	8	21	12	13	7	1	—	72
6·625	—	—	—	—	—	—	—	—	—	—	2	—	1	4	3	—	10
6·375	—	—	—	—	—	—	—	—	—	1	—	—	—	1	1	1	4
Totals	6	55	275	1129	2082	2294	1787	929	437	199	115	70	36	18	7	1	9440

FIGURE 8.4 (*continued*) **(B)**

$r = +.78$
$N = 9{,}440$ beans
X (horizontal variable) = Breadth of beans
Y (vertical variable) = Length of beans

Reference: Pretorius, S. J. Skew bivariate frequency surfaces, examined in the light of numerical illustrations. *Biometrika,* 1930, **22,** 157.

Remarks: This and other scatter diagrams have been rotated, when necessary, so that high values of both variables will be in the upper right corner.

(C)

$r = +.56$
N = 269 male members of European royal families
X = Estimated score in morality
Y = Estimated score in intellectual ability

Reference: Thorndike, Edward L. *Human nature and the social order*. Abridged edition. Cambridge, Mass.: M.I.T. Press, 1969, page 125.

Remarks: After referring to this scatter diagram, Thorndike concluded: "There is a positive correlation: the abler intellects are the more moral persons. But it is far from close: r = .56."

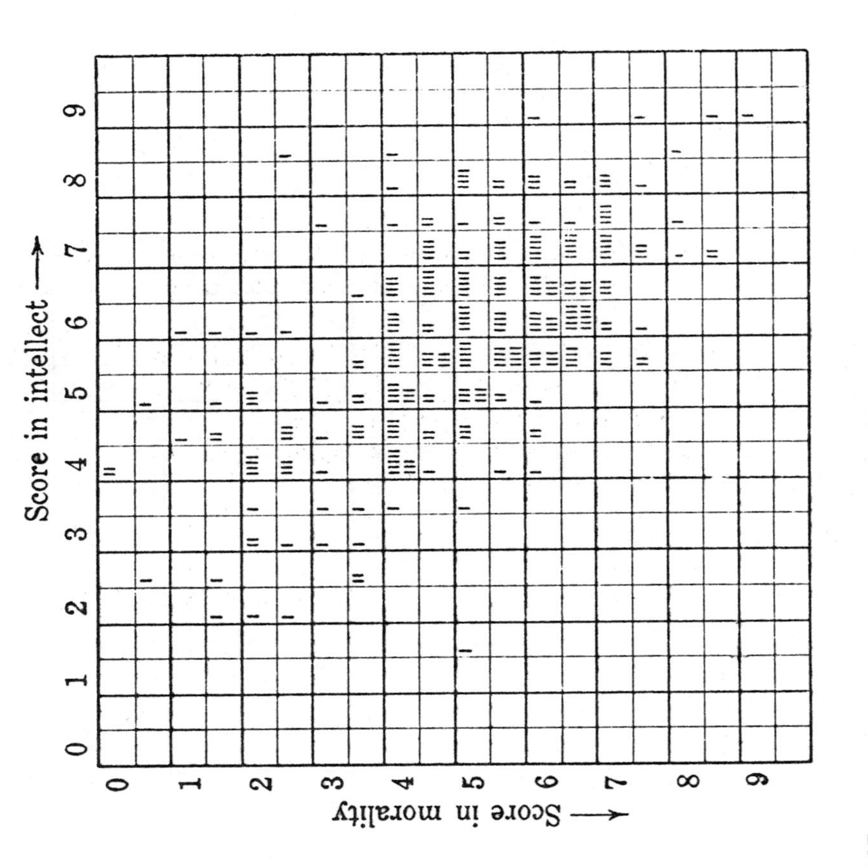

FIGURE 8.4 *(continued)* (D)

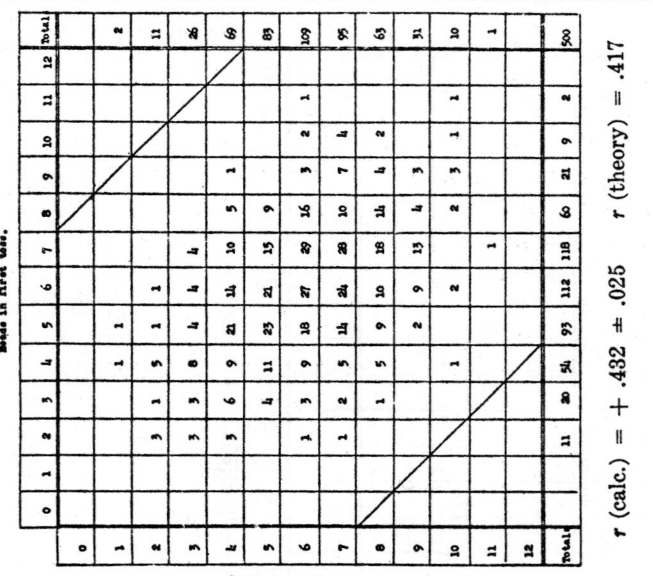

Heads in first toss	0	1	2	3	4	5	6	7	8	9	10	11	12	Total
12														
11							1				1			2
10							2	4	2		1			9
9					1		3	7	4	3	3			21
8					5	9	16	10	14	4	2			60
7				4	10	15	29	28	18	13		1		118
6			1	4	14	21	27	24	10	9	2			112
5		1	1	4	21	23	18	14	9	2				93
4		1	5	8	9	11	9	5	5		1			54
3			1	3	6	4	3	2	1					20
2			3	3	3		1	1						11
1														
0														
Total		2	11	26	69	83	109	95	65	31	10	1		500

Heads in second toss.

r (calc.) = + .432 ± .025 r (theory) = .417

$r = +.43$

N = 500 pairs of scores resulting from coin tossing

X = Number of heads showing in second toss of 12 pennies, in which 5 marked coins remain as they were in the first throw and the other 7 are tossed a second time. The heads on all 12 coins are counted.

Y = Number of heads showing in first toss of 12 pennies

Reference: Pearl, Raymond. *Introduction to medical biometry and statistics*. Philadelphia: Saunders, 1940, 3rd ed., p. 413.

Remarks: The diagonal lines cut off the cells in which events cannot possibly happen. For example, if all 12 coins had shown heads on the first toss, the fact that 5 of them remain on the table means that the second toss *must* have at least 5 heads; hence, scores of 0, 1, 2, 3, and 4 are impossible. The "r (theory) = .417" means that the population value of this correlation, computed by Formula (8.6), is .417. The computed value for these 500 tosses is .432, which is very close to this population value.

(E)

M.A. of Foster Mother	I.Q. of Foster Child 40–44	45	50	55	60	65	70	75	80	85	90	95	100	105	110	115	120	125	130	135	140	145	150	155	160
21–6													1	1		1	1	1							
21–0																									
20–6						1					1	2		1		1							1		
20–0												2	4		1	1	4	1							
19–6																									
19–0											1			1	2							1			
18–6										1	1		2	3	5	6			2						
18–0																									
17–6											1	2	3	3	5	3									
17–0									1	1	2	2	2	2	3	4		1							
16–6												2	1	2			1								
16–0												3	2	4		2	2		2					1	1
15–6									1					1		1									
15–0							1	1				2	3	2		1	1	1							
14–6										1	1		1	1		1		2							
14–0										1		2	2	1	1	1		2	1	2					
13–6							1	1			1	1		2	2		1			1					
13–0										1	1		4	1	2		1		1						
12–6												4			1		2								
12–0			1								2	2	2	1			1								
11–6	1										2		2	3		1									
11–0										1			1		1	2									
10–6										1	2	1			1		1		1						
10–0											1														
9–6												1			1										

$r = +.19$

N = 204 foster child and foster mother pairs

X = Stanford-Binet Intelligence Quotient of foster child

Y = Stanford-Binet Mental Age of foster mother

Reference: Burks, Barbara S. The relative influence of nature and nurture upon mental development: a comparative study of foster parent-foster child resemblance and true parent-true child resemblance. *Yearbook of the National Society for the Study of Education*, 1928, **27 (I)**, 315.

Age of Mother at Birth of Child	I.Q. of Child 0–9	10–19	20–29	30–39	40–49	50–59	60–69	70–79	80–89	90–99	100–109	110–19	120–29	130–39	Total	Mean
12–13							1			1					2	
14–15					1			2	1						4	75
16–17		1	1		2		5	6	7	4		2			28	
18–19		2	2	2	2	4	10	8	8	7	2			1	48	71
20–21	1		4	3	3	11	22	25	13	10	9	2			103	73
22–23		1	2	4	6	5	25	29	30	20	10	2		1	135	77
24–25	3		3	4	9	13	33	17	19	15	18	6	2		142	75
26–27	1	4	5	2	5	11	23	19	21	18	5	6		2	122	74
28–29	1	4	4	4	5	8	17	27	24	21	8	9	3		135	77
30–31		1	3	8	1	6	12	14	9	14	8	3			79	74
32–33			2	1	5	10	8	16	9	11	1	2	3		68	75
34–35		2	1	1	4	7	14	14	8	3	6		1		61	71
36–37			1	2	1	1	5	9	9	6	3	2			39	79
38–39		1	3	1	1	5	2	8	8	4	1	1			35	70
40–41	1		1		2		2	2	3	1	2				14	
42–43			2	1	4		2	1	2	2					14	
44–45				1	1		2		2	2				1	9	
46–47					1			1							2	67
48–49								2		1					3	
50–51					1										1	
52–53																
54–55						1									1	
Total	7	16	34	34	54	82	183	200	173	140	73	35	9	5	1,045	74

FIGURE 8.4 *(continued)* **(F)**

r = about zero

N = 1,045 children, most of whom came to the Institute for Juvenile Research (Chicago) because of some behavior problem

X = Age of mother at birth of child

Y = IQ of child

Reference: Thurstone, L. L., and Jenkins, Richard L. *Order of birth, parent-age and intelligence*. Chicago: University of Chicago Press, 1931, p. 46.

Remarks: Both from the size of r (practically zero) and the appearance of the scatter diagram, it is evident that there is no appreciable relationship between the two variables.

(G)

$r = -.20$

N = 234 pairs of scores resulting from coin tossing

X = Number of heads on 4 nickels and 16 pennies

Y = Number of TAILS on the same 4 nickels and of heads on 16 dimes

Source: All these scatter diagrams for negative correlation coefficients and the one for r = +1.00 are based on coins tossed by one of us. Because the coin tossing was done in Sawtell, New South Wales, the coins actually used were Australian rather than American (1977 data).

Remarks: The theoretical value of r, obtained from formula (8.6), is $-.20$, this obtained value of $-.20$ happening to coincide.

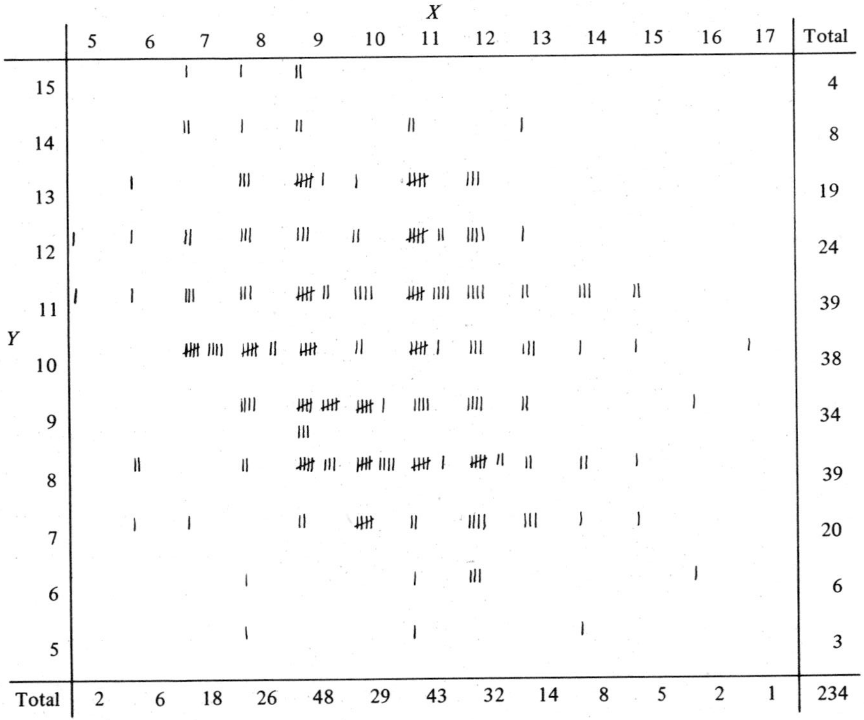

Y \ X	5	6	7	8	9	10	11	12	13	14	15	16	17	Total
15			|	|	||									4
14			||	|	||		||		|					8
13		|		|||	𝍸 |	|	𝍸	|||						19
12	|	|	||	|||	|||	||	𝍸 ||	||||	|					24
11	|	|	|||	|||	𝍸 ||	||||	𝍸 ||||	||||	||	|||	||			39
10			𝍸 ||||	𝍸 ||	𝍸	||	𝍸 |	|||	|||	|	|		|	38
9				||||	𝍸 𝍸 |||	𝍸 |	||||	||||	||			|		34
8		||		||	𝍸 |||	𝍸 ||||	𝍸 |	𝍸 ||	||	||	|			39
7		|	|		||	𝍸	||	||||	|||	|	|			20
6				|			|	|||				|		6
5				|			|			|				3
Total	2	6	18	26	48	29	43	32	14	8	5	2	1	234

FIGURE 8.4 (*continued*) **(H)**

$r = -.39$
N = 234 pairs of scores resulting from coin tossing
X = Number of heads on 8 nickels and 17 pennies
Y = Number of TAILS on the same 8 nickels and of heads on 8 dimes

Remarks: The value of r for this sample of 234 cases turns out to differ by only .01 from the population parameter, which is $-.40$.

Y \ X	6	7	8	9	10	11	12	13	14	15	16	17	18	Total
15				1										1
14														
13					2	1								3
12	1		1		1		2							5
11			1		2	4	1		1	1				10
10		1	1	4	3	5	8	7	4	3				36
9			3	6	8	2	6	5	7	5	3			45
8				3	7	3	12	9	6		3	4		47
7				2	3	5	8	6	8	5	2		1	40
6					2	4	6	4	3	3	6	2		30
5				1			1	1	2	2	3			10
4					1				3			1		5
3										1	1			2
Total	1	1	6	17	29	24	44	32	34	20	18	7	1	234

(I)

$r = -.60$

N = 222 pairs of scores resulting from coin tossing

X = Number of heads on 12 nickels and 8 pennies

Y = Number of TAILS on the same 12 nickels and of heads on 8 dimes

Remarks: This time the sample statistic agrees perfectly with the population parameter to the two recorded places. To four decimals, the population parameter is $-.6000$, while the sample statistic for these 222 cases is $-.5959$, a sampling error of only .0041. Notice that the scatter diagram shows a definite negative relationship.

Y \ X	4	5	6	7	8	9	10	11	12	13	14	15	16	Total
16			1	1										2
15		1		1	1									3
14	1	1	1	2	1	2	1							9
13				2	2	6	1	3	1		1			16
12		1	1	5	5	4	5	2		2				25
11	1		1	1	5	6	6	7	1					28
10				1	2	6	14	10	5	1	1			40
9				1	1	9	7	4	9	6				37
8					2	6	7	10	8	7				40
7						1		5	3	3	2	1		15
6						1			2	1				4
5												1	1	2
4												1		1
Total	2	3	4	14	19	41	41	41	29	20	4	3	1	222

FIGURE 8.4 (*continued*) **(J)**

$r = -.80$
N = 234 pairs of scores resulting from coin tossing
X = Number of heads on 16 nickels and 9 pennies
Y = Number of TAILS on the same 16 nickels

Remarks: Note the definite and marked change in the value of the means of successive rows as we go from one to the next. The same is true of the means of successive columns, of course, except where the number of cases is very small. This computed r of $-.80$ is identical to the parameter value of $-.80$.

Y \ X	4	5	6	7	8	9	10	11	12	13	14	15	16	17	18	19	20	21	Total
13	𝍷			𝍷															2
12				𝍷	𝍷𝍷𝍷	𝍷𝍷		𝍷	𝍷										8
11			𝍷		𝍷𝍷		𝍸𝍷		𝍷𝍷	𝍷									12
10					𝍷𝍷𝍷	𝍷𝍷𝍷𝍷	𝍸𝍸𝍷	𝍸𝍷𝍷	𝍸𝍷	𝍷𝍷	𝍷								34
9					𝍷	𝍷𝍷𝍷𝍷	𝍷𝍷	𝍸𝍸	𝍸𝍷𝍷𝍷	𝍸𝍸𝍷𝍷	𝍷𝍷𝍷								40
8							𝍷𝍷𝍷𝍷	𝍸𝍷	𝍸𝍸𝍸	𝍸𝍸𝍷𝍷	𝍸𝍷𝍷𝍷	𝍷𝍷𝍷	𝍷						44
7							𝍷	𝍷𝍷𝍷	𝍸𝍷𝍷	𝍸𝍷𝍷𝍷𝍷	𝍸𝍷𝍷𝍷𝍷	𝍷𝍷𝍷𝍷	𝍷	𝍷𝍷					36
6								𝍷	𝍷𝍷	𝍷𝍷𝍷	𝍸𝍷	𝍸𝍸	𝍸𝍷	𝍷𝍷	𝍷𝍷				32
5											𝍷𝍷𝍷	𝍷	𝍷𝍷𝍷𝍷	𝍷𝍷			𝍷		11
4											𝍷	𝍷𝍷		𝍷	𝍷𝍷				6
3														𝍷	𝍷				2
2																𝍷			1
1																		𝍷	1
Total	1		1	2	9	16	18	30	40	38	31	20	12	8	5	1	1	1	234

(K)

$r = -1.00$
N = 234 pairs of scores resulting from coin tossing
X = Number of heads on 8 nickels and 17 pennies
Y = Number of TAILS on the same 25 coins

Remarks: The X variable for this scatter diagram is the same one plotted as X on the scatter diagram for $r = -.39$. The Y variable is simply $25 - X$. When the parameter for r is either $+1.00$ or -1.00, any sample containing two or more different values of X will give the same value as the parameter. If there are no omissions, or if omissions are counted as wrong, then the number of items right on a test will correlate -1.00 with the number of items wrong on that test.

Y \ X	6	7	8	9	10	11	12	13	14	15	16	17	18	Total
19	|													1
18		|												1
17			卌 |											6
16				卌 卌 卌 ||										17
15					卌 卌 卌 卌 卌 ||||									29
14						卌 卌 卌 卌 ||||								24
13							卌 卌 卌 卌 卌 卌 卌 卌 ||||							44
12								卌 卌 卌 卌 卌 卌 ||						32
11									卌 卌 卌 卌 卌 卌 ||||					34
10										卌 卌 卌 卌				20
9											卌 卌 卌 |||			18
8												卌 ||		7
7													|	1
Total	1	1	6	17	29	24	44	32	34	20	18	7	1	234

have preferred different intervals or other score limits. However, in none of these cases are the methods that were used wrong. (If they had been wrong, the scatter diagram would not have been included here. Yes, that's right; some others *were* rejected for that reason.) Notice also that there is no reason for the number of intervals of the X variable (plotted in the horizontal direction) to be equal to the number of intervals for the Y variable (plotted in the vertical direction). If they are both between 15 and 20, as we suggested on page 18, or if there are even more intervals than this, r can be computed quite satisfactorily. Many printed correlation charts provide space for a few more than 20 intervals; and the experimenters usually try to use as many of these intervals as they reasonably can.

If one variable is being used to predict another, the first variable, usually called X, is plotted along the horizontal axis; and the other variable, usually called Y, is plotted along the vertical axis. Where the variables cannot be so classified, it is immaterial which variable is plotted in which place. Thus, in our earlier example, height may be used to predict weight, or weight may be used to predict height. Ordinarily we don't do either. If we know the heights of some people and we want to know their weights, we ask them or we measure them, since rulers and scales are readily available. Since there is no clear-cut test variable and no clear-cut criterion variable, we plotted height on the vertical scale, simply because most people (other than obstetricians and coroners) measure height that way.

Notice the shift in the position of the frequencies—at first rapid, then gradual, then rapid again—as we proceed by approximately equal units on the r scale from $+1.00$ through $.00$ to -1.00. Try to get an idea of the extent of the relationship for the various values of r, especially for the low positive ones, since those are the ones most often encountered in educational and psychological research.

What is the difference between a positive Pearson r and a negative one of the same size? First, let us note that some of these scatter diagrams have been rotated so that in all cases the high scores on one variable are plotted at the right and the high scores for the other variable are plotted at the top. This conforms to general mathematical practice, although some statisticians at times depart from this custom. If we were to reverse one of the scales (such as could be done by holding the page up to a mirror), the absolute value of the correlation coefficient would be the same, but its sign would change. Thus, if r had been $+.37$, it would change to $-.37$; if it had been $-.46$, it would become $+.46$. Of course, if both scales were reversed, the original sign would be unchanged. To illustrate this, look at the correlation of $+.56$ in Figure 8.4(C) and note that the tally marks run from the lower left to the upper right. Now turn the book upside down. Both scales are thereby reversed, but the tally marks still run from the lower left to the upper right and hence the sign is positive. The correlation is still +.56. Now compare this correlation of +.56 with the one of −.60 near the end at Figure 8.4(I). The relationship is just about equally close in both cases. In fact, if both r's had been exactly .60000 . . . 0, the degree of relationship would be exactly the same. The difference in sign simply means that: *when r is positive, high values of X and Y tend to go together, but when r is negative, high values of one variable tend to go with low values of the other variable.*

PUBLISHED CORRELATION COEFFICIENTS

The correlation coefficients given in Figure 8.4 are not typical of the correlations usually obtained in educational and psychological research. If we have a group of students take a good published test of intelligence or of school achievement, and then a few days later we retest these students with an alternate form of the same test, the correlation between the two test results will usually be around +.90 or higher. Such correlations simply tell us how accurately or how consistently the test is measuring whatever it does measure. These correlations are called *reliability coefficients*. For good tests, they are usually between +.90 and +.95 or higher for a group of students in a single grade.

Most other correlation coefficients are much lower. Although it is desirable to have higher correlations, and thus to make more accurate predictions, we seldom find tests that will correlate much higher than +.70 with academic success or +.40 with any criterion of industrial success. Negative correlations, except near zero, are also rare—partly because desirable traits tend to be positively correlated and partly because many tests are so scored as to produce positive rather than negative correlations with other variables. As a result, probably more than nine-tenths of all correlation coefficients that are reported in the psychological and educational literature lie between −.20 and +.60, even though these limits include less than half of all possible correlation coefficients.

How large are the correlation coefficients among many common variables? As stated, this question admits of no simple answer. There are too many common variables. Besides, the same two variables have sometimes been correlated hundreds or (in the case of intelligence and college grades) thousands of times.

Table 8.5 has been prepared to show correlations of various sizes, as reported in the educational and psychological literature. These Pearson r's were selected to meet several criteria:

1. Each r is based on an N of 200 or more.
2. Duplicate r's involving the same two variables were rejected.

TABLE 8.5
Selected Correlation Coefficients, Arranged from High to Low

r	Subjects	Variables	Reference	Remarks
+.993	206 eighth grade students instructed to answer the same test in two ways at the same time.	Number of right answers in parentheses on Otis Quick-Scoring Mental Ability Test, Beta Test. Number of right answers underlined on Otis Quick-Scoring Mental Ability Test, Beta Test.	Dunlap, Jack W. Problems arising from the use of a separate answer sheet. *Journal of Psychology,* 1940, **10,** 3–48.	
+.97	6,984 students in grades 6–9 in SRA achievement standardization sample of 1965.	Vocabulary subtest. Total Reading Composite score (which contains vocabulary score).	Staff, Science Research Associates. *SRA Achievement Series: Technical Report for Forms C and D.* Chicago: SRA, 1968.	This may be regarded as a *spuriously* high correlation, since one variable is contained in the other
+.92	204 college seniors and graduate students tested at 9 institutions.	Miller Analogies Test, Form L (taken first). Miller Analogies Test, Form M (taken second).	Psychological Corporation. *Manual for the Miller Analogies Test.* New York: Psychological Corp., 1970.	Corresponding *r* for 216 students, when Form M was taken first, was also +.92. When the two groups were combined, *r* became +.90.
+.91	1,414 present employees and applicants for a wide variety of jobs at several hundred companies throughout the country.	Form A of Visual Speed and Accuracy Test. Form B of Visual Speed and Accuracy Test (same session).	Ruch, Floyd L., and Ruch, William W. *Employee Aptitude Survey: Technical Report.* Los Angeles: Psychological Services, 1963.	This kind of correlation is known as an *alternate form reliability coefficient.*

+.91	221 children referred to a clinic.	Stanford–Binet IQ on first test. Stanford–Binet IQ on retest taken less than a year later.	Brown, Andrew W. The change in intelligence quotients in behavior problem children. *Journal of Educational Psychology,* 1930, **21,** 341–350.	
+.85	1709 students in Colorado.	Lorge–Thorndike verbal IQ in grade 9. Lorge–Thorndike verbal IQ in grade 11.	Hopkins, Kenneth D., and Bracht, Glenn H. Ten-year stability of verbal and nonverbal IQ scores. *American Educational Research Journal,* 1975, **12,** 469–477.	See reference for r = +.76.
+.83	920 students, constituting a 10% random sample of those tested in the National College Freshman Testing Program in 1957.	Verbal subtest of the School and College Ability Tests. Reading Comprehension subtest of the Sequential Tests of Educational Progress.	Fortna, Richard O. A factor-analytic study of the Cooperative School and College Ability Tests and Sequential Tests of Educational Progress. *Journal of Experimental Education,* 1963, **32,** 187–190.	
+.83	436 selected science students in a nationwide fellowship project.	Miller Analogies Test. Graduate Record Examination, Verbal Score.	Psychological Corporation. *Manual for the Miller Analogies Test.* New York: Psychological Corp., 1970.	

TABLE 8.5 (*continued*)

r	Subjects	Variables	Reference	Remarks
+.78	265 colleges and universities for which the needed data were available.	An estimate of Ph.D. output based on the sex, major fields, and intelligence of the students in each institution. Actual number of Ph.D.'s received by each institution's graduates about 6 years after graduation.	Astin, Alexander W. "Productivity" of undergraduate institutions. *Science*, 1962, **136,** 129–135.	
+.76	203 college freshmen.	Engineering interest on Strong Vocational Interest Blank. Engineering interest on Strong Vocational Interest Blank 19 years later.	Strong, Edward K., Jr. Nineteen-year followup of engineer interests. *Journal of Applied Psychology*, 1952, **36,** 65–74.	Note the high r indicating very stable interest scores for a long time.
+.76	1709 students in Colorado.	Lorge–Thorndike nonverbal IQ in grade 9. Lorge–Thorndike nonverbal IQ in grade 11.	Hopkins, Kenneth D., and Bracht, Glenn H. Ten-year stability of verbal and nonverbal IQ scores. *American Educational Research Journal*, 1975, **12,** 469–477.	See reference for r = +.85.
+.68	223 sixth grade pupils in the Metropolitan New York area.	Objective report card grades (in terms of grade level norms) in reading, arithmetic, and spelling. Subjective report card grades (in terms of individual pupil ability) in reading, arith-	Halliwell, Joseph W., and Robitaille, Joseph P. The relationship between theory and practice in a dual reporting program. *Journal of Educational Research*, 1963, **57,** 137–	The scatter diagram is published in the article.

	socioeconomic level (mean IQs at two testings were 114 and 117).	age 3. 1916 Stanford–Binet IQ at age 5.	stancy of the Stanford–Binet IQ from three to five years. *Journal of Psychology,* 1941, **12,** 159–181.	
+.60	271 kindergarten children who were later rated by first grade teachers and by testers in the second grade.	Combined ratings by teachers and by testers on items comprising the confident-friendly vs. anxious-withdrawn factor in kindergarten. Similar ratings on same factor in late first or early second grade.	Kohn, Martin, & Rosman, Bernice L. Cross-situational and longitudinal stability of social-emotional functioning in young children. *Child Development,* 1973, **44,** 721–727.	
+.59	1,060 males, 18 years and older, individually tested in their residences by interviewers.	20-word vocabulary test, covering a very wide range and having a reliability of .85. 12-item arithmetic reasoning test, covering a very wide range and having a reliability of .83.	Hagen, Elizabeth P. and Thorndike, Robert L. Normative test data for adult males obtained by house-to-house testing. *Journal of Educational Psychology,* 1955, **46,** 207–216.	
+.54	472 students who entered Northern Ireland grammar schools (roughly equivalent to U.S. junior high and high schools).	Arithmetic examination of the traditional type of a high standard of difficulty. Junior certificate (an equally weighted average mark in 8 subjects: English, Latin, French, History, Geography, Arithmetic, Experimental Science, and Algebra and Geometry combined) three years later.	Wrigley, Jack. The relative efficiency of intelligence and attainment tests as predictors of success in grammar schools. *British Journal of Educational Psychology,* 1955, **25,** 107–116.	The corresponding validity for an English Composition test was +.43.

TABLE 8.5 (*continued*)

r	Subjects	Variables	Reference	Remarks
+.53	361 cadets entering West Point.	Revised Minnesota Paper Form Board Test. Course grades in Military Topography and Graphics.	Psychological Corporation. *Manual for the Revised Minnesota Paper Form Board Test.* New York: Psychological Corp., 1970.	This is the highest of 27 validity coefficients, each based on an *N* of over 100, given in the Manual.
+.49	232 native-born father-child pairs in New Hampshire, Vermont, and Massachusetts, the children being 3½ to 13 years old.	Father's Army Alpha test score converted into approximate standard score for his age group. Child's Stanford-Binet score (1916 revision) converted into approximate standard score for his age group.	Conrad, H. S., and Jones, H. E. A second study of familial resemblance in intelligence: Environmental and genetic implications of parent-child and sibling correlations in the total sample. *Yearbook of the National Society for the Study of Education,* 1940, **39** (II), 97–141.	The corresponding correlation for 269 mother-child pairs was also +.49.
+.42	581 college students.	Iowa Tests of Basic Skills administered in grade 4 (about 9 years earlier). Freshman college grade point average.	Scannel, Dale P. Prediction of college success from elementary and secondary school performance. *Journal of Educational Psychology,* 1960, **51,** 130–134.	

	dents in public high schools.	musical groups, clubs, school publications, etc. participated in. Grade average relative to other students in same class.	son, James D., Hofmann, Gerhard, & Brown, William R. Determinants of students' interracial behavior and opinion change. *Sociology of Education,* 1977, **50,** 55–75.	about 1600 black students was +.19.
+.41	200 borrowers from a personal finance company.	Borrower's number of creditors. Criterion of consumer credit risk (bad = high score).	Muchinsky, Paul M. A comparison of three analytic techniques in predicting consumer installment credit risk: toward understanding a construct. *Personnel Psychology,* 1975, **28,** 511–524.	This is the highest of 8 correlations with the criterion. The article gives data for 2 more finance companies. See reference for $r = -.27$.
+.39	643 college students.	Attitude toward evolution. Attitude toward birth control.	Ferguson, Leonard W. A revision of the primary social attitude scales. *Journal of Psychology,* 1944, **17,** 229–241.	.39 = mean of 4 r's (2 forms of each scale) of .38, .38, .41, and .40.
+.36	209 second grade boys.	Social class (one background demographic variable). Reading scale of Metropolitan Achievement Tests.	Kohn, Martin, and Rosman, Bernice L. Social-emotional, cognitive, and demographic determinants of poor school achievement: Implications for a strategy of intervention. *Journal of Educational Psychology,* 1974, **66,** 267–276.	

TABLE 8.5 (*continued*)

r	Subjects	Variables	Reference	Remarks
+.33	208 children in the Guidance Study at the University of California Institute of Child Welfare.	Mother's education. Stanford-Binet IQ (1916 revision) at age 7.	Honzik, Marjorie P. Developmental studies of parent-child resemblance in intelligence. *Child Development,* 1957, **28,** 215–228.	The corresponding r with father's education was +.40.
+.31	300 twelfth grade students—100 each from large, intermediate, and small high schools.	Attitude toward marijuana, as measured by semantic differential *evaluation* dimension. A test of general knowledge of drugs.	Whiddon, Thomas, & Halpin, Gerald. Relationships between drug knowledge and drug attitudes for students in large, intermediate, and small schools. *Research Quarterly,* 1977, **48,** 191–195.	This was the highest of the r's with 7 drugs.
+.26	2,070 Ph.D. members of American Sociological Assn.	Publication output measured by points assigned to articles and books. Composite index of quality of graduate departments of sociology.	Clemente, Frank, and Sturgis, Richard B. Quality of department of doctoral training and research productivity. *Sociology of Education,* 1974, **47,** 287–299.	See reference for r = −.13.
+.26	1042 high school girls returning a questionnaire (response rate was 46%).	Academic grade average. Teacher's educational encouragement.	Snyder, Eldon E., & Spreitzer, Elmer. Participation in sport as related to educational expectations among high school	All information was supplied by the girls. See reference for r = +.19.

			girls. *Sociology of Education,* 1977, **50,** 47–55.	
+.25	239 cadets, members of the first class to enter the Air Force Academy (Class of 1959).	Academic composite, a weighted average of first-year grades in Chemistry, English, Geography, Graphics, History, Mathematics, and Philosophy. Aptitude for commissioned service, based on ratings by each cadet's peers and by his superiors.	Krumboltz, John D., & Christal, Raymond E. Predictive validities for first-year criteria at the Air Force Academy. Air Force Personnel & Training Research Center. *AFPTRC-TN-57–95,* July, 1957. Pp. v + 6.	
+.23	243 male employees of a bank in a large midwestern city.	Concern of the bank for social and environmental problems. General satisfaction with the job.	Gavin, James F., & Maynard, William S. Perceptions of corporate social responsibility. *Personnel Psychology,* 1975, **28,** 377–387.	Corresponding r for 417 female employees was +.43.
+.19	1042 high school girls returning a questionnaire (response rate was 46%).	Academic grade average. Mother's educational encouragement.	Snyder, Eldon E., and Spreitzer, Elmer. Participation in sport as related to educational expectations among high school girls. *Sociology of Education,* 1977, **50,** 47–55.	All information was supplied by the girls. See reference for $r = +.26$.
+.17	8,965 high school students taking one of the three versions of a biology course designed by the Biological Sciences Curriculum Study.	Salary of the high school biology teacher. Comprehensive final examination in biology.	Grobman, Hulda. Student performance in new high school biology programs. *Science,* 1964, **143,** 265–266.	

TABLE 8.5 (*continued*)

r	Subjects	Variables	Reference	Remarks
+.14	250 college freshmen with high school chemistry but not majoring in chemistry or chemical engineering.	Cooperative English Test (Form OM). Space relations scale of Differential Aptitude Tests (Form A).	Bennett, George K., Seashore, Harold G., & Wesman, Alexander G. *Fourth Edition Manual for the Differential Aptitude Tests* (Forms L and M), New York: Psychological Corp., 1966.	This is an example of a *discriminant validity coefficient*. The correlation is expected to be relatively low.
+.13	225 Air Force cadets.	Sentence-completion test of 24 incomplete sentences, scored on the dimension of superior-subordinate orientation. Flying training grade.	Burwen, Leroy S., Campbell, Donald T., and Kidd, Jerry. The use of a sentence completion test in measuring attitudes toward superiors and subordinates. *Journal of Applied Psychology*, 1956, **40,** 248–250.	This is the highest one of 13 correlations between the test and 13 available reputational criterion measures obtained from grades, ratings, and nominations.
+.02	225 freshman dental students.	ACE L score. Quality points for all work taken in the first year of dental school.	Webb, Sam C. The prediction of achievement for first year dental students. *Educational and Psychological Measurement*, 1956, **16,** 543–548.	This was the lowest of 15 validity coefficients for this group. The same variable also gave the third lowest (+.16) of 13 validity coefficients for 247 students in three earlier years.
+.01	522 boys measured both at 12 and 14 years of age by the Harvard Growth	Percent increase in weight in 2-year period. Gain in arithmetic score in 2-	Dearborn, Walter F., and Rothney, John W. M. *Predicting the*	

			Cambridge: Sci-Art Publishers, 1941.
−.02	300 college-age subjects.	Belief in Afterlife Scale (Osarchuk and Tatz). Templer's Death Anxiety Scale.	Berman, Alan L., and Hays, James E. Relation between death anxiety, belief in afterlife, and locus of control. *Journal of Consulting and Clinical Psychology,* 1973, **41,** 318.
−.04	957 male schizophrenics released from 12 Veterans Administration hospitals.	Percentage of unemployment in the locality to which the subject went. Number of weeks worked in the 37 weeks after exit from hospital.	Gurel, Lee, & Lorei, Theodore W. The labor market and schizophrenics' posthospital employment. *Journal of Counseling Psychology,* 1973, **20,** 450–453.
−.09	1,354 children, most of whom came to the Institute for Juvenile Research (Chicago) because of some behavior problem.	Ordinal number of birth (1 = first born). Stanford–Binet IQ.	Thurstone, L. L., and Jenkins, Richard L. *Order of Birth, Parentage and Intelligence.* Chicago: University of Chicago Press, 1931.
−.10	270 U.S. cities with an International Association of Fire Fighters local union.	Median education level in the city. Entrance salary for uniformed fire fighters.	Ehrenberg, Ronald G. Municipal government structure, unionization and the wages of fire fighters. *Industrial and Labor Relations Review,* 1973, **27,** 36–48.

TABLE 8.5 (*continued*)

r	Subjects	Variables	Reference	Remarks
−.13	2,070 Ph.D. members of American Sociological Assn.	Age at Ph.D. Composite index of quality of graduate departments of sociology.	Clemente, Frank, and Sturgis, Richard B. Quality of department of doctoral training and research productivity. *Sociology of Education,* 1974, **47,** 287–299.	The negative sign means that high quality departments tend to grant Ph.D.'s at younger ages. See reference to r = +.26.
−.13	200 high school freshmen and sophomores.	Total score on Survey of Study Habits and Attitudes. Score on Courtship, Sex and Marriage subtest of Mooney Problem Check List.	Burdt, Eric T., Palisi, Anthony T., and Ruzicka, Mary F. Relationship between adolescents' personal concerns and their study habits. *Education* (Chula Vista, Calif.), 1977, **97,** 302–304.	This was the smallest (closest to zero) of the 11 r's with subtests of the Mooney Problem Check List. See reference for r = −.38.
−.16	208 executives in a large chain grocery.	Score on test variable measuring Social Introversion-Extraversion. Job performance rating.	Guilford, Joan S. Temperament traits of executives and supervisors measured by the Guilford personality inventories. *Journal of Applied Psychology,* 1952, **36,** 228–233.	Negative correlation indicates positive relationship between criterion and sociability.
−.22	339 college freshmen.	Extent of belief (indicated on a 21-point scale) in the power of prayer. Scholastic Aptitude Test score.	Salter, Charles, A., & Routledge, Lewis M. Intelligence and belief in the supernatural. *Psychological Reports,*	This is the largest of 7 negative correlations between intelligence (SAT) and 7 supernatural beliefs.

−.23	268 cadets in a Naval School of Pre-Flight.	Authoritarianism as measured by the F-scale. Leadership scores based on leadership nomination forms filled out by other cadets in their section.	Hollander, E. P. Authoritarianism and leadership choice in a military setting. *Journal of Abnormal and Social Psychology,* 1954, **49,** 365–370.	
−.27	200 borrowers from a personal finance company.	Borrower's gross monthly salary. Criterion of consumer credit risk (bad = high score).	Muchinsky, Paul M. A comparison of three analytic techniques in predicting consumer installment credit risk: toward understanding a construct. *Personnel Psychology,* 1975, **28,** 511–524.	This was the most negative of 8 correlations with the criterion. The article gives data for 2 more finance companies. See reference for $r = +.41$.
−.29	3206 students in grades 6–12 in Maryland.	Quality of School Life Scale, measuring attitudes about school, classwork, and teachers. Cutting school.	Epstein, Joyce L., and McPartland, James M. The concept and measurement of the quality of school life. *American Educational Research Journal,* 1976, **13,** 15–30.	See reference for $r = -.43$.
−.30	441 17-year-old high school boys.	Occupational Aspiration Scale—a measure of level of occupational aspiration. Number of high school agricultural courses taken.	Miller, I. W., & Haller, A. O. A measure of level of occupational aspiration. *Personnel and Guidance Journal,* 1964, **42,** 448–455.	This is the most negative of 32 correlations with the Occupational Aspiration Scale, the latter correlating +.50 with high school grade point average.

TABLE 8.5 (*continued*)

r	Subjects	Variables	Reference	Remarks
−.33	632 children in grades 6 to 8.	Size of family. IQ obtained from National Intelligence Test.	Chapman, J. Crosby, & Wiggins, D. M. Relation of family size to intelligence of offspring and socio-economic status of family. *Pedagogical Seminary,* 1925, **32,** 414–421.	
−.33	3,540 male college freshmen.	Omnibus Personality Inventory Estheticism Scale (indicating artistic interest). Omnibus Personality Inventory Practical Outlook Scale (indicating interest in facts and material possessions).	Heist, Paul, and Yonge, George. *Manual for the Omnibus Personality Inventory*. New York: Psychological Corp., 1968.	Corresponding r for 3,743 female college freshmen was −.26.
−.38	200 high school freshmen and sophomores.	Total score on Survey of Study Habits and Attitudes. Score on Adjustment to School Work subtest of Mooney Problem Check List.	Burdt, Eric T., Palisi, Anthony T., and Ruzicka, Mary F. Relationship between adolescents' personal concerns and their study habits. *Education* (Chula Vista, Calif.), 1977, **97,** 302–304.	This was the most negative of the 11 negative r's with subtests of the Mooney Problem Check List. See reference for $r = -.13$.
−.40	4,000 cases, data for which were published in	Ectomorphy, characterized by neural development.	Humphreys, Lloyd G. Characteristics of type	When scores were transformed to produce linear

	Sheldon, W. H. *The Varieties of Human Physique*.	Endomorphy, characterized by visceral development.	concepts with special reference to Sheldon's typology. *Psychological Bulletin*, 1957, **54**, 218–228.	regression and Sheppard's correction was used, r became −.44. See reference for r = −.58.
−.42	261 students constituting the senior class of a high school.	Authoritarianism as measured by the California F-Scale. Aesthetic scale of Allport, Vernon, and Lindzey's Study of Values.	Nolan, Edward G., Bram, Paula, and Tillman, Kenneth. Attitude formation in high school seniors: A study of values and attitudes. *Journal of Educational Research*, 1963, **57**, 185–188.	
−.43	3206 students in grades 6–12 in Maryland.	Quality of School Life Scale, measuring attitudes about school, classwork, and teachers. Anxiety about school.	Epstein, Joyce L., and McPartland, James M. The concept and measurement of the quality of school life. *American Educational Research Journal*, 1976, **13**, 15–30.	See reference for r = −.29.
−.50	260 Naval Aviation Cadets instructed to turn in papers as soon as they finished, but not told they were being timed.	Working time on Guilford–Zimmerman Temperament Survey, as estimated by ranking those in each group tested, and transmuting to normal curve deviation scores. Reading speed, obtained from test routinely administered to all cadets.	Voas, Robert B. Personality correlates of reading speed and the time required to complete questionnaires. *Psychological Reports*, 1957, **3**, 177–182.	Corresponding r for MMPI for same group was −.51.

TABLE 8.5 (*continued*)

r	Subjects	Variables	Reference	Remarks
−.51	341 graduates from a national sample of teacher-training institutions.	Pretest Score on *Measurement Competency Test* while a college senior. Gain scale based on pre- and post-tests of MCT.	Mayo, Samuel T. *Pre-Service Preparation of Teachers in Educational Measurement*. (Final Report of CRP Project No. 5-0807, funded by U.S. Office of Education.) Chicago: Loyola University of Chicago, 1967.	
−.58	4,000 cases, data for which were published in Sheldon, W. H. *The Varieties of Human Physique*.	Ectomorphy, characterized by neural development. Mesomorphy, characterized by skeletal and muscular development.	Humphreys, Lloyd G. Characteristics of type concepts with special reference to Sheldon's typology. *Psychological Bulletin*, 1957, **54,** 218–228.	When scores were transformed to produce linear regression and Sheppard's correction was used, r became −.61. See reference for r = −.40.
−.61	Over 2,000 newly born human babies.	Visual evoked response latency. Conceptual age at time of birth.	Engel, Rudolf, and Benson, R. C. Estimate of conceptional age by evoked response activity. *Biologia Neonatorum*, 1968, **12,** 201–213.	This negative r means that the more premature the infant, the longer was the visual evoked response latency.
−.62	260 first grade boys in public schools.	Time required to answer all 12 items of Kagan's Matching Familiar Figures test.	Hemry, Frances P. Effect of reinforcement conditions on a discrimi-	

		Total errors on same test.	nation learning task for impulsive versus reflective children. *Child Development,* 1973, **44,** 657–660.
−.69	280 U.S. cities of population over 30,000.	Average salary of high school teachers. Infant death rate.	Thorndike, E. L. *Your City*. New York: Harcourt, Brace, 1939.
−.71	223 secondary school and college students.	Worldmindedness scale, which is called a "Social Attitudes Questionnaire." Ethnocentrism as measured by an 11-item form of the E scale of the *California Public Opinion Scale*.	Sampson, Donald L., and Smith, Howard P. A scale to measure worldminded attitudes. *Journal of Social Psychology,* 1957, **45,** 99–106.
−.76	More than 1,500 elementary schools throughout the United States.	Sixth grade verbal achievement in predominantly black grade schools. Percent of black students in school.	Mosteller, Frederick, and Moynihan, Daniel P. *On Equality of Educational Opportunity: Papers Deriving from the Harvard University Faculty Seminar on the Coleman Report*. New York: Vintage Books, 1972.

3. No more than two r's were cited from any one article or book.
4. We tried to include three or four r's within each interval of .10 such as $r = +.30$ to $r = +.39$. We were unable to get all the negative ones we wanted. (Our lowest r is $-.76$.)

In assembling the final set of r's for inclusion in the table we, of course, tried to select variables that were interesting or well known, while at the same time retaining correlations that were typical, unexpected, or unusual. The latter are so noted in the table.

The results of this process of selection and rejection are shown in Table 8.5.

Near the top of the table, many of the correlations are reliability coefficients or are correlations between very similar tests. Then follow correlations among a wide variety of variables.

Differences in certain physical traits, such as height, are generally acknowledged to be hereditary, even though it is known that certain extreme diets do have an effect on these measurements. The correlation between height of father or mother and height of adult son or daughter is usually found to be about +.50, a value which, incidentally, has significance for geneticists. Similar correlations are found for other physical measurements. Also, the correlation between intelligence of father or mother and intelligence of son or daughter is usually found to be about +.50. (See the r of .49 in Table 8.5.) This similarity led some early investigators, by erroneous reasoning, to conclude that differences in intelligence are inherited. Later, other investigators saw the logical fallacy and, by still more erroneous reasoning, concluded that since differences in intelligence had not been shown to be inherited, they must be due to environmental differences. Today, scientists realize that both heredity and environment play a part, but there is some disagreement as to the relative importance of the two. The evidence is inconclusive, but studies of identical twins reared apart show that environment does make a difference, while studies of foster children adopted at a very early age indicate that, in the United States at least, intelligence is more influenced by differences in heredity than by differences in environment.

People interested in teachers' salaries may note the correlations of +.17 and −.69. The former indicates a low but positive relation between the teacher's salary and the pupil's test score. The other, a high negative correlation, is a good illustration of the fact that correlation does not indicate causation. A high infant mortality rate does not cause salaries to drop, nor does an increase in teachers' salaries cause the infant mortality rate to drop. Both salary changes and differences in mortality have more fundamental causes, related to the general welfare of the community and the desire of its people for better conditions.

A few of the negative correlations are, to some extent, spurious. For example, ectomorphy, endomorphy, and mesomorphy are measured relative to each other, and nobody can be high on all three of them. Nor is it possible to be low on all three. Other negative correlations merely indicate that some traits, such as worldmindedness and ethnocentrism, are negatively related to each other.

We could comment at length on many of the other correlation coefficients in Table 8.5. Instead, we shall ask you to examine them closely. Pay particular attention to their size and to the changes in the nature of the pairs of variables as you go down the scale from the highest to the lowest, or most negative, correlation coefficients.

SOME CHARACTERISTICS OF PEARSON r

Now that we have seen a great many correlation coefficients, it is appropriate that we should see one of the formulas for r and learn a little about it. There are over fifty formulas, all different, but all yielding either exactly or very close to the same numerical value for r. Except for Formula (8.6), which is not a general formula, we shall give only three formulas for r in this chapter*—one formula that *looks* simple, and two that are easy to use in most situations. The formula that looks simple is:

$$r = \frac{\Sigma z_1 z_2}{N - 1} = \frac{\Sigma z_x z_y}{N - 1}, \tag{8.1}$$

in which r is the Pearson product moment coefficient of correlation, z_1 or z_x is a standard score for the first or X variable, z_2 or z_y is the corresponding standard score for the second or Y variable, and N is the number of cases.

Only in very unusual circumstances is Formula (8.1) useful as a means of computing r. It is, however, the basic product moment formula. It is useful in understanding a little about the meaning of r, just as Formula (3.1) was helpful in illustrating the meaning of the standard deviation. If we were to compute r by means of Formula (8.1), we would first have to compute $\overline{X}$ and s for the X variable; we would then have to use Formula (3.5) to compute a standard score on the X variable for every one of the N cases (This could be quite a task!); we would need to repeat all these operations

*There are also a few more in Supplement 8.

for the Y variable; we would need to compute all N of the $z_x z_y$ products (paying attention to signs and decimals); and finally we would need to add all these products and substitute in Formula (8.1) to obtain r. There are no formulas for computing r that are truly fast, accurate, and easy for the beginner, but Formula (8.1) does not even come close to attaining these objectives. It does, however, show us some important characteristics of Pearson r. Let us see what some of these are:

1. *Pearson* r *is a product moment* r. In Physics or Mechanics, a moment is a measure of the tendency to produce motion about a point or axis. Thus, the first moment about zero as an origin is the mean, since $\frac{\Sigma X}{N} = \overline{X}$. Also the first moment about the mean is always zero, since $\frac{\Sigma x}{N} = 0$. If we raise scores to the second power, we get second moments, and thus $\frac{\Sigma x^2}{N}$ is the second moment about the mean. This is very similar to the formula for s^2, and if N is large, the two are practically identical. Hence, $\frac{\Sigma x^2}{N}$ is the second moment and it is essentially equal to $\frac{\Sigma x^2}{N-1}$ or s^2, which is the square of the standard deviation.* If a moment involves the product of two variables, it is called a product moment. Thus, the fact that the product moment, $\frac{\Sigma z_x z_y}{N-1}$, leads to r, has led statisticians to refer to r as the *Pearson product moment coefficient of correlation.*

2. r_{xy} *equals* r_{yx}. For any one person, we get the same product of scores regardless of whether we multiply z_x by z_y or whether we multiply z_y by z_x. (Mathematicians call this the commutative law, and grade school pupils learn that (7)(3) = (3)(7) when they learn the multiplication table.) If each product remains unchanged by such a reversal, it follows that if we should reverse all of the products, their total would remain the same and hence r_{yx} would equal r_{xy}.

---

*In rare instances we may have a population or, even though we have a sample, we may want to obtain the standard deviation *of our sample* rather than, as we have nearly always done, to obtain an estimate of the standard deviation in the population from which our sample was drawn. In such cases, it is appropriate to use $\frac{\Sigma x^2}{N}$ rather than $\frac{\Sigma x^2}{N-1}$ in computing the standard deviation of a finite population. If standard scores were computed from such standard deviations, then Formula (8.1) would have to have N in its denominator to give the correct value of r. Thus, Formula (8.1) *is* a product moment, not just an approximation to one.

3. *r can be either positive or negative.* Standard scores are deviations from the mean, expressed in units of the standard deviation. Since their sum must be zero, ordinarily about half of them will be positive and half negative. This will be true both for z_x and for z_y. If we arbitrarily pair most of the positive z_x values with positive z_y values, and if we also pair most of the negative z_x values with negative z_y scores, nearly all of the products will be positive, $\Sigma z_x z_y$ will be positive, and hence r will be positive. On the other hand, if we pair positive z_x with negative z_y scores, and negative z_x with positive z_y scores, most of the products will be negative, and so will r. In the preceding sentences, we have been assuming that the various products would, except for sign, be roughly the same size. Ordinarily that will not be strictly true. A simpler way to set up the data to produce either a positive or a negative r, as desired, is to pair off a relatively small number of the largest deviations, either pairing off those with identical signs (to produce a positive r) or those of opposite signs (to produce a negative r). Of course, any such pairings of scores are purely of theoretical or academic interest; in actual practice, we do not rearrange scores, but we compute r from the scores as they were paired together in nature.

4. *r is confined within the limits of -1.00 and $+1.00$.* If each person had exactly the same z_x score as his z_y score, Formula (8.1) would take its maximum value and become

$$r = \frac{\Sigma z_x z_y}{N - 1} = \frac{\Sigma z_x^2}{N - 1}.$$

The last expression we recognize as the square of the standard deviation of the z_x scores. But since we learned on page 64 that the standard deviation of standard scores is 1.00, the squared standard deviation of standard scores is also 1.00 and, under our assumption, r would be equal to $+1.00$. At the other extreme, if each person's z_y score were the negative of his z_x score, Formula (8.1) would become

$$r = \frac{\Sigma z_x(-z_x)}{N - 1} = \frac{\Sigma\,(-z_x^2)}{N - 1} = -\frac{\Sigma z_x^2}{N - 1}.$$

This is the negative of the preceding value, and hence r would become -1.00.

5. *r may be regarded as an arithmetic mean.* If we call the $z_x z_y$ products by some other name such as $X_7 = z_x z_y$, it becomes clear that r is the sum of X_7 divided by $N - 1$. That is, r is approximately the mean of X_7 and hence of the $z_x z_y$ products also. In fact, if the standard deviations are computed with N instead of $N - 1$ in the denominator, the footnote on page 230 shows that r is *exactly* the mean of such standard score products.

COMPUTING r WITHOUT PLOTTING THE DATA

Formula (8.1) is, as we have just seen, helpful in enabling us to see some of the characteristics of r, but it is a very poor formula for computing r. If we were to use it, we would become involved both in decimal places and in negative signs. A formula that avoids these two very fruitful sources of error is Formula (8.2). Barring rounding and computational errors, it will give exactly the same value of r as Formula (8.1) or any of the formulas for r in Supplement 8. If anything, it is more accurate than the others, since nothing is rounded (and hence there can be no rounding errors) until the last two steps—extracting a square root and then dividing. (If we carry the square root out to several decimals, we may achieve any accuracy we desire.) There are only two things wrong with the formula: (1) it involves the multiplication of fairly large numbers; and (2) it frightens people. Here it is; get scared!

$$r = \frac{N\Sigma XY - \Sigma X \Sigma Y}{\sqrt{[N\Sigma X^2 - (\Sigma X)^2][N\Sigma Y^2 - (\Sigma Y)^2]}}. \tag{8.2}$$

That is almost the longest formula in this book. It contains only one symbol not used before, Y, a symbol denoting a score on the second variable. In no earlier formula did we compute ΣXY, but ΣX was used in computing the mean; and both ΣX and ΣX^2 were used in computing the standard deviation. Since both X and Y are treated similarly in the formula, the only new item is ΣXY. It is computed by multiplying the X score by the Y score for each person and then adding the N such products. To illustrate this process, let us consider the scores shown in Table 8.6.

TABLE 8.6
Data on Imaginary Movie Stars

Name	Number of Films in Last Two Years	Popularity Rating
Anne Arnaz	6	6
Bing Baxter	9	13
Cecil Crosby	10	24
Dale De Mille	0	3
Eddie Evans	3	13
Farley Foy	0	15
Goldie Granger	3	6
Harry Hawn	11	19
Jack James	5	11
Kay Klugman	14	21
.	.	.
.	.	.
.	.	.

TABLE 8.7
Computing r for Imaginary Movie Stars

Name	X^2	X	XY	Y	Y^2
Anne Arnaz	36	6	36	6	36
Bing Baxter	81	9	117	13	169
Cecil Crosby	100	10	240	24	576
Dale De Mille	0	0	0	3	9
Eddie Evans	9	3	39	13	169
Farley Foy	0	0	0	15	225
Goldie Granger	9	3	18	6	36
Harry Hawn	121	11	209	19	361
Jack James	25	5	55	11	121
Kay Klugman	196	14	294	21	441
.	.	.	.	.	.
.	.	.	.	.	.
.	.	.	.	.	.

We could add another column to Table 8.6 and put the products in it. However, since we shall also need the sums of the squares, as well as the sums of the scores themselves, it is better (if calculators are not available and if we do not wish to plot the data) to alter the setup slightly and proceed as indicated in Table 8.7.

Notice the very convenient arrangement of the data in Table 8.7. The X and Y columns are spaced far apart, and the appropriate scores are listed. Between these two, we insert the XY column and record the products. The X^2 column is then placed at the left, next to the X column, while the Y^2 column is placed at the extreme right, next to the Y column. The appropriate squares are then written in these columns. The next step, of course, is to add up each of the five columns. Because Table 8.7 is incomplete, listing only the first 10 stars, we cannot do this. We can, however, compute subtotals for these 10 cases. From left to right, these are 577; 61; 1,008; 131; and 2,143. When we get records on the other movie stars, we can then get the five totals that, with N, are all we need to substitute in Formula (8.2) and get Pearson r. Let us try it now with some real records.

Table 8.8 gives the heights and weights of 107 male Seniors at the University of Florida. These records were taken from reports of physical examinations given by quite a number of different M.D.'s when the men were Juniors or, in some cases, Sophomores. Among other things, the various physicians were supposed to measure and weigh these men. As is often true of such data, the measurements were not made in as scientific and accurate a manner as we might have desired. To most of the examining physicians, the nearest inch seemed good enough. The great majority of the weights

TABLE 8.8
Weight and Height of Male Seniors at University of Florida

Initials	Weight X	Height Y	Initials	Weight X	Height Y	Initials	Weight X	Height Y
AJG	164	69¾	KRB	130	64	BJT	158	68¾
ALE	156	74¾	KGR	180	68	BLP	162	70
BWS	162½	68½	KDC	156	71	BRJ	135	70
BFD	145	69¼	KJH	152	71½	CWR	164	74
BJB	149	72¾	LAE	185	72	CGF	166½	71
BWH	157	71½	LSJ	147	67½	CRF	152	67
BE	143	67	LGD	182	69	EMI	151	71
BHE	131½	67¼	MKH	161½	71	FRM	157	73
BLD	177½	72¼	MTE	176	72	FFJ	155	69
CRS	173	68	MDJ	170	72½	FCT	124½	67¼
CCG	170	72	MJB	158	71	FLB	172	72
CDC	142	69¾	NHH	149	67½	FRG	177	70½
DMJ	150	69	ODE	148	68	GI	160	72
EE	155	68	OCI	182	71	JFD	162	69
EMF	147	66	PEL	174	73¾	JR	176	75¼
EHR	157	72	POC	163	73¼	KRA	182	71
FHL	159	70½	PMM	138	66	KKD	143	67
FHR	152	70	PPT	162	71¾	LAM	163	68
FGE	141¾	69¾	PDD	174	71½	LLL	172	69
FJG	150	66	RCD	134	67½	MHM	186	71
FJM	140	65	RCR	178	71¼	MBS	175	73
GSP	145	69½	SJW	175	67	MCW	160½	70½

GJW	180	73	SHM	155½	69½	NJC	151	68
GLL	166	71¼	SHG	165	69½	OFP	138	69
GJA	199	73	SHA	166	72½	ORJ	151	70½
GRT	177	69	TAW	183	71	PRE	123	64
HPS	155	70	TJE	149	69	PTP	136	72
HRT	172	69	TWR	164½	71½	SGB	147	72
HJG	180	72	VKM	199	71	SOM	157	70
HRW	166	73¾	WSD	148	67¼	SB	142	68¼
ILJ	161	67½	WTO	133½	68½	TCH	144	69
JDL	135	69	WJY	167	68	TLC	157	66¾
JLL	172	68	ARL	148	71	TLT	161½	68¼
KBM	144	68	BJO	159	67¾	VDM	160	70
KWJ	150	70	BDL	165½	71¼	WRJ	106	60¼
KFB	154	71¼	BJF	160	67			

seem to be recorded to the nearest pound, which is satisfactory. But it might be interesting to speculate on just why one man's weight was recorded as 141¾, and why the two heaviest were recorded as tied at 199 pounds. (Incidentally, sitting pulse rates were recorded for 102 of this group and every one of the readings was divisible by two, 88 of them being 72, 76, 80, or other numbers that are divisible by 4. What does that tell you about how the measurements were made?)

Despite the apparent inaccuracy of some of the measurements, most of the heights are probably within an inch of the correct value. It seems unreasonable to assume that a man reported as 70 inches is, in reality, 68 or 72 inches tall. Because of the coarseness of the data, it is pointless to attach much significance to fractions of inches or fractions of pounds. Instead, let us take heights only to the nearest inch, taking the even height when the recorded figure ends in ½. Thus, the first six heights will become 70, 75, 68, 69, 73, and 72. Next we get rid of the few fractional pounds in the same way. It would now be possible to set up a table similar to Table 8.7 and compute r. The first two rows would read:

X^2	X	XY	Y	Y^2
26,896	164	11,480	70	4,900
24,336	156	11,700	75	5,625

What's wrong with that? You're right; it's too much work. Thank heavens, there's an easier way: group the data and measure deviations from some arbitrary origin. In other words, use x' and y' scores instead of X and Y. Fortunately, $r_{xy} = r_{XY} = r_{x'y'}$ except for minor grouping and rounding errors, so we may work with whatever scores are most convenient.

Noting that the heights run from 60 to 75 inches, an obvious and easy way to code them would be to subtract 60 from each height. The height deviations would then run from 0 to 15, giving 16 intervals. The weights range from 106 to 199. The weights are thus all over 100; and they have a range of 94. An interval width of 5 seems to be the answer. If we subtract 100 pounds from each weight (merely by ignoring the 1 in every case) and divide the residue by 5, discarding all remainders, we will get a group of coded scores ranging from 1 to 19, giving 19 intervals.

Before going on, let us note that there are many other ways to get x' and y' scores. Instead of subtracting 60, we could have subtracted 68 inches. That would have taken more time in subtraction and would later have given us smaller numbers to work with, but it would also have given us negative numbers. Sometimes it's a matter of preference. Also, we could have subtracted 105 pounds instead of 100 before dividing. That would have resulted in smaller numbers without introducing negative numbers, but the numbers wouldn't have been much smaller, and it's much more work to subtract 105 than it is to subtract 100. In dividing, we could have taken the nearest quotient, instead of discarding all remainders. One method, applied consistently, is usually as accurate as the other; the one we suggested is easier. Both would give interval widths of 5 pounds. Our method is not only easier, but it also makes the lower limit of each interval a multiple of the interval width. Thus, the same x' score (of 12) corresponds to weights of 160, 161, 162, 163, and 164, just as would be the case if we plotted the weights with $i = 5$. In fact, most of the coding problems for correlation coefficients are identical with the questions involved in selecting an interval width before computing $\overline{X}$ or s. (An easy way to determine the interval width is to use Table A.)

When the weight and height data of Table 8.8 are listed to the nearest pound and inch and then coded by the equations $x' = \frac{1}{5}(X - 100)$ and $y' = Y - 60$, in both cases discarding all remainders, the figures shown in boldface type in Table 8.9 result. Check four of the values in the x' column and four in the y' column, to make sure that you see how the inches and pounds of Table 8.8 were transformed into the little numbers in Table 8.9 that are easy to work with.

What happens next can best be done on a calculator. With *any* calculator or electronic computer, the individual products shown in the x'^2, $x'y'$, and y'^2 columns need not be recorded. They are simultaneously computed and added in the machine, only the cumulative totals and the totals being recorded. Of course, many of you may not have electronic calculators or computers readily available. But with or without them, the procedure is essentially the same. The way to compute r from the basic x' and y' data (shown in boldface type in Table 8.9) is:

1. Square each x' deviation and record its square in the x'^2 column, keeping the numbers lined up so that later they can be added easily. Thus, in the first row of Table 8.9, $(12)^2$ is recorded as 144. (Supplement 3 gives some suggestions on how to square numbers.) However, for your convenience, here are the squares of all the coded scores listed in Table 8.9, plus a few more.

x'	1	2	3	4	5	6	7	8	9	10	11	12	13	14	15	16	17	18	19
x'^2	1	4	9	16	25	36	49	64	81	100	121	144	169	196	225	256	289	324	361

x'	20	21	22	23	24	25	26	27	28	29	30	31	32	33	34	35
x'^2	400	441	484	529	576	625	676	729	784	841	900	961	1024	1089	1156	1225

2. Multiply each x' by the corresponding y' value and record the product in the $x'y'$ column. In the first row, (12)(10) is recorded as 120 in the $x'y'$ column. The multiplication table, Table J, should be of great help here,

TABLE 8.9
Coded Weight and Height of Male Seniors

	x'^2	x'	$x'y'$	y'	y'^2
AJG	144	**12**	120	**10**	100
ALE	121	**11**	165	**15**	225
BWS	144	**12**	96	**8**	64
BFD	81	**9**	81	**9**	81
BJB	81	**9**	117	**13**	169
BWH	121	**11**	132	**12**	144
BE	64	**8**	56	**7**	49
BHE	36	**6**	42	**7**	49
BLD	225	**15**	180	**12**	144
CRS	196	**14**	112	**8**	64
CCG	196	**14**	168	**12**	144
CDC	64	**8**	80	**10**	100
DMJ	100	**10**	90	**9**	81
EE	121	**11**	88	**8**	64
EMF	81	**9**	54	**6**	36
EHR	121	**11**	132	**12**	144
FHL	121	**11**	110	**10**	100
FHR	100	**10**	100	**10**	100
FGE	64	**8**	80	**10**	100
FJG	100	**10**	60	**6**	36
FJM	64	**8**	40	**5**	25
GSP	81	**9**	90	**10**	100
GJW	256	**16**	208	**13**	169
GLL	169	**13**	143	**11**	121
GJA	361	**19**	247	**13**	169
Cumulative Total	3,212	**274**	2,791	**246**	2,578
GRT	225	**15**	135	**9**	81
HPS	121	**11**	110	**10**	100
HRT	196	**14**	126	**9**	81
HJG	256	**16**	192	**12**	144
HRW	169	**13**	182	**14**	196
ILJ	144	**12**	96	**8**	64
JDL	49	**7**	63	**9**	81
JLL	196	**14**	112	**8**	64
KBM	64	**8**	64	**8**	64
KWJ	100	**10**	100	**10**	100
KFB	100	**10**	110	**11**	121
KRB	36	**6**	24	**4**	16
KGR	256	**16**	128	**8**	64
KDC	121	**11**	121	**11**	121

TABLE 8.9 (*continued*)

	x'^2	x'	$x'y'$	y'	y'^2
KJH	100	**10**	120	**12**	144
LAE	289	**17**	204	**12**	144
LSJ	81	**9**	72	**8**	64
LGD	256	**16**	144	**9**	81
MKH	144	**12**	132	**11**	121
MTE	225	**15**	180	**12**	144
MDJ	196	**14**	168	**12**	144
MJB	121	**11**	121	**11**	121
NHH	81	**9**	72	**8**	64
ODE	81	**9**	72	**8**	64
OCI	256	**16**	176	**11**	121
Cumulative Total	7,075	**575**	5,815	**491**	5,087
PEL	196	**14**	196	**14**	196
POC	144	**12**	156	**13**	169
PMM	49	**7**	42	**6**	36
PPT	144	**12**	144	**12**	144
PDD	196	**14**	168	**12**	144
RCD	36	**6**	48	**8**	64
RCR	225	**15**	165	**11**	121
SJW	225	**15**	105	**7**	49
SHM	121	**11**	110	**10**	100
SHG	169	**13**	130	**10**	100
SHA	169	**13**	156	**12**	144
TAW	256	**16**	176	**11**	121
TJE	81	**9**	81	**9**	81
TWR	144	**12**	144	**12**	144
VKM	361	**19**	209	**11**	121
WSD	81	**9**	63	**7**	49
WTO	36	**6**	48	**8**	64
WJY	169	**13**	104	**8**	64
ARL	81	**9**	99	**11**	121
BJO	121	**11**	88	**8**	64
BDL	169	**13**	143	**11**	121
BJF	144	**12**	84	**7**	49
BJT	121	**11**	92	**9**	81
BLP	144	**12**	120	**10**	100
BRJ	49	**7**	70	**10**	100
Cumulative Total	10,706	**866**	8,763	**738**	7,634

TABLE 8.9 (*continued*)

	x'^2	x'	$x'y'$	y'	y'^2
CWR	144	**12**	168	**14**	196
CGF	169	**13**	143	**11**	121
CRF	100	**10**	70	**7**	49
EMI	100	**10**	110	**11**	121
FRM	121	**11**	143	**13**	169
FFJ	121	**11**	99	**9**	81
FCT	16	**4**	28	**7**	49
FLB	196	**14**	168	**12**	144
FRG	225	**15**	150	**10**	100
GI	144	**12**	144	**12**	144
JFD	144	**12**	108	**9**	81
JR	225	**15**	225	**15**	225
KRA	256	**16**	176	**11**	121
KKD	64	**8**	56	**7**	49
LAM	144	**12**	96	**8**	64
LLL	196	**14**	126	**9**	81
MHM	289	**17**	187	**11**	121
MBS	225	**15**	195	**13**	169
MCW	144	**12**	120	**10**	100
NJC	100	**10**	80	**8**	64
OFP	49	**7**	63	**9**	81
ORJ	100	**10**	100	**10**	100
PRE	16	**4**	16	**4**	16
PTP	49	**7**	84	**12**	144
SGB	81	**9**	108	**12**	144
Cumulative Total	14,124	**1,146**	11,726	**992**	10,368
SOM	121	**11**	110	**10**	100
SB	64	**8**	64	**8**	64
TCH	64	**8**	72	**9**	81
TLC	121	**11**	77	**7**	49
TLT	144	**12**	96	**8**	64
VDM	144	**12**	120	**10**	100
WRJ	1	**1**	0	**0**	0
Total	14,783	**1,209**	12,265	**1,044**	10,826

since all the $x'y'$ products ordinarily needed in correlation work are included, most of them in the top half of the first page of the table.

3. Square each y' and record the square in the y'^2 column at the extreme right. The first entry, for $(10)^2$, is recorded as 100.

4. Check all the squares in the x'^2 column.

5. Check all the products in the $x'y'$ column. Instead of multiplying x' by y', reverse the process and multiply y' by x'. If you looked the products up in Table J by finding x' at the left and y' at the top, reverse things and find x' at the top (or bottom) and y' at the left (or right). This will eliminate nearly all errors of reading from the wrong row of the table. It is very important that $x'y'$ be *thoroughly* checked. More errors are made here than in all four of the other columns.

6. Check all the squares in the y'^2 column.

7. Add the numbers in each of the five columns, recording cumulative totals for the first 25, first 50, first 75, etc., scores. The last cumulative total is the total for all N cases. These totals are shown in the very last row of Table 8.9. In the order in which they appear there, they are

$$\begin{aligned} \Sigma x'^2 &= 14{,}783, \\ \Sigma x' &= 1{,}209, \\ \Sigma x'y' &= 12{,}265, \\ \Sigma y' &= 1{,}044, \\ \Sigma y'^2 &= 10{,}826. \end{aligned}$$

8. Check these five summations, again changing the method as much as possible. If the original adding was down each column to the first cumulative total, then down each column to the next cumulative total, etc., add the first 25 x'^2 scores going up the column from the twenty-fifth (GJA's 361) to the first (AJG's 144) and check the cumulative total. Leave that in the machine and then go up the next section of the column from the fiftieth through the twenty-sixth score and thus check the second cumulative total for x'^2. Continue until $\Sigma x'^2$ is checked. Then check the other four columns in the same way. We are now ready to substitute N and the five summations in an appropriate formula for r.

9. As was pointed out earlier in this chapter, $r_{xy} = r_{XY} = r_{x'y'}$. Our formula for r in terms of x' and y' looks exactly like Formula (8.2) except that x' is substituted for X and y' is substituted for Y. It is

$$r = \frac{N\Sigma x'y' - \Sigma x'\Sigma y'}{\sqrt{[N\Sigma x'^2 - (\Sigma x')^2][N\Sigma y'^2 - (\Sigma y')^2]}}. \qquad (8.3)$$

We now take our known value of N and the five summations obtained and checked in steps 7 and 8, and we substitute them in this formula. This gives the correct value of r, as indicated below.

$$
\begin{aligned}
r &= \frac{(107)(12{,}265) - (1{,}209)(1{,}044)}{\sqrt{[(107)(14{,}783) - (1{,}209)^2][(107)(10{,}826) - (1{,}044)^2]}} \\
&= \frac{1{,}312{,}355 - 1{,}262{,}196}{\sqrt{(120{,}100)(68{,}446)}} \\
&= \frac{+50{,}159}{\sqrt{8{,}220{,}364{,}600}} \\
&= \frac{+50{,}159}{90{,}666} \\
&= +.55.
\end{aligned}
$$

10. Check the computation of r, including the copying of all numbers substituted in the formula. Do not confuse $\Sigma x'^2$ with $(\Sigma x')^2$. The square root should not be checked by extracting it a second time. First note that the square root must be intermediate in value between the two terms enclosed in parentheses. It will always be somewhat smaller than a value halfway between them. In this case, 90,666 is intermediate in value between 120,100 and 68,446. It is also a little less than 94,273, which is halfway between 120,100 and 68,446. Note that it is given to the nearest integer; decimals are rarely needed for square roots used in Formula (8.2) or (8.3). Now check the square root; 90,666. Using method 2 on page 74 of Supplement 3, we first square 90,666 (preferably on a calculator) and obtain 8,220,323,556. This is very close, but a little less than the 8,220,364,600 we started with. So we add 90,666 to our product, obtaining 8,220,414,222 which is larger than our target. Because our product changed from being too small to being too large, our 90,666 is the correct square root. Now check all the other computations, but perform them in a different order, wherever possible.

This completes the computation of Pearson r between the weight and height of 107 students. The r is +.55. You already have some idea of the meaning of this. Further information concerning the significance and interpretation of r is given near the end of this chapter.

In computing r, we did not plot the data; we coded the scores and did our computing directly from the coded scores. An alternative method of computing r requires that the data be plotted, but eliminates the time of coding the scores. Both methods are equally accurate; sometimes one takes less time, sometimes the other does. Many people, including us, like to see the correlation plot. Hence, it is often preferable to plot the data before computing r.

PLOTTING THE DATA AND COMPUTING r FROM THE SCATTER DIAGRAM

In using a scatter diagram as an aid in computing r, we must decide upon suitable intervals, lay out our scatter diagram (or use a printed correlation chart), plot the data, check the plotting, count the frequencies in each row and column, select arbitrary origins, and do some computing. We finally substitute in Formula (8.3) to obtain r. Let us now take the data we just used and compute r from a scatter diagram. Most of our intermediate figures will be different, but we should finally get exactly the same numerator, denominator, and r (and we will).

We will take as our basic data the weight and height records listed in Table 8.8. The steps needed to get from these 107 pairs of values to the r of $+.55$ are as follows:

1. Call one variable X and decide to plot it in the horizontal direction. Call the other variable Y and decide to plot it in the vertical direction. The criterion or dependent variable, if there is one, is usually designated as Y. Here we have no criterion, so we would designate our variables arbitrarily if they were not already labeled in Table 8.8.

2. Determine suitable intervals for each variable. The lowest and highest weights are 106 and 199. Following our usual procedure for determining the proper interval width, we see that if $i = 5$, we shall have 19 intervals, the lowest being 105–109 and the highest 195–199. Similarly, since the heights vary from 60 to 75 inches, we shall use $i = 1$ and have 16 intervals for our second variable. (Fractional inches and fractional pounds will, of course, be rounded to the nearest integer, for the reason given on page 236. If all heights had actually been recorded to the nearest ¼ inch, we would have set up our intervals as 60–60¾, 61–61¾, etc., but since most are given only to the nearest inch, this would be inappropriate.)

3. Get two pieces of paper, preferably graph paper with pale blue rulings about ¼ to ⅓-inch apart, and a piece of carbon paper. Line up the rulings on the two sheets with the carbon paper between them and fasten them with paper clips. Lay out a rectangle to include all possible scores, and label the intervals in X above the rectangle, and those in Y to the left of it. (Be sure to write clearly the lower limit of every interval, and the upper limits of the two lowest and of the two highest intervals. It is not necessary to write in the other upper limits, unless you feel they will help you in plotting the data.) Leave enough space at the right and below the rectangle for subsequent computations.

4. Separate the sheets and plot the data on the original. For example, the first student weighed 164 and was 69¾ inches tall. As noted in step 2, we

regard these scores as 164 and 70. Put a small tally mark near the left side of the cell found at the intersection of the 160–164 column and the 70 row. The next case, with scores of 156 and 74¾, is similarly plotted at the intersection of the 155–159 column and the 75 row. This is one of the two scores in the top row of the completed scatter diagram shown as Figure 8.10. Notice that each tally mark represents one person and is located so that it indicates his score on *both* variables. Continue until all 107 pairs of scores are plotted. *Be extremely careful to do your plotting accurately.* More errors are made in plotting than in any other phase of the computation of r by this method.

5. Check the plotting. Unfortunately, there is no satisfactory way to be sure of finding *all* plotting mistakes. Check by replotting the same data on the carbon copy of the scatter diagram. Compare the two plots. If you have used black ink, one way to do this is to put one plot on top of the other, hold

FIGURE 8.10
Correlation between Weight (X) and Height (Y), Based on Data from Table 8.8

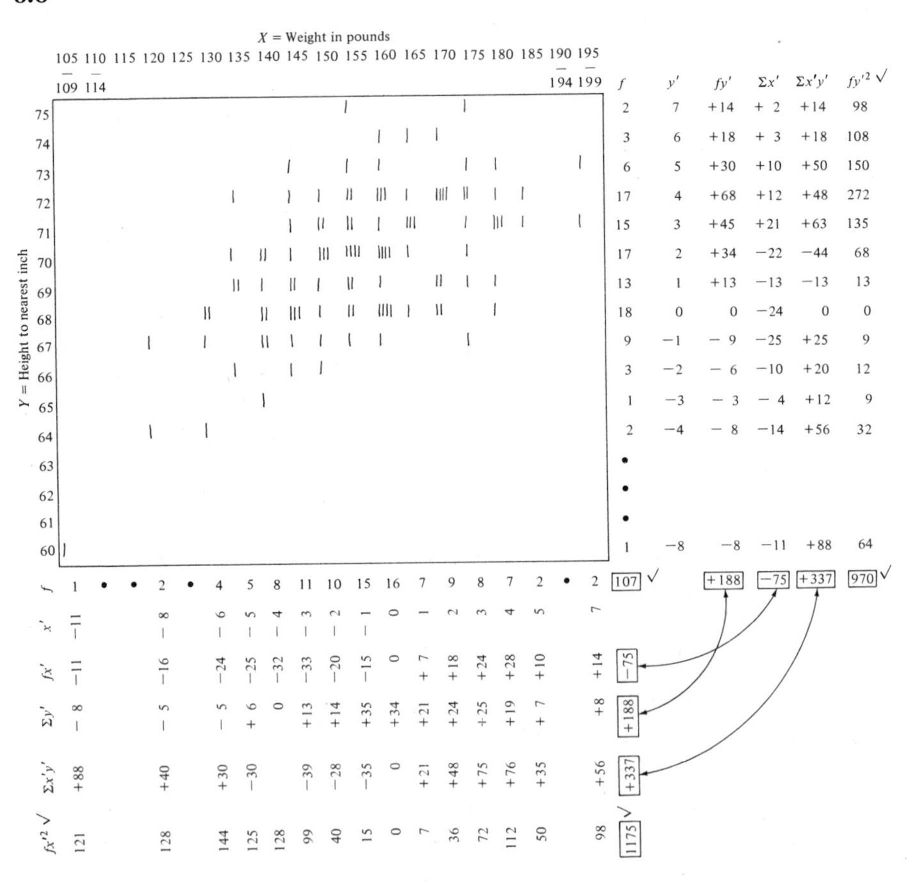

them in front of a strong light, and move the paper a little. If *all* tally marks agree perfectly, either both plots are right or, nasty as it may sound, you have made the same mistake both times. (It is quite possible for an inaccurate plotter to have two plots agree even though both are wrong.) If there is a discrepancy anywhere, plot *all* the data a third time. If this third plot agrees with either of the first two, it is probably correct. If it does not, keep plotting until you get 3 identical plots out of 5 attempts, or at least 4 identical plots out of 6 or more plottings. If there is any disagreement between two plots, *do not try to find and correct the mistake*. Here's why: If we know that at least one mistake has been made, and we do, we should realize that it is quite possible that there are others. If one of these others was made on both plots, it can be located only by replotting the entire set of scores. With a little practice, you can greatly improve your plotting accuracy if you wish to do so. Once you start, don't let anybody interrupt you.

6. Count the tallies in each row and record these frequencies in a column headed f immediately to the right of the ruled rectangle. The sum of these frequencies should equal N—in this case, 107. Write it at the foot of this column. Count the tallies in the columns and record these frequencies in an f row immediately below the ruled rectangle. The sum of these frequencies should also equal N. If it does, check the N. If it does not, find the mistake and correct it.

7(A). If you are going to use a *calculator*, label the next row x' and write 0, 1, 2, 3, . . . in it. Also label the next column y' and, beginning at the bottom, write 0, 1, 2, 3, . . . in it. All subsequent calculations will be positive until you substitute in the formula for r. As a result, you may forget about signs; you can easily add such things as fx' and fx'^2 simultaneously. You will need to record all the entries in the $\Sigma y'$ row and in the $\Sigma x'$ column, but elsewhere in steps 8 through 11 following, you need not record the individual products unless you wish to do so; just let them cumulate in the machine.

7(B). If you are *not* going to use a *calculator*, label the next row x' and select an arbitrary origin, just as was done in Chapter 2: Put 0 opposite the largest frequency (in this case, 16); and run the negative numbers to the left and the positive ones to the right. Write them down sideways, as illustrated in Figure 8.10. Also label the next column y' and select its arbitrary origin. Put 0 opposite the largest frequency (here, 18); and run the positive numbers up and the negative ones down from this arbitrary origin.

8. Label the next row fx' and record all the products of f multiplied by x'. Label the next column fy' and record all the $(f)(y')$ products in it. All products to the right of or above the arbitrary origin are positive. All products to the left of or below the arbitrary origin are negative. Be sure to

indicate all signs clearly and to line up the numbers so they will be easy to add.

9. Label the next row $\Sigma y'$. In each column, add the y' scores corresponding to all the tally marks in that column and record their sum in the $\Sigma y'$ row. (Hint: Onto the left edge of another piece of graph paper, copy the numbers in the y' column. Put horizontal lines just above the top number (7) and just below the bottom number (-8). These aid in aligning the paper. Now slide this paper alongside any column. Multiply the number of tally marks by y' and add. Do the same for all the other columns.) For example, the total for the column in which $x' = -2$ is $4 + 2(3) + 3(2) + 1 + 0 - 1 - 2 = +14$. Note that the y' score for every tally mark is added and that we must pay close attention to signs. Label the next column $\Sigma x'$. In each row, add the x' score corresponding to every tally mark in that row; record the total in the $\Sigma x'$ column.

10. We are through referring to the tally marks. Label the next row $\Sigma x'y'$. We are going to get the entries for this row by multiplying each $\Sigma y'$ by the corresponding x' (or x' by $\Sigma y'$, if that is easier). First record the sign of every one of the products. Then record the products, lining them up carefully so they will be easy to add. Similarly, label the next column $\Sigma x'y'$. First record the sign of each of the $(y')(\Sigma x')$ products. Then record the products themselves.

11. Label the next row fx'^2. To get the entries for this row, mentally square x' and then (mentally or otherwise) multiply this value of x'^2 by f. All products are positive, so there is no need to record signs. For example, for $x' = -11$, $x'^2 = 121$ and $fx'^2 = 121$. For $x' = -8$, $x'^2 = 64$, and since $f = 2$, $fx'^2 = 128$. Label the next column fy'^2. Mentally square each y' and multiply this square by the corresponding f.

12. You may have noted that we haven't been doing much checking of the computations that have been done subsequent to the plotting. We will now start checking these figures, starting with the last ones. To check the fx'^2 row, we multiply each number in the fx' row by the corresponding number in the x' row. (Since all signs will be positive, we may ignore them.) Thus, we multiply 11 by 11 to check the previously recorded 121; then $(16)(8) = 128$; $(24)(6) = 144$; etc. (Table J will be useful here.) If you find any discrepancies, recompute both ways and correct the figures. Then put a check mark after fx'^2. Similarly, check the fy'^2 column by multiplying each fy' by its y'. When this column is correct, put a check mark after fy'^2.

13. We now check all the columns and rows. We add every row except x' and we add every column except y'. Three of our five summations are

determined twice. $\Sigma x'$ should have the same value, whether calculated by adding the fx' row or the $\Sigma x'$ column. In this problem, both equal -75. Similarly, $\Sigma y'$ is computed both by adding the row so labeled and by adding the fy' column. Here, both sums are $+188$. The hardest sum to compute is $\Sigma x'y'$. The two computed values of it should agree. They do; both add to $+337$. If any of these three pairs of sums fail to agree, find the mistake or mistakes and make all necessary corrections.

14. Because every figure in the fx'^2 row and the fy'^2 column has been computed by two different methods, check $\Sigma x'^2$ and $\Sigma y'^2$ by adding twice, starting the addition first at one end of the row or column and then at the other. Put a check mark after these sums when you have checked them.

15. We are now ready to substitute our sums in Formula (8.3) for r. (Incidentally, some people would rewrite the formula for grouped data, inserting an f immediately after each summation sign. So long as we add, for example, all 107 of the x'^2 scores, it makes no difference whether we call the resulting total $\Sigma x'^2$ or $\Sigma fx'^2$. We feel that the f's are both unnecessary and confusing in a formula for r, but if it will make you happy to write in the f's, go ahead.) Take our known value of N and the five sums we have just checked and substitute them in Formula (8.3):

$$\begin{aligned} r &= \frac{N\Sigma x'y' - \Sigma x'\Sigma y'}{\sqrt{[N\Sigma x'^2 - (\Sigma x')^2][N\Sigma y'^2 - (\Sigma y')^2]}} \\ &= \frac{(107)(337) - (-75)(+188)}{\sqrt{[(107)(1175) - (-75)^2][(107)(970) - (+188)^2]}} \\ &= \frac{36{,}059 + 14{,}100}{\sqrt{(120{,}100)(68{,}446)}} \\ &= \frac{+50{,}159}{\sqrt{8{,}220{,}364{,}600}} \\ &= \frac{+50{,}159}{90{,}666} \\ &= +.55. \end{aligned}$$

16. Check the computation of r. First check the copying of N and the five sums from the correlation chart. Pay close attention to parentheses and brackets, noting for example that $\Sigma x'^2$ is quite different from $(\Sigma x')^2$. The three numerator terms other than N may each be either positive or negative, so watch the $+$ and $-$ signs carefully. In the denominator, each term within square brackets will always be a positive product minus a smaller (positive) square, ending up as a positive number. The denominator is the square root of the product of these two numbers; hence it must be intermediate in value between them. Check all computations, reversing the order in

which multiplications are carried out, and checking the square root by one of the four methods given in Supplement 3.

This completes the second computation of the Pearson product moment coefficient of correlation between the weight and height of 107 male seniors. At the end of this and the next chapter, we will give 15 ways of interpreting a correlation. Temporarily, one of the best of these ways is simply to say that an r of $+.55$ is the r that depicts the extent of the relationship when two variables are related as indicated by the tally marks in Figure 8.10.

It is worth noting that whether or not we use a scatter diagram, the computed value of r is not just about the same, but is absolutely identical in the two methods. None of the five summations are the same (unless the same arbitrary origins are selected, which is unlikely because when scores are not plotted, positive numbers are best; when scores are plotted, small numbers usually are best), but note that the numerators (+50,159) are identical in both solutions, so are both denominator terms, and, therefore, all subsequent computations. Both methods are accurate.

The next question is to determine whether or not an obtained r is merely a chance fluctuation from zero and, if not, to get some idea of about how far it may differ from the true correlation in the universe or population from which our sample was taken.

THE z′ TRANSFORMATION AND ITS STANDARD ERROR

Up till now, we have given a formula for the standard error of nearly every statistic we have presented. We shall now deviate from that pattern. It would be possible to give a formula for s_r. In fact, we shall do so in Supplement 8, but there is a better method. It sounds like going around in circles and, in fact, it almost is: We transform r into z'. We find $s_{z'}$. We add and subtract various multiples of $s_{z'}$ to z' and get some other values of z'. We transform these other values back into r's. Why do we do all of this? The reason is simply that successive values of r obtained from a large number of samples of, say, N cases are not normally distributed. But if all these r's are transformed into z''s, the z''s are normally distributed. Thus, we can use the normal probability curve when dealing with z', although it is not appropriate for use with r. Because we can very rapidly transform an r to a z' or a z' to an r, there is usually no particular reason for not using this more exact procedure.

The transformation equation may be written in several forms, but unless your teacher says otherwise, you need not memorize any of them.

$$
\begin{aligned}
z' &= \tanh^{-1} r \\
&= \frac{1}{2} \log_e \frac{1+r}{1-r} \\
&= \frac{1}{2} [\log_e (1+r) - \log_e (1-r)] \qquad (8.4) \\
&= 1.15129 \log_{10} \frac{1+r}{1-r} \\
&= 1.15129 [\log_{10}(1+r) - \log_{10}(1-r)] \\
&= \coth^{-1} \left(\frac{1}{r}\right) \\
&= r + \frac{1}{3}r^3 + \frac{1}{5}r^5 + \frac{1}{7}r^7 + \frac{1}{9}r^9 + \cdots .
\end{aligned}
$$

From the last of these equations, we can draw several conclusions: (1) z' and r always have the same sign. (2) When r equals zero, z' also equals zero. (3) Except when $r = z' = .00$, z' is always greater in absolute value than r. (4) When r is small, the cube, fifth power, etc. of r are very small indeed, and z' does not differ much from r. (5) Since $1 + \frac{1}{3} + \frac{1}{5} + \cdots$ is a divergent series, z' becomes infinite when $r = \pm 1.00$.

The formula for the standard error of z' is extremely simple. It is

$$s_{z'} = \frac{1}{\sqrt{N-3}}. \qquad (8.5)$$

Let us now see how z' and $s_{z'}$ are used in such a practical problem as the weight–height correlation we have just computed. We found that r (to 3 decimals)* was $+.553$ and that N was 107. We should like to set up confidence limits for this correlation coefficient, so that we can say that if the true r were within those limits, the chance would be only 1 in 100 of obtaining an r as far away from the true r as our r of $+.553$. Thus, we can get an idea of where the true r may reasonably be expected to lie.

First, we must find the z' corresponding to an r of $+.553$. The easiest way to do this is to use Table K and read the answer in the body of the table as

---

*This correlation was, following our usual 2-decimal rule, given to 2 decimals as $+.55$ when we computed it. It is customary to compute Pearson r's to 2 decimals, but now we need 3. As was pointed out on page 10, you may need more decimals to look something up in a table such as Table K, because that table allows for values of r to 3 decimal places. When we give our r and its confidence limits we will, of course, round the figures at the end of Table 8.11 to 2 decimals.

+.6227. Next, one might suppose that we should compute $s_{z'}$ and then multiply it by 2.58. This would be all right, but Table L not only gives the value of $s_{z'}$, but even multiplies it by 2.58 for us. For $N = 107$, this product is .2526. Taking $z' \pm 2.58\, s_{z'}$ or, in this case, $+.6227 \pm .2526$, will give the desired limits, which obviously are +.3701 and +.8753. But these are z' units. To transmute them back to r's, we find the values in the body of Table K that are closest to .3701 and .8753 and read the desired values of r from the side and top (or bottom) of the table. These values of r are +.354 and +.704. Thus, our .99 confidence limits for r, obtained with the aid of the z' transformation, are +.35 and +.70. Our result is consistent with the hypotheses that our obtained r of +.55 is a chance fluctuation from any true value between +.35 and +.70. But if the true correlation were outside these limits, then a correlation as extreme as ours could have arisen by chance less than once in a hundred times. Since this is unlikely, we reject all such hypotheses and usually add that we do so at the .01 level of significance.

But there are other confidence limits: $s_{z'}$ itself gives one of these since .6827, or roughly two-thirds, of the cases are included within the limits of $\pm 1\ s$ so the other .3173 must be beyond these limits, and $\pm 1\ s$ could be regarded as the 68% confidence limits. Still two other confidence limits are shown in Table L. Sometimes if we have no previous or *a priori* hypothesis, we may want to use all of them. It isn't much work and it sometimes gives us some useful information. Table 8.11 shows one way to set this up in systematic fashion.

The first two rows are, in effect, column headings serving to identify the figures that appear below them. The next row simply repeats the value of z' (.6227) corresponding to the given r (.553), as given in Table K. The four entries in the next row are the four figures in Table L immediately to the right of our given value of N (107). To obtain the next two rows, we first subtract this row from the preceding row and then we add it to that preceding row. These rows give us the lower and upper confidence limits

TABLE 8.11
Table for Determining Confidence Limits When $r = .553$ and $N = 107$

Confidence limits	.68	.95	.99	.999
k	1.00	1.96	2.58	3.29
z'	.6227	.6227	.6227	.6227
$ks_{z'}$	.0981	.1922	.2526	.3227
$z' - ks_{z'}$	.5246	.4305	.3701	.3000
$z' + ks_{z'}$	.7208	.8149	.8753	.9454
Lower confidence limit of r	.481	.406	.354	.291
Upper confidence limit of r	.617	.672	.704	.738

for z' as specified at the tops of the various columns. Finally we convert these eight z' values to r's and record them in the last two rows of the table. This is done by finding the figure in the body of Table K which is nearest to our z', then reading the corresponding r from the side and top or bottom of the table. For example our first value of .5246 is not in the body of the table, but the two adjacent figures of .5243 and .5256 do appear. The former is clearly closer, and since it is the z' for an r of .481, it is obvious that this r of .481 also corresponds to a z' of .5246. Hence .481 is recorded in the same column two lines below .5246 in Table 8.11.

The last two rows of Table 8.11 enable us to make the following interpretative statements.

1. We do not know the true r between weight and height for all such people as those included in our sample of 107 cases, but it is entirely reasonable that this r is between .48 and .62, since if it were any value between these limits, at least .32 of the time we would expect to get an obtained r further from the true r than our .55 is. The "at least" means this: If the true r were either .48 or .62, 32% of the time obtained r's would be further away than our own .55; if the true r were any intermediate value, more of them would be further away. In the extreme case of the true r being .55, practically all of the obtained r's would be further away than our .55.
2. The .95 confidence limits for r are .41 and .67. It is reasonable to expect the true r to lie within these limits because if it does not, then an r as extreme as our obtained .55 could have occurred only 5 times in 100. This is possible, but since the odds are 19 to 1 against it, we say that our r of .55 is significantly different (at the .05 level) from all r's below .41 or above .67.
3. Our obtained r of .55 is highly significantly different from all correlations outside the range of .35 to .70. If the true r had been either of these figures, a random sample of 107 cases would have given an r as far* away as our obtained .55 only once in 100 such experiments. Since this is highly unlikely, our r is highly significantly different from both of these values and from all more extreme values of r.
4. Only rarely are experimenters so conservative that they demand odds of greater than 99 to 1. The statistician who wishes to be extremely sure of a conclusion can set up confidence limits at the .999 level. If the true

*Technically speaking, distances from the true r to obtained r's are measured in terms of the z' scale, although it is more convenient to speak of them in terms of r. Notice that this range extends from an r .20 below .55 to one only .15 above it. The distances are, however, equal in terms of z'.

> r were .29 or .74, only once in 1000 times would a random sample yield an r as far away from the true value as our .55. The confidence limits for r, at the .999 level, are .29 and .74. Notice that these limits are far apart. We are quite positive that the weight–height correlation for such students is between .29 and .74, but this is not a very surprising conclusion. Nearly anyone would have felt that it would be positive and somewhere within this range before we ever computed it. If we want to make more definite statements, we must either use less extreme confidence limits or increase the size of our sample. Most people are quite willing to settle for statements based on confidence limits of .99.

We did not consider signs in the preceding discussion, since they were all positive. If r is lower or if $s_{z'}$ is larger, some of the figures in such a table as Table 8.11 may be negative. Careful attention will then need to be paid to all signs.

Sometimes we are not interested in confidence limits. We may merely wish to determine whether or not our obtained r is significantly different from zero. If this is our purpose, all we need to do is use Table K to transform our r into z' and then see whether this z' exceeds any of the values in Table L in the row corresponding to our N. Thus, if $r = .22$ and $N = 111$, we first transform r and obtain $z' = .2237$. Next we compare .2237 with the figures in the row corresponding to $N = 111$ in Table L. Since .2237 is greater than the .1886 in the .95 column, but less than the .2479 in the .99 column, our r of .22 is significantly different from zero at the .05 but not at the .01 or .001 levels.

ASSUMPTIONS UPON WHICH PEARSON r IS BASED

Arguments sometimes arise concerning the assumptions involved in computing the Pearson product moment coefficient of correlation. People sometimes say that a given r doesn't mean anything because some of the assumptions don't hold. The critics are usually wrong in such cases because very little is assumed in the derivation of r. Only two assumptions are made and one of them is assumed so commonly in other situations that we scarcely think of it as a restriction. Let us consider it first.

1. The units of measurement are equidistant throughout the range of scores involved in the correlation. This must, of course, be true for both variables. This assumption ordinarily offers no difficulties. It seems obvious that a difference in height from 4 feet 10 inches to 5 feet is the same as the difference from 5 feet 10 inches to 6 feet. We are also correct in assuming that a difference of 15 IQ points in a well-constructed test means the same thing, whether the change is from Jeff's IQ of 45 to Christine's 60 or from Jenny's IQ of 100 to Bobby's 115.

(Notice that we are referring to equivalence of scale units only—not to the importance of a 15-point change in terms of its psychological significance nor in terms of its contribution to society.) This assumption of equality of units is not so simple as it may appear, however. On the 1916 revision of the Stanford–Binet test, for example, a person over 16 years of age could not get an IQ above 122. Therefore, many brighter people were recorded as 122 and hence the difference between a recorded 122 and 112 was much greater than the same difference of 10 points anywhere else on the scale. Again, the difference in selling ability of a person who sells $600,000 and one who sells $800,000 of life insurance in one year is negligible, while there is a large difference between the people who sell $100,000 and those who sell $300,000 a year. (That is one reason why transformations such as those suggested at the end of Chapter 7 are sometimes used with such data.) Most psychological and educational records for which *r*s are computed have units that are sufficiently nearly equal to meet the requirements of this assumption. Incidentally, we make this same assumption in computing $\overline{X}$, s, and AD.

2. In computing r, we assume that regression is linear. *Linear regression* is a concept that will be clearer after the reader has finished the next chapter. In a scatter diagram, regression is linear if, except for chance fluctuations, the means of the successive columns lie on a straight line (and if the means of the successive rows lie on another straight line). If the line connecting the means of the columns (and the line connecting the means of the rows) does not deviate systematically from a straight line, we have linear regression. Such a line may have a slope which is positive, negative, or even zero. If each time the X score increases by 1 unit, the Y score increases (or decreases) by a constant amount, regression is linear. Thus, if each time we select a group of men who are one inch taller, we find that their average weight is 4 pounds higher than for the preceding group, the regression is linear. On the other hand, if the average weight of males increases about 8 pounds per year from age 7 to 17, but only 1 pound per year from age 25 to 45, and *decreases* after age 60, as it does, the regression is not linear and Pearson r is inapplicable. Except for age, other functions of time, and some variables involving money, most of the variables that educators and psychologists correlate show linear, or substantially linear, regression.

What happens if an investigator computes Pearson r between two variables that do not show linear regression? Different statisticians will give different answers to this question; our answer is, "Nothing much." Technically, of course, r is inapplicable and should not have been used. More realistically, the value shown by r is an underestimate; the relationship is at least as high as r indicates (actually, it is higher) and no one can accuse the scientist who uses r when he shouldn't of overestimating his results. For example, if a clinical psychologist reports a Pearson r of .40 between her test and a measure of marital happiness, it is still true that she can predict that criterion to the extent of an r of .40, regardless of whether or not regression is linear. If it is not linear, the correlation would have been higher if she had transformed one or both variables, or if she had taken account of the absence of linearity in some other way. Such a person may justly be accused of carelessness or ignorance of advanced statistics, but not of misrepresentation.

3. Pearson r does *not* assume normality. Many people believe that both variables must be normally distributed in order to justify the computation of r. But many people also believe that they can drive better after they have been drinking. Others believe that the average person can come out ahead by devoting a lifetime to betting on the ponies. The facts don't back up these beliefs. In some interpretations of r, we assume normal distributions within each column (or row), but in most interpretations, no such assumptions are made. Finally, no matter what the shape of the distribution of either variable, we are always justified in computing r to determine the strength of the (linear) relationship between the two variables.

4. Pearson r does *not* assume that an increase in the score on one variable is or is not related to an increase or a decrease on the other variable. Pearson r *measures* the strength or amount of the relationship *if it exists*. That is, we do not assume anything about either the direction or the amount of the relationship. Instead, we use r to determine the sign and size of the existing correlation, which may be either positive, zero, or negative.

To summarize, aside from the assumption of equality of units throughout the scale, the only other assumption needed to justify the computation of Pearson r is the assumption of linearity of regression, and even that assumption may be waived in some cases.

INTERPRETATION OF PEARSON r

There are a number of different ways in which a Pearson r of any given size may be interpreted. Although some interpretations have already been used, it seems desirable to make a formal listing of all fifteen of them. Some will be more readily understood after we know about regression lines, so they will follow in Chapter 9. The first eight methods of interpreting r now follow. The first two of these methods are somewhat different from the others; in testing agreement with hypotheses, they are concerned with chance fluctuations. All the other methods ignore or minimize the effect of chance and base their interpretation on the size of r, treating the obtained value in the sample as though it were a true value in the population. Here are the interpretations.

1. A correlation coefficient is significantly different* from zero if and only if its transformed value of z' exceeds the appropriate value shown in Table L, corresponding to the number of cases used in computing it. The reference to "appropriate value" is purposely vague. Some statisticians are quite willing to regard an r as significant if it exceeds the value shown in the column for the .05 level; others insist on the .01 level. There are some superconservative people who insist on the .001 level, or even on a value equal to 3.00 times its standard error (about the .0027 level if N is large, less significant if N is small). The safe thing to do is to state that the obtained coefficient of correlation is "not significant at the .05 level," "significant at the .05 but not at the .01 level," or "significant beyond the .01 level," as the case may be. Then no matter how liberal or conservative your readers are, they cannot quarrel with your interpretation. Let us clarify the meaning of these three categories and eliminate one common misconception. If r is not significant at the .05 level, we have failed to prove that the obtained r is not just a chance fluctuation from a true r of zero. Notice that if we obtain an r of .25 with an N of only 50, we have failed to show more than a chance relationship; but we have *not* proved that there is no correlation. The true correlation may be .00, but it may also be .25; in fact, it may be higher than .25 or it may even be a small negative correlation. Stated differently, we don't know much about it. The reason for this awkward situation is that we don't have enough cases; that is why N should usually be over 100 if we wish to be able to draw reliable conclusions. When r is significant at the .05 level, we know that if no actual correlation existed chance alone would produce an r that high only 5 times in 100. That's not very often, so, while such an r *may* be due to chance, it probably isn't. Hence, we usually conclude that there is more than a chance relationship between the two variables. Finally, when r is significant at the .01 level, the value is one that could be given by chance fluctuations from a true r of zero only once in a hundred times. This is very unlikely, so we conclude that our

---

*Going back to Table 8.11, notice that since the .95 confidence limits for our obtained r are .406 and .672, it follows that our obtained r is significantly different from all values outside these limits at the .05 level of significance. Thus, the same point, as .406, is the boundary both of the .95 confidence limits and also of the .05 level of significance. In this interpretation of r, we will use the figures in the .95 confidence limit column as though that column were labeled the .05 level of significance.

obtained r is highly significant. The correlation in the universe from which we selected our N cases isn't zero; its sign is the same as that of our obtained r. This method of interpretation is ideal if we are concerned merely with the presence or absence of relationship; if we wish to know the degree of the relationship, we should use the next method.

2. Closely related to the preceding method is that of testing the significance of the difference between r and some hypothetical value other than zero. Many scientists have obtained correlations of .50 between various physical measurements of fathers and corresponding measurements of their adult daughters. Suppose we correlate the knee height of the fathers with that of their daughters and, with 75 cases, obtain an r of .642. Now we want to know whether our finding differs significantly from the usual .50. We transform our r's of .500 and .642 into z''s of .5493 and .7616 by the use of Table K. The difference, .7616 − .5493, is .2123. Table L shows us that when N is 75, $s_{z'}$ will be .1179, and the significance values at the .05 and .01 level will be .2310 and .3036. Since the obtained difference is not significant, even at the .05 level, we are forced to conclude we have no evidence that our result is significantly different from the .50 obtained by other investigators. Thus our results are consistent with the hypothesis that knee height measures also give an r of .50. Notice that this does not prove that the correlation is .50, because this r of .642 is also consistent with the hypothesis that the true correlation between knee heights is .55, .60, .65, .70, or .75.

The two preceding interpretations differ in principle from those that follow. These first two interpretations are concerned with testing hypotheses, with determining whether the obtained correlations are chance fluctuations from fixed values. These interpretations of r are markedly affected by the number of cases upon which r is based. The remaining methods are concerned only with the degree of association or the strength of the relationship between two variables. If, for example, $r = .64$, the interpretation will be the same regardless of whether $N = 75$ or 75,000. Stated crudely, these methods take a correlation of a given size and interpret the correlation for what it *is*—not in terms of the probability that the correlation *might be* something else. That is, they ignore standard errors and deal only with the observed size of r.

3. If you want to know what an r of any given size looks like, find the one or two correlations closest to it in Figure 8.4 and look at the scatter diagrams. An examination of the scatter diagrams should go far toward clarifying the meaning of any Pearson r.
4. It is often very useful to compare our r with the r's that have been obtained by other investigators. As a person acquires experience with coefficients of correlation, their sizes take on new meaning, but past experience may be utilized, even when a person has had no previous

knowledge whatever of correlations as such. Thus, one of us once explained to a group of life insurance executives that a new set of tests would predict the success of their sales managers about as well as you can predict the height of an adult son from the height of his father. This statement is far more meaningful to such an audience than it would be to tell them that $r = +.50$. In an attempt to give just the background needed to achieve such understanding, Table 8.5 was prepared. It should be of considerable value in interpreting r, especially until your experience enables you to build up your own frame of reference.

5. Sometimes, r may be interpreted in terms of the number of common elements, if the scores on X and Y are determined by the presence or absence of a number of independent elements, each of which has the same effect on the score. For example, suppose we toss 4 pennies, 12 nickels, and 13 dimes, and count the number of heads. Let us do this 200 times and then determine 200 X and 200 Y scores that we will correlate. If $X =$ the number of heads on the pennies and nickels, and if $Y =$ the number of heads on the dimes and nickels, then

$$r = \frac{n_c}{\sqrt{n_a + n_c}\sqrt{n_b + n_c}}, \tag{8.6}$$

in which n_a is the number of factors (i.e., pennies) unique to X, n_b is the number of factors (i.e., dimes) unique to Y, and n_c is the number of factors (i.e., nickels) common to both X and Y. In this case

$$r = \frac{12}{\sqrt{4 + 12}\sqrt{13 + 12}} = \frac{12}{\sqrt{16}\sqrt{25}} = +.60.$$

Since we rarely know the elemental composition of the variables that we correlate, this method of interpretation is of very limited applicability. In using it we do not assume equality of units, linearity of regression, nor make any of the other assumptions we have been discussing. We must, however, assume that both X and Y scores are determined by a number of elemental factors, each with the same probability of being present, each having the same effect as any other element when it is present, and each having a correlation of zero with all the others. Seldom can we justifiably make all these assumptions with respect to the variables we work with in educational and psychological research.

6. As we already noted on pages 229–230, r may be regarded as an arithmetic mean, the mean of the $(z_x z_y)$ products, if each z is calculated from a standard deviation computed with N rather than $N - 1$ in its denominator. With most distributions, there will be about the same number of positive and of negative values of z. If the positive and negative values of z_x and z_y pair off at random, about half the products will be positive, half negative, and their sum, and hence r, will approach zero. If, however, the positive z_x values go with positive z_y

values, while negative z_x scores go with negative z_y scores, the cross-products will all be positive, and r will be plus. If in addition, the large values of z_x are paired with large z_y scores, r will be very high indeed. This interpretation of r as a mean is quite legitimate for any r, but since most people do not have a clear understanding of the nature of the $(z_x z_y)$ products that are averaged, it is not likely to be very meaningful. The only assumption involved is that of equality of units, a very trivial restriction indeed.

7. The next interpretation involves three-dimensional space. It is in terms of what is called the normal correlation surface. We can get an approximate idea of this fairly simply. Put this book on the floor and look at one of the Figure 8.4 scatter diagrams. Imagine that a little square or round tower is erected on each cell of that scatter diagram. The height of each tower will be proportional to the number of tally marks in that cell. (If there aren't any tally marks, there won't be any tower either.) Such a three-dimensional diagram for the correlation of +.78 in Figure 8.4(B) was constructed with discs and is reproduced as Figure 8.12. Because N was so large (9,440), each disc represents 100 cases and every fifth disc is white. Unfortunately, actual frequencies never quite fit into the smooth pattern that one would like. Despite a few erratic fluctuations, however, most of the major characteristics of a normal correlation surface are present. By making a rough estimate of the mean of each of the rows (or columns), we can see that these means do not depart much from a straight line, except at the ends where only a few cases are involved. Hence the regression is nearly linear. Next we can approximate the standard deviations of these same rows (and columns). The easiest way to do this is to estimate 2 s (like the mathematician who counted the legs and divided by 4 to find the number of sheep in a field) by noting the scores in each row that include approximately the middle two-thirds of the cases in that row. Again, except where the number of cases in a row is very small, the (double) standard deviations are all about the same size. There is no marked tendency for them to get much larger or smaller as we go across the scatter diagram. Even with our large N, there really aren't enough cases in most of the rows to tell whether the distributions within the rows are normal, but they seem to be. Look across the successive rows in Figure 8.12 and you will see this. Then reorient the page and look across the columns. They, too, all seem to be about equally variable and nearly normal in shape. They certainly do not give either a rectangular or a markedly skewed appearance. If we had still more cases, it seems that they would probably be normal or pretty close to it. These characteristics, plus the marginal frequencies also being normal—unfortunately not true in this case—correspond to the assumptions that are made in deriving the normal correlation surface. Such a surface is the limiting value that would be approached if we greatly increased N while also making the intervals smaller. An approximate

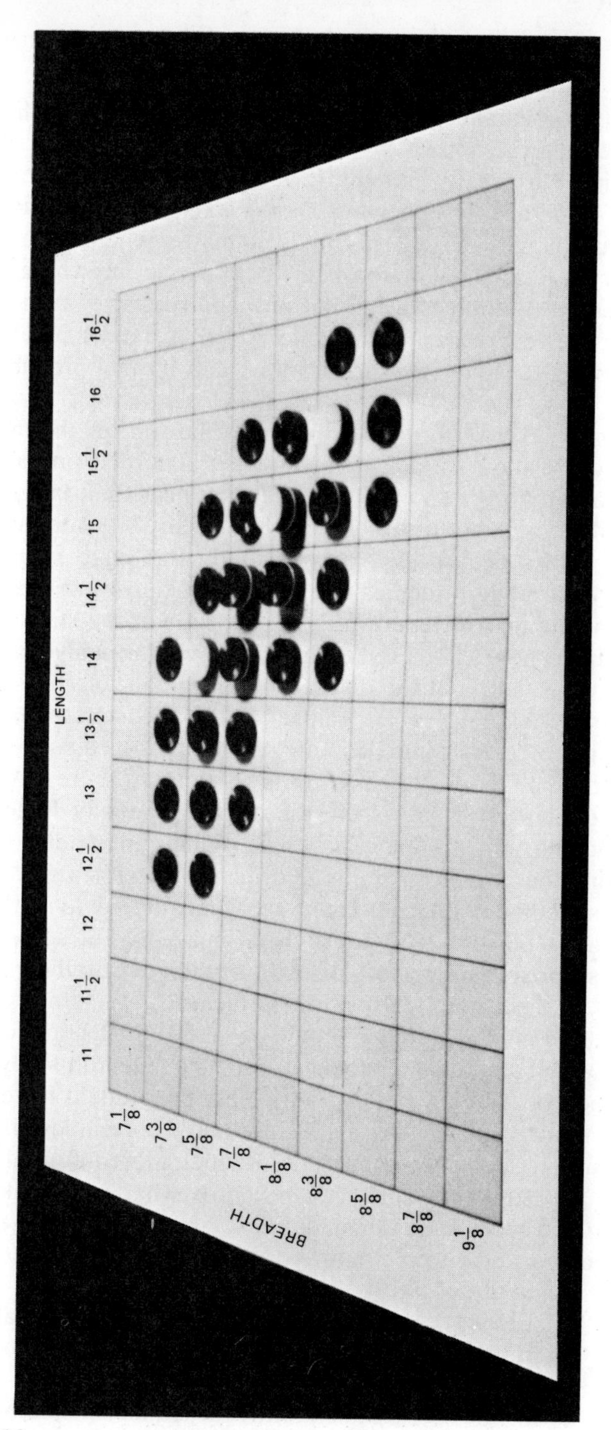
LENGTH
11
11 1/2
12
12 1/2
13
13 1/2
14
14 1/2
15
15 1/2
16
16 1/2
BREADTH
7 1/8
7 3/8
7 5/8
7 7/8
8 1/8
8 3/8
8 5/8
8 7/8
9 1/8

FIGURE 8.12

idea of its appearance could be obtained by throwing a clinging cloth over the material shown in Figure 8.12. If we then cut the normal correlation surface by passing a vertical plane (or a crosscut saw) through it, parallel to either the X or Y axis, the cut portion will be a normal probability curve. This holds no matter whether we cut through the mean, a little above it, markedly below it, or anywhere else. Every such vertical section is a normal probability curve. And this is true whether the sections are cut parallel to the X axis or parallel to the Y axis. No matter how you slice it, it's still a normal probability curve.

Now let us take a sharp sword and cut the top off the model of the normal correlation surface. We can't get a normal probability curve this time; we do get an ellipse. If we slice near the top we get a little ellipse; if near the bottom, a big one; but for any one normal probability surface, all its ellipses will be of exactly the same shape. A crude attempt was made to depict a few of these ellipses in Figure 8.12 by using a white disc for every fifth disc in each of the stacks. Thus, the ellipse that is now 5 units above the base is very roughly outlined by the outside white edges of the discs. An ellipse can, of course, have any shape from a circle to an ellipse that is so stretched out that it approaches a straight line. When $r = .00$, all the ellipses will become circles if the plotting is done in terms of standard scores. As r changes toward either -1.00 or $+1.00$, the circles gradually flatten out until they become straight lines at ± 1.00. This can readily be seen by examining the scatter diagrams in order from zero to $+1.00$ in Figure 8.4. This method of interpreting r is very meaningful to mathematicians and to most persons with good visual imagery, but some people are merely surprised, interested, or even bewildered by the various cross sections of the normal correlation surface.

8. This method of interpreting r should be taken with a grain (or a carload) of salt. When a graduate student conducts a research study and gets a correlation of .57 between two variables, the student is very likely to want to know whether that is a high or a low correlation. It is impossible to answer such a question without much more information. Nevertheless, the student wants an answer to what seems like a simple question. To meet such demands, here are some suggested answers, classified according to the nature of the variables correlated:

A. Reliability of a test

.95 to .99	Excellent; satisfactory for individual prediction
.90 to .94	Passable; good enough to justify some individual predictions
.80 to .89	Low; seldom regarded as satisfactory
.00 to .50	Unsatisfactory; test should be discarded or its reliability should be improved

B. Reliability of an industrial criterion

.80 to .99	Excellent; very unusual
.60	Quite good

.40 Permissible if it cannot be improved
.00 The criterion is worthless; get another one

C. Prediction of success in industry
+.80 to +.99 or −.80 to −.99 Fantastically high; validities so reported should be viewed with suspicion
+.60 or −.60 Very high
+.40 or −.40 High for most occupations
+.30 or −.30 Satisfactory if nothing better available, especially if the selection ratio (See page 305) can be kept low
+.20 or −.20 Low; useful only with very low selection ratios

D. Prediction of college grades
+.70 or −.70 High
+.50 to +.60 or −.50 to −.60 Moderate
−.30 to +.30 Low

E. Intercorrelations of factor analyzed tests (which are supposed to be independent of each other)
+.40 or −.40 Too high
+.10 to +.20 or −.10 to −.20 Permissible
−.10 to +.10 Excellent

F. All other correlations
+1.00 Perfect positive correlation
+.01 to +.99 "It depends."
.00 No relation between the two variables
−.01 to −.99 "It depends."
−1.00 Perfect negative correlation

There it is. We hope this crude classification may do more good than harm. The next statistician you ask may well disagree with at least a fourth of the statements listed under A to F. Interpretation in terms of such verbal descriptions is not at all satisfactory. Until you build up your own understanding of the kinds of correlations you work with or read about, these suggestions may be better than nothing.

SUMMARY

The Pearson product moment coefficient of correlation is usually the best method of measuring the strength of the relationship between two variables. Scatter diagrams and selected correlation coefficients were used to illustrate r's of various sizes from +1.00 through .00 to −1.00. It is possible to compute r either with or without a scatter diagram, either using the raw scores, coding them, or grouping them into intervals. It is usually best to transform r into z' before computing standard errors. To justify the use of r we need to satisfy only two requirements: equality of measuring units throughout the scale, and linearity of regression. Eight different ways of interpreting r have been given so far, the first two having to do with the probability that the size of the obtained r is due to chance, the others assuming that the obtained value is correct. Those in the early part of the list are generally the most useful and satisfactory. Since it is fundamental to all predictions of human behavior, r is probably the most important statistic with which psychologists work and one of the most important for educators, sociologists, and many other scientists.

SUPPLEMENT 8

OTHER FORMULAS FOR PEARSON r

The choice of a method for computing Pearson r depends upon the kind of data one has at hand and the availability of a machine (calculator or computer).

If one has no machine and there are a large number of scores, and especially if the scores have large values or decimals, then it would be most feasible to group on both variables and use a correlation computing chart (a modified scatter diagram) such as the one on page 244.

If one has a machine, the raw score Formula (8.2) would usually be the easiest.

For conditions departing from the ones above, there are a number of alternative formulas. They have some disadvantages, such as more computational labor or less accuracy due to rounding errors. However, there may be a time and place for them.

It will be recalled from the text of this chapter that the basic formula for Pearson r is

$$r_{xy} = \frac{\Sigma z_x z_y}{N - 1}.$$

Let us call this the "z score formula." Although this might be an easy formula to use if we already had the z scores, only rarely will this be the case. We recall from earlier in the chapter that this is not a good computing formula. Suffice it to say that z scores involve decimals and negative numbers.

If we remember that

$$z_x = \frac{x}{s_x} \quad \text{and} \quad z_y = \frac{y}{s_y},$$

then

$$r_{xy} = \frac{\Sigma\left(\frac{x}{s_x}\right)\left(\frac{y}{s_y}\right)}{N - 1},$$

$$r_{xy} = \frac{\Sigma xy}{(N - 1)s_x s_y}. \tag{8.7}$$

Let us call this the "original deviation score formula." Starting from scratch, it would be more laborious than the raw score formula (assuming a machine, of course). However, if we already had the two standard deviations and the deviation scores available, it wouldn't involve much work.

A variation of the "original deviation score formula" is the following:

$$r_{xy} = \frac{\Sigma xy}{\sqrt{(\Sigma x^2)(\Sigma y^2)}}. \tag{8.8}$$

Let us call this formula the "modified deviation score formula." If we had a situation such as a small number of whole number scores, and if both means were whole numbers, you could conceivably work out r with this formula by longhand. Since this practically never happens with real data, this formula is generally not useful for computing r.

The "raw score formula" (8.2) from the text of Chapter 8 was

$$r_{xy} = \frac{N\Sigma XY - \Sigma X \Sigma Y}{\sqrt{[N\Sigma X^2 - (\Sigma X)^2][N\Sigma Y^2 - (\Sigma Y)^2]}}.$$

It is relatively easy to use on a calculator and gives maximum accuracy, since it defers division and extraction of square roots as long as possible, thereby minimizing rounding errors. It can easily be derived from what we could call "the original raw score formula," which is

$$r_{xy} = \frac{\frac{\Sigma XY}{N} - \overline{X}\overline{Y}}{\sqrt{\frac{\Sigma X^2}{N} - \overline{X}^2}\sqrt{\frac{\Sigma Y^2}{N} - \overline{Y}^2}}. \tag{8.9}$$

If we have the two regression coefficients from a set of data or a published report on a study, but the correlation has not been reported, we can calculate r accurately from the following formula:

$$r_{xy} = \sqrt{b_{yx} b_{xy}}. \tag{8.10}$$

The b's in the above formula have not been covered as yet. They are covered in Chapter 9. This formula is an application of the geometric mean, which was described in Supplement 2.

If, for any reason, you want a larger number of formulas for computing r, 52 are given by Symonds.*

ALTERNATIVE WAYS TO TEST THE SIGNIFICANCE OF AN OBTAINED PEARSON r

As was pointed out in the text of Chapter 8 (p. 248), the Fisher z' transformation and the accompanying standard error

$$s_{z'} = \frac{1}{\sqrt{N - 3}}$$

is probably the best method of testing significance of r. Its disadvantage is having to make the transformation.

One of the oldest alternatives was a formula for the standard error of r for large samples and small r's, which is as follows:

$$s_r = \frac{1 - r^2}{\sqrt{N - 1}}. \tag{8.11}$$

A formula which is suitable for relatively small samples was developed by R. A. Fisher to test the hypothesis that the true correlation is zero in the population. It is given in terms of Fisher's t, which can be interpreted by means of Table G, which is the t table. The Fisher formula for testing r by means of the t table is

$$t = \frac{r\sqrt{N - 2}}{\sqrt{1 - r^2}}. \tag{8.12}$$

To save the labor of computing t and looking up t each time, Table M was prepared. We enter at the left with N, the sample size, and then looking across to the right on that row, we see the minimum size r that would be significant at either the .05 or .01 level.

RELIABILITY AND VALIDITY

Briefly, *reliability* tells how well a test measures whatever it does measure; while *validity* tells how well it predicts whatever we would like it to predict.

Whenever we give a test to anybody for any purpose, we assume or hope that the test can be depended upon to give the same result, regardless of whether it is given on Monday morning or Tuesday evening, regardless of who gives it, regardless of whether the person taking the test is happy or

*Symonds, Percival M. Variations of the product-moment (Pearson) coefficient of correlation. *Journal of Educational Psychology,* 1926, **17,** 458–469.

tired, and so on. That is, we want the test to give consistent results rather than for the test scores to vary widely for irrelevant reasons. Of course, tests aren't just consistent or not. They vary; and we need to measure the extent or degree to which they do give consistent results. This is measured by the reliability coefficient. One of the best ways to measure reliability is to test a reasonably large group of people with Form A of a test, wait a couple of weeks, retest them with Form B, and then compute Pearson r between the two sets of scores.

Of course, consistency isn't everything. We may have a very reliable test of height, of visual acuity, or of attitude toward Communism. But if we are using any one of these three measures to predict college success, we should not be surprised if we fail. We have to get a test whose scores are actually related to the variable (here, college success) which we are trying to predict. The extent to which any test measure correlates with a criterion of success is called the test's *validity coefficient*. Ideally, one of the best ways to measure validity is to test a reasonably large group of people otherwise qualified for employment, hire all of them, wait, get production records, and compute Pearson r between the test scores and success on the job. Practical situations, however, often dictate that different methods be used. There are, of course, other kinds of reliability and other kinds of validity.

Reliability Computed from Alternate Forms. If a measuring device has two forms or entirely different ways of arriving at scores, and if it is feasible to measure the individuals on two separate occasions, we can thus get two scores for each individual and correlate them. The result is a *reliability coefficient*. It is, of course, assumed that the two forms are measuring the same basic concepts, but the items should not be identical, or nearly so, in the two forms. It is further assumed that the nature of the function tested remains the same on both testings. This assumption generally holds, but may not in puzzlelike items in which insight plays a part. The time between testings should be great enough to eliminate day to day variability, and to diminish greatly any advantage that otherwise might accrue on the second testing. Ordinarily an interval of a week to a month seems about right.

This is a simple and straightforward method of determining reliability. It is perfectly consistent with the definition of reliability. When it is feasible, it is usually the best method of determining reliability. But it is not widely used because (a) many tests do not have two forms and (b) it is frequently impractical to assemble people twice to take the tests. As a result, other methods have been developed.

The Kuder–Richardson Formulas. Some time ago, a common procedure for determining the reliability coefficient was to split a test into comparable halves, to correlate scores on these halves, and then by the Spearman–

Brown prophecy formula—given later in this Supplement as Formula (8.17)—to estimate what the correlation would be between whole tests rather than between half tests. It was recognized that different ways of splitting a test into halves gave different results, but nobody did anything about it until Kuder and Richardson independently saw how to solve the problem and wrote up the solution as a joint article.* In the years since their article appeared, as the formulas they derived became more and more common, the odd-even and other split-half correlations just faded away. This is as it should be. All the Kuder–Richardson formulas are better and most of them are easier to compute than are the split-half coefficients.

Let us now give a passing reference to the split-half method of estimating reliability before we turn to the more modern method—the Kuder–Richardson formulas. In using the split-half method, the investigator obtains a total score for each person on two supposedly comparable halves of a test. Frequently, one half consists of the odd numbered items and the other, the even. These two sets of scores are correlated. This gives the correlation between scores on one half of the test and scores on the other half, or the reliability of a half test. This figure is then substituted in the Spearman–Brown formula and a somewhat larger figure is obtained as an estimate of the reliability of the whole test. In doing this we are, of course, assuming that the correlation obtained between the two halves is a representative value.

Often it is not, and although it is generally believed that this procedure tends to give estimates that are too high, Kuder and Richardson base their criticism of it on the fact that there are many ways of splitting a test. Hence any one reliability coefficient is not a unique value. Their formulas do give unique values.

In deriving their set of formulas for estimating reliability coefficients,† Kuder and Richardson first produced a formula that was very nice except for one thing: some of the terms in it "are not operationally determinable." English translation: There was no way to use it. So they made an assumption and got formula K–R (8). It can be solved but it takes a long time. So they made another assumption and got formula K–R (14). Then they made

*Kuder, G. F., and Richardson, M. W. The theory of the estimation of test reliability. *Psychometrika,* 1937, **2,** 151–160.

†Four of the Kuder–Richardson formulas have been so widely used and so often referred to by their original numbers that, even though we call them (8.13) through (8.16) in order to fit into our numbering system, we also give their original numbers and all our references to them are in terms of these original numbers: K–R (8), K–R (14), K–R (20), and K–R (21). That's the way they are almost always cited in both the educational and the psychological literature.

another assumption and got the still simpler formula K–R (20). Finally, they made still another assumption and got formula K–R (21).

The more assumptions they made, the easier the formulas were to solve. If the assumptions are justified in any particular problem, all their formulas will give the same answer. But if they are not, then the more work you are willing to do, the greater will be your reward in terms of getting a high reliability coefficient. In practice, this means that statisticians usually compute K–R (21) first. If the resulting reliability coefficient is high enough to satisfy them, they quit while they are ahead. If it is not high enough, they turn to formula K–R (20). If this is sufficiently high, fine; if not, they may go to K–R (14), but they usually stop there. Formula K–R (8) may give a higher reliability coefficient but, unless people have access to a programmed large scale computing system, it is so much work to compute it that they rarely do so. What happens is that, except for K–R (8), each of these formulas simply sets a lower limit to the size of the reliability coefficient.

First, Kuder and Richardson make the assumption that each item measures exactly the same composite of factors as does every other item. That is, if some of the items involve verbal ability, then all of them must. The test must not be composed of some arithmetic computation items, some memory items, and some inductive reasoning items, for example. This is not much of an assumption. Except for tests that deliberately try to cover several fields (as many intelligence tests do), most tests ought to be sufficiently homogeneous to justify our making this assumption. The resulting formula is

$$r_{11} = \frac{\sigma_N^2 - \Sigma pq}{2\sigma_N^2} + \sqrt{\frac{\Sigma r_{it}^2 pq}{\sigma_N^2} + \left(\frac{\sigma_N^2 - \Sigma pq}{2\sigma_N^2}\right)^2}, \tag{8.13}$$

[also known as K–R (8)]

where σ_N^2 is the squared standard deviation of the total score on the test, computed by Formula (3.5) for a finite population, p is the proportion of persons answering the item correctly, $q = 1 - p$, and r_{it} is the item-test correlation. These r_{it}'s are point biserial coefficients of correlation, computed by Formula (10.5), (10.6), or (10.7) (though they could be computed by one of the formulas for Pearson r if one does not mind doing extra work). In each case, the summation signs run from 1 to n, the number of items, although σ_N^2 is, as usual, based on the scores of the N people tested. Because there are n item-test correlations to be computed, this formula is rarely used. It is important, primarily because it is the first one in the series of four Kuder–Richardson formulas that have revolutionized the methods of computing reliability coefficients.

The next formula is much simpler. It makes the original assumption that all items measure the same factors and further assumes that all inter-item correlations are equal. This assumption seems very stringent, but in practice, the inter-item correlations usually are sufficiently similar that the assumption causes no difficulty. If the assumption is false, formula K–R (14) merely gives a value lower than would otherwise be the case. The formula is

$$r_{11} = \left(\frac{\sigma_N^2 - \Sigma pq}{(\Sigma\sqrt{pq})^2 - \Sigma pq}\right)\left(\frac{(\Sigma\sqrt{pq})^2}{\sigma_N^2}\right) \tag{8.14}$$

[also known as K–R (14)].

Even less work is involved in the next formula. We still need the pq products, but we do not need to bother with $\sqrt{pq}$. To the assumptions already made, we further assume that all items have the same squared standard deviation or variance, pq. In practical situations, this assumption is always false, but frequently the item variances do not differ very much.

A minor deviation from equal variances (or even a much larger departure from such equality) does not seem serious enough to affect the formula appreciably. In any case, the only effect is a slight lowering of the estimated reliability coefficient. The elimination of the necessity for including $\sqrt{pq}$ is generally regarded as a sufficient reason for preferring the next formula, K–R (20):

$$\begin{aligned} r_{11} &= \left(\frac{n}{n-1}\right)\left(\frac{\sigma_N^2 - \Sigma pq}{\sigma_N^2}\right) \\ &= \left(\frac{n}{n-1}\right)\left(\frac{\sigma_N^2 - \overline{X} + \dfrac{\overline{X}^2}{n} + n\sigma_p^2}{\sigma_N^2}\right) \end{aligned} \tag{8.15}$$

[also known as K–R (20)].

It does not take long to solve formula K–R (20), yet it usually gives a good approximation to the values computed by the more rigorous formulas presented earlier in this chapter. For these reasons, Kuder and Richardson have indicated a preference for it. Nevertheless, each practitioner will have to decide how to weight the higher accuracy and higher reliability coefficients given by the lower-numbered formulas against the additional assumptions but greatly lowered computing time of the higher-numbered formulas.

Interestingly enough, since 1937, at least four different authors have derived formulas that are mathematically identical with K–R (20) but are based on different assumptions. Most of these other authors feel that their assumptions are much less stringent than those made in the original derivation. Perhaps the extreme in this respect is the statement by one of these others that "the only assumption made in deriving this equation was that the average covariance among *non-parallel* items was equal to the average covariance among *parallel* items." (A covariance is a correlation coefficient multiplied by the two standard deviations. It thus is $\frac{\Sigma xy}{N}$.) While this is a less restrictive assumption, there are no comparable derivations for the other formulas. Hence such a derivation fails to make clear the interrelationships existing among the K–R formulas.

The final reliability formula in the original article makes only one more assumption—that all items have the same difficulty—and it leads to the extremely simple formula:

$$r_{1I} = \left(\frac{n}{n-1}\right)\left(\frac{\sigma_N^2 - \overline{X} + \frac{\overline{X}^2}{n}}{\sigma_N^2}\right) \tag{8.16}$$

[also known as K–R (21)].

Thus, if all you have are total scores on a test, you can readily compute $\overline{X}$ and σ_N^2, count the number of items, n, and substitute in the formula. This very simple formula always gives an underestimate unless all items are exactly equal in difficulty, because it is just like formula K–R (20) except that the intrinsically positive term, $n\sigma_p^2$, is omitted from the numerator. It has been stated by Ledyard Tucker, who derived the second formula for K–R (20), that the results obtained from formula K–R (21) are seldom more than 10% less than those for the more accurate formula K–R (20). Thus if you compute K–R (21) and get a low reliability coefficient, you can ordinarily expect a higher one (but not much higher) if you compute K–R (20), and, of course, a possibly still higher r_{1I} from K–R (14) and, finally, from K–R (8). But if K–R (21) gives a satisfactorily high value, the research worker has, with very little effort, been saved the necessity of computing any of the more complicated formulas.

We shall now present in Table 8.13 some of the characteristics of these various methods of determining the reliability coefficient and comment briefly on a few of them.

Most educators and psychologists know that the split-half method is not applicable to speed tests. Some of them do not realize that it is not applicable to tests that are not called speed tests, but which a few students do not have time to finish. Further, knowledge of the relevance of the effect of speed on the K–R formulas is rather limited. The answer is simple: In

TABLE 8.13
A Comparison of Reliability Formulas

	Alternate Forms	Split Half	K–R (8)	K–R (14)	K–R (20)	K–R (21)
A. Characteristics						
Appropriate for test involving speed	Yes	No	No	No	No	No
Gives a unique value	—	No	Yes	Yes	Yes	Yes
Items must be scored 1 for right and 0 for wrong	No	No	Yes	Yes	Yes	Yes
Approximate calculating time in minutes ($N = 100$; $n = 50$)	—	45	600	60	30	10
Size of reliability coefficient	Varies	Varies	Highest	High	Low	Lowest
May underestimate the reliability	Yes	Yes	Yes	Yes	Yes	Yes
May overestimate the reliability	Yes	Yes	Yes	No	No	No
B. General assumptions						
Response to one item does not depend on how the subject answers other items	No	Yes	Yes	Yes	Yes	Yes
The two halves have equal standard deviations	—	Yes	—	—	—	—
The correlation between the two halves is representative	—	Yes	—	—	—	—
C. Specific K–R assumptions						
Each item measures exactly the same composite of factors as every other item	—	—	Yes	Yes	Yes	Yes
All inter-item correlations are equal	—	—	No	Yes	Yes	Yes
All items have the same variance	—	—	No	No	Yes	Yes
All items have the same difficulty	—	—	No	No	No	Yes

Note: When one of the situations described at the left does not arise with respect to a particular formula, — is inserted in the table.

any circumstance in which some subjects are prevented from doing their best on a test because of insufficient time, no satisfactory determination of reliability can be made. (It is difficult to measure a person's knowledge by the use of an item that he has not read.) If we must determine the reliability of such a test, two procedures are available: (1) use alternate forms or (2) break the test into two comparable and separately timed halves, then estimate the reliability of the whole test by the Spearman–Brown formula (8.17).

The restriction about scoring items 1 for right and 0 for wrong means that formulas such as R − W and R − $\frac{1}{4}$W must not be used if Kuder–Richardson reliabilities are to be computed. Actually, modifications of the K–R formulas have been derived for such situations, but it is usually much easier to ask students to answer all items, to score papers by counting the number right, and to use the original K–R formulas.

The calculating times shown in Table 8.13 are approximate and will vary greatly from person to person. These estimates are all based on the assumption that the tests have already been given and scored.

Effect of Test Length on Reliability. We have already referred to the fact that the correlation between the two halves of a test, or the reliability of a half test, is lower than the reliability of a whole test. In fact, so long as additional items are measuring the same function, the greater the number of items, the higher the reliability. This may not be obvious, but it is usually apparent that if we keep shortening a test until finally we have only one or two items, chance will play a very large part in determining whether a person's score is 0, 1, or 2 and the correlation between such a "test" and another equally short one is bound to be low. As we increase the length of the test, chance plays a less and less important role. The correlation between scores on one test and an alternate form of it will then be high. The formula depicting this relationship is

$$r_{AA} = \frac{Ar_{1I}}{1 + (A - 1)r_{1I}}, \tag{8.17}$$

where r_{1I} is the reliability of a test and r_{AA} is the reliability of a test A times as long as the original test.

This is the Spearman–Brown prophecy formula. In deriving the formula, it is assumed additional forms of the test exist or could exist that (1) all have equal standard deviations and that (2) all correlate equally with each other. In practice, it seems not to matter much whether or not these assumptions hold. A large number of studies have shown that the Spearman–Brown formula gives a very close estimate of the reliability of a lengthened test.

TABLE 8.14
Reliability of a Lengthened Test, as Estimated by the Spearman–Brown Formula

r_{1I}	r_{2II}	r_{3III}	r_{4IV}	r_{5V}	$r_{(10)X}$
.00	.00	.00	.00	.00	.00
.10	.18	.25	.31	.36	.53
.20	.33	.43	.50	.56	.71
.30	.46	.56	.63	.68	.81
.40	.57	.67	.73	.77	.87
.50	.67	.75	.80	.83	.909
.60	.75	.82	.86	.88	.938
.70	.82	.88	.903	.921	.959
.80	.89	.923	.941	.952	.976
.90	.947	.964	.973	.978	.989

Reference: The above values were taken, by permission, from the more extensive table in Dunlap, Jack W., and Kurtz, Albert K. *Handbook of statistical nomographs, tables, and formulas*. Yonkers-on-Hudson, N.Y.: World Book Company, 1932.

Some sample values of r_{AA} are shown in Table 8.14 for various values of r_{1I}. For example, if a test has a reliability of .40, then a comparable test four times as long will have a reliability of .73.

If the reliability of our test is known, but we wish to know how much it would have to be lengthened to give any desired reliability, we can simply substitute in Formula (8.17) and solve for A. Thus if a short test has a reliability of .61, but we would like to raise that reliability to .90, simply substitute .61 for r_{1I}, .90 for r_{AA}, and solve for A:

$$.90 = \frac{(A)(.61)}{1 + (A - 1).61},$$

$$A = 5.75.$$

Hence the test will need to be made about six times its present length in order to give the desired reliability. Formula (8.17) provides all the information we need to decide whether it is feasible to lengthen the test or whether it should be revised or abandoned.

Standard Error of the Spearman–Brown Prophecy Formula. It is apparent that even if the two assumptions basic to the Spearman–Brown formula are fulfilled exactly, the formula can still not be counted on to give us the *exact* reliability of a lengthened test. Our original r_{1I} was based on a finite sample of N people and is subject to error. So, therefore, are all subsequent figures based upon it. The formula for the standard error of r_{AA} is

$$s_{r_{AA}} = \frac{A(1 - r_{1I}^2)}{(\sqrt{N - 2})[1 + (A - 1)r_{1I}]^2} \tag{8.18}$$

in which N is the number of cases, and the other symbols have the same meanings as in the preceding formula. Notice that the expression in brackets does not have to be computed. We already have it as the denominator of Formula (8.17); now we just square it.

Following up on a hint in the original derivation, we have derived the following alternative formula:

$$s_{r_{AA}} = \frac{(1 - r_{AA})[A - (A - 2)r_{AA}]}{\sqrt{N - 2}}. \tag{8.19}$$

What's it good for? Well, since the two terms in parentheses or brackets are very easy to compute, it isn't much work to calculate the numerator, so this formula *may* be a little easier to compute than the previous formula. But it has a *much better* use. We have said many times that "redoing it" is a very poor way of checking. In this case we have a superb check. When you compute r_{AA}, carry the final division out to four decimals. Then compute $s_{r_{AA}}$ by both formulas, keeping six decimals in the final division. If these two results agree or differ by no more than about .000005 or .000010, you have checked all three computations—both values of $s_{r_{AA}}$, of course, but also r_{AA} itself.

Reliability Standards. In interpretation (8) of Pearson r near the end of Chapter 8, we have already summarized our viewpoint on reliability. That brief summary now requires a little explanation and elaboration.

Reliability coefficients cannot be greater than +1.00 and, except by chance, cannot be negative. A negative reliability is a contradiction in terms, because if a test is no good at all, it should show no agreement within itself and have a reliability of zero. It cannot consistently give scores that are the opposite of the scores it gives. Hence, in setting standards, we need concern ourselves only with positive reliability coefficients.

An attempt has been made to bring about a reasonable compromise between what we would like and what it is practicable to attain. Thus, there is no theoretical reason why a test should have any higher or lower reliability than the criterion it is intended to predict. But every industrial psychologist knows that it is often extremely difficult to secure a criterion with a reliability as high as what we routinely expect from reasonably good tests. Thus, on p. 260 a test with a reliability of .80 to .89 was described as "Low; seldom regarded as satisfactory," while an industrial criterion with a reliability anywhere from .80 to .99 was described as "Excellent; very unusual."

There has been a trend toward insisting on higher reliability coefficients. This is entirely reasonable. As our test construction techniques have improved, so have our tests. In part, this improvement was promoted on an unsound basis. People have advocated higher reliability on the argument that a test can't correlate with anything else any higher than it correlates with itself. However reasonable this may sound, it isn't true. A very poor test of vocabulary, for instance, not only can but will correlate higher with an excellent vocabulary test than it will correlate with itself, regardless of whether we measure its reliability by alternate forms, by one of the K–R formulas, or in some other way. Actually, a test with a reliability of r_{1I} can correlate as highly as $\sqrt{r_{1I}}$ with an ideal test of what it is attempting to measure. Thus, a test with a reliability of .36 has a maximum possible correlation with anything else of .60. We have no objection to increasing the reliability of such a test—it certainly needs improving—but let us not employ fallacious reasoning to justify such action.

In setting standards for test reliability, no differentiation was made by type or purpose of the test. We know, for instance, that it is difficult to attain high reliabilities with certain types of performance tests, but until their reliabilities are improved, such tests will continue to be less useful than other more reliable tests that are used for the same or similar purposes.

It has been assumed throughout that all reliabilities to which we have referred have been based on an appropriate range. In the case of educational tests, this nearly always means that the reliability coefficient must be based on pupils in a single grade. It would be satisfactory to give reliabilities based on pupils with a one-year age range, but since it is much simpler administratively to test, for example, fifth-graders than to test eleven-year-olds, reliability coefficients are usually computed for the successive school grades. What happens if we use a much wider range, such as all pupils from the first through the sixth grade? The sixth graders will obtain very high scores and the first graders very low scores, not only on the test as a whole but on practically all the individual items. As a result, the computed reliability coefficient will be extremely high, but it won't mean anything. (Formulas are available for estimating reliability or other correlation coefficients in one range from those obtained in a different range, but they are not very satisfactory. The simple method is the best: compute the correlation coefficient for a group with the desired range and other characteristics.) The only fair reliability coefficients for authors of educational tests to present are those based on a one grade (or a one-year age) range.

For tests designed for use in industry, the group for whom reliability coefficients are to be computed is the group for which it is intended that the test be used. This statement is a little ambiguous since it could refer to (1) the group of applicants, (2) the persons hired, or (3) those who were hired and remained long enough to be included in some other study. The first of these groups would usually give the highest reliability coefficient, and the last the lowest, but all are appropriate. The test constructor or research worker may use whichever seems most proper for the purpose.

For other tests, for criteria, and for all other measuring devices, the group used in the computation of reliability coefficients should be any proper group. In addition, the group should be briefly, but suitably, described. Nearly always, unduly wide ranges give spuriously high reliabilities (and other correlations as well), while unduly restricted ranges of talent result in spuriously low figures.

At the present time, we cannot characterize the reliability of an industrial criterion as excellent unless it is at least .80; we cannot so characterize a test unless its reliability is at least .95. The standard for most other measurements should be much nearer the latter than the former figure.

Validity. A *criterion* is a variable that we wish to predict. In industrial psychology, it is a measure of the degree of success attained by an employee. Such a criterion may be amount of work produced, but since, oddly enough, records of this are often not available, it is more likely to be supervisor's ratings or rankings, current earnings, job satisfaction, freedom from accidents, or merely whether or not the employee remains on the job for a year. In other branches of Psychology and Education, analogous criteria are used; in predicting academic success, school grades frequently constitute the criterion.

A validity coefficient is simply the correlation between the criterion and whatever other variable is used to predict it. Such correlations may have any value from -1.00 through .00 to $+1.00$. In practice, the great majority of published validity coefficients are between $-.40$ and $+.40$. (The unpublished ones, of course, tend to be lower in absolute value than the published ones.) For predictive purposes, it makes no difference whether the validity coefficient is positive or negative. A test that correlates $-.43$ with a criterion will predict it more accurately than will another that correlates $+.32$ with it.

As stated in the tabulation on page 261, validity coefficients in industry of .40 (or $-.40$) are high. Unless a better test is available for that particular occupation, we are certainly justified in using a test with a validity of .40. Higher validities do occur, but tests with validities as high as .60 are extremely rare. They are even rarer if we adhere to our usual standard of insisting upon an N of at least 100.

Considerably better success has been encountered in the prediction of academic success. This holds in grade school, in high school, and in college. Validities of .50 to .60 can usually be obtained by the use of a good intelligence or achievement test. In a few instances, a properly weighted composite of several tests has consistently given validities of around .70, and occasionally even higher, in the prediction of college grades. Perhaps it is unfortunate that we can predict how well students will do in school so much better than we can predict how well they will succeed in the world after they get out of school, but that's the way it is.

We sometimes distinguish between concurrent validity and predictive validity. *Concurrent validity* is what we obtain if we test a group of presently employed workers and correlate their test scores with production. If the validity is high, it is assumed, or at least hoped, that it will still be equally high when the test is used on new applicants at the time they are hired. In different situations, there are good reasons for expecting a change in either direction, but usually the validity will be lower on the applicant group.

The *predictive validity* of a test is obtained when we test applicants, wait until sufficient time has elapsed for criterion records to become available, and then compute the correlation coefficient between test scores and the criterion. This is the best way to conduct test validation studies, but it is seldom used because it takes so much time. It is best because it is so similar to the situation that will exist if the test actually is found to have predictive validity and is adopted as a part of the regular employment procedure.

Concurrent validity and predictive validity are the only types of validity that can be expressed in terms of the coefficient of correlation. Ordinarily when the word validity is used, it refers to one or the other of these two kinds of validity. There are, however, some other uses of the term validity. Let us look at them very briefly.

The *content validity* of a test is determined by finding out how well the test content represents the subject matter and situations upon which the test is supposed to be based. This is clearly a subjective decision, and there is no such thing as a content validity coefficient. It is, of course, important that a test should properly sample some universe of content, but this is properly an early phase of test construction and is entirely separate and distinct from concurrent and predictive validity. In tests of school subjects, the term *curricular validity* is sometimes preferred to content validity.

A test's *construct validity* is evaluated by investigating what psychological qualities it measures. If a test is based upon a certain theory or hypothesis, and if the theoretical predictions concerning individuals or groups are confirmed by the test results, then the test has construct validity.* This type of validity is important in developing measures of personality traits and attitudes.

---

*An important paper treating this concept is that by Cronbach, Lee J., and Meehl, Paul E. Construct validity in psychological tests. *Psychological Bulletin,* 1955, **52,** 281–302.

The misleading term, *face validity,* is not validity at all. If a test looks practical, especially if it looks practical to the people who are tested, to executives, and to others who know little or nothing about tests, it is said to have face validity. This is sometimes of some importance, but face validity bears the same relation to validity that quicksilver and nickel silver bear to silver.

Although both content validity and construct validity may be of great interest and value to the test constructor, the typical test user will be directly concerned either with a test's predictive validity or with its concurrent validity used as an estimate of its predictive validity. For him, the *only* way to determine the validity of a test or other measuring device is to compare it (preferably by means of Pearson *r*) with some criterion of success. When we realize that this definition is simply a more precise way of stating that validity is how well a test predicts whatever it is supposed to predict, it should be clear that *validity is the most important single characteristic of a test.*

A Preview of the Next Chapter:

Kermit Schafer: *The Blunderful World of Bloopers*

Child Bride

The Bob Newhart Show told about his parents coming for Thanksgiving dinner. Several times throughout the show it told how his parents had been married for forty-five years, but within two minutes of the end of it, said that his mother was fifty-six years old. This would mean that she was married at the age of eleven!

CHAPTER 9 REGRESSION EQUATIONS

A *regression equation* is an equation for predicting the most probable value of one variable from the known value of another. Regression equations are used, for instance, to predict such different things as college success, a life insurance agent's average weekly commissions, or even marital happiness from the score on an appropriate objective test.

THE PURPOSE OF A REGRESSION EQUATION

If we can either establish the fact of correlation or determine the degree of the relationship between two variables, that is sometimes sufficient. For many purposes, however, we often want something more than a mere statement of the amount of correlation between two variables. One function of science is to predict and control. The regression equation provides the best possible way to predict the score on a second variable from the knowledge of a person's score on the first variable.

Thus, the purpose of a regression equation is simply to tell us what score we may expect to attain on one variable if we already know that person's score on another variable. We frequently use regression equations to predict job success from the score on an aptitude test. They are also used to predict college grades by means of intelligence test scores or high school record. In fact, if there is a correlation other than zero between *any* two variables, a regression equation can be used to predict either one of them from the other.

Usually we wish to predict in one direction but not in the other. Thus, we may wish to use a biographical inventory blank to predict combat success in naval aviation, as was done in World War II, but we would not be at all interested in using a flier's combat success to predict biographical information. Similarly, weight is used to predict the outcome of prize fights, but we never arrange to have two boxers engage in a slugging match so that we may estimate their relative weights. We *could* do these things, and, incidentally, we can always predict just as accurately in one direction as in the other. But common sense tells us that there are simpler and better ways to find out that a boy once won a bicycle race and that a boxer now weighs 188 pounds. In some other situations, it may be just as logical to predict in one direction as in the other. Thus, if we know somebody's attitude toward God, we can predict the corresponding attitude toward evolution ($r = -.47$).* Or, if we know an individual's attitude toward evolution, we can, of course, predict the attitude toward God. Similarly, we can predict grades in History from those in Geography just as readily as we can predict grades in Geography from those in History. The direction in which we predict is determined by what we already know and what we wish to know.

FORMULAS FOR REGRESSION EQUATIONS

The fundamental formula for estimating a standard score on one variable from an individual's standard score on any other variable is given by Equations (9.1):

$$\begin{aligned} \tilde{z}_y &= r_{yx} z_x, \\ \tilde{z}_x &= r_{xy} z_y, \\ \tilde{z}_1 &= r_{12} z_2, \\ \tilde{z}_0 &= r_{01} z_1. \end{aligned} \tag{9.1}$$

There are many other ways to write the formula for a regression equation in terms of standard scores, but they differ only in the subscripts. In each case, the symbol on the left consists of the letter z with a tilde above it (to indicate that the value under the tilde is being predicted) and a subscript (to identify the variable being predicted). The z at the right signifies a standard score in the variable upon which the prediction is based. The subscripts of r are those of the two variables involved. They may be written in either order. Thus the first of the equations listed as (9.1) says: The predicted standard score of a person on variable Y is equal to the correlation between

*Ferguson, Leonard W. A revision of the primary social attitude scales. *Journal of Psychology,* 1944, **17,** 229–241. The quoted figure is the mean of four r's (2 forms of each scale) of $-.44$, $-.46$, $-.48$, and $-.50$.

the variables Y and X multiplied by the standard score on variable X. If X = weight, Y = height, $r_{xy} = .55$, and Percy's weight is $1.23s_x$ below $\overline{X}$, this equation becomes: Percy's standard score in height is predicted to be .55 times -1.23, or $-.68$. Thus, we predict that he will be .68 standard deviations below the mean height. To reverse the situation and predict weight from height we would use the next equation. Notice that however superior or inferior an individual is on one variable, we never expect that person to be any further away from the mean on the second variable. In fact, if r has any value other than ± 1.00, our best prediction always is that any individual will be closer to the mean on the second variable than that person was on the first, when we use standard scores. This phenomenon is called *regression toward the mean*. Note that we always predict that a person's standard score in the predicted variable will be only r times as great as that same person's known standard score in the other variable.

Like the first formula in some of our other chapters, this formula is primarily of theoretical interest. Ordinarily we do not have our data expressed as standard scores. Hence we need a more practical formula—one based on raw scores (X) rather than on deviations (x) or on standard scores (z). To derive such a formula, we shall first need two other formulas. The first of these is Formula (3.6), which is

$$z = \frac{X - \overline{X}}{s}.$$

The second is a formula for what is called a regression coefficient. Regression coefficients are defined as follows:

$$\begin{aligned} b_{yx} &= r_{yx}\left(\frac{s_y}{s_x}\right), \\ b_{xy} &= r_{xy}\left(\frac{s_x}{s_y}\right), \\ b_{12} &= r_{12}\left(\frac{s_1}{s_2}\right), \\ b_{01} &= r_{01}\left(\frac{s_0}{s_1}\right). \end{aligned} \tag{9.2}$$

Notice that the subscripts of b always occur in the same order as do those of the two s's. Although $r_{yx} = r_{xy}$, b_{yx} does not equal b_{xy}.

Let us now start with a regression equation in terms of standard scores and derive one in terms of raw scores. It is easy. First, let us take one of the equations in (9.1):

$$\tilde{z}_y = r_{yx}z_x.$$

Substituting the appropriate values from (3.6), we obtain

$$\frac{\tilde{Y} - \overline{Y}}{s_y} = r_{yx}\frac{X - \overline{X}}{s_x}.$$

Multiplying both sides by s_y, the equation becomes

$$\tilde{Y} - \overline{Y} = r_{yx}\left(\frac{s_y}{s_x}\right)(X - \overline{X}).$$

Noting from Formula (9.2) that $r_{yx}\left(\frac{s_y}{s_x}\right) = b_{yx}$, this becomes

$$\tilde{Y} - \overline{Y} = b_{yx}X - b_{yx}\overline{X}.$$

Finally, adding $\overline{Y}$ to both sides, we get

$$\tilde{Y} = b_{yx}X + (\overline{Y} - b_{yx}\overline{X}).$$

Similar operations applied to all the equations numbered (9.1) will give

$$\begin{aligned} \tilde{Y} &= b_{yx}X + (\overline{Y} - b_{yx}\overline{X}), \\ \tilde{X} &= b_{xy}Y + (\overline{X} - b_{xy}\overline{Y}), \\ \tilde{X}_1 &= b_{12}X_2 + (\overline{X}_1 - b_{12}\overline{X}_2), \\ \tilde{X}_0 &= b_{01}X_1 + (\overline{X}_0 - b_{01}\overline{X}_1). \end{aligned} \tag{9.3}$$

As before, the tilde indicates a predicted variable. $\overline{X}$ or $\overline{Y}$ with any subscript indicates the mean of the corresponding raw scores. Thus, $\overline{Y}$ is the mean of the Y's; $\overline{X}_1$ is the mean of the X_1's; $\overline{X}_0$ is the mean of the X_0's, etc.

The parentheses were included in each of equations (9.3) for a special reason. They would be unimportant or even undesirable if we were going to make only one prediction, but when we predict a number of scores, they help because the material within the parentheses reduces to a single constant. Thus, the first of the equations (9.3) might become $\tilde{Y} = 98.76X - 543.21$. It is obviously much more convenient to compute the value -543.21 once than it would be to use a slightly different formula and go through the entire computation for every prediction. In fact, some texts use the letter C or K for this parenthetical material, thus writing two equations such as $\tilde{Y} = b_{yx}X + C$ and $C = \overline{Y} - b_{yx}\overline{X}$ for each one of those we have written in (9.3).

The four equations given as Formula (9.3) are not the only regression equations. A regression equation may be written between *any* two variables. For example, if we wish to predict scores on the fifth variable, X_5, from the known scores on the third variable, X_3, we simply substitute the subscripts 5 and 3 for 0 and 1, respectively, in the last of the equations (9.3) and obtain the correct equation:

$$\tilde{X}_5 = b_{53}X_3 + (\overline{X}_5 - b_{53}\overline{X}_3).$$

As we noted on page 275 in Supplement 8, the variable we wish to predict is called the criterion. It is often designated by the symbol Y or X_0. The criterion is a rating, production record, or other measure of success that is being predicted by one or more tests. The tests are usually designated by the letter X, sometimes with appropriate subscripts. However, as just stated, *any* two variables may be connected by a regression equation. We need to know the two means, the two standard deviations, and the correlation between the variables. Then, by means of Formulas (9.2) and (9.3), we can write an equation such that if we know anybody's score on only one variable, we can estimate what that person's score is apt to be on the other variable.

Such predictions are very important. Some testing in schools is designed solely to measure what has been accomplished, but the overwhelming majority of printed tests used in the schools are given in order to predict what students will do in the future. All tests used for educational or vocational guidance have as their prime objective the prediction of future success. The same is true of aptitude tests. Industry's widespread use of tests in hiring workers is based almost exclusively on the belief (usually justified, sometimes not) that the test scores can be used to predict success on the job. Thus, nearly all educational and psychological testing is definitely concerned with the predictive value of tests. Pearson r and the regression equation are needed to make such predictions. Let us now see exactly how regression equations are used in prediction.

THE USE OF REGRESSION EQUATIONS

Once r has been computed, it is very easy to write the appropriate regression equation (or even both regression equations if we sometimes wish to predict in one direction and sometimes in the other). There are three simple steps. First, we substitute the known values of r and the two s's in Formula (9.2) in order to obtain the regression coefficient. Second, we substitute this regression coefficient and the two means in Formula (9.3) to obtain the equation to be used for our predictions. Finally, we take any score in which we are interested, substitute it on the right-hand side of our equation, and obtain the desired predicted score. Once we have the prediction equation, it can be used to predict one score, several scores, or a hundred. (If we wish to make a large number of predictions, the work can be shortened by using a method explained in Supplement 9. For only a few predictions, we simply substitute each time in the formula.)

Suppose we wish to estimate the heights of two of the 107 male seniors whose weights are 177 and 133 pounds. In the previous chapter, Figure 8.10 gave us $r = +.55$ and it also gave us the basic data which we can readily substitute in Formulas (2.2) and (3.3) in order to obtain the means and standard deviations we need. These basic figures are:

$$\begin{aligned}
\overline{X} &= i\left(\frac{\Sigma fx'}{N}\right) + M' \\
&= 5\left(\frac{-75}{107}\right) + 162 \\
&= 158.50, \\
\overline{Y} &= i\left(\frac{\Sigma fy'}{N}\right) + M' \\
&= 1\left(\frac{+188}{107}\right) + 68 \\
&= 69.76, \\
s_x &= i\sqrt{\frac{N\Sigma f(x')^2 - (\Sigma fx')^2}{N(N-1)}} \\
&= 5\sqrt{\frac{(107)(1{,}175) - (-75)^2}{(107)(106)}} \\
&= 16.25, \\
s_y &= i\sqrt{\frac{N\Sigma f(y')^2 - (\Sigma fy')^2}{N(N-1)}} \\
&= 1\sqrt{\frac{(107)(970) - (188)^2}{(107)(106)}} \\
&= 2.46.
\end{aligned}$$

We need an equation for predicting height (Y) from weight (X). The first one of equations (9.3) fills the bill. It is

$$\tilde{Y} = b_{yx}X + (\overline{Y} - b_{yx}\overline{X}).$$

We also need something else—one of equations (9.2) for evaluating b_{yx}. It is

$$\begin{aligned}
b_{yx} &= r_{yx}\left(\frac{s_y}{s_x}\right) \\
&= .55\left(\frac{2.46}{16.25}\right) \\
&= .0833.
\end{aligned}$$

Notice, in passing, that we retained four decimals. To have retained our usual two would have given us only one significant figure, and our answer would have been quite inaccurate, since .08 could mean anything from .0751 through .0849, the latter being 13% larger than the former. We are now ready to substitute in the equation for $\tilde{Y}$. It becomes

$$\begin{aligned}\tilde{Y} &= .0833\,X + [69.76 - (.0833)(158.50)]\\ &= .0833\,X + (69.76 - 13.20)\\ &= .0833\,X + 56.56.\end{aligned}$$

Substituting the weight of 177 pounds for X, this regression equation gives a predicted height, $\tilde{Y}$, of

$$\begin{aligned}\tilde{Y} &= (.0833)(177) + 56.56\\ &= 14.74 + 56.56\\ &= 71.30 \text{ inches}.\end{aligned}$$

Thus, the regression equation tells us that the most probable height for this man who weighs 177 pounds is 71.30 inches, or just a trifle over 5 feet 11¼ inches. This is just about 1½ inches above the average height of 69.76 inches.

Similarly, we very easily predict the height of the 133-pound senior as

$$\begin{aligned}\tilde{Y} &= (.0833)(133) + 56.56\\ &= 11.08 + 56.56\\ &= 67.64 \text{ inches}.\end{aligned}$$

Thus, our best estimate of the height of this 133-pound man is 67.64 inches, or approximately 5 feet 7⅝ inches.

Since r is positive, all persons (such as the 177-pounder) who are above the average weight (158.50 pounds) will be predicted as being above the average height (69.76 inches); and all who are below the average weight will have estimated heights below average also.

Now let us write an equation for predicting weight (X) from height (Y). First we must solve one of equations (9.2) for b_{xy}:

$$\begin{aligned}b_{xy} &= r_{xy}\left(\frac{s_x}{s_y}\right)\\ &= .55\left(\frac{16.25}{2.46}\right)\\ &= 3.63.\end{aligned}$$

Next we select the appropriate equation from (9.3) and solve it for $\tilde{X}$:

$$\begin{aligned}\tilde{X} &= b_{xy}Y + (\overline{X} - b_{xy}\overline{Y})\\ &= 3.63\,Y + [158.50 - (3.63)(69.76)]\\ &= 3.63\,Y + (158.50 - 253.23)\\ &= 3.63\,Y - 94.73.\end{aligned}$$

How much should a six-footer weigh? As stated, the question is ambiguous. What do we mean by "should"? If we say a man should weigh what the average man of his height weighs, then the preceding equation will give us the answer; all we have to do is substitute 72 inches for Y and solve:

$$\begin{aligned}\tilde{X} &= (3.63)(72) - 94.73 \\ &= 261.36 - 94.73 \\ &= 166.63 \\ &= 167 \text{ pounds.}\end{aligned}$$

What about a man 5 feet, 5 inches tall? We simply substitute his height of 65 inches for Y and get:

$$\begin{aligned}\tilde{X} &= (3.63)(65) - 94.73 \\ &= 235.95 - 94.73 \\ &= 141.22 \\ &= 141 \text{ pounds.}\end{aligned}$$

Thus we see that we can predict either height or weight if we know the other. Using our last regression equation, we could predict weights for heights of 66, 67, 68, etc. inches. This is done as Table 9.10 in Supplement 9 which is a height-weight table similar to those appearing on some scales. In fact, that is exactly the way that some of the height-weight tables are computed, except that the people used in constructing them are supposed to be representative of the population at large rather than University of Florida male seniors.

What would happen if we were to take a student who is exactly $1s$ above the mean height and predict his weight? This student's height would be $\overline{Y} + 1.00s_y$, or $69.76 + 2.46 = 72.22$. Substituting, we get:

$$\begin{aligned}\tilde{X} &= 3.63Y - 94.73 \\ &= (3.63)(72.22) - 94.73 \\ &= 262.16 - 94.73 \\ &= 167.43 \\ &= 167 \text{ pounds.}\end{aligned}$$

This does not differ much from the value already obtained for the six-footer. It's the same to the nearest pound. But let's see how far above the mean weight this man probably is, in terms of standard scores. We'll keep fractions to reduce rounding errors. This predicted standard score in weight will then be

$$\begin{aligned}\tilde{z}_x &= \frac{\tilde{X} - \overline{X}}{s_x} \\ &= \frac{167.43 - 158.50}{16.25}\end{aligned}$$

$$= \frac{8.93}{16.25}$$
$$= +.5495$$
$$= +.55.$$

Please note that this value is the same as the value of the correlation coefficient. It had to be. Why? Look back at equations (9.1) and notice that the predicted score is always r times as far above the mean as is the known score from which the prediction is made. So when we start with a score 1.00 standard deviation above the mean, our predicted score will be r (or, in this case, .55) times 1.00, or .55 standard deviations above the other mean.

This is another illustration of regression toward the mean. We can now describe regression toward the mean more precisely. A person ns above the mean of one variable is expected to be rns above the mean of the other variable. Similarly, if he is ns below the mean of either variable, his predicted score on the other variable will be rns below the other mean. We have just seen this illustrated for $n = 1.00$ and $r = +.55$. As a second example, if $r = +.59$ and a person is $2.22s$ below the mean on one variable, his predicted score will be (.59)(2.22) or $1.31s$ below the mean of the other variable. What difference would it make if r were negative? The predicted distance from the mean would be the same, but its sign would be reversed, and the original score and the predicted score would be on opposite sides of their means. Thus if r were $-.55$ instead of $+.55$, the person $1.00s$ above one mean would have a predicted score of $-.55s$, which would, of course, be $.55s$ *below* the other mean. Also, if the r of .59 were negative, the person with a standard score of -2.22 would have a predicted standard score of $+1.31s$ instead of $-1.31s$.

The Relation of the Regression Coefficients to the Correlation Coefficient. For any scatter diagram, there are two regression coefficients. If the variables are called X_1 and X_2, then the regression coefficient used in the equation for predicting X_1 from X_2 is

$$b_{12} = r_{12}\left(\frac{s_1}{s_2}\right).$$

The other regression coefficient, used in the regression equation for predicting X_2 from X_1, is

$$b_{21} = r_{21}\left(\frac{s_2}{s_1}\right).$$

Let us multiply these equations by each other. We then get

$$b_{12}b_{21} = r_{12}\left(\frac{s_1}{s_2}\right) r_{21}\left(\frac{s_2}{s_1}\right)$$
$$= r_{12}r_{21}.$$

Since $r_{12} = r_{21}$, this may be written

$$b_{12}b_{21} = r_{12}^2.$$

Or, taking the square root,

$$r_{12} = \sqrt{b_{12}b_{21}}.$$

Thus the correlation coefficient, with appropriate sign, is equal to the square root of the product of the two regression coefficients. (The sign is very simply determined since r and the two regression coefficients always have the same sign. If any one of them is positive, all three are positive; if any one is negative, so are all three.) This relationship between the b's and r holds regardless of the units of measurement. We used pounds and inches, but the equation would still hold if we used kilograms and centimeters, for example, or if we just left all measurements in terms of x'_1 and x'_2 deviations. In every case, the standard deviations would always cancel out.

For a little more discussion of the relationship between the b's and r, read the tenth interpretation of r on pages 298–299 near the end of this chapter.

The Standard Error of Estimate. The appropriate regression equations enable us to predict, for example, that the six-foot man will weigh 167 pounds. Will he? Will his weight be the even more exact figure of 166.63 pounds that we obtained before rounding? We all know that this is not at all likely. The 167 pounds (or 166.63 pounds) is merely our best estimate of the man's weight. We are well aware of the fact that all six-foot students do not weigh the same amount. The regression equation simply estimates the mean weight of a large group of such men. It would help if we also knew the variability of the individual weights about this estimated mean. For instance, what is the standard deviation of the weight of these six-footers whose mean weight is estimated to be 167 pounds? This question is answered by computing a statistic known as the *standard error of estimate*. Its formula is

$$\begin{aligned} s_{y \cdot x} &= s_y\sqrt{1 - r_{yx}^2}, \\ s_{x \cdot y} &= s_x\sqrt{1 - r_{xy}^2}, \\ s_{1 \cdot 2} &= s_1\sqrt{1 - r_{12}^2}, \\ s_{0 \cdot 1} &= s_0\sqrt{1 - r_{01}^2}. \end{aligned} \qquad (9.4)$$

In each symbol to the left of the equal sign, the subscript to the left of the dot indicates the variable that is being predicted. The subscript to the right denotes the one from which the prediction is made. Thus, for the data of Table 8.8, the second of equations (9.4) becomes

$$
\begin{aligned}
s_{x \cdot y} &= s_x\sqrt{1 - r_{xy}^2} \\
&= 16.25\sqrt{1 - .55^2} \\
&= 16.25\sqrt{1 - .3025} \\
&= 16.25\sqrt{.6975} \\
&= (16.25)(.8352) \\
&= 13.57.
\end{aligned}
$$

For men 6 feet tall, it follows that the average weight is nearly 167 pounds with a standard deviation of about 13½ pounds. The usual interpretations apply: About two-thirds of such men may be expected to weigh between 153 and 180 pounds; 95% of them should be between 139 and 194. Do these things really happen? You may be interested to look now at row 72 of Figure 8.10 in order to see how well these predictions are fulfilled. The agreement is quite high for the 17 men in that row.

An interesting characteristic of the standard error of estimate is the fact that its value remains constant, regardless of whether the predicted score is high or low. Earlier we predicted the weight of a 5 foot 5 inch man as 141 pounds. This has exactly the same standard error of estimate as does the 167 pound prediction for the taller man. Hence, about two-thirds of the 5 foot 5 men should be within 13½ pounds of the predicted weight of 141 pounds. We cannot illustrate this in Figure 8.10 because there is only one man of that height. The characteristic of having equal variability in all arrays is called *homoscedasticity,* a long word which your instructor may or may not wish you to remember. There are a few circumstances in which it may not be justifiable to assume homoscedasticity, but ordinarily the assumption that all rows are equally variable (and that all columns are) is essentially correct and it is ever so much more reasonable than it would be to assume that there was any real significance in the erratic changes in the size of the standard deviation from one row to the next. The mere fact that in Figure 8.10 the standard deviation of the weights of 70-inch men is smaller than that of 69-inch or 71-inch men does not impress us as anything more than the chance fluctuation it undoubtedly is. We should not be at all surprised to find the situation reversed if another similar group were weighed and measured.

There are, of course, two standard errors of estimate for every scatter diagram or correlation coefficient. If we wish to predict height (Y) from weight (X), the standard error of estimate, $s_{y \cdot x}$, will give us a measure of the variability of the heights about the other regression line. Using the first of equations (9.4), we can compute the standard deviation of the errors made in predicting the heights of the male seniors from their weights. This gives

$$\begin{aligned} s_{y \cdot x} &= s_y\sqrt{1 - r_{yx}^2} \\ &= 2.46\sqrt{1 - .55^2} \\ &= (2.46)(.8352) \\ &= 2.05. \end{aligned}$$

Thus, going back to page 284, we may say that if one of these men weighs 177 pounds, we predict his height as 71.30 inches with a standard error of 2.05 inches, or about two-thirds of the time, the 177-pound seniors will be between 5 feet 9¼ inches and 6 feet 1¼ inches. This same standard deviation of predicted heights, 2.05 inches, is used regardless of what the student's weight is. Hence, the predicted height of a 133-pound student is 67.64 inches with the same standard error of estimate of 2.05 inches.

Just as the regression line does not go exactly through the mean of every row, so the standard error of estimate is not exactly equal to the standard deviation of that row. The correspondence is close, however, and as N increases, the means of successive rows come closer to the regression line, and the standard deviations of the rows come closer to the standard error of estimate. The position of the regression line in any particular row is definitely influenced by the position of the mean of that row. It is also influenced by the means of all the other rows, but most strongly, as it should be, by those nearby. The standard error of estimate is also influenced by every tally mark in the scatter diagram. It is, in fact, the standard deviation of all the scores, each expressed as a deviation from the value predicted by means of the regression equation. To illustrate, if we were to predict the weights of 75-inch men and subtract this value from the actual weights of the two men of that height, we would get two deviations. In the next row of Figure 8.10, there would be three deviations from the estimated weight for a man 74 inches tall. Similarly, the 73-inch row would give us six more deviations from the weight estimated by the regression equation. Continuing, we would finally obtain 107 deviations. Some would be positive, some negative, and their mean would be zero. Their standard deviation would turn out to be 13.57 pounds. That would be one way to determine the standard error of estimate. A much easier way is to substitute the appropriate values of s and r in Equation (9.4). Both procedures give exactly the same answer.

The standard error of estimate is thus the standard deviation of the differences between the actual values of the predicted variable and those estimated from the regression equation. More briefly, it is simply the standard deviation of the errors of estimate.

We should call attention to two limitations on the use of regression equations and the standard error of estimate. First, if regression is not linear, at least for the direction in which we are predicting, we are not entirely justified in using the regression equation. While we can still make predictions and the standard error of estimate will still tell us the size of our errors in prediction, we will not be doing as well as we might. It is usually best to try to attain linear regression by transforming one or both variables, using one of the methods given in Chapter 7. We shall then probably get both a higher r and better predictions, as evidenced by smaller errors of estimate.

Second, when we are using our data to predict scores in a second group (the usual situation), the errors of estimate may be somewhat larger than those given by Formula (9.4), especially for the high and low scores, primarily because if the two correlation coefficients are not identical the regression lines will differ. This effect diminishes, of course, when N is large.

THE GRAPHIC REPRESENTATION OF PREDICTION

With the aid of the four sets of numbered equations given earlier in this chapter, we have now discussed the fundamental concepts of regression both algebraically and verbally. Because these equations have interesting geometrical properties also, it seems desirable to show some of the relationships graphically. This can be done by means of Figure 9.1, which is based on Figure 8.10.

The line labeled $\hat{Y}$ is used for predicting Y from X. It is a plot of the equation on page 284,

$$\hat{Y} = .0833X + 56.56.$$

This line may be plotted from any two points, but its plotting is both easier and more accurate if we select X values near the ends of the scale. We can then check our work by plotting a third point anywhere else. We already have $\hat{Y}$ values for $X = 177$ and $X = 133$. These are fairly far apart, but since the scale goes down to 105, let us compute another value for the midpoint of the lowest interval. When $X = 107$, our equation gives $\hat{Y} = 65.47$. We now plot the three points ($X = 177$, $\hat{Y} = 71.30$), ($X = 133$, $\hat{Y} = 67.64$), and ($X = 107$, $\hat{Y} = 65.47$), and draw a straight line. Our work checks because the line goes through all three points. This is the line labeled $\hat{Y}$ in Figure 9.1. Notice how well it fits the midpoints of the columns. These midpoints are designated by small circles except when there are fewer than 8 cases in the column.

Similarly, the line labeled $\hat{X}$ is used for predicting weight, X, from height. It is a plot of the equation on page 284,

$$\hat{X} = 3.63Y - 94.73.$$

We have already solved this for heights of 6 feet and of 5 feet 5 inches. This gave us the points ($\hat{X} = 167$, $Y = 72$) and ($\hat{X} = 141$, $Y = 65$). We could plot

FIGURE 9.1
Regression Lines for Predicting Weight ($\tilde{X}$) from Height and for Predicting Height ($\tilde{Y}$) from Weight

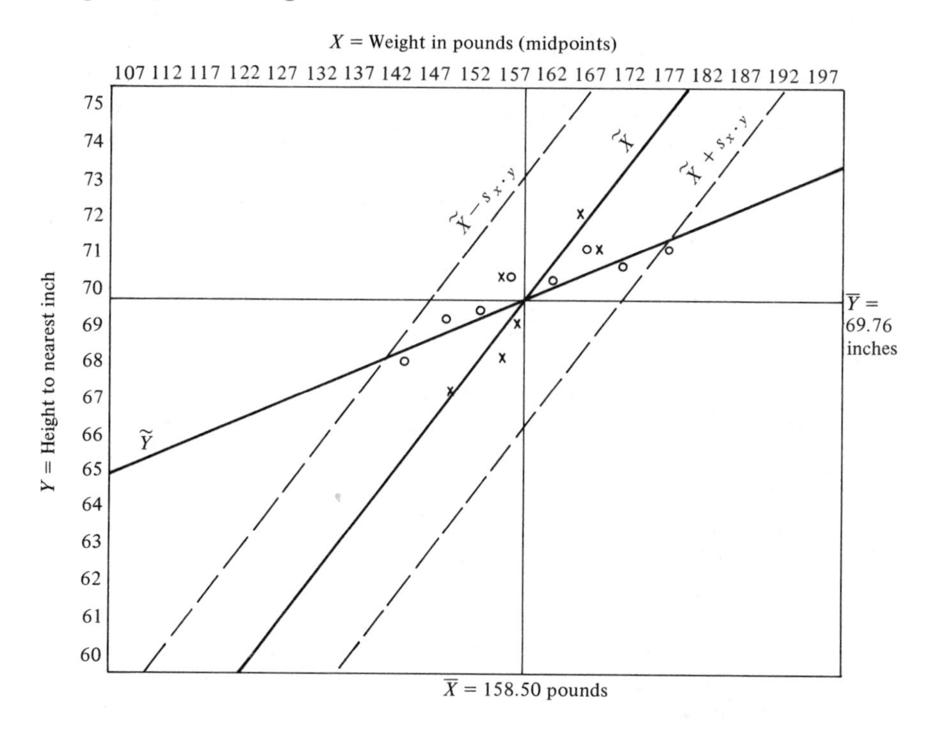

the regression line from these two values, but we can do it more easily and more accurately if we first compute a more extreme value. This will also give us our third point as a check. Assuming $Y = 60$ and solving for $\tilde{X}$, we obtain 123 pounds. Plotting these three points, we draw the regression line, noting that our work checks because the same straight line goes through all three points.

In the extremely rare situation in which $r = \pm 1.00$, the two regression lines coincide. For all other values of r, they do not, but they intersect at the point corresponding to the two means. In other words, a vertical line going through $\overline{X}$, a horizontal line going through $\overline{Y}$, and both of the regression lines will all meet in a single point, as they do in Figure 9.1.

The Graphic Representation of the Standard Error of Estimate. It is also possible to represent the standard error of estimate graphically. We simply measure it off above and below (or to the right and left of) the appropriate regression line. On page 288, we found the standard deviation of the errors made in estimating weight from height to be 13.57. This is the standard error of estimate, $s_{x \cdot y}$. This distance has been measured off to the right and to the left of the line labeled $\tilde{X}$ in Figure 9.1. It is indicated by the two dotted lines. These limits will ordinarily include about two-thirds of the cases plotted on the graph. In fact, they do in this case. Now if the arrays are homoscedastic, we shall have essentially equal standard deviations in all the rows. This means that roughly two-thirds of the cases in every row should be included between the dotted line at the left and the dotted line at the right. If you actually compare these lines on Figure 9.1 with the frequencies in Figure 8.10, it will be seen that this is approximately true. Any excess in one row is apt to be balanced by a deficiency in an adjacent row. The agreement between the expected and the actual distribution of cases is quite close.

As we also saw in Chapter 4, about 95 out of 100 cases are included within the limits of $\pm 2\sigma$. These relationships should hold true, at least approximately, for $s_{x \cdot y}$, as well as for σ. Imagine two more lines, just twice as far from the $\tilde{X}$ line as the two dotted lines are. Our knowledge of the normal probability curve would lead us to expect all but about 5 of our 107 cases to be included within these limits of $\pm 2s_{x \cdot y}$. Again the prediction is closely fulfilled by our data: Allowing for fractions when the line cuts through a cell containing a tally mark, there are between 5 and 6 cases outside these limits. This is very close to the theoretical expectation.

Thus the standard error of estimate (or multiples of it) may be represented by a zone or area around the regression line. Such a zone contains, at least approximately, the proportion of cases expected in accordance with the normal probability curve.

A SECOND ILLUSTRATION OF REGRESSION EQUATIONS

We will now show how regression equations can be used to predict the ages of brides or grooms. During a recent year in Florida, 53,848 persons were married. In one case neither party gave age information, in fifteen the groom didn't tell, and in the one other instance the bride didn't tell her age. Omitting these 17 marriages, the remaining 26,907 are shown in Figure 9.2.

Notice the high frequencies. In 6 of the rows and 7 of the columns, there are over a thousand marriages. Two cells contain over 4,000 apiece. As a result, any statistics (such as r_{xy}, s_x, and s_y) we compute from these records will be highly accurate; consequently, their standard errors will be unusually small.

The figures just below the scatter diagram show the number of bridegrooms in each of the 5-year age groups from 15–19 years on up to 90–94. It is obvious that this distribution is markedly skewed. At the right-hand side of the scatter diagram, we can see at a glance that the brides' figures are skewed too. It would be possible to eliminate the skewness by transforming the age scale. This is not necessary since, as we pointed out in Chapter 8,

FIGURE 9.2
Ages of Partners in Florida Marriages

Y = Age of Bride	X = Age of Groom																
	15–19	20–24	25	30	35	40	45	50	55	60	65	70	75	80	85–89	90–94	
85–89												1			1		2
80–84														4	1		5
75	1		1								3	5	6	7	3		26
70							1	1	1	6	18	32	25	13	1		98
65						2		3	13	26	67	60	50	13	3	1	238
60				1			3	14	28	83	98	72	42	6	1		348
55			1		2	9	20	52	101	106	116	50	23	5	1		486
50	1		2	5	7	27	53	142	145	117	85	36	21	4	1		646
45		7	4	16	53	102	202	227	154	94	72	20	8	1	1		961
40	1	5	27	75	161	352	306	216	97	57	32	8	2	1			1340
35		31	120	299	467	428	287	136	68	30	13	4	1				1884
30	2	108	431	676	515	330	181	73	33	8	5	1	2				2365
25	17	588	1345	875	480	213	107	44	13	7	2						3691
20	234	4072	2248	718	266	118	36	15	4	4	1	2	1				7719
15–19	1739	4169	878	174	56	24	2	2	2	1							7047
10–14	31	17	3														51
	2026	8997	5060	2839	2007	1605	1198	925	659	539	512	291	181	54	13	1	26907

the computation of Pearson r does not assume normal—or even symmetrical—distributions. If regression is linear, as it is here, we may use our regression equations also. The standard error of estimate is, however, inapplicable, as we shall see later.

There is nothing new in computing r. When N is 26,907, the amount of arithmetic is greater than it is when N is 100, as we found when we computed r for Figure 9.2. It is not, however, anything like 269 times as much work. It takes much longer to enter 26,907 than only 100 frequencies in a scatter diagram, of course, and the chances of error are greatly increased. After the frequencies are entered in the scatter diagram, it takes perhaps two to five times as long to finish computing r for 26,907 cases as for 100. This is true whether the calculations are done by machine or by hand. The numbers are large, the numerator for r computed by Formula (8.3) being 4,031,713,695 and the denominator 4,604,617,100. Dividing, we find that r is + .88. Because of the large number of cases, we know that this obtained r of .88 cannot differ much from the true r. How much? First, let us carry out the computation of r to two more decimal places. The figure is +.8756. Transmuting this into z' by interpolating in Table K, we find that $z' = 1.3566$. Ordinarily we would next use Table L to obtain an appropriate multiple of the standard error of z', but since our N is so far beyond the range of the table, we shall compute $s_{z'}$ by using Formula (8.5):

$$\begin{aligned} s_{z'} &= \frac{1}{\sqrt{N-3}} \\ &= \frac{1}{\sqrt{26{,}907 - 3}} \\ &= .0061. \end{aligned}$$

We can afford to be ultraconservative, so we shall use the .001 significance level and multiply this value by 3.29, obtaining .0201. Thus, the .999 confidence limits for our z' are $1.3566 \pm .0201$. These limits are 1.3365 and 1.3767. Using Table K in the reverse direction and interpolating, we obtain limiting values for r of +.8708 and +.8802. We can, then, be quite sure that the true value of r in such situations as this lies between .87 and .88. These are narrow limits indeed.

Let us digress a moment to a technicality in the computation of the means. Ask a girl who was born 10 years, 10 months, and 10 days ago how old she is. She will probably say "10" and not "11." Let us compute the average age of 100 16-year-old girls and 100 17-year-old girls. We could get the wrong answer by thinking that

$$\overline{X} = \frac{(100)(16) + (100)(17)}{200} = 16.50.$$

Why is this wrong? The girls' ages vary all the way from 16 years and 0 days to 17 years and 364 days. The mean is clearly near the middle of this range—at 17 and not at 16½. Thus, although one of our intervals is correctly labeled as 20–24, since it includes only those people who stated they were 20, 21, 22, 23, or 24 years of age, this interval actually refers to people of all ages from 20 years and 0 days to 24 years and 364 days. Its midpoint, then, is 22.50 years and not 22, which would ordinarily be the midpoint of a 20–24 interval. This digression shows that we must be aware of the nature and meaning of our data; statistical methods should not be applied blindly.

By the use of formulas previously given in Chapters 2, 3, 8, and 9, we can use the data on Florida grooms and brides to compute the statistics needed for our regression equations. The correct values of the various statistics are shown in Table 9.4.

Since r is positive, it follows that both regression coefficients are also positive. Because s_x is a little larger than s_y, it follows that b_{xy} is also a little larger than r_{xy}. The other regression coefficient, b_{yx}, must be correspondingly smaller since the value of r is always between the values of the two regression coefficients. (In fact, r is equal to the square root of their product.) The value of b_{xy} is probably the first we ever computed that turned out to be exactly 1.00. This is very convenient, because the regression equation for predicting the groom's age becomes simply: Add 3.95 years to the age of the bride. This regression line is labeled $\hat{X}$ and runs practically from one corner to the other corner of Figure 9.3.

As is well known, grooms are usually older than their brides. This is true, even though one of these 76-year-old women married a man who stated his age as exactly one-fourth that of his perhaps-blushing bride. While not a prevalent practice, the pairing of pensioners with premature partners—whether predestined, premediated, or merely precipitous—does occur, though rarely in Florida, for the average groom (at 32.15 years) is almost 4 years older than the average bride (28.20 years). The variability of the grooms' ages is also a little larger than that of their brides.

Let us solve this regression equation for a few values. Suppose Adelaide decides to marry just as soon as she gets to be 18 years old. How old will her groom be? He might, of course, be any age, but our best guess is 18 + 3.95, or 21.95 years. Belle didn't realize that in over 30% of the marriages, the bride is over 30. Because of her anxiety over the thought of reaching 30 while still single, she married What's his name at 29. What is the age of her groom? The regression equation predicts his age as 3.95 more than Belle's 29.50, or 33.45. When Constance and Ernest became engaged, they decided to wait a few days (or months) and get married on her fiftieth birthday. Our best estimate of her bridegroom's age is 50.00 + 3.95 or 53.95 years.

How accurate are these estimates? Ordinarily, this question might be answered in two quite different ways. Each row of Figure 9.2 shows the age distribution of grooms whose brides are all in the same interval. Starting at the bottom with the 10–14 year child brides, the standard deviation of their grooms' ages can be computed as 3.05, but such standard deviations increase steadily until they reach 9.50 for the grooms of 45–49 year brides. The same is true of the columns. Thus, the standard deviation of brides' ages increases from 2.55 for the youngest grooms to 9.30 for the brides of 60–64 year men. Both for the grooms and for the brides, the standard deviation of successive arrays keeps increasing to over 3 times its initial value, and then behaves somewhat erratically at the oldest ages. This is not the ordinary situation, but it did happen with this particular scatter diagram. Hence, because this scatter diagram is not homoscedastic, the standard error of estimate is inapplicable here, and we shall not use it with these predictions.

FIGURE 9.3
Regression Lines for Predicting Age of Other Partner in Florida Marriages

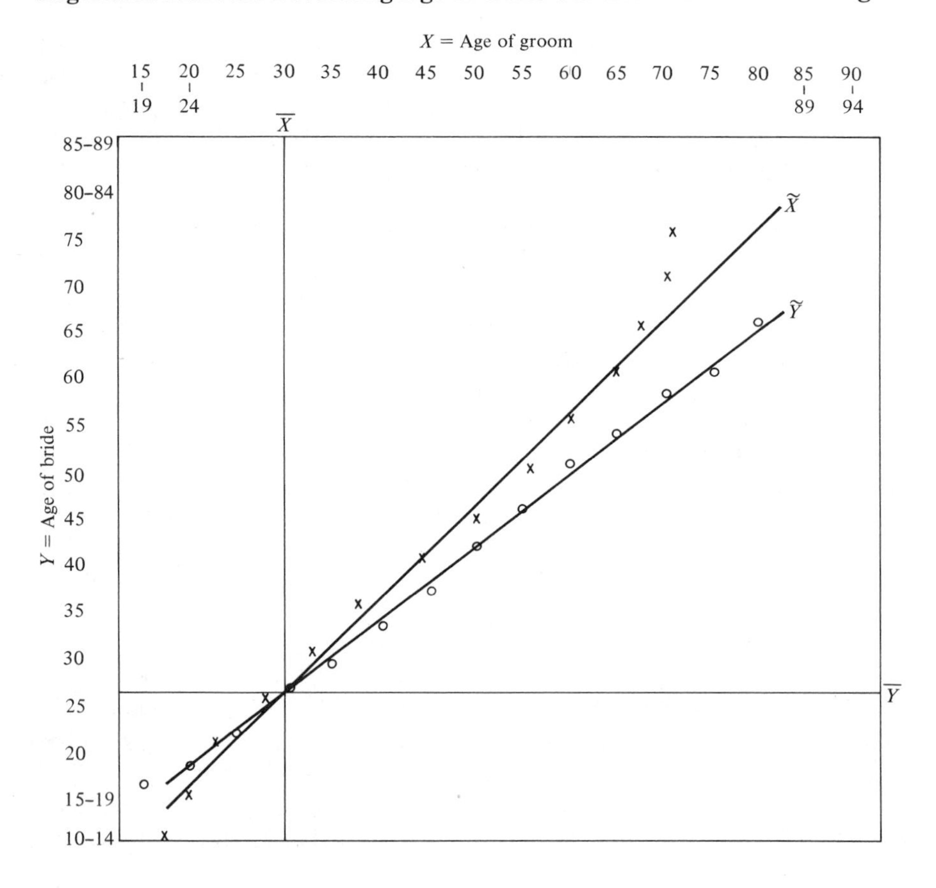

TABLE 9.4
Basic Statistics Computed from Scatter Diagram for Age of Groom and Age of Bride

Groom	Bride
$N = 26{,}907$	
$\overline{X} = 32.15$	$\overline{Y} = 28.20$
$s_x = 13.45$	$s_y = 11.85$
$r_{xy} = +.88$	
$b_{xy} = +1.00$	$b_{yx} = +.78$
$\tilde{X} = 1.00Y + 3.95$	$\tilde{Y} = .78X + 3.12$

How else can we tell how accurate these estimates are? In terms of how much or how little the $\tilde{X}$ predicted for a given row differs from the actual mean of that row. The regression line, labeled $\tilde{X}$, can be seen to pass very close to the means of the rows of Figure 9.3. These means are shown by small crosses (×) for each row containing more than 20 cases. The only discrepancies of any consequence are in the few rows containing relatively small numbers of cases at the extreme ages. It is obvious that a straight line fits the data very well.

To be specific, let us consider the brides Debbie and Elizabeth, who give their ages as 22 and 42, respectively. The ages of the two grooms as estimated from the regression equation are 26.45 and 46.45. The average ages of grooms of the 20–24 and 40–44 year brides, as computed from these two rows of Figure 9.2 are 25.80 and 46.50. These two sets of values are very close together. Thus, this regression equation is quite satisfactory.

The other regression equation is

$$\tilde{Y} = .78X + 3.12.$$

In other words, to estimate the age of the bride, add 3.12 years to $\frac{78}{100}$ of the age of the groom. Let us continue through the alphabet and predict the ages of the brides of several grooms. Frank married on his twenty-first birthday; his bride's age was probably close to (.78) (21.00) + 3.12, or 19.50 years. Gerald marries when his age reaches exactly one-third of a century; his spouse's age is predicted as 29.12 years. Harry's head was not hairy when he married the girl of his dreams. They decided to attempt a double celebration on his eightieth birthday. Our regression equation predicts his bride's age as 65.52 years.

Let us assume that Ike and Jimmy both marry when their ages coincide with the midpoints of two of the columns of Figure 9.2. We can then get two estimates of their brides' ages—one from the regression equation and one from the actual mean of the column. If Ike was 22.50 years old when he married, the regression equation predicts his bride's age as 20.67. The average age of the 8,997 brides of 20–24 year old men is almost identical: 20.70 years. Similarly, if Jimmy married at the age of 42.50, the corresponding estimate of his bride's age is 36.27 years. This is practically the same as the 35.90 average for that column. These two examples merely illustrate what can be seen in Figure 9.3. The regression line (labeled $\hat{Y}$) for predicting the age of the bride gives a very good fit to the small circles (○) which mark the means of all columns containing more than 20 cases.

We have now seen that regression is linear in both directions, but that the standard error of estimate is inapplicable. The latter is very unusual, but unusual things do happen—not only to California weather and Florida sunshine, but also to statistical data. In the ordinary problems dealt with in Psychology and Education, regression is usually linear or essentially so (though we rarely have data on thousands of subjects to provide such convincing proof of it), but ordinarily there is no systematic change in variability as we go from one column (or row) to the next. Consequently, the standard error of estimate is usually a very good indicator of the accuracy of our predictions.

FURTHER INTERPRETATIONS OF *r*

In Chapter 8 we gave eight methods of interpreting r. The other seven were delayed because they required an understanding of concepts related to regression. Here they are.

9. Statisticians are often bothered by the fact that some people seem to have a very strong desire to interpret r as a proportion or percent. Such people are practically always wrong, but there is one sense in which r may be regarded as a proportion. We have already referred to this on pages 280 and 285–286. When we know how many standard deviations above or below the mean a person is on one variable, r_{xy} tells us at what proportion of this distance he is likely to be on the other variable. Thus, if his standard score on X is $+1.5$, and if $r_{xy} = +.60$, then his predicted standard score on Y is .60 times $+1.5$, or $+.9s_y$. The correlation coefficient tells us by what proportion we need to multiply z_x in order to obtain $\tilde{z}_y$. It shows what proportion the predicted standard score, $\tilde{z}_y$, will be of the known standard score on the first variable, z_x.
10. The correlation coefficient may be regarded as the geometric mean of the two regression coefficients, as we saw on page 287. (The geometric mean of two numbers is the square root of their product.) If r is either $+1.00$ or -1.00, the two regression lines will coincide, as can be seen from the first and last scatter diagrams in Figure 8.4, and as is

also shown in the last diagram of Figure 9.5. If $r = .00$, one regression line will be horizontal and the other vertical. Thus the slopes of these two lines are obviously related to the size of r. The slope of one regression line is determined by measuring the number of units change in the vertical scale corresponding to a change of one unit in the horizontal scale. That is, its slope is the amount of vertical change in the value of Y corresponding to a horizontal change of 1 unit in the value of X; the slope is a measure of the extent of the departure from horizontal. The slope of the other regression line is similarly measured by the extent to which it departs from being a vertical line. Now r is the square root, with appropriate sign, of the product of these two slopes. If we plot standard scores, z_x and z_y, instead of raw scores, X and Y, or if in some other way we make the standard deviations equal in the horizontal and vertical directions, two interesting things happen: (1) both regression lines will have the same slope and (2) the slope of either of them will be equal to r. Then for a change of any amount in either variable, we expect a change r times as great in the other variable, just as we found in interpretation (9). This is a very useful interpretation and it involves very few assumptions, nothing but equality of units and linear regression.

11. Instead of dealing with the slopes of the regression lines, we could shift our attention to the angle between them. Figure 9.5 shows the regression lines for predicting X (labeled $\hat{X}$) and for predicting Y (labeled $\hat{Y}$) for correlations from .00 to 1.00 when $s_x = s_y$. A glance shows that the larger r is, the smaller is the angle between the regression lines $\hat{X}$ and $\hat{Y}$.

If r is zero, then there is no relation between the two variables; and knowledge of a man's score on one variable will be of no value in predicting his score on the other variable. For instance, there is a correlation of just about zero between the height and the pulse rate (sitting) of the 107 male seniors. Now estimate the heights of three of the students whose pulse rates are recorded as 64, 80, and 96, respectively. Which one is the tallest? Clearly we have no information that will help us to predict their heights. The best we can do is to guess *all three* at the mean. And that is exactly what the regression equation and the regression line do. If $r = .00$, b_{yx} will also equal zero; and the regression equation for predicting Y from X will become simply $\tilde{Y} = \overline{Y}$. This may appear to be unsatisfactory or even discouraging, but if $r = .00$, the plain fact is that high or low scores on one variable are related only in a strictly chance manner to high or low scores on the other variable. Hence, when $r = .00$, the regression line for predicting Y from X is simply a horizontal line going through $\overline{Y}$. This is shown as the line labeled $\tilde{Y}$ in the small diagram in the upper left corner of Figure 9.5. Similarly, in such a situation, the other regression line becomes the vertical line shown in the same diagram and labeled $\tilde{X}$.

FIGURE 9.5
Selected Regression Lines for Predicting X and Y for Correlations from .00 to 1.00 when $s_x = s_y$

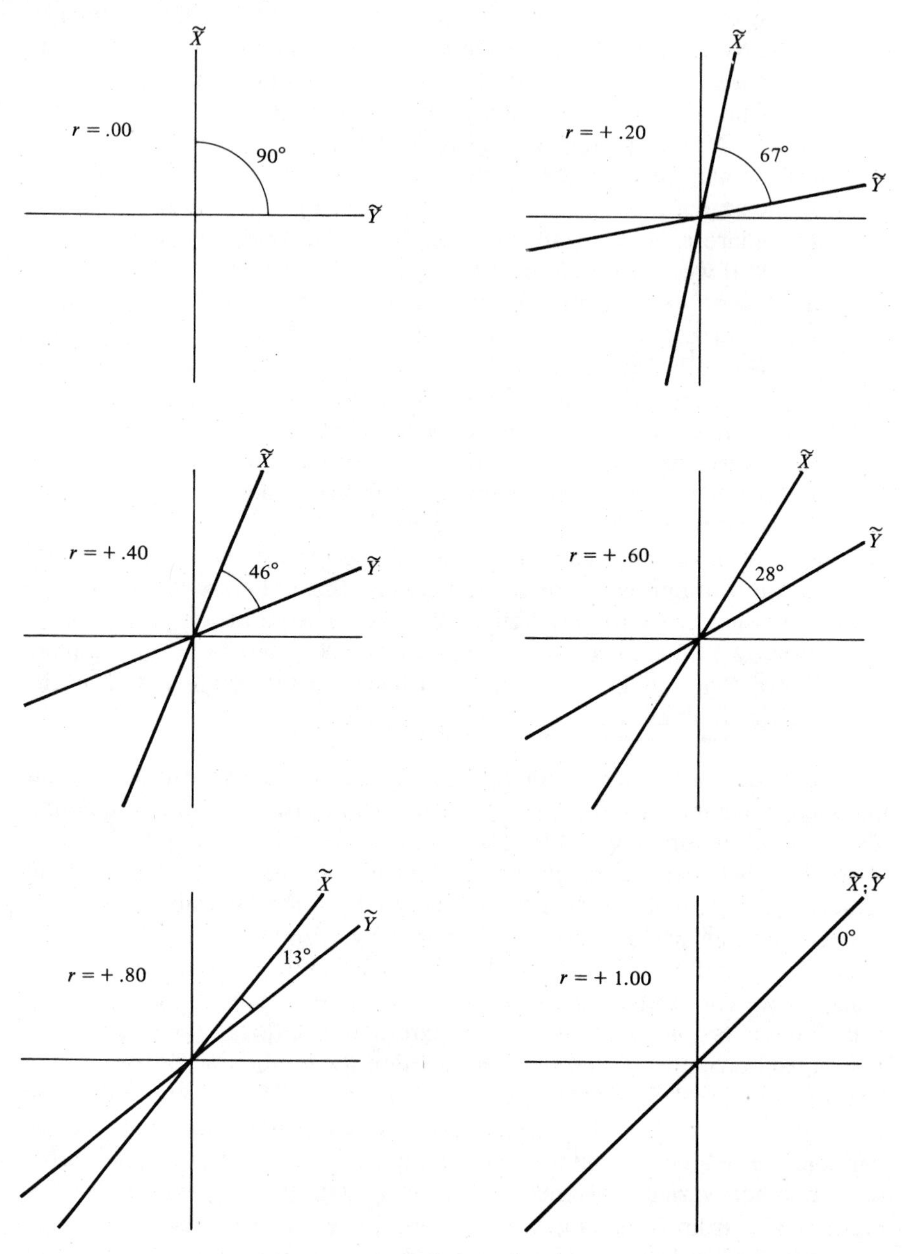

The remaining small diagrams of Figure 9.5 show that as r increases from .00 to +1.00, the angle between the two regression lines* keeps getting smaller and finally becomes zero when $r = +1.00$. This relationship is also depicted clearly in the left half of Table 9.6. Notice that in both of these places we are assuming equality of the two standard deviations.

What happens if the two standard deviations are not equal? If the correlation is plotted, it will rarely happen that the standard deviations are *exactly* equal, but usually the difference will not be great. Without giving any special attention to it, the interval widths will usually be so chosen that the larger standard deviation (measured in inches on the scatter diagram) will be less than twice as large as the smaller one. The right-hand side of Table 9.6 shows the relation between r and θ in this relatively extreme situation.

The two sides differ all right, but not very much. And in both of these tables the sum of θ in degrees plus r in hundredths is nearly a constant for most of the common correlation coefficients. Hence, the rule: For all correlation coefficients from .00 to .70, **r (in hundredths) is equal to 85 minus the angle** (in degrees) **between the two regression lines.** This rule never produces an error of more than .05 (or 5°) if the plotted standard deviation of X is anywhere from half to double that of Y—and these limits include nearly all the correlations you'll ever plot. If the two standard deviations are equal, as is the case when standard scores are used, then an even more precise rule may be given: ***r* is equal to 88 minus *θ*.** In this case, the error will not exceed .02 for any r from .00 to .70.

Figure 9.5 and Table 9.6 thus demonstrate the simple inverse relationship between θ and r for practically all positive correlations we encounter except reliability coefficients. What about negative correlations? There are two equally good alternatives. (1) Simply subtract θ from 180°, proceed as before, and put a minus sign before r, or (2), use the smaller of the two angles between the regression lines, proceed as before, and put a minus sign before r.

---

*The exact relationship between r and this angle is a complex one that can be completely understood only by those familiar with trigonometry. It can be expressed by either of two equations

$$\tan \theta = \frac{1 - r^2}{2r}$$

or

$$r = \sqrt{1 + \tan^2 \theta} - \tan \theta.$$

It is not necessary, however, to know, or even to understand, these formulas. Forget them. If you need to know the value of r or of the angle, θ, between the regression lines, just look it up in the left half of Table 9.6.

TABLE 9.6
The Effect of the Closeness of the Regression Lines Upon the Size of Pearson *r*

When Standard Deviations are Equal		When One Standard Deviation is Twice as Large as the Other	
Angle Between Regression Lines (to nearest Degree)	*r*	*Angle Between Regression Lines (to nearest Degree)*	*r*
0°	1.00	0°	1.00
3°	.95	2°	.95
6°	.90	5°	.90
9°	.85	7°	.85
13°	.80	10°	.80
16°	.75	13°	.75
20°	.70	16°	.70
24°	.65	20°	.65
28°	.60	23°	.60
32°	.55	27°	.55
37°	.50	31°	.50
42°	.45	35°	.45
46°	.40	40°	.40
51°	.35	45°	.35
57°	.30	51°	.30
62°	.25	56°	.25
67°	.20	62°	.20
73°	.15	69°	.15
79°	.10	76°	.10
84°	.05	83°	.05
90°	.00	90°	.00

Aside from the stated restriction that neither s_x nor s_y as plotted must be more than double the other, this method of interpreting r involves no assumptions whatever.

12. When we use the knowledge of a person's score on one variable to predict his score on a second variable, we use a regression line to make our prediction. This gives us the best estimate of the individual's score on that second variable. Our tenth method of interpreting r was in terms of such regression lines. However, such a best estimate is subject to error. Except when $r = \pm 1.00$, there is a scatter of cases about the regression line. The smaller this scattering, the higher the correlation, and the greater the scattering, the lower the correlation. So instead of interpreting r in terms of the regression lines, we may interpret it in terms of deviations from one of these regression lines.

From Formulas (9.4), we learned that the standard deviation of the points about a regression line is equal to the standard deviation of the variable being predicted multiplied by $\sqrt{1 - r^2}$. Thus the quantity, $\sqrt{1 - r^2}$, called the *coefficient of alienation,* gives us a measure of the extent to which the errors in our estimates are reduced. When $r = \pm 1.00$, the coefficient of alienation is .00 and we make no errors in predicting one score from the other. When $r = .00$, the coefficient of alienation is 1.00 and hence the standard error of estimate is the same as the standard deviation of the marginal frequencies. When $r = .50$, the coefficient of alienation is .87, which means that our errors of estimate are .87 as large as, or 13% smaller than, they would be if there were no correlation between the two variables. Values of the coefficient of alienation for selected values of r are shown in Table 9.7. The table shows that if we wish to cut our error of estimate to .80 of what it would have been if there were no correlation, we need an r of .60. To cut the standard error of estimate to half the standard deviation, r must be .8660. An excellent way to get a clearer understanding of this interpretation is through the use of the eleven scatter diagrams in Figure 8.4. Notice that when $r = +.19$, zero, or $-.20$, the standard deviation within the columns is practically as great as the standard deviation of the marginal frequencies. (The same is true of the rows, of course.) When r is equal to $+.56$ or $-.60$, the variability within columns is, on the average, eight-tenths as great as the standard deviation obtained from the marginal frequencies. This within-column variation is reduced to $.60s$ when r is $+.78$ or $-.80$; and it is

TABLE 9.7
Relation between Coefficient of Correlation, r, and the Coefficient of Alienation, k

r	k
.00	1.0000
.20	.9798
.40	.9165
.50	.8660
.60	.8000
.7071	.7071
.80	.6000
.8660	.5000
.90	.4359
.95	.3122
.99	.1411
1.00	.0000

nonexistent when r reaches the most extreme values in the table, $+1.00$ and -1.00. The coefficient of alienation, k, is related to the coefficient of correlation, r, in the same way that the sine is related to the cosine of an angle. Those who have studied trigonometry will probably remember that

$$(\text{sine})^2 + (\text{cosine})^2 = 1.00.$$

The corresponding relationship, which is still true whether or not you have ever seen a trigonometry book, is

$$r^2 + k^2 = 1.00. \tag{9.5}$$

This relation is also shown in Figure 9.8. Notice that as r increases, the decline in k is extremely gradual—almost negligible—at first, but that slight increases in r result in large decreases in k when r is in the neighborhood of .80 to 1.00. The curve is an arc of a circle; it is horizontal at $r = .00$ and vertical at $r = 1.00$.

This interpretation of r in terms of k or the reduction in the standard error of estimate makes several assumptions. Besides equality of units and linear regression, it assumes that the scores in each column (and in each row) are essentially normally distributed about the regression line, and it further assumes that the standard deviations in the various columns (and in the

FIGURE 9.8
Relation between r and k where $k = \sqrt{1 - r^2}$

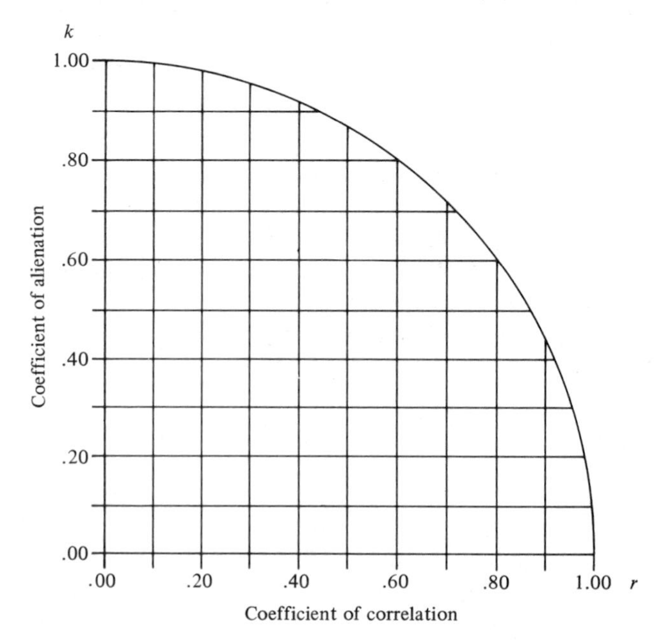

various rows) are equal. Although we do not have to assume that the X and Y variables are normally distributed, it is often true that when they are, the assumptions just listed will be approximated if not completely satisfied.

13. The interpretation of r in terms of k gives big numbers when correlation is low and little numbers when the correlation is high. This reversal bothers some people. A minor modification of the preceding method gives a value, E, called the *coefficient of forecasting efficiency* or the predictive index or the improvement over chance. The formula for E is

$$E = 1 - k = 1 - \sqrt{1 - r^2}. \tag{9.6}$$

The use of this formula adds very little to the understanding of r that is obtained through the use of k alone. We recommend method (12) rather than method (13). This method is listed solely because other people use it and you should know what they mean if they refer to it in a book or an article. Both methods are similar and, of course, involve exactly the same assumptions.

14. Method (12) gives a rather discouraging picture. If a clerical test correlates .60 with success on the job, most industrial psychologists would say that the test is giving an excellent prediction of future success, since most such correlations are lower than .60. Nevertheless, when r is .60, k is .80, and E is .20. Telling an executive that we can now make predictions that enable us to miss a clerk's actual production only .80 as much as we would have missed it if we had used no test at all and simply guessed everybody at the mean is accurate, all right, but the executive is not apt to be impressed. To tell an executive that this is an improvement of 20% over chance could prompt consideration of taking the psychologist off the payroll. These figures are correct, but they are sometimes misinterpreted.

To give a more favorable but absolutely correct picture of the meaning of a given value of r, when used to predict the success of newly hired employees, Taylor and Russell* published their famous tables, a portion of which is reproduced as Table 9.9. The *selection ratio* is the ratio of the number of persons hired to the number who were tested and available for placement. If we are going to hire everybody we can get for some job, clearly it doesn't matter whether we use a test with high validity or one with low validity, or whether we use any test at all. But if we can select only the top half, or the

*Taylor, H. C., and Russell, J. T. The relationship of validity coefficients to the practical effectiveness of tests in selection: discussion and tables. *Journal of Applied Psychology*, 1939, **23**, 565–578.

TABLE 9.9
Proportion of Above Average Employees for Various *r*'s and Selection Ratios

Correlation Coefficient, *r*	Selection Ratio .05	.10	.20	.30	.40	.50	.60	.70	.80	.90	.95	1.00	
.00	.50	.50	.50	.50	.50	.50	.50	.50	.50	.50	.50	.50	**.00**
.05	.54	.54	.53	.52	.52	.52	.51	.51	.51	.50	.50	.50	**.05**
.10	.58	.57	.56	.55	.54	.53	.53	.52	.51	.51	.50	.50	**.10**
.15	.63	.61	.58	.57	.56	.55	.54	.53	.52	.51	.51	.50	**.15**
.20	.67	.64	.61	.59	.58	.56	.55	.54	.53	.52	.51	.50	**.20**
.25	.70	.67	.64	.62	.60	.58	.56	.55	.54	.52	.51	.50	**.25**
.30	.74	.71	.67	.64	.62	.60	.58	.56	.54	.52	.51	.50	**.30**
.35	.78	.74	.70	.66	.64	.61	.59	.57	.55	.53	.51	.50	**.35**
.40	.82	.78	.73	.69	.66	.63	.61	.58	.56	.53	.52	.50	**.40**
.45	.85	.81	.75	.71	.68	.65	.62	.59	.56	.53	.52	.50	**.45**
.50	.88	.84	.78	.74	.70	.67	.63	.60	.57	.54	.52	.50	**.50**
.55	.91	.87	.81	.76	.72	.69	.65	.61	.58	.54	.52	.50	**.55**
.60	.94	.90	.84	.79	.75	.70	.66	.62	.59	.54	.52	.50	**.60**
.65	.96	.92	.87	.82	.77	.73	.68	.64	.59	.55	.52	.50	**.65**
.70	.98	.95	.90	.85	.80	.75	.70	.65	.60	.55	.53	.50	**.70**
.75	.99	.97	.92	.87	.82	.77	.72	.66	.61	.55	.53	.50	**.75**
.80	1.00	.99	.95	.90	.85	.80	.73	.67	.61	.55	.53	.50	**.80**
.85	1.00	.99	.97	.94	.88	.82	.76	.69	.62	.55	.53	.50	**.85**
.90	1.00	1.00	.99	.97	.92	.86	.78	.70	.62	.56	.53	.50	**.90**
.95	1.00	1.00	1.00	.99	.96	.90	.81	.71	.63	.56	.53	.50	**.95**
1.00	1.00	1.00	1.00	1.00	1.00	1.00	.83	.71	.63	.56	.53	.50	**1.00**
	.05	**.10**	**.20**	**.30**	**.40**	**.50**	**.60**	**.70**	**.80**	**.90**	**.95**	**1.00**	

top tenth, of the applicants, more than 50% of such employees will be above average, the exact proportion being determined both by the selection ratio and by the validity coefficient (which is merely the *r* between the test score and a criterion of success on the job). To illustrate, the column of the table corresponding to a selection ratio of 1.00 shows that if we hire everybody, .50 of those we hire will be above average no matter what the validity of our test may be. If our test has a validity of .40 and we hire half the applicants, the figure of .63 at the intersection of the .40 row and the .50 column shows that 63% of such new employees will be above average. If through advertising or other methods we can get many more *equally good* people to apply so that our selection ratio can be reduced to .20, Table 9.9 shows that 73% of those hired will be above average. This is just as good selecting, from the point of view of above-average employees, as could have been achieved by a test with the much higher validity of .65 if half those tested had been hired.

Let us now assume a selection ratio of .05 and a validity of only .30 for the testing procedure. Table 9.9 again shows 73 or 74% above average. Let us summarize:

Validity	Selection Ratio	Percent of Those Hired Who Are Above Average
.65	.50	73%
.40	.20	73%
.30	.05	74%

This little tabulation shows clearly that, in a very real sense, a decrease in the selection ratio can be used as a substitute for increased validity. The tables prepared by Taylor and Russell cover a wide range of standards from only .05 of those employed being considered satisfactory through .50 satisfactory (which is the table we quoted) on up to .90. The tables are very effective in showing the improvement in the selection of groups of individuals under any specified conditions with respect to the three variables involved—selection ratio, validity coefficient, and proportion considered satisfactory. They are of primary use in industry but can be used equally well in selecting students, either at the time of admitting them to college or in selecting from the applicants for various advanced degrees. The assumptions involved here are similar to but not exactly the same as those made in methods (12) and (13), even though this method of interpreting r is a direct outgrowth of the principles involved in method (12). In addition to equality of units and linear regression, we assume that the scores in each column of the Y or criterion variable are normally distributed, that the standard deviations in the various columns are equal, and that the marginal frequencies of both the X and the Y distributions are normally distributed.

15. This last method of interpreting r is somewhat technical, but we shall skip most of the details. First, we recall that the variance is simply the square of the standard deviation. The variance of the scores on a variable Y may then be divided into two parts: the part that can be predicted from knowledge of a person's score on another variable, X, and the part that cannot be so predicted. The part that cannot be so predicted is the variance of a column in the scatter diagram. We noted in method (12) that the standard deviation of a column is $\sqrt{1 - r^2}$ times the standard deviation of the Y variable. The variance of a column will then be $(1 - r^2)$ times the variance of the Y scores. Thus $(1 - r^2)$ corresponds to the proportion of the variance of the Y scores that cannot be predicted from X. Consequently, $1 - (1 - r^2)$, or simply r^2, corresponds to the part of the variance of Y that *can* be predicted from X. Hence, r^2 gives the proportion of the variance of either variable that can be predicted from the other variable. If in some other way we know that variations in one of the variables are the cause of variability in the other, then r^2 tells what proportion of the variance of the one variable is caused by variations in the other. Whether or not we know about the direction of causation, r^2 tells us the proportion of the variance of either variable that is determined by the variance of the other variable. For this reason r^2 is sometimes called the *coefficient of determination*. It should be noted that although r^2 can be interpreted in terms of changes in s^2, there is no corresponding interpretation of r in terms of changes in s. In making this final interpretation in terms of variance we make few assumptions—only the basic ones of equality of units and linearity of regression.

SUMMARY

Regression equations enable us to perform one of the most important scientific functions—that of predicting. These equations may be written in terms of standard scores or of deviations from the mean, but the most practically useful are those in terms of raw scores, as given in Equations (9.3). The regression equations may be plotted as regression lines. There are two regression lines for every scatter diagram. When r is near zero, they will be far apart—approximately at right angles to each other. As r increases, the regression lines come closer together until, when $r = 1.00$, they coincide. Pearson r may be interpreted in terms of their slopes, or of the angle between the regression lines, or in many other ways.

If there is any correlation at all, either positive or negative, between two variables, we can always make a better estimate of the score on a second variable by using a regression equation than we could without it. That is, our predicted score is better than a blind guess of the mean for everyone. Unless $r = \pm 1.00$, however, these predicted scores are seldom *exactly* right and are subject to some error. The standard error of estimate is ordinarily the best measure of this error. The higher the absolute value of r, the smaller will be the standard error of estimate.

SUPPLEMENT 9

MAKING A LARGE NUMBER OF PREDICTIONS

In Chapter 9, several examples were given of predictions based upon Equations (9.3). Suppose, though, that rather than making a few individual predictions we wish to make dozens or even hundreds of predictions. Instead of making each one individually, we will be much better off to operate on a wholesale basis.

We shall illustrate this with the data of Table 8.8 and Figure 8.10 on the height and weight of college seniors. (If you don't feel that those University of Florida men are typical, get some other data—preferably on a few thousand such students—and proceed in exactly the same way.) We have already used the second one of Equations (9.3) in order to get a regression equation for predicting weight from height. That equation was

$$\hat{X} = 3.63\,Y - 94.73.$$

This equation clearly shows that every time height (Y) increases by one inch, the predicted weight ($\hat{X}$) increases by 3.63 pounds. We shall limit our predictions to the range from 64 to 74 inches and shall make them for people measured to the nearest half inch.

Substituting $Y = 64$ in the equation, we get $\hat{X} = 137.59$. Substituting $Y = 64.5$, we get $\hat{X} = 139.40$. Substituting $Y = 65$, we get $\hat{X} = 141.22$. Clearly, each change of ½ inch results in a change in the predicted weight of 1.815 pounds. So all we need to do is keep adding 1.815 pounds each time the height goes up. Doing this, rounding to the nearest pound, and putting the

TABLE 9.10
Predicted Weights for Various Heights, Based on Data from University of Florida Senior Men

Height	Predicted Weight
6′2	174
6′1½	172
6′1	170
6′0½	168
6′0	167
5′11½	165
5′11	163
5′10½	161
5′10	159
5′9½	158
5′9	156
5′8½	154
5′8	152
5′7½	150
5′7	148
5′6½	147
5′6	145
5′5½	143
5′5	141
5′4½	139
5′4	138

tallest men at the top of the page, we get Table 9.10. To check the table, start at the tall end, for $Y = 74$, get $\tilde{X} = 173.89$ which is rounded to 174. Then from this 173.89, successively subtract 1.815 and thus check every value in Table 9.10. This table can then be used to predict the weight of any senior if you think he is reasonably typical of the group on whom these norms or standards are based.

Notice that if the predictions had been made for heights differing by whole inches, the subtractive factor would have been the 3.63 given in the regression equation, rather than 1.815 which was half of it (for half inches). Note also that the weight predictions we made in Chapter 9 could readily have been obtained by simply looking them up in Table 9.10. It's a real time-saver if you wish to make many predictions—no matter what the variables are.

CHAPTER 10 MORE MEASURES OF CORRELATION

Under most circumstances, if it is physically possible to compute it, the Pearson product moment coefficient of correlation is the best measure of relationship and should be used. There are, however, at least twenty other measures of correlation. Except for the measurement of curvilinear relationship, all the others are merely imitations of the Pearson r we studied in Chapter 8.

WHY OTHER CORRELATIONS

In view of what has just been said, what reason is there for the existence of all these other correlation coefficients? Quite aside from the fact that some statisticians just love to derive formulas, there are at least four reasons: (1) Sometimes the available records are not in such condition as to make it possible or feasible to compute Pearson r. (2) Some methods have been proposed as time-savers. Their sponsors realize that these procedures are less accurate, but they are willing to sacrifice accuracy for time. It must be admitted that there actually are situations in which such a decision is justifiable; ordinarily it is not. (3) In some situations, the use of Pearson r would be inappropriate and other methods are available for properly measuring the amount of relationship between the two variables. Such a circumstance is, of course, an entirely appropriate justification for the use of the other measure. (4) Under certain conditions, the formula for r will simplify, and hence simpler computational methods become available for certain types of data. These different methods are sometimes given new names, even though they are mathematically equivalent to Pearson r and give the same numerical answers.

Several of these four reasons may apply to a given coefficient. Thus, the use of biserial r and tetrachoric r can be justified by reasons (1), (2), or (3); while the use of point biserial r and phi can be justified by reasons (2), (4), and perhaps (1). The four correlation methods just mentioned are certainly among the best and the most useful measures of correlation other than r. They constitute the subject matter of this chapter. Three other methods are named and briefly described in Supplements 10 and 11. Pearson r and the seven methods about to be discussed will adequately care for the great majority of situations.

One general restriction should be mentioned, which applies both to Pearson r and to all four of the methods to be discussed in this chapter. We assume that all of our sample is present. For example, we do not use just the students who make A's and E's, rejecting the middle ones. Nor do we eliminate those who make high or low scores. These five correlation methods are applicable to random samples from a larger population, but not to biased or curtailed portions of such samples.

BISERIAL *r*

The biserial coefficient of correlation is usually called *biserial r*. It is designated by the symbol r_{bis}. Biserial r is one of two measures of relationship that can be used when we have two scores—one continuous and one dichotomous—for each individual. There is nothing unusual about the continuous score; it can be any variable that we might use in computing Pearson r. A dichotomy is a division of a variable into two parts. A dichotomous variable, then, is one in which all individuals are classified into one or the other of two classes. Thus, the scatter diagram, or 2 by n table, for biserial r will look something like:

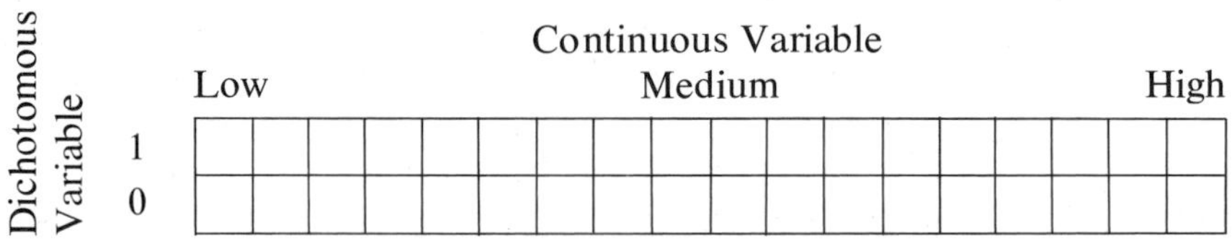

Dichotomous Variable	Continuous Variable: Low								Medium									High
1																		
0																		

When we compute biserial r, we indirectly admit that we feel it is unfortunate that our records are so incomplete. We are sorry we have merely a plus-or-minus (or yes-or-no, or more-or-less, etc.) classification on the dichotomous variable. We feel that, even though we have only two possible scores, the dichotomous variable is fundamentally continuous and normally distributed. The trouble is that this ideal normal distribution is cut into only two parts: those above and those below some critical point. If only we had many more scores on this dichotomy, so that we had 15 or 20 equal intervals, we could compute Pearson r. But we can't do this. We can, however, make the best use of the data we do have and use these records to estimate what value Pearson r would have if we were able to compute it. This is exactly what biserial r does. In technical language, we may define biserial r as a coefficient of correlation between a two-categoried and a continuous variable, assuming that the two-categoried variable is in reality continuous and normally distributed, although it is expressed in only two degrees. Notice that we do not need to make any assumptions whatever regarding the distribution of the continuous variable. The only necessary assumptions are that (1) the two-categoried variable is, theoretically, at least, continuous and normally distributed and (2) if the two-categoried variable actually were subdivided, we would have linear regression. These assumptions are met, or nearly satisfied, for most of the variables we use in Psychology and Education if the dichotomous variable is one that we can regard as capable of further subdivision.

Thus, we may use biserial r if for each individual we have two scores. One such score is on a variable that is continuous but not necessarily normally distributed. The other is a score indicating whether each person is above or below a point that dichotomizes a trait that would give a normal distribution if we were able to measure it in appropriate units. We further assume that if we had such measures, the regression line for predicting scores on the continuous variable from those on what is now the dichotomy would be essentially a straight line. These restrictions do not greatly limit the usefulness of biserial r. If a variable is continuous, it usually is either normally distributed or it is possible to imagine that some transformation of the scores (square root, logarithm, reciprocal, or some other alteration) would result in a normal distribution. If so, biserial r gives an estimate of the correlation between the continuous variable and the dichotomous variable when so transformed.

As to linearity of regression, the plain facts are that (1) regression is usually essentially linear and (2) if it is not, we ordinarily have no way of knowing it. Hence, as a practical matter, most people ignore that requirement. What is left? To be technically correct, we should try to see that our data satisfy both the assumptions listed in the preceding paragraph. In practice, if we know only that the two-categoried variable can be regarded as fundamentally continuous, we often compute biserial r. Nearly always, this is all right. Some samples of dichotomous variables that may appropriately be used in computing biserial r are given in sections A and C of Table 10.6.

Computation of Biserial r What is the relation between swimming ability and running speed? At the University of Florida, freshmen men are required to take tests in both. Four sections were selected as probably

TABLE 10.1
Running and Swimming Times of Florida Freshmen

100 Yard DASH (Seconds)	40 Yard SWIM (Seconds)	100 Yard DASH (Seconds)	40 Yard SWIM (Seconds)
11.6	26.5	12.7	26.3
14.1	27.2	11.7	32.4
11.9	24.7	12.5	21.5
12.2	27.6	11.8	33.0
12.9	22.0	12.2	29.1
11.7	24.4	12.9	28.8
12.8	22.8	11.6	25.1
12.1	34.3	12.5	29.0
14.3	24.4	12.7	31.4
11.8	36.5	13.0	27.7
12.6	39.4	12.6	37.1
13.2	27.4	12.6	27.2
13.7	27.8	13.4	22.8
12.8	37.0	12.5	29.5
12.9	27.4	14.0	27.5
14.1	28.5	13.4	24.6
13.9	26.9	13.8	27.6
13.3	N.S.	13.5	27.4
13.3	25.4	13.5	32.4
13.5	31.0	13.0	43.9
13.6	32.6	13.2	46.0
13.6	26.0	13.3	39.5
13.7	28.0	13.2	33.7
14.0	35.6	14.0	31.1
13.9	36.4	14.0	38.0
13.6	N.S.	13.5	38.8
12.5	24.9	13.4	33.6
14.1	34.6	14.8	28.2
13.8	24.7	14.4	31.5
15.0	N.S.	14.8	39.5

Note: N.S. indicates a Non-Swimmer.

being typical of the class. The results for these 115 men are given in Table 10.1. Everyone could run 100 yards, but 11 in this group could not swim 40 yards or would not try. These men were called non-swimmers and hence are designated by "N.S." in the table.

Using the data of Table 10.1, how can we determine the extent of the relationship between running and swimming proficiency? At first it might appear that Pearson r is entirely appropriate. Before computing Pearson r,

TABLE 10.1
(*continued*)

100 Yard DASH (Seconds)	40 Yard SWIM (Seconds)	100 Yard DASH (Seconds)	40 Yard SWIM (Seconds)
12.7	28.5	12.4	26.4
11.8	27.0	12.4	23.7
12.6	27.0	12.9	23.1
12.4	27.0	12.1	35.8
12.0	27.5	12.2	27.4
11.5	27.9	11.9	31.6
13.5	N.S.	11.8	30.1
13.0	26.9	13.4	31.4
13.0	21.2	12.5	N.S.
13.3	38.2	12.8	28.0
12.5	26.2	13.4	32.4
12.5	29.8	13.5	33.6
12.7	32.4	12.5	N.S.
13.3	28.1	12.5	30.8
13.2	34.5	12.5	35.9
13.5	27.6	13.3	30.8
13.0	34.0	15.5	N.S.
13.4	26.0	14.0	N.S.
13.8	30.1	13.3	28.0
14.0	28.9	13.4	23.3
13.8	36.8	14.0	34.7
13.5	27.5	13.3	31.0
13.3	N.S.	14.6	31.0
15.0	34.4	14.9	N.S.
13.4	26.0	15.5	32.5
13.1	39.0		
15.4	N.S.		
14.5	25.9		
12.9	24.0		
15.8	38.4		

however, we must raise one little, but highly important, question: What are we going to do about the 11 non-swimmers? That's a tough one! We might just omit them. That choice often appeals to naive people. When inexperienced people throw away some of their data, they are usually wrong in doing so. When anyone throws away cases that are particularly high or low on any variable, he is bound to be wrong. In this situation, the 11 cases we are considering are obviously extremely poor swimmers—they can't swim at all! If we were to omit them and compute Pearson r, we would then have the correlation between swimming proficiency and running ability, but only for the restricted group who possess at least a certain minimum swimming ability. Such an r could be computed, but it would be misleading. It would not show the relationship that we seek.

Another possibility would be to assign some arbitrary time score to each of these cases. Such a score might be 50 seconds (which is slower than the slowest swimmer in the group), but it also might be 100 seconds or 1000. This method is not always as bad as it sounds. If there is any logical basis for assigning scores to the missing cases, it is sometimes reasonable to do so. Although we think it would not be too bad to assign arbitrary scores of 50 seconds in this particular problem, some other statisticians might be horrified at the thought of doing so. We will be better off if we can find a more defensible and more acceptable method. If it didn't make much difference what score we assigned to the non-swimmers, then we would have no problem. In some cases it doesn't.

Just as automobile times became speeds by the reciprocal transformation, we could try transforming swimming times into speeds by dividing 1000 by the time in seconds. Thus, the first time in Table 10.1 of 26.5 seconds becomes a speed score of 38; the fastest swimming time of 21.2 seconds becomes a speed score of 47; and the slowest swimming time of 46.0 seconds becomes a speed score of 22. If we then assume that the non-swimmers are something like 50-second swimmers, their speed scores would be 20, but if we assume that they resemble 200-second swimmers (or people who would take infinitely long), their speed scores would be 5 (or 0). When most of the people have scores between 22 and 47, it is obvious that unless we have a reason for believing that these people really resemble those who take 50 seconds rather than 200 seconds or forever, then we have no basis for assigning them scores of 20 or 5 or 0. We have no such reason and what we assume makes a real difference, so that system won't work here. It does work in many other situations, including the example on pages 378–383.

There is a way out. Even though swimming *time* is not distributed normally, it is not too unreasonable to assume that the underlying swimming *proficiency* is normally distributed with the non-swimmers, of course, at the low end of the scale. Biserial r is appropriate for just such a situation. We can correlate swimming proficiency (dichotomized as able versus unable to swim 40 yards) against 100 yard dash time.

To compute biserial r, it is first necessary to tabulate two distributions of the scores made on the continuous variable. One is for those with low scores on the two-categoried variable. The other is for those with high scores on the two-categoried variable. From the records given in Table 10.1, we make two tabulations of 100 yard DASH time. The first, for the non-swimmers, is shown in column (2) of Table 10.2. There were only 11 non-swimmers (N.S. in Table 10.1), so there are only 11 tally marks in column (2) of Table 10.2. Similarly, the 100 yard DASH times of the 104 other freshmen are tabulated in column (6). This tabulating should be checked by doing it again. The tally marks in columns (2) and (6) are then added and the sums are recorded in column (9). You may check this adding, but you do not have to; any mistakes will be discovered when we compare the sums of the various columns.

We now have figures for non-swimmers, swimmers, and for the total group. The next step is to select an arbitrary origin opposite the largest total frequency (or, if desired, at some other place) and fill in the x' deviations in column (3). Then we compute the three sets of fx' and fx'^2 entries. If we are not using a calculator, we compute the fx'^2 entries in column (11) by whichever method was *not* used in computing columns (5) and (8). This procedure minimizes the chance of having an undetected error arise through making the same mistake twice (as by multiplying 36 by 3 in the last row and consistently getting 118). If we are using a calculator, columns (4), (5), (7), (8), (10), and (11) need not be filled in; only the sums need to be recorded.

We now add every column of the table except columns (1) and (3) and record the sums. These sums are checked by noting that the sum of f, fx', or fx'^2 for non-swimmers plus the corresponding figure for swimmers must equal that for the total group. For example, using the sums of columns (2), (6), and (9), we see that $11 + 104 = 115$. Similarly, $+23 + (-58) = -35$. Next, we note that $183 + 900 = 1{,}083$. Since all three checks hold, we then place a $\surd$ between the parentheses at the right. If any of these checks ever fails to hold, recompute the sums and correct the error or errors before proceeding further.

We need one other value, y, before we can substitute in a formula and obtain biserial r. To get this, we first compute p and q, check them, and then find y in Table P. We compute $p = \frac{104}{115} = .904$ and $q = \frac{11}{115} = .096$. (Checking, $.904 + .096 = 1.000$.) From Table P, we find $y = .1703$.

There are three formulas for biserial r, all having identical denominators:*

$$r_{\text{bis}} = \frac{N_q \Sigma x'_p - N_p \Sigma x'_q}{N y \sqrt{N \Sigma x'^2 - (\Sigma x')^2}}, \tag{10.1}$$

*Each of these formulas can be adapted for use with raw score data by merely substituting X for x' throughout.

TABLE 10.2
Biserial r Between Swimming Proficiency (Dichotomous) and Running Time (Continuous)

100 Yard DASH Time *(1)*	Non-Swimmers *(2)*	x' *(3)*	fx' *(4)*	fx'^2 *(5)*	Swimmers *(6)*	fx' *(7)*	fx'^2 *(8)*	Total *(9)*	fx' *(10)*	fx'^2 *(11)*	
15.6–15.8		+8			|	8	64	1	8	64	
15.3–15.5	||	+7	14	98	|	7	49	3	21	147	
15.0–15.2	|	+6	6	36	|	6	36	2	12	72	
14.7–14.9	|	+5	5	25	||	10	50	3	15	75	
14.4–14.6		+4			|||	12	48	3	12	48	
14.1–14.3		+3			||||	12	36	4	12	36	
13.8–14.0	|	+2	2	4	𝍸 𝍸 ||	24	48	13	26	52	
13.5–13.7	||	+1	2	2	𝍸 𝍸 |	11	11	13	13	13	
13.2–13.4	||	0			𝍸 𝍸 𝍸 ||||			21			
12.9–13.1		−1			𝍸 𝍸 |	−11	11	11	−11	11	
12.6–12.8		−2			𝍸 𝍸 |	−22	44	11	−22	44	
12.3–12.5	||	−3	−6	18	𝍸 𝍸 |	−33	99	13	−39	117	
12.0–12.2		−4			𝍸 |	−24	96	6	−24	96	
11.7–11.9		−5			𝍸 |||	−40	200	8	−40	200	
11.4–11.6		−6			|||	−18	108	3	−18	108	
Sum	11		+23	183	104	−58	900	115	−35	1083	(√)

$$r_{\text{bis}} = \frac{N\Sigma x'_p - N_p\Sigma x'}{N\mathcal{y}\sqrt{N\Sigma x'^2 - (\Sigma x')^2}}, \tag{10.2}$$

$$r_{\text{bis}} = \frac{N_q\Sigma x' - N\Sigma x'_q}{N\mathcal{y}\sqrt{N\Sigma x'^2 - (\Sigma x')^2}}. \tag{10.3}$$

The subscript p refers to the passing (or higher) group, while q refers to the failing (or lower) group. Where no subscript is used, the figures are for the total group. All the symbols, except r_{bis} for biserial r, have been previously used.

It makes no difference which formulas we use, so we shall take the first two in order to get a check on our work. First using Formula (10.1), we obtain

$$\begin{aligned} r_{\text{bis}} &= \frac{N_q\Sigma x'_p - N_p\Sigma x'_q}{N\mathcal{y}\sqrt{N\Sigma x'^2 - (\Sigma x')^2}} \\ &= \frac{(11)(-58) - (104)(23)}{(115)(.1703)\sqrt{(115)(1{,}083) - (-35)^2}} \\ &= \frac{-3{,}030}{19.5845\sqrt{123{,}320}} \\ &= \frac{-3{,}030}{(19.5845)(351.17)} \\ &= \frac{-3{,}030}{6{,}877.49} \\ &= -.44. \end{aligned}$$

To check this, we use the next formula, but in recomputing the denominator, we pay special attention to decimals and signs, we multiply in a different order, and we check the square root rather than extracting it again. This gives

$$\begin{aligned} r_{\text{bis}} &= \frac{N\Sigma x'_p - N_p\Sigma x'}{N\mathcal{y}\sqrt{N\Sigma x'^2 - (\Sigma x')^2}} \\ &= \frac{(115)(-58) - (104)(-35)}{(115)(.1703)\sqrt{(115)(1{,}083) - (-35)^2}} \\ &= \frac{-3{,}030}{6{,}877.49} \\ &= -.44. \end{aligned}$$

Since the values of the numerators, denominators, and hence the quotients turn out to be identical, our work is checked. If the numerators had differed, we would have used Formula (10.3) and then traced down the error. If the denominators had differed, the error would have been easy to find and correct.

This computation of r_{bis} shows us that there is a fairly high negative correlation between ability to swim and 100-yard-dash time. The negative correlation should not surprise us, since the greater the time, the slower the runner. The correlation, then, is positive between ability to swim and ability to run rapidly. We may, accordingly, say that in this group of freshmen,there is a biserial correlation of +.44 between ability to swim and speed in running the 100-yard dash.

The formulas we have used for r_{bis} are time saving, but they do not give us any means or standard deviations. If we want these, we can compute them from the sums given in the last row of Table 10.2.

Interpretation of Biserial r Except for the obvious fact that one of the variables is dichotomous, biserial r has about the same meaning as a Pearson product moment coefficient of correlation of the same size. Actually, biserial r gives us the best available estimate of what the size of Pearson r would be if the continuous variable remained as it is, while the dichotomous variable was divided into a number of intervals that gave a normal distribution. That is, if we were correct in our original assumption that the dichotomous variable is in reality continuous and normally distributed, then biserial r tells us what Pearson r would be, if we were able to reclassify our dichotomous variable into one with a large number of intervals.

There are a few ways in which biserial r differs from Pearson r which have a bearing on its interpretation:

1. Where several points of dichotomy are possible, the various values of biserial r will vary somewhat among themselves.
2. Biserial r is ordinarily more accurate if the dichotomized variable is cut near the median. If there is a choice among several points at which the variable may be dichotomized, unless other considerations are involved, preference should go to that point nearest the median. A procedure that usually is even better is to compute Jaspen's multiserial correlation, which is referred to on page 328.
3. Biserial r is not strictly confined between the limits of $+1.00$ and -1.00, as Pearson r is. In rare instances, biserial r may take a value greater than $+1.00$ (or less than -1.00). Usually, however, values that are computed and reported to be greater than $+1.00$ contain computational errors. (*Hint:* If you ever get a biserial r beyond the limits of -1.00 to $+1.00$, get two other people to take your original data and independently compute biserial r before you report your result publicly.)
4. The standard error of biserial r is considerably larger than the standard error of Pearson r. Hence, for any given number of cases, biserial r has

to be larger than Pearson r in order to be equally likely to be significant. This will be discussed in the following section.

Standard Error of Biserial r If our assumption about the normality of the dichotomous variable is correct, and if neither p nor q is less than about .07, the standard error of biserial r is, approximately,

$$s_{r_{\text{bis}}} = \frac{\left(\frac{\sqrt{pq}}{y}\right) - r_{\text{bis}}^2}{\sqrt{N-1}}. \tag{10.4}$$

All the symbols have their usual meanings. Notice that the square root sign applies to pq only and not to y. It is not necessary to compute the function, $\frac{\sqrt{pq}}{y}$; it has been computed and tabled in Table Q. The table contains a built-in warning. If either p or q is less than .07 (the other being greater than .93), the entries are printed in italics. This is our attempt to dissuade people from computing r_{bis} and its standard error for dichotomies more extreme than .07 and .93.

Let us now apply Formula (10.4) to the data on running and swimming times. We have already found p, the proportion of swimmers, to be .904. We have just computed r_{bis} and found it to be $-.44$. We know that N was 115. Next we look at Table Q. In the row labeled .900 and in the column headed .004, we find the entry 1.7296. This means that when $p = .904$ the corresponding value of $\frac{\sqrt{pq}}{y}$ will be 1.7296. Substituting these values in Formula (10.4), we readily obtain

$$\begin{aligned} s_{r_{\text{bis}}} &= \frac{\left(\frac{\sqrt{pq}}{y}\right) - r_{\text{bis}}^2}{\sqrt{N-1}} \\ &= \frac{1.7296 - (-.44)^2}{\sqrt{115-1}} \\ &= \frac{1.7296 - .1936}{10.6771} \\ &= \frac{1.5360}{10.6771} \\ &= .14. \end{aligned}$$

This means that if the true biserial r in an infinite population of similar students were $-.44$ between swimming ability and running time, and if we were to draw a number of groups of 115 students at random from this theoretically infinite population, the obtained values of r_{bis} would center around $-.44$ but would have a standard deviation of .14. Roughly speaking, about two-thirds of such correlations would be between $-.58$ and $-.30$, and about 95% of them would be between $-.72$ and $-.16$. These are not exact statements for several reasons: The true r probably is not exactly $-.44$; it might be $-.45$, $-.50$ or even $-.30$. (All of these values are reasonable, insofar as we can judge from our data.) This means that we do not know the true point from which we should measure our deviations. This, however, would not matter much if the standard error had the same value for all values of r_{bis}. This is not the case. For our data, if the true value of r_{bis} were .00, then $s_{r_{\text{bis}}}$ would equal .16, while if the true value of r_{bis} were $-.60$, then $s_{r_{\text{bis}}}$ would equal .13. Further, although these distributions of correlations obtained from successive samples would cluster around $-.44$, .00, $-.60$ or whatever the true mean might be, they would not be symmetrical.

Let us see how serious these restrictions are in various practical situations. They are most serious when r is high, when N is small, and to a lesser extent, when either p or q is very large or very small. Thus, if r_{bis} were .90, if N were 20, and if p were .95, the standard error of biserial r would be .21. It would be rather absurd to state that if the true biserial r were .90, two-thirds of such correlations based on 20 cases would be expected to lie between $+.69$ and $+1.11$, with one-sixth of them being greater than $+1.11$. The actual situation in the case of samples of 20 being drawn from a universe in which the true biserial r were .90, and in which 95% of the cases were in one of the two categories, would be that the distribution of biserial r's would be markedly skewed, some values being .80, .70, or even .60 with very few reaching values as high as .99 or 1.00.

First, what happens to the standard error formula if p is near .50 rather than near .05 or .95? A glance at Table Q shows that the standard error will be smaller because the function, $\frac{\sqrt{pq}}{y}$, will take smaller values (such as 1.25) rather than larger ones (such as 2.11).

Second, what happens to the standard error formula when biserial r is large rather than small? Since r_{bis} is squared, its sign is immaterial. For all values of r_{bis} between $-.35$, .00, and $+.35$, the standard error of biserial r is within 10% of what it would be if r were zero. (This comes about because the smallest possible value of the function, $\frac{\sqrt{pq}}{y}$, is 1.2533, and no correlation of .35 or less has a square that is greater than .12533.) Similarly, the change is still under 20% when r_{bis} gets to $-.50$ or $+.50$. Thus, changes in the size of the correlation have very little effect when the absolute value of the correlation is below .35 or .50. This, of course, is where many of our correlations are found. Only when biserial r reaches a value of something over .79 (depending on the point of dichotomy), will the standard error be cut in half.

Third, what happens to the standard error formula when N changes? This is easy: the standard error is cut in half every time $N - 1$ is quadrupled. Sometimes we can alter or control our points of dichotomy and thus reduce our standard error a bit. We cannot, of course, raise our correlation in order to reduce its standard error; the value of r_{bis} is determined by computation, not by desire. Usually, however, we can alter N. If we will just gather data on 21% more cases, we shall reduce our standard error by 10%; if we double our number of cases, we cut it down by about 30%; and if we use $4N$ instead of N (or, more precisely, $4N - 3$ instead of N), we reduce it to only 50% of its previous value. Of course, there comes a point when it is just not worth while to increase N further in order to decrease standard errors. This point is hard, yes, impossible, to determine, but practical people need practical answers, so here it is: Try to plan your study so that you will wind up with something over a hundred usable cases. Not only will this give you a smaller standard error than if you had used 10 or 20 cases, but also the formulas will be more directly applicable, and, most important of all, your values of r_{bis} and all other statistics will be much closer to the parameters in the population from which your sample is drawn than would otherwise be the case.

A Second Biserial r Example It really doesn't take long to compute biserial r even without electronic computers. With the aid of older nonelectronic machines, biserial r's have been turned out by the hundreds in less than 5 minutes each; with electronic equipment the time is, of course, measured in seconds. But even with a desk or pocket calculator, the time will ordinarily be far less than it has taken to read the preceding description of biserial r.

Our second example is from the field of industrial psychology. By the use of multiple correlation (an advanced technique which is referred to but not explained in Chapter 15), three tests were selected that, when appropriately weighted, had given excellent results in predicting the success of long distance telephone operators. Some time later, Rusmore and Toorenaar*

---

*Rusmore, Jay T., and Toorenaar, Gerald J. Reducing training costs by employment testing. *Personnel Psychology*, 1956, **9**, 39–44. The weighted test battery scores were necessarily in decimal form, but were rounded to the nearest whole number. Because the interval width of 2 used in the Rusmore and Toorenaar article resulted in 35 intervals, we secured the original data through the kind cooperation of these authors and have retabulated the scores, as shown in columns (1), (2), and (6) of Table 10.3.

studied the relation between these test battery scores and a dichotomous summary evaluation made by the supervisor at the time of termination. Their study is based on 135 operators who were hired and later terminated during the same year. The results of the study are tabulated as Table 10.3. This tabulating was checked and biserial r was computed by carrying out the following steps:

First, we add the tally marks in each row of column (2) to those in column (6) and record the total in column (9). This gives us our basic data.

We now select our arbitrary origin and put a zero in column (3) to indicate it. Here there are two intervals tied with highest frequencies of 20, so we take the one nearer the middle of the distribution and write a 0 in column (3) opposite the 40 (which obviously means 40–44) in column (1).

We fill in the other x' values, compute all the figures in the body of the table, being sure to obtain fx'^2 in column (11) by the method that was *not* used for columns (5) and (8).

We now add all the columns of the table except columns (1) and (3) and record the sums. We check these nine values by noting that corresponding figures for the unsuccessful and for the successful must add to those for the total. If this is true for N, for $\Sigma x'$, and for $\Sigma x'^2$, place a $\surd$ in the space at the right. If it is not true, find the error and correct it.

Next we obtain p, q, and $\mathcal{y}$. Here, $p = \frac{75}{135} = .556$, $q = \frac{60}{135} = .444$, and they check since $.556 + .444 = 1.000$. From Table P, we find that $\mathcal{y} = .3950$.

This time, let us skip (10.1) and use Formulas (10.2) and (10.3). Formula (10.2) gives us

$$\begin{aligned} r_{\text{bis}} &= \frac{N\Sigma x'_p - N_p\Sigma x'}{N\mathcal{y}\sqrt{N\Sigma x'^2 - (\Sigma x')^2}} \\ &= \frac{(135)\,(237) - (75)\,(237)}{(135)\,(.3950)\sqrt{(135)\,(1507) - (237)^2}} \\ &= \frac{14{,}220}{53.3250\sqrt{147{,}276}} \\ &= \frac{14{,}220}{(53.3250)\,(383.77)} \\ &= \frac{14{,}220}{20{,}464.54} \\ &= .69. \end{aligned}$$

TABLE 10.3

Biserial *r* between Degree of Success of Terminating Telephone Operators (Dichotomous) and Test Battery Score (Continuous)

(1)	Unsuccessful *(2)*	x' *(3)*	fx' *(4)*	fx'^2 *(5)*	Successful *(6)*	fx' *(7)*	fx'^2 *(8)*	Total *(9)*	fx' *(10)*	fx'^2 *(11)*	
85–89		+9			I	9	81	1	9	81	
80–84		+8			IIII	32	256	4	32	256	
75–79		+7			卌 I	42	294	6	42	294	
70	I	+6	6	36	III	18	108	4	24	144	
65	I	+5	5	25	卌 I	30	150	7	35	175	
60	II	+4	8	32	卌 卌 I	44	176	13	52	208	
55	卌 I	+3	18	54	卌 IIII	27	81	15	45	135	
50	III	+2	6	12	卌 卌 IIII	28	56	17	34	68	
45	卌 II	+1	7	7	卌 卌	10	10	17	17	17	
40	卌 卌 II	0			卌 III			20			
35	卌 卌 卌 II	−1	−17	17	III	−3	3	20	−20	20	
30	IIII	−2	−8	16				4	−8	16	
25–29	IIII	−3	−12	36				4	−12	36	
20–24	II	−4	−8	32				2	−8	32	
15–19	I	−5	−5	25				1	−5	25	
Sum	60		0	292	75	+237	1215	135	+237	1507	(√)

We now check the denominator, first seeing that the proper figures are substituted in the formula, then checking rather than recomputing the square root, and then performing all the other computations in a different order. Since everything checks perfectly, we can very readily use this denominator in solving Formula (10.3):

$$
\begin{aligned}
r_{\text{bis}} &= \frac{N_q \Sigma x' - N \Sigma x'_q}{N_{\mathscr{Y}} \sqrt{N \Sigma x'^2 - (\Sigma x')^2}} \\
&= \frac{(60)\,(237) - (135)\,(0)}{20{,}464.54} \\
&= \frac{14{,}220}{20{,}464.54} \\
&= .69.
\end{aligned}
$$

Let us now compute the standard error. As usual, we do not know the true value of r_{bis}, but since N is over 100, it will not matter much whether we substitute our obtained value of .69 or a value of .60 or .80. Using our p of .556, Table Q gives us 1.2578 as the value of $\frac{\sqrt{pq}}{\mathscr{Y}}$. Substituting this and our known values of r_{bis} and N in Formula (10.4), we obtain

$$
\begin{aligned}
s_{r_{\text{bis}}} &= \frac{\left(\frac{\sqrt{pq}}{\mathscr{Y}}\right) - r^2_{\text{bis}}}{\sqrt{N - 1}} \\
&= \frac{1.2578 - (.69)^2}{\sqrt{135 - 1}} \\
&= \frac{.7817}{11.5758} \\
&= .07.
\end{aligned}
$$

For such a correlation of .69 with a standard error of .07, the interpretation is extremely simple. First of all, we notice that if the true correlation really were .69, about ⅔ of the time we should expect obtained correlations between .62 and .76; and 95 times out of 100, they would be between .55 and .83. Referring back to the eighth interpretation of Pearson r on page 260, we see that correlations anywhere in either of these ranges are very high for validity coefficients in an industrial situation.* Unless there is something spurious about it, of which the authors are unaware, their weighted test battery score is giving a highly satisfactory prediction of the

*In fairness, it should be conceded that technical reasons related to sampling and the size of the standard errors of the various r's would lead most statisticians to feel that a biserial (or other) r of .60 or .70 is really not so good or dependable as a Pearson r of the same size.

success of telephone operators at the time of their termination. There remains, of course, the question of how well it will predict the degree of success of those who remain on the job and how well it will differentiate them from terminators. But these are other questions that their records are not intended to answer. Insofar as predicting the stated criterion in this situation is concerned, the test battery is doing an excellent job. Note, for instance, that very few of the successful telephone operators scored below the mean (42.00) of the unsuccessful; and also that very few of the unsuccessful operators scored above the mean (57.80) of the successful ones.

With a biserial r of .69 based on 135 cases, there is no point in asking whether it is significantly different from zero. It obviously is. But we might want to know whether or not we can be confident that this correlation is really different from .40 (which we referred to on page 261 as being "high for most occupations" in an industrial situation), or from .50 (a value that often interests people), or from .80 (a correlation nearing the fantastic), or from other values. Substituting these correlations in Formula (10.2), and then dividing the differences by the appropriate standard errors, we obtain Table 10.4. This table readily enables us to answer the questions just put before us. All three differences are significant at the .05 level or better. If the true biserial r were .40, we would get a correlation as far away from .40 (in either direction) as our .69 less than 1% of the time. Next, if the true correlation were .50, a correlation differing as much as .19 or more from it would occur less than 5% of the time. Similarly, we can reject the hypothesis that the true correlation is .80 because if it were .80, a difference as large as our obtained difference of .11 would occur in less than 5 instances out of 100.

TABLE 10.4
Significance of a Biserial r of .69 Based on N = 135

Assumed True Correlation	Obtained Minus True Correlation	Standard Error of True Correlation	Difference Divided by Its Standard Error	Level of Significance
Any value from .81 to 1.00	−.12 to −.31	.05 or less	2.4 or more	Beyond .05, at least
.80	−.11	.05	2.2	Beyond .05
.70	−.01	.07	.1	Not significant
.60	+.09	.08	1.1	Not significant
.50	+.19	.09	2.1	Beyond .05
.40	+.29	.09	3.2	Beyond .01
Any value from .39 to .00	+.30 to +.69	.10 or .11	3.0 to 6.3	Beyond .01

Notice the top and bottom rows of the table. When we have found out that our obtained r_{bis} is significantly different from any hypothesized true value, it is even more significantly different from all other assumed values beyond that one. This is obvious in the top row, since as we assume higher values, the difference gets larger while the standard error gets smaller, thus forcing the quotient to increase rapidly. It is less obvious at the bottom, both because of a rounding error and because the difference is increasing much more rapidly than the standard error is. (The latter is rather slowly approaching its maximum of .11 which is attained when $r_{\text{bis}} = .00$.)

In conclusion, our r_{bis} of .69 is not a chance fluctuation (at the .05 level of significance) from any of the three values with which we compared it. The true correlation must, therefore, be rather closely in the vicinity of our obtained .69; and this is a very high correlation between a test battery and an industrial criterion.

MULTISERIAL CORRELATION

When we correlated a continuous variable with another variable that was fundamentally continuous but divided into two categories, we got biserial r. If this other variable had been divided into three categories, we could have computed triserial r; if it had been divided into six categories, we could have computed sexiserial r; etc. In 1946, Jaspen* derived formulas applicable to such situations. See his article if you have data for which it might be appropriate to use one of his formulas. For instance, triserial r would be appropriate for use in test validation if the workers were placed in three groups according to their proficiency.

POINT BISERIAL r

The description of point biserial r might start out very much like the description of biserial r, even including the use of the diagram on page 312 showing the procedure used in classifying individuals on the basis of their scores, both on a continuous variable and on a dichotomous variable. There is only one important difference between the two: In working with biserial r, we assumed that the dichotomous variable was fundamentally continuous and normally distributed. Sometimes this assumption is not legitimate. Thus, if our continuous variable is intelligence and if the dichotomous variable is sex, Richardson and Stalnaker have pointed out that to assume that sex is a continuous normally distributed variable is sanctioning a biological absurdity. The difference between men and women is of such a nature that, whether or not we agree with the Frenchman who said, "Vive la différence!", it is clear that we have two distinct categories of people—

---

*Jaspen, Nathan. Serial correlation. *Psychometrika,* 1946, **11,** 23–50.

men at one point on the scale and women at the other. Point biserial r is applicable to this situation.

Thus, point biserial r expresses the relationship existing between a continuous variable and a dichotomous variable which is regarded as composed of two categorically distinct groups. There are no particular restrictions on the continuous variable. As was true of Pearson r, any old continuous variable may be used. It need not even be normally distributed. The dichotomous variable, however, should be composed of two discrete classes or types of individuals. Sex is a nice example. So far as point biserial r is concerned, there are no gradations of maleness or femaleness.

Computation of Point Biserial *r* Because the formula for point biserial r is similar to that for biserial r, the computational details in the two situations are extremely similar. In fact, the design of the basic table for the tabulation and the bulk of the computation is identical for the two measures. (Compare Table 10.2 or 10.3 for r_{bis} with Table 10.5 for r_{pb}.) If we wish to compute the correlation between the correct answer to a test item and subsequent grade point average, point biserial r is well adapted to determining this relationship. Hence, such data will be used to illustrate the computations involved. The item chosen* is:

Which of the following is due entirely to heredity?

1. Eye color
2. Insanity
3. Body weight
4. Cancer

The subjects were a random selection (the others taking Form B) of freshmen who entered a state university. We are interested in seeing how those who passed the item (by marking response 1) and those who failed it differ in their first semester grade point average.

In this university, and in this group, grades ranged from 0.00 to 5.00, thus creating a minor difficulty in selecting the interval. If $i = .20$, we would have 26 classes; if $i = .50$, we would have 11 intervals. Neither of these preferred interval sizes gives us the 15 to 20 intervals we want. We could take the one ($i = .50$) that would come closest to giving us 15 to 20 intervals, but 11 intervals seem to be too few, so we used $i = .30$ (slightly more difficult to tabulate), which gives us 17 intervals.

*This is item 12 of Test I (Information) of Form A of the College Qualification Tests. This copyrighted item is reproduced here by special permission of The Psychological Corporation, who also supplied the needed data on the item.

TABLE 10.5

Point Biserial r between Passing or Failing Item 12 (Dichotomous) and First Semester Grade Point Average (Continuous)

(1)	Fail *(2)*	x' *(3)*	fx' *(4)*	fx'^2 *(5)*	Pass *(6)*	fx' *(7)*	fx'^2 *(8)*	Total *(9)*	fx' *(10)*	fx'^2 *(11)*	
4.80–5.09		+10			IIII	40	400	4	40	400	
4.50–4.79		+9			卌 I	54	486	6	54	486	
4.20–4.49		+8			卌 III	64	512	8	64	512	
3.90	I	+7	7	49	卌 I	42	294	7	49	343	
3.60		+6			卌 IIII	54	324	9	54	324	
3.30		+5			卌 卌 卌 I	80	400	16	80	400	
3.00	I	+4	4	16	卌 卌 II	48	192	13	52	208	
2.70	II	+3	6	18	卌 卌 I	33	99	13	39	117	
2.40	I	+2	2	4	卌 III	16	32	9	18	36	
2.10	I	+1	1	1	卌 卌 I	11	11	12	12	12	
1.80	IIII	0			卌 卌 III			17			
1.50	III	−1	−3	3	卌 I	−6	6	9	−9	9	
1.20	II	−2	−4	8	IIII	−8	16	6	−12	24	
.90	I	−3	−3	9	IIII	−12	36	5	−15	45	
.60–.89	II	−4	−8	32	II	−8	32	4	−16	64	
.30–.59	II	−5	−10	50	I	−5	25	3	−15	75	
.00–.29	II	−6	−12	72				2	−12	72	
Sum	22		−20	262	121	+403	2865	143	+383	3127	(√)

A glance at the tally marks in columns (2) and (6) of Table 10.5 suffices to show that the item is easy (22 failed while 121 passed) and that there is a very marked and obvious difference in the mean grade point average of these two subgroups; it's easy and it's a good item.

We now compute and check the figures in Table 10.5. The numbers are different, but the procedure is nearly identical to that explained earlier for Tables 10.2 and 10.3. We perform the following specific steps.

1. Check the tallies in columns (2) and (6) by retabulation and a very careful comparison of the two sets of figures. If there are any discrepancies, retabulate *both* columns until you are sure you know what the correct tabulation is.
2. Add the tally marks, one row at a time, in columns (2) and (6) and record the sums in column (9).
3. If you are not using a calculator, place a zero in column (3) opposite the highest frequency in column (9). In this case, write 0 in the row labeled 1.80, which obviously means 1.80–2.09.
4. Fill in the positive and negative deviations in column (3) and inspect the column to check them.
5. Fill in columns (4), (5), (7), (8), (10), and (11). In filling in column (11), use a method of computation which was *not* used in columns (5) and (8). For instance, use $(f)\ (x'^2)$ for columns (5) and (8) and $(x')\ (fx')$ for column (11). You may, but you need not, check these computations.
6. Add all columns except columns (1) and (3) and record the sums. You may, but you need not, re-add any of these columns.
7. Check to see that each of the three column sums for the Fail group, plus the corresponding sum for the Pass group, equals the corresponding sum for the Total group. Specifically,

   To check N: $22 + 121 = 143$
   To check $\Sigma x'$: $-20 + 403 = +383$
   To check $\Sigma x'^2$: $262 + 2865 = 3127$

   If any of these equations fails to check, find the error by redoing the relevant parts of steps (5) and (6). When all three checks hold, put a check mark between the parentheses at the right.

We now have all the figures needed to substitute in the formula and compute r_{pb}, the point biserial r. There are three formulas, all very similar in appearance, and all having identical denominators.*

*Each of these formulas can be adapted for use with raw score data by merely substituting X for x' throughout.

$$r_{pb} = \frac{N_q\Sigma x'_p - N_p\Sigma x'_q}{\sqrt{[N_pN_q][N\Sigma x'^2 - (\Sigma x')^2]}}, \tag{10.5}$$

$$r_{pb} = \frac{N\Sigma x'_p - N_p\Sigma x'}{\sqrt{[N_pN_q][N\Sigma x'^2 - (\Sigma x')^2]}}, \tag{10.6}$$

$$r_{pb} = \frac{N_q\Sigma x' - N\Sigma x'_q}{\sqrt{[N_pN_q][N\Sigma x'^2 - (\Sigma x')^2]}}. \tag{10.7}$$

Except for r_{pb} itself, all the symbols have been used before, the subscript p referring to the passing (or higher) group, the subscript q referring to the failing (or lower) group, and no subscript referring, as usual, to the total group. It does not matter which of these formulas we use, so we shall take the first one, then check its denominator, and then use the second formula to complete the check on our work. From Formula (10.5), we readily obtain

$$\begin{aligned} r_{pb} &= \frac{N_q\Sigma x'_p - N_p\Sigma x'_q}{\sqrt{[N_pN_q][N\Sigma x'^2 - (\Sigma x')^2]}} \\ &= \frac{(22)\,(403) - (121)\,(-20)}{\sqrt{[(121)\,(22)][(143)\,(3{,}127) - (383)^2]}} \\ &= \frac{11{,}286}{\sqrt{[2{,}662][300{,}472]}} \\ &= \frac{11{,}286}{\sqrt{799{,}856{,}464}} \\ &= \frac{11{,}286}{28{,}282} \\ &= .40. \end{aligned}$$

To check the denominator, we perform all the multiplications in a different order and then check the square root, not by extracting it again, but by one of our checking methods. For example, we square 28,282 and obtain 799,871,524. This is close, but it is a little larger than the 799,856,464 we started with. Since our square was too big, we subtract 28,282 from it, getting 799,843,242. This is now smaller than the number we started with, so the square root we obtained is correct. This checks the denominator, and the same denominator is used in all three formulas.

We now use Formula [10.6] for checking:

$$\begin{aligned} r_{pb} &= \frac{N\Sigma x'_p - N_p\Sigma x'}{\sqrt{[N_pN_q][N\Sigma x'^2 - (\Sigma x')^2]}} \\ &= \frac{(143)(403) - (121)(383)}{28{,}282} \end{aligned}$$

$$= \frac{11,286}{28,282}$$
$$= .40.$$

This checks all our computations and shows us that the correlation between this single item about heredity and first semester grade point average is .40. Since many entire tests of 50 to 200 items give correlations with grade point average of only .50 to .60, any one item that will correlate .40 with college success is doing very well indeed.

Interpretation of Point Biserial r Point biserial r is not an estimate of what Pearson r *might be* under some circumstances; instead, it is a statement of what Pearson r *is* between the two variables—the continuous variable being scored in the usual manner and the dichotomy being scored 0 for one of the responses and 1 for the other. (It is customary to state the situation this way and to assign 1 to the higher or more desirable category, but any two numbers may be used as scores.) If the favorable degree of the dichotomy is assigned a lower number, the sign will be reversed, but r_{pb} will have the same absolute value, whether it is computed by one of the three formulas just given, or whether it is computed by a formula for Pearson r with any arbitrarily assigned numbers for the two levels of the dichotomous variable.

On the same scale used by Pearson r, r_{pb} tells us to what extent the two variables are related and how well the continuous variable can be predicted from the dichotomy. Thus, although the Pearson r between the 75-item Information Test and grade point average was .52 for a group somewhat larger than this, our item 12 (definitely one of the better items) alone correlates .40 with this same criterion of college success. If we had reversed the situation and used the continuous variable (grade point average) to predict whether students would get the item right or wrong, again r_{pb} of .40 tells us the extent of the relationship. (In section C of Table 10.6, we shall learn that if we are interested in the theoretical or philosophical question of the relation between the continuous variable and *such information* as that needed to answer the item, r_{bis} would be appropriate but such circumstances are rare.)

Unlike r_{bis}, r_{pb} actually is a Pearson r; hence it varies between the limits of -1.00 and $+1.00$. It can never have a value outside this range. Like r_{bis}, when several points of dichotomy are possible, the various values of r_{pb} will vary somewhat. In such a situation, it is usually preferable to select the point of dichotomy nearest the median, thus making the two groups as nearly equal as possible.

Confidence Limits for Point Biserial r Even though r_{pb} really is a Pearson r, its standard error is different from that of Pearson r. The reason is that the several assumptions made in the derivation of the standard error of Pearson r definitely do not hold when one of the two variables being correlated is a dichotomy. Furthermore, the z' transformation is not strictly applicable, although it gives results that are often fairly similar to those obtained by the more exact procedure, which we shall now present.

Usually, when we compute a standard error of any statistic, we are not interested in the standard error in and of itself. Rather, we then use it to give us an idea of some confidence limits within which the true value probably lies. Often such limits extend the same or nearly the same distance in each direction. With high correlations, we might expect that such limits would extend further in the low than in the high direction. This, in fact, is exactly what happens. Thus, if we obtain $r_{pb} = .90$ in a sample of 31 individuals, the .99 confidence limits of .77 and .94 obviously go from .13 below our obtained r_{pb} to only .04 above it. This means that if the parameter or true value of r_{pb} in the population were .77, the probability of getting a value of .90 or more in random samples of $N = 31$ is .005; and that if the true value were .94, the probability of occurrence of a value of r_{pb} of .90 or less is also .005. Taken together, these two probabilities add up to .01 and, of course, give the .99 confidence limits. These limits are shown in Table R in the row for $N = 31$ and the column for obtained $r_{pb} = .90$. Just below them are the figures .81 and .93, which are the .95 confidence limits.

For every value of N and obtained r_{pb} shown in the table, there are four figures. Those in the upper row are the .99 confidence limits; those in the lower row are the .95 confidence limits. This copyrighted table, somewhat abridged and rearranged, is reproduced here through the kind permission of the authors, Norman C. Perry and William B. Michael, and the publisher, *Educational and Psychological Measurement*. The authors point out that, due to the method of computation, there may in rare instances be an error of .01 in the table.

Let us now determine confidence limits for our r_{pb} between item 12 and grade point average. N was 143 and r_{pb} was .40. The .99 confidence limits in the upper row are .19 and .56 for $N = 127$; and they are .21 and .55 for $N = 152$. The very simple interpolation gives .20 and .55 for $N = 143$. (If r_{pb} had been a value such as .46, not given in the table, we would do what we just did, similarly interpolate for $r_{pb} = .50$, and then interpolate between these two values to get the right figures for $N = 143$ and $r_{pb} = .46$. Usually, the interpolation is very easy.)

Using the two sets of figures in the lower row, we interpolate between the .24 and .52 for $N = 127$ and the .26 and .52 for $N = 152$, seeing at a glance that the desired .95 confidence limits are .25 and .52 for our N of 143.

We are now in a position to comment on just about any hypothesis we may have started out with concerning the relation between Item 12 and grade point average. For example, are our results consistent with the hypothesis that the population correlation is .30? Yes. Are they consistent with the hypothesis that it is .24? Maybe—the hypothesis is rejected at the .05 level of significance but not at the .01 level. Are they consistent with the view that it is .18? No. This hypothesis is rejected at the .01 level (and, of course, also at the .05 level). We may express these and other results, if desired, in tabular form:

Hypothesis about Population r_{pb}	Decision
−1.00 to +.19	Reject at .01 level
+.20	Barely significant at .01 level
+.21 to +.24	Reject at .05 level, but not at .01 level
+.25	Barely significant at .05 level
+.26 to +.51	Do not reject the hypothesis
+.52	Barely significant at .05 level
+.53 to +.54	Reject at .05 level, but not at .01 level
+.55	Barely significant at .01 level
+.56 to +1.00	Reject at .01 level

We would not ordinarily print such a tabulation, since we would usually have started out with one, or possibly as many as two or three hypotheses. We would then comment only on them, as was done just before this tabulation with respect to three hypothetical values.

CLASSIFICATION OF DICHOTOMOUS VARIABLES

We have learned that certain dichotomous variables are appropriate for use with biserial r, while others are appropriate for use with point biserial r. We will soon learn that the dichotomies regarded as continuous and normally distributed are appropriate both for biserial r and for tetrachoric r, whose symbol is r_t. Analogously, we will find that categorically distinct dichotomies are suitable for use both with point biserial r and with phi, whose symbol is ϕ. From a clear understanding of this distinction, it might be thought that the application of deductive logic would enable one to determine whether a proposed variable is suitable for use with biserial r or with point biserial r. This is almost true but, unfortunately, the situation is complicated a bit: there are some variables for which neither is applicable and others for which either one is applicable, depending on the circumstances.

Accordingly, Table 10.6 has been prepared. The variables in it were obtained from statistics texts, journal articles, and other sources.

Part A of the table lists variables that may be regarded as continuous and normally distributed. In most cases, more refined measures of these varia-

TABLE 10.6
Classification of Dichotomous Variables

A. Continuous dichotomies (suitable for r_{bis} and r_t)		
Above the mean on a test	or	Below the mean on a test
Active (church goer)	or	Inactive
Answering 50 or more items correctly on a test	or	Answering 49 or fewer items correctly on a test*
(Child) with developmental defect	or	Without developmental defects
Discharged (employee)	or	Not discharged
Favorable attitude (toward anything)	or	Neutral and unfavorable*
Fleshy (build)	or	Medium and slender
Good (motion picture)	or	Poor
Graduated from university	or	Not graduated*
Healthy	or	Unhealthy*
Literate	or	Illiterate
(Manufactured part) accepted by inspector	or	Rejected by inspector
Normal	or	Abnormal
Normal	or	Retarded
Pass (in school or on a test)	or	Fail
Problem child	or	Not a problem child
Radical (or liberal)	or	Conservative (or reactionary)
Socially mature (child)	or	Socially immature
Socially well adjusted	or	Not socially well adjusted
Success (in technique)	or	Failure*
Successful (worker)	or	Unsuccessful
Swimmer	or	Non-swimmer
Wife aged 15–59	or	Wife aged 60–99

B. Categorical dichotomies (suitable for r_{pb} and ϕ)		
Adenoids removed (from a child)	or	Adenoids not removed
Blue-eyed	or	Not blue-eyed†
Catholic	or	Protestant
Dead	or	Alive†
Divorced (woman)	or	Not divorced†
First lieutenant in Army Medical Department	or	First lieutenant in Army, but not in Medical Department†
Have brothers or sisters	or	Have no siblings
Live in Switzerland	or	Do not live there
Male	or	Female†
Married	or	Not married
Own a television set	or	Not own one
Parents	or	Not parents
Preference for pie	or	Preference for cake
Presence of vaccination scar	or	Absence†
Recover (from smallpox)	or	Die†

TABLE 10.6
Classification of Dichotomous Variables (*continued*)

	C. Ambiguous or equivocal dichotomies	
Variable	*Circumstance in which* r_{bis} *or* r_t *is more appropriate*	*Circumstance in which* r_{pb} *or* ϕ *is more appropriate*
Alcoholic (person) or Nonalcoholic	If alcoholics are regarded simply as people who drink more than a certain (very large) amount; or if the relation of amount of drinking is important	If alcoholics are regarded as a distinct type
Experienced (worker) or Inexperienced	If eventually to be used as a continuous variable; or if the amount (rather than presence or absence) of experience is important (though such a distribution may not be normal)	If a small amount of experience makes a big difference and additional experience has little influence; or if used to predict future success
Jew or Gentile	If regarded as a political or social continuum	If regarded as distinct races; or if regarded as religious types
Like (an interest item) or Not like	If theoretical relation of the continuum covered by the item is wanted	If item is used to predict any criterion; or if liking and not liking are regarded as distinct types
Negro or White	If regarded as a continuum, the term Negro including 100% to 1% Negro blood (though such would not be a normal distribution)	If regarded as distinct races
Pass an item or Fail an item	If theoretical significance of field measured by the item is of interest	If item is used to predict any criterion (as part of a test or otherwise)
Republican or Democrat	If regarded as degrees on a political continuum	If regarded as distinct types; or if used to predict viewpoint toward political issues
Socially minded or Mechanically minded	If dichotomy will be replaced by a continuous test score for future use	If used to predict a criterion
Yes on a questionnaire or No on a questionnaire	If theoretical relation of the continuum covered by the item is wanted	If item is used to predict any criterion; or if yes and no are regarded as distinct types

*At some time or other, at least one statistician has (erroneously) used this variable with point biserial r or with phi. The student is referred not to authority, but to logic to decide how a variable should be used.

†At some time or other, at least one statistician has (erroneously) used this variable with biserial r or with tetrachoric r. The student is referred not to authority, but to logic to decide how a variable should be used.

bles might, at least in theory, be expected to be normally distributed. In other cases, this is not true, but if some function of the variable would yield a normal distribution, then biserial r or tetrachoric r is an appropriate measure of the relation between that function of the variable and whatever else it is compared with. Thus, a more refined measure of the Radical or Conservative dichotomy might not yield normally distributed scores, but since some simple function of these scores probably would be normal, this variable is appropriately included under A. In the case of some other dichotomies, such as Wife aged 15–59 or Wife aged 60–99, the assumption of even a transformed normal distribution may be quite dubious with some sets of data; with others it would be all right. For example, if all the women in a city were used, the distribution of their ages would be far from normal, but if those belonging to a certain club were used, a normal distribution might be quite reasonable.

Part B of Table 10.6 lists the variables that constitute true categorical dichotomies with all the individuals in one category being alike and all those in the other category also being alike insofar as this variable is concerned. Variables yielding only two types of individuals fit nicely in here. Such variables are ideally adapted for use with point biserial r or with phi.

Unfortunately, Parts A and B do not exhaust the list of possible dichotomous variables. There are a few variables that should sometimes be used with biserial r (or tetrachoric r), while at other times they should be used with point biserial r (or phi). A typical selection of these is given in Part C of Table 10.6. For these variables, the appropriate correlational procedure is a matter of purpose and interpretation. For each such variable, at least one situation is described that would justify the use of the variable with biserial r (or r_t) and at least one other that would justify its use with point biserial r (or ϕ).

The most important variable in Table 10.6 is undoubtedly Pass an item or Fail an item. More biserial r's have been computed with this variable than with all the others put together. But it is also true that more point biserial r's have been computed with it than with all other variables put together. Item analysis is big business. Let us consider this very simple true–false test item: "San Francisco is closer than Los Angeles is to Detroit." This item clearly involves a number of geographical concepts, the most relevant of which are the relative positions of these three cities with respect to each other. Now if you know that Detroit is in the midwest, near the Great Lakes, and that the other cities are both in California, next to the Pacific Ocean, with San Francisco being several hundred miles further north than Los Angeles, you will probably have enough information to lead you to answer the item rather than to omit it. But you may pass it or you may fail it. To be reasonably sure of getting it right, you must know and apply still more information. You will probably miss the item unless you realize that Los Angeles is also east of San Francisco. The fact that it is over 200 miles east of San Francisco makes it closer to practically all the big cities of the United States—regardless of whether we measure by airline miles, auto miles, railroad miles, or by the way no crow ever flew but all crows are supposed to. Most people do not know that Los Angeles is so far east of San Francisco; still fewer know that even the western boundary of its current city limit is over 60 miles *east* of Reno, Nevada. You do not need to know all this in order to answer the item correctly, but you certainly need some of this information and this is exactly the type of information on which the item is based.

Suppose we are interested in the relation between this kind of information and some other continuous variable. Biserial r would be an appropriate measure. Except for chance, people who have more than a certain minimum amount of such information will get the item right; those who have less than this necessary minimum will miss it. Biserial r thus gives us an estimate of what the Pearson r would be, if we were able to compute it, between the other variable and the amount of geographical knowledge of the kinds (chiefly directional in space) needed to answer such items. Biserial r, then, gives us an estimate of what the Pearson r would be between our other variable and the amount of relevant information possessed by those people who answered the item. We are estimating what the correlation would be if, instead of having two scores (0 for wrong and 1 for right), we had a dozen or more degrees of information or scores and could compute Pearson r in the usual way.

We hope we have not made too good a case for the use of biserial r in item analysis, because point biserial r is appropriate under many of the more usual circumstances. Biserial r is considerably older and much more widely known. It was derived and first published by Karl Pearson (after whom Pearson r was named) in 1909. The derivation of point biserial r (for item analysis, it happens) was published by Richardson and Stalnaker in 1933. Although point biserial r is gaining both in absolute and in relative frequency of use, it is still not as widely used as it should be.

A very important function of tests and of test items is to predict some future performance. This is the purpose of nearly all tests used in industry or in military situations. It is, by definition, the sole reason for the existence of aptitude tests—whether the aptitude be to sell life insurance, to receive radio code, to make good grades in college, or to do first grade work. Even with achievement test items, it frequently happens that the aim of such testing is not just to find out what has already been learned (for historical purposes) but, rather, to use past performance as an indicator of future potential (for educational or vocational guidance or placement). If, now, we plot a 2 by n table (such as was shown on page 312) to show the relation between passing or failing the item, and score on any criterion of industrial success, we are clearly doing so for the purpose of predicting industrial success from the item response. Notice that no matter how many degrees of extent of geographical (or other) knowledge there may be, all we will ever know about any respondent's information about the item is that the person passed it or failed it. We cannot know that he almost passed it or that he failed miserably. For predictive purposes, any other considerations are irrelevant. We must make the best prediction with what we have. For this purpose, point biserial r is ideal.

If we are correlating our item response with total score on the test containing the item in order to get measures of the internal consistency of the test items, a case can be made out for either measure, but, in our opinion, it is usually better to use point biserial r. Thus, we regard biserial r as appropriate for use with test items only when the theoretical nature of an idealized relationship is of primary concern; in most situations, point biserial r is either preferable or is the only correct procedure.

It is appropriate at this time to emphasize that biserial r tells us what Pearson r *would be* if we were able to substitute continuous normally distributed scores for our dichotomy, while point biserial r tells us what Pearson r *is* when we score our dichotomy 1 for the correct alternative and 0 for the others. In the next two sections of this chapter, we will see that much of what has been said about biserial r applies with equal force to the two dichotomies of tetrachoric r and that much of the discussion of point biserial r applies also to phi.

TETRACHORIC r

Sometimes we wish to correlate two dichotomous variables with each other. Each of these variables may, in reality, be continuous and normally distributed, but recorded in only two categories. For many reasons, it may be desirable to know the relationship between the two variables. But the relation we want is not between the two dichotomies, but rather is the correlation between the two continuous and normally distributed variables that we know only as dichotomies.

For instance, if for a number of students we know whether each is above or below 5 feet 4 inches and we also know whether each one's score on a test is above or below the mean of 32.10, we would have data of a form that could be used to compute the tetrachoric correlation coefficient. In one sense, it would be appropriate to compute tetrachoric r from such a fourfold table. Both variables are continuous, both are normally distributed, both are dichotomized. Tetrachoric r can appropriately be computed. It will give us an estimate of what Pearson r would be if we took the actual measurements of height and correlated them with the test scores. In theory, there is nothing wrong with this use of tetrachoric r; in practice, too, it is all right except perhaps for one little thing: Why use tetrachoric r as a means of estimating what Pearson r would be? Why not compute Pearson r directly? In such a situation as this, that is exactly what we recommend. Most statisticians would, we believe, agree. Those who would use tetrachoric r in such cases do so solely as a timesaving measure. There is no secret about the facts that (1) it is more accurate to compute Pearson r than it is to use only part of the data to estimate what it would be; and (2) the standard error of tetrachoric r is much greater than that of Pearson r.

There are, however, many situations in which it is impossible to compute Pearson r. If our records are available only as dichotomies, we have no alternative—regardless of how continuous and normal the underlying distributions may be. Let us consider an example.

Example of Tetrachoric r Mandell and Shultz* give data for a group of 258 overseas U.S. Government employees in all occupational levels from clerical through professional and administrative. One variable is based on a medical examination. The high group have no or only slight medical defects, and no history of serious illness. The low group have more serious disabilities or illnesses. The other variable is a rating by supervisors on adjustment to the overseas environment. They give their data in the following form:

	Adjustment		
Medical Examination	Very Satisfactory	Satisfactory	Unsatisfactory
High	72	77	45
Low	18	26	20

This is not a fourfold table; it has six categories. The three classes of adjustment to the overseas environment are hardly enough to justify us in computing biserial r, so we will group two of them together and compute tetrachoric r. There are two ways we might do this. The better way always is to group our data so that the dichotomic line comes as near the median as possible. More simply stated, we combine so that our two remaining groups

*Mandell, Milton M., and Shultz, Meyer. Validity information exchange report No. 9-1. *Personnel Psychology,* 1956, **9**, 103–104.

(on this variable) will be as nearly equal in size as possible. This gives us the following table:

	Satisfactory or Unsatisfactory Adjustment	Very Satisfactory Adjustment	Total
High Medical Examination	122	72	194
Low Medical Examination	46	18	64
Total	168	90	258

The columns have been rearranged so that the ++ cell is in the usual upper right corner and, of course, the −− cell is in the lower left corner. The correlation is obviously low, but since 72 is 4 times 18 and 122 is less than 4 times 46, it is clearly positive. (Only if two such ratios are equal, will r_t be zero. If the one on the left is greater, r_t will be negative.)

One might expect the next step to be the presentation of a formula for tetrachoric r and its solution. This is not desirable, however. The formula for tetrachoric r is of the general type

$$a = r_t + br_t^2 + cr_t^3 + dr_t^4 + er_t^5 + \cdots,$$

where a, b, c, d, and e are coefficients derived from the data in a complicated fashion and where the series is infinite in length. When r_t is low, the first few terms give an accurate answer, because the small coefficients multiplied by high powers of r_t give negligible quantities. When r_t is high, many terms (such as 6 or 8) may need to be retained. Take our word for it—we've computed such r_t's—it takes a long time to solve such an equation.

Today there are three main ways to compute the tetrachoric correlation coefficient: by using one of the several computer programs now available, by using the Chesire–Saffir–Thurstone Computing Diagrams, or by using a method of approximating it. We shall briefly consider all three of these methods.

Using a Computer Program for Computing r_t If you have access to a computer, there are at least three programs for obtaining r_t.

The Educational Testing Service at Princeton, NJ, has developed a Fortran subroutine for computing r_t, using a formula developed by Ledyard Tucker. For information regarding the availability of this subroutine, please write to Director, Office of Data Analysis Research, ETS, Princeton, NJ 08540.

IBM has a program which will give both r_t and its standard error. It is described on pages 168–171 of IBM's Program Reference Manual, SH20-1069. It is alluded to in their Program Product Manual GH-20-1027-3, which is entitled "STAT/BASIC for System/3 Model 6, ITF, and VM/370-CMS General Information Manual." In this manual, tetrachoric correlation is listed under nonparametric statistics. This is an interactive program. In order to use it, your IBM computer should have installed either TSO or ITF or VM370. Both these references and more information are available at IBM offices.

The University of California at Los Angeles has just completed a major revision in their program BMDP1F—Two-way Contingency Tables. Included in the revision is a routine to compute tetrachoric *r* and its standard error, which will be available as part of the BMDP series of programs from the Health Sciences Computing Facility, UCLA, Los Angeles, CA 90024.

With the information given in the three preceding paragraphs, anyone familiar with data processing and computers should be able to help you compute tetrachoric correlation coefficients and their standard errors. Since, however, not every one has access to expensive computers, the next best method is the use of a book of computing diagrams which, hopefully, you can obtain at your college library.

Using the Chesire–Saffir–Thurstone Computing Diagrams For several years it was believed that these computing diagrams were out of print. They are now available and will continue to be available.* To use them, it is first necessary to prepare a new table, similar to the preceding, but based on proportions. To do this, each number in the table based on frequencies is divided by *N*, in this case 258, resulting in a table of proportions:

	Satisfactory or Unsatisfactory Adjustment	Very Satisfactory Adjustment	Total
High Medical Examination	.47	.28	.75
Low Medical Examination	.18	.07	.25
Total	.65	.35	1.00

*The original publisher was University of Chicago Book Store which in 1976 had a few copies available. Arrangements have now been made for republication of xerographic paper bound copies, of cloth bound copies, and of microfilm copies. For ease of use, the full size copies

Then, in accordance with the directions accompanying the *Computing Diagrams,* four separate estimates of r_t are obtained. The average of these four determinations is recorded (usually to two decimals) as r_t. In our example, we obtain $r_t = +.15$. Even a novice can make these four estimates and average them in less than five minutes. The obtained r_t is very accurately determined. In our example, the value obtained from the *Computing Diagrams* was $r_t = +.149$, while that obtained by applying the formula to these same proportions was $r_t = +.151$. This is much closer agreement than can usually be expected, although the authors state that, "Only rarely do the discrepancies exceed .01."

Using Methods of Approximating r_t Without either a computer program or the Computing Diagrams just referred to, the computation of r_t is straightforward but extremely time-consuming. There are several methods of obtaining approximations to it, the best of which is probably the cosine π correlation which is discussed in Supplement 10. If both dichotomies are at the medians, it is not just an approximation—it is exact. Any time that p, q, p', and q' are all between .40 and .60, the approximation will be very close (errors of about .01 or .02). If, however, the correlation is high (.80 or more) and the splits depart appreciably from the medians, the cosine π correlation may overestimate r_t by relatively large amounts (.05 or .10).

Hence, if it is not feasible for you to use a computer program, and if the four proportions are all between .40 and .60, use the cosine π correlation coefficient as an approximation to r_t. If the proportions are not all within these limits, use the Computing Diagrams.

Interpretation of Tetrachoric r As already indicated, tetrachoric r gives us an estimate of the correlation we would obtain if (1) the data were more finely divided on both variables and (2) Pearson r were computed. It is further necessary to assume that both variables are, in reality, continuous and normally distributed and that regression is linear. Usually, we do not know whether or not all these assumptions are satisfied, but if we are correlating variables that are truly continuous and normally distributed (See Part A of Table 10.6), the chances are good that regression will be essentially linear.

Tetrachoric r is subject to exactly the same type of interpretation that applies to Pearson r, since it is always an estimate of Pearson r. The only

---

are recommended. At the present time, orders may be placed with the Book Store. As soon as their copies are gone, the citation will be: Chesire, Leone, Saffir, Milton, and Thurstone, L. L. *Computing diagrams for the tetrachoric correlation coefficient*. Ann Arbor: University Microfilms International. (Reprint of 1938 edition).

complication lies in the fact that its standard error is larger—much larger, in fact.

Standard Error of Tetrachoric *r* The formula for the standard error of tetrachoric r is not as complicated as the formula for r_t itself, but it is not simple. It is usually given as

$$s_{r_t} = \frac{\sqrt{pqp'q'}}{yy'\sqrt{N-1}} \sqrt{(1 - r_t^2)\left[1 - \left(\frac{\sin^{-1} r_t}{90°}\right)^2\right]}, \tag{10.8}$$

in which p and q are the marginal proportions for one variable and p' and q' are those for the other variable. By the use of two tables, this formula can be greatly simplified. First note that for any value of p (or q), Table Q gives $\frac{\sqrt{pq}}{y}$ and, of course, for any value of p' it gives $\frac{\sqrt{p'q'}}{y'}$. The function of r_t, comprising a little more than the right half of the formula, is given in Table S for all values of r_t. This function has the same value for positive and for negative values of r_t, so the sign may be ignored. Using F as a symbol for an appropriate tabled function of the variable indicated by its subscript, we may rewrite the formula for the standard error of r_t as

$$s_{r_t} = \frac{F_p F_{p'} F_{r_t}}{\sqrt{N-1}}. \tag{10.9}$$

The values of F_p and $F_{p'}$ are obtained from Table Q, the value of F_{r_t} from Table S, and the value of $\sqrt{N-1}$ from Table Z.

For our overseas health data, $p = .35$; $p' = .75$; $r_t = +.15$; and $N = 258$. First we note that since p and q enter symmetrically into the formula, F_p has the same value for $p = .35$ as that tabled for p (or q) $= .65$. Substituting the tabled values in Formula (10.9), we obtain

$$\begin{aligned} s_{r_t} &= \frac{F_p F_{p'} F_{r_t}}{\sqrt{N-1}} \\ &= \frac{(1.2877)\,(1.3626)\,(.9841)}{16.0312} \\ &= \frac{1.7267}{16.0312} \\ &= .1077 \\ &= .11. \end{aligned}$$

If the true r_t were $+.15$, this tells us that about two-thirds of the time, our obtained r_t's would fall between $+.04$ and $+.26$, while 95% of the time they would be between $-.07$ and $+.37$. Since one of these values is very close to zero, it is obviously appropriate to see whether or not this r really differs from .00. (Some statisticians routinely perform this step first.) To do so, we use the same formula. But note that since $F_{r_t} = 1.0000$ when $r_t = .0000$, we simply write 1.0000 for the last numerator term and find that s_{r_t} is a little larger but, to two decimal places, it remains .11 as before. Because +.15 is less than 1.96 times this value, we conclude that our correlation is not significantly different from zero, even at the .05 level of significance.

Is it different from +.40? Substituting again in the formula, we find that s_{r_t} now is .10. The .99 confidence limits are $+.40 \pm (2.58)(.10)$, or +.14 and +.66; the .95 confidence limits are $+.40 \pm (1.96)(.10)$, or +.20 and +.60. Our r_t of +.15 is outside these .95 limits and almost on the boundary of the .99 limits. It is therefore unlikely that it arose as a chance fluctuation from a true correlation of +.40 or higher.

Thus, we conclude our analysis by noting that we have failed to demonstrate any significant relationship, positive or negative, between health (as measured by a medical examination) and adjustment to the overseas environment (as rated by supervisors). Because the N of 258 is fairly large, our sample statistic of .15 and its standard error enable us to see that the population parameter may be slightly negative, zero, or some positive value no higher than about +.40. In brief, we have shown that there is little or no relationship between the two variables.

In order to use r_t appropriately, must both variables be normally distributed? The answer is a qualified yes. If the variables would be normally distributed if we had more exact measures, r_t is obviously appropriate. If either or both are not, but some function of them is normally distributed, r_t may be computed and interpreted as showing the relationship between these functions of the variables. Since any continuous variable can have its scores transmuted into a normalized distribution, and since a relatively simple transformation (such as taking the square root of the scores) will give essentially normal distributions in many cases, it follows that r_t is quite widely applicable if we regard it as an estimate of the Pearson r between these *transformed* variables. The same principle applies, of course, to the dichotomous variable in biserial r.

This extension of the applicability of r_t is not quite so general as it may appear. Although appropriate for many, if not most, continuous variables, it does not apply to discrete variables at all; it cannot apply even to continuous variables if one plans to use them for predictive purposes in their present dichotomous state. For such situations, r_t is not appropriate, but ϕ is, as can be seen in the next section of this chapter.

PHI

In Part B of Table 10.6, a number of truly categorical dichotomies are listed. Often one wishes to measure the degree of correlation between such variables. This is one of the two purposes served by ϕ. Its other purpose is to provide a measure of the amount of correlation that *now exists* between any two categorical variables *as dichotomies,* rather than to give an estimate, as r_t does, of what Pearson r *would be* under other circumstances.

Derivation of Phi When we compute Pearson r, we usually have a dozen or more intervals for each variable. What would happen if we cut down the number of intervals until we had only two for each variable? We would then have our familiar fourfold table. If we denote the frequencies in the four cells by the first four letters of the alphabet, the table would look like

B	A
D	C

. If we include the marginal totals, and remember that $A + B + C + D = N$, it would become

B	A	$A + B$
D	C	$C + D$
$B + D$	$A + C$	N

We may now regard the right column (containing A and C) as corresponding to higher scores than the left column on the X variable. (If this is not so, reverse the columns.) Similarly, the top row is regarded as the higher on the Y variable. With no loss of generality, but with considerable saving of arithmetic, we may assign scores of 1 to the higher groups and 0 to the lower. (This is the same as assigning x' and y' scores.)

What is the value of ΣX? Each person's X score is either 0 or 1. Those in cells B and D have X scores of 0, while those in cells A and C have X scores of 1. ΣX is, then, the number of people who have X scores of 1. Hence $\Sigma X = A + C$. If $X = 0$, then $X^2 = 0$; and if $X = 1$, then $X^2 = 1$. Oddly enough, in this situation, therefore, $\Sigma X^2 = \Sigma X = A + C$. Similarly, $\Sigma Y = \Sigma Y^2 = A + B$. ΣXY is even easier to compute. If either X or Y is equal to 0, the product XY will be zero. Hence the only XY products not equal to zero are those in cell A and each of them is equal to 1. Since there are A of them, $\Sigma XY = A$. Let us substitute these values in Formula (8.2) for Pearson r:

$$r = \frac{N\Sigma XY - \Sigma X \,\Sigma Y}{\sqrt{[N\Sigma X^2 - (\Sigma X)^2][N\Sigma Y^2 - (\Sigma Y)^2]}}$$

$$= \frac{NA - (A + C)(A + B)}{\sqrt{[N(A + C) - (A + C)^2][N(A + B) - (A + B)^2]}}$$

$$= \frac{NA - A^2 - AB - AC - BC}{\sqrt{[(A + C)(N - A - C)][(A + B)(N - A - B)]}}$$

$$= \frac{A(N - A - B - C) - BC}{\sqrt{[(A + C)(B + D)][(A + B)(C + D)]}},$$

$$\phi = \frac{AD - BC}{\sqrt{(A + C)(B + D)(A + B)(C + D)}}. \qquad (10.10)$$

Notice that although we have written ϕ at the left of the equal sign, as is customary, ϕ is really nothing but the Pearson r between two categorical distributions. We derived it by simply substituting the correct values for all the summations and noting that $N = A + B + C + D$. We made no further assumptions beyond our original one that each of our variables had all its scores concentrated at one or the other of two points.

Computation of Phi We have heard a person referred to, usually in a derogatory fashion, as a pie-face, cake-eater, or cookie-pusher. Is there any relation between sex and preference for various desserts? On one summer day, a university cafeteria had available chocolate and spice cake, and apple, chocolate chiffon, lemon meringue, and peach pie. People who chose ice cream, watermelon, or no dessert were ignored. The others were tabulated:

	Female	Male	Total
Cake	10	32	42
Pie	55	113	168
Total	65	145	210

There were more men than women (even in summer school!) and many more pie-eaters than cake-eaters. Rousseau refers to "a great princess, who, on being informed that the country people had no bread, replied, 'Let them eat cake.'" Because he wrote this before Marie Antoinette's arrival in France, she may not have been the one who said it but, in any case, there seems some tendency for the men rather than the women to follow this advice. How strong is this relationship?

To compute ϕ, all we need do is substitute the four cell frequencies in the numerator and the four marginal frequencies in the denominator of the formula and solve. Thus,

$$\phi = \frac{AD - BC}{\sqrt{(A + C)\,(B + D)\,(A + B)\,(C + D)}}$$

$$= \frac{(32)\,(55) - (10)\,(113)}{\sqrt{(145)\,(65)\,(42)\,(168)}}$$

$$= \frac{1{,}760 - 1{,}130}{\sqrt{66{,}502{,}800}}$$

$$= \frac{630.}{8,155.}$$
$$= + .08.$$

This correlation is very low. Even if it were statistically significantly different from zero (which it is not), it would be too low to be of any practical value. We have no evidence, then, for believing that there is a correlation between sex and preference for cake rather than pie. Note that we have not proved that there is *no* relation. This is impossible to do. Because our N was reasonably large, we have, however, proved that the correlation cannot be very high—either positive or negative. (This will be discussed further in connection with the standard error of phi.) Note that we have by no means established that there is little or no relation between sex and dessert preferences in general, or even between sex and preference for different kinds of cake or pie. Quite the contrary. In fact, here is a further breakdown of the pie data:

	Female	Male	Total
Apple or Peach Pie	14	66	80
Chocolate Chiffon or Lemon Meringue Pie	41	47	88
Total	55	113	168

The two pies with chunks of fruit in them are preferred by the men and the two fluffy pies are preferred by the women. Let us compute the correlation:

$$\phi = \frac{AD - BC}{\sqrt{(A + C)(B + D)(A + B)(C + D)}}$$
$$= \frac{(66)(41) - (14)(47)}{\sqrt{(113)(55)(80)(88)}}$$
$$= \frac{2,706 - 658}{\sqrt{43,753,600}}$$
$$= \frac{2,048}{6,615}$$
$$= +.31.$$

Thus, among the pie-eaters, there is a definite correlation between preference for the two kinds of pie and sex.

The sign of phi is usually arbitrary. We have here regarded both males and apple-or-peach-pie as positive. If we reversed one of these (as by tabulating the males at the left), ϕ would have the same absolute value, but would have its sign reversed. If we reversed both variables (regarding females and fluffy pies as positive), ϕ would retain the same value we computed for it in the first place.

Alternative Formula for Phi If we divide both numerator and denominator of Formula (10.10) by N^2, the formula becomes

$$\phi = \frac{ad - bc}{\sqrt{(a + c)(b + d)(a + b)(c + d)}},$$

where the lower-case letters represent proportions rather than the frequencies which were represented by the corresponding capital letters. If we further substitute the symbols p and q for the proportions above and below the point of dichotomy in the first variable, and p' and q' in the second variable, the formula simplifies to

$$\phi = \frac{ad - bc}{\sqrt{pqp'q'}}, \tag{10.11}$$

where all the symbols refer to proportions in the fourfold table:

	$X = 0$	$X = 1$	Total
$Y = 1$	b	a	p'
$Y = 0$	d	c	q'
Total	q	p	1.00

Formula (10.11) is frequently encountered. It looks simpler than Formula (10.10), but actually is not. One is based on the raw data, while the other is based on the proportions resulting when all the table entries are divided by N. Formula (10.11) has the slight disadvantage of introducing small rounding errors in all eight of the proportions entering into it. It is, however, sometimes useful in checking a computation carried out by Formula (10.10) although, of course, it can provide no check on the original tabulation. Formula (10.11) is also useful when the research worker is computing a number of ϕ's based on different N's and wishes to have his data expressed in somewhat comparable form (although the ϕ's might well be computed by either formula—or better by both—in such cases). Some persons who have either a somewhat artistic or a mathematical frame of mind may prefer Formula (10.11) for still another reason. The formula for s_ϕ is best written in terms of proportions rather than frequencies. Hence, there may be some slight aesthetic appeal in using proportions both times. We have a slight (uncultured) preference for Formula (10.10) on the basis of a trivial increase in accuracy and the elimination of decimals as a source of error in the computations, but either formula is entirely acceptable.

Standard Error of Phi If by any chance you thought the formula for s_{r_t} looked complicated, just take a deep breath and look at the one for s_ϕ. Perhaps this is one of the reasons that the formula is seldom published and rarely used. It can be written in a number of different ways. One of the simplest* of these is

$$s_\phi = \sqrt{\frac{1 - \phi^2 + \left(\phi + \frac{1}{2}\phi^3\right)\left(\frac{p - q}{\sqrt{pq}}\right)\left(\frac{p' - q'}{\sqrt{p'q'}}\right) - \frac{3}{4}\phi^2\left[\frac{(p - q)^2}{pq} + \frac{(p' - q')^2}{p'q'}\right]}{N - 1}} \tag{10.12}$$

In certain situations, this formula will simplify greatly: If both dichotomic lines are at the medians, it becomes simply $\sqrt{\frac{1 - \phi^2}{N - 1}}$. Or, regardless of where the dichotomic lines are, if we are testing the null hypothesis that $\phi = 0$, the formula becomes $\frac{1}{\sqrt{N - 1}}$. The latter situation is of fairly frequent occurrence, but the former is not. Hence, if we wish to determine whether our obtained ϕ is significantly different from any value other than zero, the entire formula must nearly always be used.

Let us look the formula over. The portion on the right contains 32 letters and numbers, and even more than 32 signs and other symbols. Yet it contains only four independent variables: ϕ, p, p', and N. We can't do anything to make this formula as simple as that for $s_{\bar{X}}$, but we can fairly easily get it down to a range where it becomes quite reasonable. With the aid of a couple of tables and a calculator, it can then be computed in less than 5 minutes. Without the calculator, the time should still be well under 15 minutes.

To simplify Formula (10.12), we will first define three functions of ϕ, two more of p, and the same two of p'. Then with the aid of tables of these five (or seven) functions, we can rewrite the formula. We do not claim that our new version of it is simple, but it is at least simpler than Formula (10.12). The functions we need to tabulate are:

$$F_{\phi_1} = 1 - \phi^2,$$

$$F_{\phi_2} = \phi + \frac{1}{2}\phi^3,$$

$$F_{\phi_3} = \frac{3}{4}\phi^2,$$

*That is not a typographical error. You should see the others!

$$F_{p_4} = \frac{p - q}{\sqrt{pq}}, \qquad F_{p'_4} = \frac{p' - q'}{\sqrt{p'q'}},$$
$$F_{p_5} = \frac{(p - q)^2}{pq}, \qquad F_{p'_5} = \frac{(p' - q')^2}{p'q'}.$$

Substituting all these functions in Formula (10.12), it becomes

$$s_\phi = \sqrt{\frac{F_{\phi_1} + F_{\phi_2} F_{p_4} F_{p'_4} - F_{\phi_3}(F_{p_5} + F_{p'_5})}{N - 1}}. \tag{10.13}$$

The three functions of ϕ and the four functions of p and p' are found, respectively, in Tables U and T. To solve this formula, we enter Table U with the value of ϕ we are testing and substitute the three functions of ϕ in the formula. We then enter Table T with the obtained value of p and substitute its two functions in the formula. We then do the same for p'. Finally we substitute N in the formula and solve. Notice that the three functions of ϕ are numbered in the order in which they appear in the formula and are given in that same order in Table U. The two functions of p also appear in the formula and in the table in the same order. The functions of both p and p' are taken from the same table. It happens that $F_{p_5} = (F_{p_4})^2$, but one does not need to know this to use the formula.

Let us compute the standard error of our pie-sex correlation of .31. First, is it significantly different from zero? For $N = 168$, $\frac{1}{\sqrt{N - 1}} = .08$. Since our correlation of .31 is well over (2.58)(.08), or .21, it is clearly significant beyond the .01 level of significance. Is it significantly different from .15 or .50? To determine this we first compute and check p and p', which turn out to be $p = .67$ and $p' = .48$. Then we enter Table U with $\phi = .15$ and enter Table T, first with $p = .67$ and then with $p' = .48$. Substituting the tabled values and N in Formula (10.13) we obtain

$$\begin{aligned}
s_\phi &= \sqrt{\frac{F_{\phi_1} + F_{\phi_2} F_{p_4} F_{p'_4} - F_{\phi_3}(F_{p_5} + F_{p'_5})}{N - 1}} \\
&= \sqrt{\frac{.9775 + (.1517)(.7231)(-.0801) - .0169(.5228 + .0064)}{168 - 1}} \\
&= \sqrt{\frac{.9775 - .0088 - .0089}{167}} \\
&= \sqrt{\frac{.9598}{167}} \\
&= \sqrt{.005747} \\
&= .0758 = .08.
\end{aligned}$$

A similar computation for an assumed ϕ of .50 gives

$$s_\phi = \sqrt{\frac{.7500 + (.5625)(.7231)(-.0801) - .1875(.5228 + .0064)}{168 - 1}}$$

$$= \sqrt{\frac{.7500 - .0326 - .0992}{167}}$$

$$= .06.$$

Of the three terms in the numerator, the first is always positive, and the third is always negative. The second has the same sign as ϕ if both p and p' are greater than .50, or if they are both less than .50. If either p or p' is greater than .50, and the other is less than .50, the sign of the second term will be opposite that of ϕ.

Interpretation of Phi Phi is the Pearson product moment coefficient of correlation between two variables, each of which has all its cases concentrated at one or the other of two discrete points. Any obtained value of ϕ should first be tested to see whether or not it could reasonably have arisen as a chance fluctuation from zero. If it could, then the true correlation is either zero, low, or (if we used a very small N) unknown. If our obtained value of ϕ is statistically significantly different from zero, we then test it against any theoretical or other expected values of ϕ. If N is large, say over 200, we may without much error simply substitute our obtained ϕ in the formula in order to give us a basis for estimating where the true value might lie.

Our pie-sex correlation is certainly not zero. The obtained ϕ of +.31 is just barely significantly different from .15 at the .05 level of significance, but it is significantly different from .50 beyond the .01 level. Substituting our obtained ϕ of +.31 in the formula, we obtain $s_\phi = .07$. We thus may say that if the true ϕ had been +.31, .95 of the time we would obtain values of ϕ between +.17 and +.45. Since these are relatively narrow limits, we have pretty well established the fact that in similar campus cafeterias the luncheon preferences for such fruity or fluffy pies correlate about +.17 to +.45 with sex.

The interpretation of ϕ is thus essentially the same as that of any other Pearson r. The only difference is that our variables are, and are expected to remain, available only as categorical dichotomies.

INTERRELATIONS AMONG r_{bis}, r_{pb}, r_t, AND ϕ

We are sure that some of our mathematician friends will object to this analogy, but we will give it anyway. It is not mathematically true if numbers are substituted for the symbols, but it depicts very clearly the fundamental similarities and differences upon which these four measures of correlation are based:

Here is the analogy:

$$\begin{array}{ccc} r_{\text{bis}} & \text{is related to} & r_{pb} \\ & \text{as} & \\ r_t & \text{is related to} & \phi \end{array}$$

Interpreting, r_{bis} and r_{pb} have many things in common (both are measures of relationship, both can be either positive or negative, both are based on a 2 by n table) but they differ in that the former assumes the dichotomous variable is continuous and normally distributed, while the latter assumes it is categorical. Analogously, r_t and ϕ have many things in common, but they, too, differ in that the former assumes both dichotomous variables are continuous and normally distributed, while the latter assumes both are categorical.

In a proportion of the type $a/b = c/d$, the two middle terms may be reversed and the relationship $a/c = b/d$ will still be true. Making this transposition, our analogy would then read

$$\begin{array}{ccc} r_{\text{bis}} & \text{is related to} & r_t \\ & \text{as} & \\ r_{pb} & \text{is related to} & \phi \end{array}$$

This shows that r_{bis} and r_t have many things in common (both are estimates of what Pearson r would be, both assume dichotomous variables are continuous and normally distributed), but the former has only one dichotomous variable while the latter has two. Similarly, r_{pb} and ϕ have many things in common, but, again, the former has only one dichotomous variable while the latter has two.

Since both r_{pb} and ϕ are special cases of r, it is possible to derive the formula for either of them from that for Pearson r. Going even further, ϕ is a special case of r_{pb} and the formula for ϕ can also be derived from that for r_{pb}. ϕ is also related to χ^2, as will be seen in Supplement 11.

As was pointed out earlier, there are rare instances in which r_{bis} exceeds $+1.00$ or is less than -1.00. The three other measures, and Pearson r as well, *always* stay within the limits of ± 1.00.

Because of their rather extensive use in item analysis, it often happens that one or more of these measures of relationship is used on a wholesale basis. Such is the case if one wishes to correlate every item of one or more tests with a criterion of industrial or academic success. This is apt to involve the computation of several hundred r_{pb}'s (or possibly other measures). If the interrelationships among the items, or among what appear to be the best items, are desired, the number of ϕ's (or possibly other measures) can easily run into the hundreds or thousands. For such situations, quite a number of tables and graphs designed to facilitate the computation or approximate estimation of these four measures are now available; others are constantly being derived and written up. Many of these have been published in *Psychometrika* and in *Educational and Psychological Measurement*. You are referred to these journals and to *Psychological Abstracts* for other references if you contemplate wholesale computations. For a dozen or less, it will probably be better to use the procedures described in this chapter.

SUMMARY

With most educational and psychological data, Pearson r is the best measure of relationship to use. When it is impossible or definitely not feasible to compute Pearson r, dozens of other procedures are available. For most such data, one of four measures will be appropriate, depending upon the characteristics of the data. The relevant characteristics and the proper correlational procedure to use are shown in Table 10.7.

When our two variables are a continuous variable and a dichotomy, we use *biserial r* if we wish to know what the correlation *would be* between the continuous variable and the fundamental continuous, normally distributed variable underlying the dichotomy.

In a similar appearing situation, we use *point biserial r* if we want to know what the Pearson r *is* between the continuous variable and a true dichotomy scored as 0 or 1.

When both variables are dichotomies, we use *tetrachoric r* if we desire to estimate what the relationship *would be* between the two

TABLE 10.7
Conditions Determining the Appropriateness of Four Measures of Correlation

		Fundamental Nature of Dichotomy(ies)	
		Really continuous and normally distributed	*Two distinct categories; types; discrete classes*
MECHANICAL FORM OF DATA	$2 \times n$ table	biserial r	point biserial r
	2×2 table	tetrachoric r	phi

variables—both being regarded fundamentally as continuous normally distributed variables artificially forced into dichotomies.

When both variables are dichotomies, we use *phi* when we want to know what Pearson r *is* if both dichotomies are regarded as categorically distinct and are scored as 0 or 1.

Table 10.6 gives over 40 specific examples, and the text explains the basis for distinguishing between the two kinds of dichotomies and classifying them for use with the appropriate ones of these four measures of correlation.

SUPPLEMENT 10

Chapter 8 was devoted to the best measure of correlation, Pearson r. In Chapter 10, we have just presented four more excellent measures which can be used when Pearson r is inapplicable. We have alluded to the fact that there are many measures of correlation, most of which are of very limited usefulness and most of which are not used much anyway. A list of formulas for 14 of these (excluding the five just referred to) is given in the Dunlap and Kurtz Handbook,* along with citations to books that discuss each of them. Most of them are not widely used, which is probably as it should be, but there are two exceptions. They are discussed in this Supplement. One was referred to earlier as a means of approximating r_t; the other is the rank correlation coefficient, ρ. We shall discuss them in that order, starting with the cosine π correlation coefficient which we recommend, under the conditions described, for getting a close approximation to r_t.

COSINE PI CORRELATION COEFFICIENT

Using the same type of table that was used in computing r_t and ϕ, it is possible to compute still a third measure of relationship. This *cosine pi correlation coefficient* is so named because its value is a function, called the cosine, of an angle whose size is expressed as a fraction of 180°, which mathematicians sometimes refer to as π. The formula for the cosine pi correlation coefficient is

$$r_{\cos\pi} = \cos \frac{1}{1 + \sqrt{\frac{AD}{BC}}} 180°. \qquad (10.14)$$

---

*See citation in footnote on page 272.

TABLE 10.8
Table for Estimating $r_{\cos\pi}$

$r_{\cos\pi}$	$\frac{AD}{BC}$	$r_{\cos\pi}$	$\frac{AD}{BC}$	$r_{\cos\pi}$	$\frac{AD}{BC}$
.00	0–1.00	.35	2.49–2.55	.70	8.50–8.90
.01	1.01–1.03	.36	2.56–2.63	.71	8.91–9.35
.02	1.04–1.06	.37	2.64–2.71	.72	9.36–9.82
.03	1.07–1.08	.38	2.72–2.79	.73	9.83–10.33
.04	1.09–1.11	.39	2.80–2.87	.74	10.34–10.90
.05	1.12–1.14	.40	2.88–2.96	.75	10.91–11.51
.06	1.15–1.17	.41	2.97–3.05	.76	11.52–12.16
.07	1.18–1.20	.42	3.06–3.14	.77	12.17–12.89
.08	1.21–1.23	.43	3.15–3.24	.78	12.90–13.70
.09	1.24–1.27	.44	3.25–3.34	.79	13.71–14.58
.10	1.28–1.30	.45	3.35–3.45	.80	14.59–15.57
.11	1.31–1.33	.46	3.46–3.56	.81	15.58–16.65
.12	1.34–1.37	.47	3.57–3.68	.82	16.66–17.88
.13	1.38–1.40	.48	3.69–3.80	.83	17.89–19.28
.14	1.41–1.44	.49	3.81–3.92	.84	19.29–20.85
.15	1.45–1.48	.50	3.93–4.06	.85	20.86–22.68
.16	1.49–1.52	.51	4.07–4.20	.86	22.69–24.76
.17	1.53–1.56	.52	4.21–4.34	.87	24.77–27.22
.18	1.57–1.60	.53	4.35–4.49	.88	27.23–30.09
.19	1.61–1.64	.54	4.50–4.66	.89	30.10–33.60
.20	1.65–1.69	.55	4.67–4.82	.90	33.61–37.79
.21	1.70–1.73	.56	4.83–4.99	.91	37.80–43.06
.22	1.74–1.78	.57	5.00–5.18	.92	43.07–49.83
.23	1.79–1.83	.58	5.19–5.38	.93	49.84–58.79
.24	1.84–1.88	.59	5.39–5.59	.94	58.80–70.95
.25	1.89–1.93	.60	5.60–5.80	.95	70.96–89.01
.26	1.94–1.98	.61	5.81–6.03	.96	89.02–117.54
.27	1.99–2.04	.62	6.04–6.28	.97	117.55–169.67
.28	2.05–2.10	.63	6.29–6.54	.98	169.68–293.12
.29	2.11–2.15	.64	6.55–6.81	.99	293.13–923.97
.30	2.16–2.22	.65	6.82–7.10	1.00	923.98 . . .
.31	2.23–2.28	.66	7.11–7.42		
.32	2.29–2.34	.67	7.43–7.75		
.33	2.35–2.41	.68	7.76–8.11		
.34	2.42–2.48	.69	8.12–8.49		

SOURCE: M. D. Davidoff and H. W. Goheen. A table for the rapid determination of the tetrachoric correlation coefficient. *Psychometrika,* 1953, **18,** 115–121. Reprinted with the permission of the authors and publisher.

That's not hard to compute, but you don't even need to do that. Just perform two multiplications, one division, find this quotient in Table 10.8, and read the value of $r_{\cos \pi}$ from the table. In case your quotient isn't given, divide the other way, getting $\frac{BC}{AD}$, find the corresponding value of $r_{\cos \pi}$ and put a minus sign in front of it.

It does not matter whether you compute $\frac{AD}{BC}$ or $\frac{ad}{bc}$. Except for small rounding errors, you will get exactly the same value in either case. Let us use the same data we used to compute r_t and find the cosine pi correlation coefficient between adjustment and medical examination. First, we compute

$$\frac{AD}{BC} = \frac{(72)(46)}{(122)(18)} = \frac{3{,}312}{2{,}196} = 1.51.$$

In Table 10.8 we find that this quotient of 1.51 corresponds to .16, so $r_{\cos \pi} = +.16$.

You may recall that we found that r_t for those same records was +.15. This neatly illustrates the chief use of $r_{\cos \pi}$. It is an easy-to-get approximation to r_t. When the correlation is low, the agreement, as here, is good. When both dichotomies are split at or near the median, the agreement is good. In fact, if the split is exactly at the median so that $p = q = p' = q' = .50$, the two coefficients have exactly the same value.

RANK CORRELATION COEFFICIENT

The statement that statisticians do not agree on the merit or lack of merit of the rank correlation coefficient as a measure of relationship will probably get more agreement than almost anything else that could be said about it.

For example, the writers of this text are not in complete agreement regarding its usefulness. We spent what we both agree is a highly excessive amount of time in writing and rewriting several versions of this Supplement, with this writing ranging all the way from conciliatory and cooperative to adamant and dictatorial. Here's our final draft, guaranteed to be tolerable to both but not *wholly* satisfactory to either of us. (It is quite possible that your instructor's attitude toward it may differ from ours. Find out what it is before your next test.) This topic is present because although only one of us regards it as highly meritorious, we both agree that you should be familiar with it because you are almost certain to run into it in your reading of articles published in psychological or educational journals. Besides, some people just love it and so we should at least mention it for them. The rank correlation coefficient is often referred to by its symbol, ρ, which is the Greek letter rho.

What happens if instead of having scores for people we have rankings? We may wish to find out if there actually is a relationship between friendliness of service station managers and the quality of the service rendered by them. After all, one oil company used the theme, "I can be very friendly," in their advertising for quite a while. Were they friendly? More importantly, was there a relation between friendliness and the quality of the service rendered? Suppose we visit a number of stations in our own area, make notes, and then rank the managers from the friendliest, Gerry, to Chris who was the least friendly of the 123. Similarly, we rank all the managers from the best service rendered by Fran to the poorest service by Dana.

When we list all our managers and their rankings, we get Table 10.9. From these rankings it is possible to make a pretty good estimate of what Pearson r would be if we had actual scores instead of just rankings on friendliness and on quality of service. Let us do this. The process appears to be very simple but, as Longfellow said, "things are not what they seem."

First, we need to subtract the pairs of rankings to get a column of differences, which we will designate as D. Then we square these differences, add them up, and substitute in formula (10.15) to obtain the *rank difference correlation coefficient*, ρ. This is also called rank order correlation, Spearman's rho, and, of course, rho, or ρ. Let us compute it for our service station managers. We first obtain the differences in rank shown in the last column of Table 10.9. Then we square each difference and add these

TABLE 10.9
Rankings of Service Station Managers

Name	Ranking on Friendliness	Ranking on Service	Difference
Adrian	13	5	8
Bobby	17	63	46
Chris	123	104	19
Dana	88	123	35
Ellis	101	76	25
Fran	48	1	47
Gerry	1	98	97
Harley	2	39	37
Isa	39	28	11
Jacky	42	94	52
Kelly	14	17	3
Lee	69	89	20
Merle	107	62	45
.	.	.	.
.	.	.	.
.	.	.	.

squared differences to obtain ΣD^2. Thus, for the first 13 managers, $\Sigma D^2 = 8^2 + 46^2 + 19^2 + \cdots = 64 + 2{,}116 + 361 + \cdots = 22{,}637$.

When we get the ranks, not just of these 13, but of all 123 managers with the boy-or-girl names, we then add all the squared differences, getting $\Sigma D^2 = 123{,}456$.

The formula for ρ is

$$\rho = 1 - \frac{6\Sigma D^2}{N(N^2 - 1)}. \tag{10.15}$$

Substituting our known values of $N = 123$ and $\Sigma D^2 = 123{,}456$, we obtain

$$\begin{aligned}\rho &= 1 - \frac{(6)(123{,}456)}{(123)(15{,}128)} \\ &= 1 - \frac{740{,}736}{1{,}860{,}744} \\ &= 1 - .40 \\ &= +.60.\end{aligned}$$

How is this ρ of $+.60$ to be interpreted? Essentially the same way you would interpret a Pearson r of $+.60$, except that you should realize that its standard error, though not precisely known except when $\rho = .00$, is larger than that of Pearson r. This obviously means that we should ordinarily have a larger N when we compute ρ than when we compute r. Obvious or not, this is the exact opposite of what usually happens. When people have a large N, they tend (correctly) to compute r. When people have a small N, they sometimes (correctly) decide that their N is too small to justify computing r and then they (incorrectly) decide to compute ρ. Why do they do this? Let's look at it from their point of view—temporarily, at least.

Suppose they have two sets of scores for a large group (say, 200) of people. It is reasonable to compute r, because if they decided to compute ρ they would have to (1) rank the people in order and assign ranks running from 1 to 200 on the X variable, (2) rank them on Y, (3) get 200 D's, (4) obtain ΣD^2, and (5) substitute in the formula. Pearson r could almost certainly be computed in less time than it would take to perform either step (1) or (2) even if there were no tied scores; with ties, it would take much less time to compute r. So either the temptation to do it the easy way or the desire to do it the right way would lead to the same (correct) conclusion: compute r.

Now suppose they have a small N (say, 12). Take a look at Formula (8.2) for Pearson r and take a look at Formula (10.15) for ρ. Clearly, (8.2) looks worse than (10.15). So here the temptation to do it the easy way may triumph even though with a calculator the difference in time will probably be 15 minutes or even less. Although some statisticians regard the use of ρ

in such a circumstance as acceptable, many others feel that it is completely pointless to compute any measure of relationship with such a small N.

Why don't people compute ρ when they have a large N? We just did with 123 imaginary service station operators, but practically nobody ever does such a thing. The reason is simple. It is virtually impossible to visit 123 service stations, observe the two traits, and rank the people in order from 1 to 123. You'd have a real tough time doing it even with a much smaller group.

But let's face it: it is perfectly all right to compute ρ rather than r for a group of 123 people if you wish to do it. You'll probably get an answer close to the true value in the population from which your sample came in either case. When N is small, it isn't that easy to decide. Clearly, r is better, but ρ may be quicker. If you feel that your records aren't worth much anyway, don't try to decide *which* to use; discard them and don't compute anything. If, however, you feel that time is money and you can't conscientiously devote much time to analyzing the records, go ahead and use ρ.

We conclude that there is no reason for not using either ρ or r if it suits your purpose to do so, but we feel strongly that your decision should usually be based on the purpose and appropriateness of your analysis rather than on the basis of how much time the competing methods may require.

CHAPTER 11 CHI SQUARE

In Chapters 8 and 10, we discussed five of the best measures of the degree of relationship existing between two variables. All five are excellent measures of the *degree* of association between the two variables. We employed them, not primarily to find out whether there was or was not any (nonchance) relationship between the variables, but rather to measure the amount, degree, or extent of such relationship. For such a purpose, they are eminently proper procedures.

NATURE OF CHI SQUARE

Sometimes, however, this may not be our purpose. Suppose we have a collection of observations classified in some manner or other. Suppose further that we have or can get some corresponding set of expected or theoretical frequencies. Then it is very reasonable to compare the observed and theoretical frequencies and to ask whether the difference could or could not reasonably be ascribed to the operation of chance. The answer to this question is given by a statistic known as *chi square*.

The basic concepts behind the use of χ^2 are that (1) we have a theory of any kind concerning how our cases should be distributed; (2) we have a sample showing how the cases actually are distributed; and (3) we wish to know whether or not the differences between the theoretical and the observed frequencies are of such size that these differences might reasonably be ascribed to the operation of chance.

The phrase "theory of any kind" means exactly that. For instance, in an attitude survey of 60 people, we may assume equal frequencies for unfavorable, neutral, and favorable responses (giving expected frequencies of 20 20 20); or we may assume only half as many neutral as unfavorable or favorable responses (24 12 24); or a concentration in the middle (15 30 15); or successive increases toward the favorable end of the scale (10 20 30). In a table with a few columns and a few rows, the expected frequencies are often obtained by the hypothesis of independence, as will be illustrated in obtaining the theoretical or expected frequencies in Table 11.2 from the actual frequencies in Table 11.1. Of course, any other hypothesis may be tested. The only restriction is that the sources of error described on pages 366–369 must be avoided.

Chi square appears to be very simple, clearly understood, and easy to use. It has been very widely used, but it has also been very widely misused. Although many of the better statisticians were well aware of some of the crimes being perpetrated in the name of chi square, such awareness on the part of the majority of educational and psychological statisticians dates from the 1949 publication of an article entitled, "The use and misuse of the chi-square test."* While the first sentence of this paragraph is still true, from 1949 on it became increasingly apparent to statisticians that, regardless of appearances, (1) chi square is not simple, but it is highly technical; (2) it has not been and is not clearly understood; and (3) although it is very easy to compute, it is easily and frequently used incorrectly. More specifically, Lewis and Burke selected a journal, based on their belief that the authors who published in it were better versed in statistics than those in other journals. They examined three consecutive years and found 14 articles using chi square. They concluded that in 9 of the 14 papers the applications of chi square were clearly unwarranted, in 2 there was some doubt, and in only 3 of the 14 articles were the applications of chi square judged to be acceptable.

Before considering the hazards to be avoided, let us briefly show a simple but correct application of chi square. We will then list and discuss the pitfalls to be avoided, following this with other examples, applications, and uses of chi square.

---

*Lewis, D. and Burke, C. J. The use and misuse of the chi-square test. *Psychological Bulletin*, 1949, **46**, 433–489. See also replies by three statisticians and further discussion by Lewis and Burke in the same journal, 1950, **47**, 331–355.

TABLE 11.1
Preferences of Male and Female Students for Male or Female Personal Confidants

Sex of Student	Prefer Male	Prefer Female	No Preference	Total
Male	100	25	70	195
Female	22	38	48	108
Total	122	63	118	303

ILLUSTRATION OF CORRECT USE OF CHI SQUARE

For our first illustration, we will use data obtained from 303 University of Texas reading program students who had never had counseling.* The subjects were asked to state their preferences if a nonprofessional person (friend or personal confidant) were chosen for discussion of a personal problem. Table 11.1 shows, for instance, that 100 of the men preferred a male confidant, while 25 preferred a female, and the other 70 expressed no preference as to the sex of the confidant. The hypothesis to be tested is that there are no sex differences in such preferences and hence the sex of the student has no bearing on the student's preference (or lack of it) for a male or for a female confidant.

If the preferences were independent of the sex of the student, then the 195 males would distribute their preferences in a manner strictly proportional to those in the total row. Since $\frac{63}{303}$ of all students preferred a female personal confidant, then $\frac{63}{303}$ of the 195 male students, or 40.5 students, should also show such a preference if there is no relation between the student's sex and preference. Such figures as this 40.5 are referred to as expected frequencies. They are shown, for all six cells, in Table 11.2. The exact computed figures of 75.9 and 42.1 in the No Preference column were changed to 76.0 and 42.0. The reason for this small change will be given in section 4 on page 367 of this chapter. Notice that although no one of these expected frequencies is the same as the observed frequency given earlier in Table 11.1, all figures in the Total column, the Total row, and the two grand totals (303), are identical in the two tables.

*Fuller, Frances F. Preferences for male and female counselors. *Personnel and Guidance Journal,* 1964, **42,** 463–467.

TABLE 11.2
Expected Frequency of Preferences for Male or Female Personal Confidants

Sex of Student	Prefer Male	Prefer Female	No Preference	Total
Male	78.5	40.5	76.0	195.0
Female	43.5	22.5	42.0	108.0
Total	122.0	63.0	118.0	303.0

We now have all the information we need for computing chi square and determining its significance. The formula for χ^2 is

$$\chi^2 = \sum \frac{(f_o - f_e)^2}{f_e}, \tag{11.1}$$

where χ is the Greek letter chi, which rarely appears in statistics without being squared, where f_o is an observed frequency, and where f_e is the frequency that would be expected upon the basis of whatever hypothesis is being tested.

Substituting in this formula we obtain

$$\begin{aligned}
\chi^2 &= \sum \frac{(f_o - f_e)^2}{f_e} \\
&= \frac{(100 - 78.5)^2}{78.5} + \frac{(25 - 40.5)^2}{40.5} + \frac{(70 - 76.0)^2}{76.0} \\
&\quad + \frac{(22 - 43.5)^2}{43.5} + \frac{(38 - 22.5)^2}{22.5} + \frac{(48 - 42.0)^2}{42.0} \\
&= \frac{(21.5)^2}{78.5} + \frac{(-15.5)^2}{40.5} + \frac{(-6.0)^2}{76.0} + \frac{(-21.5)^2}{43.5} \\
&\quad + \frac{(15.5)^2}{22.5} + \frac{(6.0)^2}{42.0} \\
&= 5.89 + 5.93 + .47 + 10.63 + 10.68 + .86 \\
&= 34.46.
\end{aligned}$$

We now know that $\chi^2 = 34.46$. Is this big or little? We answer this by looking in Table V. But first we must know the number of degrees of freedom for chi square. It is not 302; it is only 2. In working with chi square, unlike all the situations discussed earlier in this book, *df* is equal to the number of frequencies that can be filled in arbitrarily before all the others can be determined automatically. Since the marginal totals are regarded as fixed, we may arbitrarily write in two frequencies in, say, the Male row. Then, since the three frequencies in that row must add up to 195, the third is automatically determined. That ends it. All the frequencies in the Female row are obtained by subtracting the corresponding figure in the Male row from that in the Total row. Hence, $df = 2$. Looking in the second row (for $df = 2$) of Table V, we note that our χ^2 of 34.46 is significant at the .001

level. We reject the hypothesis of independence and conclude that there is a definite sex difference in the expressed preferences for a personal confidant.

SOURCES OF ERROR IN CHI SQUARE

Now that we have seen a correct application of chi square, let us briefly consider the 9 principal sources of error in the papers that Lewis and Burke characterized as having used chi square in ways that were clearly unwarranted. We shall then be in a good position to join the small minority who use chi square correctly.

1. *Lack of independence among the single events or measures.* Chi square may be correctly used only if all N observations are made independently and do not influence each other. Thus, if we have only 15 subjects, and we decide to build our N up to 150 by measuring each subject 10 times, it should be clear that there is no conceivable universe of subjects of which these 150 observations could possibly constitute a random sample. (Although we can get a somewhat more reliable score for each subject by adding the 10 scores, we still have an N of 15 and not 150.) This lack of independence was the commonest of the incorrect applications of chi square noted by Lewis and Burke.
2. *Small theoretical frequencies.* There is disagreement among the experts here. Some say no expected frequency should be less than 5, but about as many others hold out for 10. A few, not surprisingly, show various forms of confusion. A small minority, which we join, will accept 5 as an absolute minimum but demand 10 whenever possible. Note that there is no minimum for observed frequencies; these minima apply to the expected frequencies only. Further, in the case of a 2×2 table, Yates' correction for continuity is usually appropriate, and an exact method of computing the probability without using χ^2 is feasible with small N's. Both of these procedures will be discussed later in this chapter.
3. *Neglect of frequencies of non-occurrence.* Suppose someone notices (or thinks he notices) that heads has been coming up more often than it should in the last 10 or 20 times that a coin has been tossed. We set up a simple experiment to test the hypothesis that the coin is biased. We toss the coin 120 times and get 45 heads. Naively and erroneously, we compute χ^2 by comparing our observed frequency, 45, with our expected frequency, 60, getting

$$\begin{aligned}\chi^2 &= \sum \frac{(f_o - f_e)^2}{f_e} \\ &= \frac{(45 - 60)^2}{60} = \frac{(-15)^2}{60} = \frac{225}{60} \\ &= 3.75.\end{aligned}$$

For $df = 1$, this would not be significant, even at the .05 level. We have, however, made two errors. The more serious is that we have failed to take any account of the 75 times that the coin did not turn up heads. The other error is that we did not use Yates' correction for continuity. This will be explained on pages 372–374. But in this problem we simply diminish the absolute value of each difference by .5 before squaring it. The correct solution, then is

$$
\begin{aligned}
{}_c\chi^2 &= \sum \frac{(|f_o - f_e| - .5)^2}{f_e} \qquad (11.2)\\
&= \frac{(|45 - 60| - .5)^2}{60} + \frac{(|75 - 60| - .5)^2}{60}\\
&= \frac{(14.5)^2}{60} + \frac{(14.5)^2}{60}\\
&= \frac{210.25}{60} + \frac{210.25}{60}\\
&= 7.00.
\end{aligned}
$$

For 1 df, this is now significant, not only at the .05 level but even at the .01 level. Note, incidentally, that this happens to flatly contradict the original assumption that heads was coming up more often.

4. *Failure to equalize Σf_o and Σf_e.* This was unintentionally illustrated in the erroneous part of the preceding paragraph in which $45 \neq 60$. A better illustration is afforded by a problem in which a die is rolled 70 times and we are interested in possible bias on any of the six faces. Our data might be:

Face	1	2	3	4	5	6
f_o	13	14	12	9	13	9

It would be incorrect to use $f_e = \frac{70}{6} = 12$, rounding to the nearest integer. If we did this, we would get $\Sigma f_e = 72$, which does not equal Σf_o, which is 70. If we carry several decimals, we reduce but do not eliminate the discrepancy. The rule we shall use is to keep one decimal in our expected frequency, f_e, and in the difference, $f_o - f_e$, keeping two decimals, as usual, in all subsequent calculations involving χ^2 but we shall "adjust" one or more of the f_e's in order to make $\Sigma f_o = \Sigma f_e$ exactly. Ordinarily, we adjust the one or two largest f_e's, or we make the entire adjustment in the miscellaneous category, if there is one. Here, neither of these can be done, so we adjust the last few values. Thus, our data, showing both f_o and f_e, would become:

Face	1	2	3	4	5	6
f_o	13	14	12	9	13	9
f_e	11.7	11.7	11.7	11.7	11.6	11.6

The high importance placed on getting exact equality can be illustrated by the fact that one author was criticized because 199.1 did not equal his observed 200, and because in another problem theoretical frequencies of 438.4 were adjusted so that they became 439.0.

5. *Indeterminate theoretical frequencies.* This objection is technical and does not often arise. If, however, either (1) the observed frequencies are in any way related, or (2) mutually contradictory assumptions regarding the f_e's can be made with about equal justification, this source of error is present, so don't use χ^2.
6. *Incorrect or questionable categorizing.* If our data must be subjectively classified or placed in one of several categories, a real possibility of error exists. This is especially true in such situations as classifying drawings according to subject-matter categories, or trying out new methods for grouping the responses made to the Rorschach, TAT, or any other projective test. We should guard against so grouping our categories as to lump together several which all show deviations in the same direction. Perhaps the best way to avoid or at least minimize this error when working with such data is to set up the categories on the basis of external criteria, if possible, but in any case to set them up in advance. If, later, some categories must be combined in order to avoid some unduly small expected frequencies, this should be done in as objective and unbiased a manner as possible.
7. *Use of nonfrequency data.* Although χ^2 is more general, when it is computed by Formula (11.1) or (11.2), as it nearly always is, it may legitimately be applied only to data in the form of frequencies. It is not correct to insert test scores or other measurements in the cells and compute chi square.
8. *Incorrect determination of* df. In a table with 2 or more columns and 2 or more rows,

$$df = (n_c - 1)(n_r - 1), \tag{11.3}$$

when n_c is the number of columns, and n_r is the number of rows. In more formal language, df is always equal to the number of cells in the table minus the number of restrictions imposed during the calculation of the theoretical frequencies. In less stilted English, df is the number of cells that may be filled in arbitrarily. Thus, in a table with 3 columms and 4 rows, we may write 2 numbers in the first row, but the third is determined automatically (since all row totals, all column totals, and N must agree for both the observed and the expected frequencies). Similarly, we may write any 2 numbers in the second row, the third being found automatically. There are also 2 numbers to be assigned arbitrarily in the third row. But none may be so inserted in the fourth or last row because each column total is also restricted. Thus, we obtain $df = 6$. An easier way is to use Formula (11.3), getting

$$\begin{aligned} df &= (n_c - 1)(n_r - 1) \\ &= (3 - 1)(4 - 1) = (2)(3) \\ &= 6, \end{aligned}$$

or, more formally, we have 12 cells, but we have restrictions on all but one of the column totals, on all but one of the row totals, and on N. Thus we have

$$12 - (3 - 1) - (4 - 1) - 1 = 12 - 2 - 3 - 1 = 6$$

as before. If we are using a 1 by n_r or 1 by n_c table, as happens in fitting frequency curves, our more formal definition still applies, but Equation (11.3) no longer gives the right answer. This case will be discussed on pages 378–383.

9. *Incorrect computations.* There are many ways to make arithmetical blunders. In computing χ^2, some of the most common mistakes result from the use of formulas based on percentages or proportions, or from an incorrect understanding of Yates' correction for continuity. In this book we shall give no formulas for χ^2 based on percentages or proportions*—that takes care of that one—and we shall try to explain Yates' correction clearly.

The nine preceding paragraphs consider what Lewis and Burke regarded as the principal sources of error in the use of χ^2, somewhat modified by us to take account both of replies to the Lewis and Burke article and of discussions of χ^2 by other statisticians. With this background, let us now consider some of the *proper* uses of chi square.

CHI SQUARE IN THE GENERAL CONTINGENCY TABLE

A contingency table is any table such as Table 11.1 or 11.3 in which the frequencies are classified according to two possibly related variables. An appropriate test of independence for such a table can be provided by χ^2. It is appropriate to use chi square in contingency tables of any size, but because of the need for avoiding small theoretical frequencies, tables as large as those commonly used in plotting Pearson r are almost nonexistent. A 15×15 table, even with the most favorable distribution of marginal totals, would have to have an N of well over 2,000 in order that each f_e might be 10 or more. Let us consider a table that is fairly representative of some of the larger tables for which χ^2 might be computed. In a study of

---

*If you ever need to compute chi square for such data, simply multiply the proportions by N or the percentages by $\frac{N}{100}$ and then compute chi square from the resulting frequencies.

various snails* that transmit disease to man, monthly counts of young and older snails (P. globosus) were made over a period of three years at Chonyi, Kenya, near the Indian Ocean. Because interest was in the breeding season, the figures for the three years were combined, yielding figures for each month. Since three of the months (March, June, and August) would have given f_e's below 10, we grouped all the data in two-month intervals, as shown in Table 11.3.

Subject only to the restriction that we get the same marginal totals (the figures in the right column and in the bottom row), there are any number of hypotheses that we could test. The most usual one in such an instance as this is the hypothesis of independence, in which the expected frequencies are derived from the observed data. Here, each f_e is obtained by assuming that each bimonthly total is distributed between Young and Older in the same ratio that the overall total of 819 is distributed. Thus, $\frac{395}{819}$ of the 356 Jan.–Feb. snails, or 171.7 snails, would be expected to be Young. Note two things: (1) We would get exactly the same answer if we said that we expect $\frac{356}{819}$ of the 395 Young snails to be in the Jan.–Feb. row. (2) We can very easily obtain the f_e for any cell by multiplying together its row total and its column total, and dividing by N. Doing this we obtain Table 11.4.

Since Table 11.3 gives the 12 f_o's, and Table 11.4 gives the corresponding f_e's, we are now ready to compute χ^2. One way to do it is to make the indicated substitutions in Formula (11.1). This gives

$$\begin{aligned}\chi^2 &= \sum \frac{(f_o - f_e)^2}{f_e} \\ &= \frac{(158 - 171.7)^2}{171.7} + \frac{(198 - 184.3)^2}{184.3} + \frac{(25 - 19.8)^2}{19.8} + \cdots \\ &= 1.09 + 1.02 + 1.37 + \cdots \\ &= 60.81.\end{aligned}$$

A somewhat better procedure, because errors are less likely to occur, is illustrated in Table 11.5. Notice that whenever χ^2 is computed for a table with two columns (or with two rows), there will be pairs of differences in column (4) that are identical except for sign. Furthermore, the corresponding figures in column (6) will be inversely proportional to the column (or row) totals. In this case, in column (6), all the figures for Older are just a little smaller than those for Young because 424 (the Older total) is just a

*Teesdale, C. Ecological observations on the molluscs of significance in the transmission of bilharziasis in Kenya. *Bulletin of the World Health Organization,* 1962, **27,** 759–782.

TABLE 11.3
Snails at Chonyi

Months	Young	Older	Total
Jan.–Feb.	158	198	356
Mar.–Apr.	25	16	41
May–Jun.	28	43	71
Jul.–Aug.	26	21	47
Sep.–Oct.	99	32	131
Nov.–Dec.	59	114	173
Total	395	424	819

little larger than 395 (the Young total). After columns (2) and (3) are filled in from the preceding tables, they should be added. If the totals do not agree *exactly,* find out why. If there are any computational errors, correct them. If there are not, and the errors are due to rounding, change the f_e's as suggested in error section 4 on page 367. Here no adjustment is needed, but if one had been necessary, say of .3, we would have changed each of our 3 largest f_e's. The difference column should then be added if its total is not obviously zero, as it is here. Even if it does add to zero, there may be errors. We might, for instance, have obtained -25.3 as the difference for ND Young and then have automatically recorded $+25.3$ in the next row. Regrettably, there is no very satisfactory way to check the last three columns. Column (4) can be checked by adding it to column (3) to obtain column (2). Columns (5) and (6) can be checked by doing them over; no better check is practical, but watch the decimal points carefully. Finally, add column (6) once in each direction to check the final operation in obtaining χ^2.

Here, $df = (n_c - 1)(n_r - 1) = (2 - 1)(6 - 1) = 5$. Since our χ^2 of 60.81 is vastly in excess of the 20.515 required at $5df$ for significance at the .001 level, we may be very sure that the bimonthly variation in the numbers of

TABLE 11.4
Expected Frequencies of Snails at Chonyi

Months	Young	Older	Total
Jan.–Feb.	171.7	184.3	356
Mar.–Apr.	19.8	21.2	41
May–Jun.	34.2	36.8	71
Jul.–Aug.	22.7	24.3	47
Sep.–Oct.	63.2	67.8	131
Nov.–Dec.	83.4	89.6	173
Total	395	424	819

TABLE 11.5
Computation of Chi Square for Snails at Chonyi

Category (1)	Observed Frequency f_o (2)	Expected Frequency f_e (3)	Difference $f_o - f_e$ (4)	$(f_o - f_e)^2$ (5)	$\frac{(f_o - f_e)^2}{f_e}$ (6)
JF Young	158	171.7	−13.7	187.69	1.09
JF Older	198	184.3	+13.7	187.69	1.02
MA Young	25	19.8	+5.2	27.04	1.37
MA Older	16	21.2	−5.2	27.04	1.28
MJ Young	28	34.2	−6.2	38.44	1.12
MJ Older	43	36.8	+6.2	38.44	1.04
JA Young	26	22.7	+3.3	10.89	.48
JA Older	21	24.3	−3.3	10.89	.45
SO Young	99	63.2	+35.8	1281.64	20.28
SO Older	32	67.8	−35.8	1281.64	18.90
ND Young	59	83.4	−24.4	595.36	7.14
ND Older	114	89.6	+24.4	595.36	6.64
Total	819	819.0	0		60.81

Young and Older snails is not a chance phenomenon. Whether these differences are great enough to justify any conclusions about the breeding season is another matter. Teesdale, the author of the article, made no statistical analysis of these figures and he said, "Young snails were recorded throughout the year but there appeared to be no clear distribution pattern." Whether or not we agree with this statement, the observed fluctuations are clearly not due to chance.

Yates' Correction for Continuity The theoretical sampling distribution of χ^2 may be plotted as a continuous curve for any given *df*. There are a whole collection of χ^2 distributions—one for every *df*—just as was true of the t distribution. While each of these theoretical χ^2 distributions is continuous, actual samples of real data may show marked discontinuity. This is particularly true for 2 × 2 tables, which, of course, have only one degree of freedom. Some correction for continuity thus becomes necessary, simply because we are using the continuous theoretical distribution of χ^2 to approximate the discrete distribution obtained with actual data. In theory, this is done by regarding each observed frequency as occupying an interval, the exact lower limit being a half unit below its expressed value, while the exact upper limit is a half unit above it. In practice, we diminish each $|f_o - f_e|$ difference by .5 before squaring it. Let us illustrate the advantage of such a correction by supposing that we have a 2 × 2 table with f_o's such as

14	16	30
12	12	24
26	28	54

To see whether chance will account for these results, we compute the corresponding table of f_e's:

14.4	15.6	30.0
11.6	12.4	24.0
26.0	28.0	54.0

How are we doing? It's true that every one of the four cells has a discrepancy of .4, but let's face it, our f_o's couldn't possibly have come closer.

Suppose, now, that our observed figures had been

13	17	30
13	11	24
26	28	54

This table would give the same set of f_e's, and all the discrepancies would now be 1.4. But is it fair to figure them in at that size? Since the observed frequencies must be integers, it may seem reasonable to allow them a leeway of .5 before starting to measure the discrepancy. This is exactly what Yates' correction for continuity does. Formula (11.2) for ${}_c\chi^2$, or chi square corrected by Yates' correction for continuity, has already been given, but is repeated here

$$ {}_c\chi^2 = \sum \frac{(|f_o - f_e| - .5)^2}{f_e}. $$

Yates' correction is applicable only when $df = 1$. The most common such situation is the 2×2 table, but the correction may also be used in a 1×2 table, as was done on page 367.* It should never be used in 1×3 or 2×3 or any larger tables. Also, it should not be used if several values of χ^2 are going to be combined. (This is a process that is rarely necessary; we will not discuss it in this text. Thus, in practically all 1×2 and 2×2 tables, Yates' correction should routinely be applied.) How should Yates' correction be used?

First, look at the 2×2 or 1×2 table and see whether all the discrepancies are .5 or less. If so, call ${}_c\chi^2$ zero and do not reject the null hypothesis at any

*There are also other tables with $df = 1$, but Yates' correction is not applicable to them. An example is a 1×4 table in which we fit a normal curve to a distribution of grades of *A*, *B*, *C*, and *D*.

level of significance. Next, if all differences in the 2 × 2 or 1 × 2 table are greater than .5, subtract .5 from the absolute value of each, computing ${}_c\chi^2$ by Formula (11.2).

There is a common but mistaken belief that the use of Yates' correction for continuity somehow makes it all right to work with small theoretical frequencies. This notion probably arose because the correction is less important when the frequencies are large than it is when they are small. Hence, some texts ignore it for large frequencies and say it should be used when dealing with small frequencies. From this and the fact that it is called a "correction," it is a small, though illogical, step to conclude that it in some way corrects for the smallness of the frequencies. Because the correction is easy to use and is appropriate for use with samples of all sizes, we recommend its use without regard to the size of N.

What do we do if we have some extremely small theoretical frequencies—less than 10 or maybe even smaller than 5? There are at least three answers. One good one is to combine them with adjacent ones, as was done with the data from which Table 11.3 was obtained. Another good answer is to get more data. This is always a desirable procedure and, if we already have a 2 × 2 table so that no further combining is possible, it may be the *only* desirable procedure. A third possibility, if we already have a 2 × 2 table and if it is not possible or feasible to gather more data, is to use what is referred to as the exact method for determining the probability of getting a distribution as extreme or more extreme than the one we did get. This is the subject of our next section.

THE EXACT TEST OF SIGNIFICANCE IN 2 × 2 TABLES

Instead of computing χ^2 and determining its significance by means of Table V, it is possible to compute the probability directly. Since the probability obtained via χ^2 is actually an approximation, the direct calculation of exact probabilities might seem preferable. Sometimes it is, but usually far too much work is involved. It is fortunate that when the calculation of exact probabilities is most laborious (when frequencies are large), the use of χ^2 provides an excellent approximation to the exact method; hence χ^2 is almost universally preferred in such circumstances. Especially when the frequencies are small, we have seen that Yates' correction for continuity provides better results. When all f_e's in the 2 × 2 table are at least 10, the probabilities obtained from ${}_c\chi^2$ are ordinarily quite satisfactory for most research work. But if the obtained ${}_c\chi^2$ is very close to the value of χ^2 given in Table V for the significance level in which we are interested, it still may be desirable to use the exact method. The other situation in which the exact method is commonly preferred is when one or more of the f_e's is very small. Then χ^2 would be completely inapplicable, but the exact method may still be used (with caution, since no amount of statistical manipulation will compensate for inadequate data).

Let us use the exact test with a small portion of the data on the effectiveness of influenza vaccination in Iceland.* These figures are based on responses to a questionnaire for families in which the vaccinated members of the family had received a second dose at least one week before the first case of influenza occurred in that household. Instead of using the data on all 433 persons, let us consider the group of females aged 15–19. The figures were:

	Not Attacked	Attacked	Sum
Vaccinated	2	1	3
Not Vaccinated	2	10	12
Sum	4	11	15

If the data had not already been so arranged that the smallest f_0 was in the upper right (or ++) cell, we would rearrange the data so that this would be true.

We will let A, B, C, and D represent the observed cell frequencies in our 2 × 2 table just as we did in connection with ϕ in Chapter 10. The table will then be

B	A	$A+B$
D	C	$C+D$
$B+D$	$A+C$	N

In this instance, $A = 1$, $B = 2$, $C = 10$, $D = 2$, and $N = 15$. We will use these figures and others derived very simply from them in order to determine the exact probability of obtaining figures deviating this much or more from those based on the null hypothesis of a pure chance relation between the two variables. We do this by, in effect, considering all possible 2 × 2 tables with these marginal totals, finding how probable each is, summing the appropriate probabilities for deviations in one direction, and then doubling this value to correspond to the usual two-tailed test for deviations in either direction.

The probability for any particular set of observed frequencies is

$$P = \frac{(A+B)!\,(C+D)!\,(A+C)!\,(B+D)!}{N!A!B!C!D!}. \qquad (11.4)$$

*Sigurjónsson, J., *et al.* Experience with influenza vaccination in Iceland, 1957. *Bulletin of the World Health Organization,* 1959, **20,** 401–409.

For our vaccination data, this becomes

$$P = \frac{3!\ 12!\ 11!\ 4!}{15!\ 1!\ 2!\ 10!\ 2!}$$
$$= \frac{(3)(2)(1)(11)(4)(3)(2)(1)}{(15)(14)(13)(1)(2)(1)(2)(1)}$$
$$= \frac{66}{455}.$$

(Notice that after we cancelled 12! from 15! leaving (15)(14)(13), after we cancelled 10! from 11! leaving 11, and after we wrote out all remaining factorials as products, we had exactly the same number of terms in the numerator as in the denominator. This is a general rule and will check some, but not too many, errors.)

Now we want the probability for the next more extreme table with the same marginal totals. We reduce A by 1, getting $A = 0$; the new set of figures is obtained by getting B, C, and D by appropriate subtractions from the marginal totals. This gives us the new table

3	0	3
1	11	12
4	11	15

The probability for this set of frequencies is going to involve multiplying nearly the same things that we just did multiply. The numerator is exactly the same, but there will be four changes in the denominator. Notice that in the second table, the new A and D are 1 unit smaller, while the new B and C are 1 unit larger than they were in our first table. This means that we can get the second probability from the first by multiplying the first probability by the old A and old D and dividing by the new B and new C. If we use the subscript 1 to refer to the first table and 2 to refer to the second, the relation between the second probability and the first one* is

$$P_2 = P_1\left(\frac{A_1D_1}{B_2C_2}\right). \tag{11.5}$$

In this case,

$$P_2 = \left(\frac{66}{455}\right)\left[\frac{(1)(2)}{(3)(11)}\right]$$

*This method of getting successive probabilities from the preceding ones is not well known, although it was given in Freeman, G. H., and Halton, J. H. Note on an exact treatment of contingency, goodness of fit and other problems of significance. *Biometrika,* 1951, **38,** 141–149.

$$= \frac{4}{455}.$$

We continue writing and evaluating more extreme tables with the same marginal totals until what had been our smallest f_o (and is now called A with an appropriate subscript) has been reduced to zero. Since we have reached that state, we have evaluated all the needed probabilities. If our smallest f_o (or A_1) had been 4, we would have computed the first probability, P_1, by Formula (11.4). We would then have used Formula (11.5) to compute P_2 for $A = 3$, P_3 for $A = 2$, P_4 for $A = 1$, and P_5 for $A = 0$. Whatever the number of terms needed, we check the last one by using Formula (11.4). In our problem on vaccinated and attacked females in Iceland, the check is

$$P = \frac{(A + B)!\,(C + D)!\,(A + C)!\,(B + D)!}{N!A!B!C!D!}$$

$$P_2 = \frac{3!\;12!\;11!\;4!}{15!\;0!\;3!\;11!\;1!}$$

$$= \frac{(4)(3)(2)(1)}{(15)(14)(13)(0!)(1)}$$

$$= \frac{4}{455}.$$

The rule that the number of terms remaining in the numerator and denominator must be equal does not work when one of our terms is 0! This, as is well known, is arbitrarily defined as equal to 1, but the number of terms in the numerator and denominator, after cancelling out factorials, will be equal only if we ignore 0! After further cancellation, we obtain the same value of P_2 as we did by Formula (11.5). This checks *all* the probabilities. In this case, it checks two, but if our smallest f_o had been 4, the final computation by Formula (11.4) would check all five P's.

Adding our probabilities, we get

$$\frac{66}{455} + \frac{4}{455} = \frac{70}{455} = .15$$

Doubling this, we see that the probability is .30 of getting a difference as extreme as this in either direction if there were no relation between getting vaccinated and getting attacked. We definitely do not reject the null hypothesis, but, once again, this does not mean that we can say we have proved there is no relation between the two variables. We just didn't have enough data to get the answer to our question, and we most certainly have not proved there is no relation. In fact, whether we take the data for all 234 females, for all 199 males, or for all 433 people, χ^2 turns out to be significant at the .001 level. We didn't prove it with our 15 young females, but there most definitely is a relation between vaccination and influenza attacks. Just to illustrate this, the complete table is

	Not Attacked	Attacked	Sum
Vaccinated	94	24	118
Not Vaccinated	96	219	315
Sum	190	243	433

and ${}_c\chi^2$ turns out to be 82.26, while Table V shows that only 10.827 is needed to be significant at the .001 level.

USE OF CHI SQUARE IN CURVE FITTING

Sometimes we want to fit a curve to a given collection of scores. Sometimes we may not necessarily wish to fit the curve, but we want to know whether it would be possible to fit it. In one peculiar sense, we can always fit any type of curve to any set of data. For example, we can always draw a normal probability curve having the same *N,* mean, and standard deviation as our sample. This, of course, is not much of an achievement, especially if the sample departs markedly from the ''normal'' shape. What we really need is (1) a procedure for fitting the desired curve, and (2) some indication of how good the fit is. This would solve both of the problems mentioned at the beginning of this paragraph.

From any frequency distribution, it is easy to see the observed frequencies. The expected frequencies are computed on the basis of whatever hypothesis or procedure happens to interest us. For instance, if we had 150 scores tabulated in 15 classes, we could see whether the obtained distribution was significantly different from a rectangular distribution with 10 scores expected in each class. This would be a simple χ^2 problem with a 1×15 table having 14*df.*

Most educational and psychological distributions are not rectangular, but are essentially normally distributed. Hence, we may wish to determine whether or not the obtained scores are consistent with the hypothesis that the sample was drawn from a normally distributed population. We shall take as our illustrative material a set of scores that have been shown not to be normally distributed. The basic scores are the response times of 55 college students to the Wechsler–Bellevue profile assembly task, as reported by Tate and Clelland.* The scores ranged from 12.4 to 73.4

*See page 59 of Merle W. Tate and Richard C. Clelland. Nonparametric and Shortcut Statistics. Danville, Ill.: Interstate Printers and Publishers, 1957.

seconds; and when those authors correctly used i = 5.0 and tabulated the scores, the following distribution (with low scores at the left) resulted:

3 21 9 10 5 4 1 0 0 0 0 1 1

They fitted a normal probability curve to these 55 scores, got $\chi^2 = 30.49$ with $4df$, and concluded, "Since the corresponding P is less than .005, the assumption that the data are drawn from a normal population is strongly discredited." That sounds definite enough, but in this book we will now find out whether or not the speed with which people complete the profile assembly test is normally distributed. We will find that it very well may be so distributed.

What gives? Nearly every statistician has been bored many times by hearing someone quote the series, "Lies, Damned Lies, and Statistics." Sometimes the victim retorts by adding another term to make it read Lies, Damned Lies, Statistics, and "Cold Facts." We all know that in criminal trials the defense is often able to hire a psychiatrist who will testify the accused was insane when he committed the crime; while the prosecution is able to hire another psychiatrist who will testify the accused was sane. Are statisticians that way, too, so that you can easily buy normality or nonnormality? Well, possibly a very small number of them are, but that's not the source of the confusion here. A very similar situation came up with respect to the time taken to swim 40 yards. There, as noted on page 316, we transformed time scores into speed scores by dividing 1000 by the time in seconds. Let's try it again. The shortest and longest times (12.4 and 73.4 seconds) will transform into the fastest and slowest speeds (80.6 and 13.6, respectively). The first two columns of Table 11.6 show the distribution of the 55 speed scores. The distribution of speeds is obviously much more nearly normal than the skewed distribution of times shown in the preceding paragraph.

In curve fitting, we wish the figures for our sample to be the same as those for the theoretical curve. Hence, we are not interested in estimating the standard deviation in some universe. What we want is simply the standard deviation of these scores, regarding our sample as the entire population.

Computing σ_N by Formula (3.5), $\overline{X}$ by Formula (2.2), and noting that $N = 55$, we have the three basic figures needed for fitting a normal probability curve to our data. These figures are $N = 55$, $\overline{X} = 46.09$, and $\sigma_N = 15.20$. We are ready to proceed with our curve fitting, but we shall now use areas rather than ordinates as we did in Chapter 4.

Write the exact limits as shown in column (3) of Table 11.6, noting that these limits are to be written halfway between the intervals shown in the

TABLE 11.6
Use of Chi Square in Evaluating Fit of a Normal Curve to W-B Profile Assembly Task Speed Scores

X (1)	f_o (2)	Exact Limits (3)	Standard Score (4)	Normal Curve Area (5)	Area in Interval (6)	Ungrouped f_e (7)	Final f_e (8)	$f_o - f_e$ (9)	$\frac{(f_o - f_e)^2}{f_e}$ (10)
			$(+\infty)$	(1.0000)					
80.0–84.9	1								
		79.95	+2.23	.9871					
75.0–79.9	1				.0582	3.2			
		74.95	+1.90	.9713			10.0	−1.0	.10
70.0–74.9	1								
		69.95	+1.57	.9418					
65	3				.0493	2.7			
		64.95	+1.24	.8925					
60	3				.0739	4.1			
		59.95	+.91	.8186					
55	8				.0996	5.5			
		54.95	+.58	.7190			12.1	+2.9	.70
50	7				.1203	6.6			
		49.95	+.25	.5987					
45	5				.1306	7.2			
		44.95	−.08	.4681			14.0	−5.0	1.79
40	4				.1235	6.8			
		39.95	−.40	.3446					
35	8				.1119	6.2			
		34.95	−.73	.2327			11.0	+2.0	.36
30	5				.0881	4.8			

		29.95	−1.06	.1446					
25	6				.0623	3.4			
		24.95	−1.39	.0823					
20.0–24.9	1								
		19.95	−1.72	.0427					
15.0–19.9	1				.0823	4.5	7.9	+1.1	.15
		14.95	−2.05	.0202					
10.0–14.9	1								
			(−∞)	(.0000)					
Total	55				1.0000	55.0	55.0	0.0	3.10

preceding columns. Compute the standard score corresponding to each exact limit, using σ_N instead of s in Formula (3.6). A good way to do this on a calculator is to compute the top, bottom, and one other standard score by the formula in the usual manner, keeping about 6 decimal places. Then compute $\frac{1}{\sigma_N}$ to six decimal places. Subtract this from the top standard score, record the answer to the usual 2 decimal places, subtract again, record, etc. The last value should check with an error of no more than about 10 in the sixth decimal place. In this case, the top value is +2.227,632, the subtractive factor is .328,947, and the bottom value by subtraction is −2.048,679, which is sufficiently close to the value of −2.048,684 obtained by direct calculation. We check by starting at the bottom and adding until we obtain the top value. (In this case, we get −.07 one time and −.08 the other for the z corresponding to $X = 44.95$. Direct calculation shows this value to be exactly −.075000.)

Table E gives the values in column (5). Since the normal curve goes from $-\infty$ to $+\infty$, these values and the appropriate areas are included at the ends of the distribution. Subtraction of adjacent values in column (5) gives the areas shown in column (6). Note that these areas correspond to the rows in the first two columns, so they are no longer centered between the rows, as columns (3), (4), and (5) were. Column (6) must add to 1.0000 exactly.

We multiply column (6) by our N, 55, in order to get column (7). Check the multiplication and then add. If the sum is not *exactly* equal to N, adjust one or more of the figures in column (7), as explained on page 367, until it is equal. In this case, no adjustment is needed.

Note that some grouping was done in column (6), since it was obvious that the extreme f_e's were going to be very small. In column (8), we do whatever additional grouping is necessary. The grouping shown here gives one f_e below 10, but it isn't much below 10, and we get a better test of our curve fitting when we have more groups. We did not feel justified in getting still more groups by using all those shown in column (7), however; that would have given far too many very small f_e's. Column (8), too, must add up to N exactly.

Column (9) is obtained by subtracting column (8) from the appropriate group of entries in column (2). Column (9) should always add to zero.

Column (10) is obtained by squaring column (9) and dividing these squares by the corresponding figures in column (8). The most satisfactory check on this operation is to repeat it.

The sum of column (10) is χ^2, in this case 3.10.

There is a small problem concerning the number of degrees of freedom. We forced our data and the theoretical curve to agree exactly on N, $\overline{X}$, and σ_N. Because of this, there are three restrictions, and df will be the number of classes (here, 5) minus three. Thus, $df = 5 - 3 = 2$.

Table V shows that for $df = 2$, χ^2 must be 5.991 to be significant, even at the .05 level. Our 3.10 falls far short of this. Hence we conclude that the fit is good, and that there is no reason to cast any doubt on the assumption that the speed with which college students complete the profile assembly task is normally distributed. Although perhaps this statement about speed does not flatly contradict Tate and Clelland's statement about time quoted above, we believe that most people would draw entirely different conclusions from the two statements. This merely points to the desirability of using statistics judiciously rather than mechanically.

ADVANTAGES AND DISADVANTAGES OF CHI SQUARE

As we noted early in this chapter, chi square is, for all practical purposes, restricted to the use of data in the form of frequencies. Subject to this limitation, chi square may be used with practically any kind of data, classified in either one or two (or more) dimensions. The categories may be arranged in order (as the speed groupings were in Table 11.6), or there may be no logical arrangement (as nationalities, hair color, or food preferences).

The order in which the categories are arranged has no effect on the size of chi square. Thus, we would get the same value of χ^2 from the data of Table 11.3 if we decided to rearrange the bimonthly periods in any other order, such as Mar.–Apr., Sep.–Oct., May–Jun., Nov.–Dec., Jan.–Feb., and Jul.–Aug. This may be a disadvantage in tables larger than 2×2 in which there is a logical arrangement for one or both variables.

Another disadvantage in many circumstances is that chi square merely tells us how certain we can be that there is a difference between our sample results and those called for by our theory. If we are testing the hypothesis of independence for a table classifying our cases on two variables, χ^2 merely tells us whether or not the classification on one variable is related in a nonchance manner to the classification on the other variable. It does not tell us anything about the strength of that relationship.

Also, χ^2 is useful in curve fitting, giving us a definite figure that tells us whether or not we may regard any given fit of a theoretical curve to a set of data as a good one.

Perhaps the greatest advantage of χ^2 is that it may be used to test the consistency between any set of frequency data and any hypothesis or set of theoretical values, no matter what the latter may be.

The biggest disadvantage of χ^2 is that, perhaps because of its simple appearance, so many people misuse rather than use it. This is a disadvantage that the listing and explanation of Lewis and Burke's nine sources of error may, to some extent, diminish.

SUMMARY

Chi square is often used to determine whether a set of observed frequencies is such that it might well have arisen as a result of chance fluctuations from the frequencies that would be expected on the basis of a theory of any kind.

After illustrating the use of chi square in the problem of preferences for a personal confidant, nine principal sources of error in the use (or misuse) of chi square were listed and described. This summary of errors to be avoided is followed by several correct examples of the use of chi square.

Yates' correction for continuity adjusts for the fact that truly continuous distributions cannot be accurately represented by frequencies which are necessarily whole numbers. This correction is applicable only when $df = 1$.

The exact test of significance sometimes replaces chi square when the frequencies are small.

The use of chi square in curve fitting is illustrated with assembly times after they have been transformed to speed scores. The normal probability curve gave a good fit to the data.

The chapter ends with a brief review of the advantages and disadvantages of chi square.

SUPPLEMENT 11

UNIQUE CHARACTERISTICS OF CHI SQUARE

Six topics will be considered in this Supplement. They are:

Unusually good agreement
Alternative formula for computing χ^2

Computation of χ^2 by reciprocals in 2 × 2 tables
Computing χ^2 in 2 × 2 tables without using expected frequencies
Relation between χ^2 and ϕ
Relation between chi square and the contingency coefficient

Unusually Good Agreement Suppose we gave someone one of our dice, asked that it be rolled 60 times and that the results be reported to us. If the report was

Face	1	2	3	4	5	6
f_o	0	0	0	20	20	20

we would certainly be suspicious. We could compute χ^2, find it to be 60, see that P is less than .001, and have our suspicion confirmed. But suppose these results had been reported

Face	1	2	3	4	5	6
f_o	10	10	9	11	10	10

We would have to be extremely naive not to feel there was something fishy about such alleged results; the agreement seems just too good to be true. Let us use χ^2 to see whether something went haywire. We quickly obtain $\chi^2 = .20$. For $5df$, Table V shows that this gives a P with a value greater than .99. That is, more than 99 times in 100 we should expect to get discrepancies greater than this if we had an honest die that was rolled by an honest individual who tabulated the results correctly and reported them accurately. The agreement is much too close to lead us to believe that it occurred by chance. Perhaps it did, but the odds are better than 99 to 1 that it didn't. Partly because scientists rarely accuse other persons of fraud, even when the evidence is overwhelming, and partly because unlikely events do sometimes happen, we should look for other possible explanations when we encounter a χ^2 that yields a P greater than .99 or .95.

The first two columns of Table V are not often needed, but when we do get observed frequencies showing an unduly high degree of agreement with the expected frequencies, we have an excellent basis for rejecting the null hypothesis that chance alone accounts for the discrepancies, just as we do in the more common case in which the observed and expected frequencies show large differences.

Alternative Formula for Computing χ^2 Formula (11.1) for χ^2 may be changed to another form that is often easier to compute. The derivation is very simple. We square the binomial, perform the indicated divisions, sum the separate terms, and notice that $\Sigma f_o = \Sigma f_e = N$. We start with Formula (11.1).

$$
\begin{aligned}
\chi^2 &= \sum \frac{(f_o - f_e)^2}{f_e} \\
&= \sum \frac{f_o^2 - 2f_o f_e + f_e^2}{f_e} \\
&= \sum \left[\left(\frac{f_o^2}{f_e} \right) - 2f_o + f_e \right] \\
&= \sum \left(\frac{f_o^2}{f_e} \right) - 2\Sigma f_o + \Sigma f_e \\
&= \sum \left(\frac{f_o^2}{f_e} \right) - 2N + N, \\
\chi^2 &= \sum \left(\frac{f_o^2}{f_e} \right) - N. \qquad (11.6)
\end{aligned}
$$

Let us use this formula with the data on personal confidants shown in Table 11.1 and 11.2.

$$
\begin{aligned}
\chi^2 &= \frac{100^2}{78.5} + \frac{25^2}{40.5} + \frac{70^2}{76.0} + \frac{22^2}{43.5} + \frac{38^2}{22.5} + \frac{48^2}{42.0} - 303 \\
&= 127.39 + 15.43 + 64.47 + 11.13 + 64.18 + 54.86 - 303 \\
&= 34.46.
\end{aligned}
$$

This value of χ^2 is necessarily identical to that of 34.46 obtained on page 365. There are advantages and disadvantages to both formulas:

Formula (11.1) gives a better understanding of the meaning of χ^2.
Formula (11.1) shows which cell discrepancies contribute the most to χ^2.
Formula (11.1) is much more commonly used.
Errors are usually easier to locate if Formula (11.1) is used.
There is less arithmetic if Formula (11.6) is used.
The two formulas give exactly the same answer—not even differing in rounding errors.

Probably the best procedure is to compute all f_e's, check them by adding the rows and columns, and then compute χ^2 once by each formula. If the two agree, you have a wonderful check on the computations. If they differ in the last retained decimal place by any amount at all, the difference cannot be due to any allowable causes, so you should find and correct the mistake.

A formula analogous to (11.6) but including the correction for continuity can be derived, but it contains five terms instead of the two in Formula (11.6), so it is not practical for use. Besides, the correction for continuity is usually used only with 2 × 2 tables, and the next two sections of this supplement will give formulas for ${}_c\chi^2$ in such tables.

Computation of χ^2 by Reciprocals in 2 × 2 Tables If we ignore signs, since we are going to square the discrepancies anyway, whatever value of $(f_o - f_e)$ is found in any one cell of a 2 × 2 table will also be found in each of the other cells. This fact enables us to save some time in computing χ^2 from such a table.

Because $(f_o - f_e)^2$ must also be the same for all four expected frequencies, Formula (11.1), for a 2 × 2 table, becomes

$$\chi^2 = (f_o - f_e)^2 \sum \left(\frac{1}{f_e}\right). \tag{11.7}$$

For the same reason, Formula (11.2), for a 2 × 2 table, becomes

$${}_c\chi^2 = (|f_o - f_e| - .5)^2 \sum \left(\frac{1}{f_e}\right). \tag{11.8}$$

Let us illustrate the use of one of these formulas by computing ${}_c\chi^2$ for the Iceland vaccination data. We compute any one of the expected frequencies, obtain the other three by subtraction, and check the last one by recomputation. Our table of expected frequencies will then be

51.8	66.2	118.0
138.2	176.8	315.0
190.0	243.0	433.0

Each $(f_o - f_e)$ is either +42.2 or −42.2. Taking the absolute value and subtracting .5, we get 41.7. Substitution in Formula (11.8) gives

$$\begin{aligned}
{}_c\chi^2 &= (|f_o - f_e| - .5)^2 \sum \left(\frac{1}{f_e}\right) \\
&= (41.7)^2 \left(\frac{1}{51.8} + \frac{1}{66.2} + \frac{1}{138.2} + \frac{1}{176.8}\right) \\
&= (1738.89)(.019305 + .015106 + .007236 + .005656) \\
&= (1738.89)(.047303) \\
&= 82.25.
\end{aligned}$$

This value agrees, except for a rounding error of .01, with that obtained earlier by the traditional method of calculation.

This shortcut method is particularly advantageous if one has a table of reciprocals available. The chief caution, of course, is to make sure of the decimal points.

Computing χ^2 in 2 × 2 Tables Without Using Expected Frequencies If we start with Formula (11.1) for χ^2, substitute A, B, C, and D for the four values of f_o, and substitute the corresponding values of f_e, it is obvious that there are four terms to be summed in order to get χ^2. Each of these terms will have a different value for f_o, these values being the frequencies A, B, C, and D in such a table as the unnumbered one shown on page 347 and again on page 375. The expected frequencies, based on the hypothesis of independence, can also be expressed in terms of the letters A, B, C, and D. Thus, f_o for cell A is A and f_e for cell A is $\frac{(A+B)(A+C)}{A+B+C+D}$. If we make such substitutions for each of the four terms and do a large amount of simple algebraic manipulation, we finally obtain

$$\chi^2 = \frac{N(AD - BC)^2}{(A+C)(B+D)(A+B)(C+D)}. \tag{11.9}$$

We will illustrate the use of this simple formula with the pie-sex data given on page 349 in Chapter 10, and we will purposely not use the correction for continuity. (Our reason for this will be explained in the next section of this supplement.)

$$\begin{aligned}
\chi^2 &= \frac{N(AD - BC)^2}{(A+C)(B+D)(A+B)(C+D)} \\
&= \frac{168\,[(66)(41) - (14)(47)]^2}{(113)(55)(80)(88)} \\
&= \frac{168(2{,}048)^2}{43{,}753{,}600} \\
&= \frac{704{,}643{,}072}{43{,}753{,}600} \\
&= 16.14.
\end{aligned}$$

If a correction for continuity is appropriate, the formula for ${}_c\chi^2$ may be written as

$${}_c\chi^2 = \frac{N\left(|AD - BC| - \dfrac{N}{2}\right)^2}{(A+C)(B+D)(A+B)(C+D)}. \tag{11.10}$$

We will use the Iceland vaccination data to compute ${}_c\chi^2$ again.

$$\begin{aligned}
{}_c\chi^2 &= \frac{433\left(|(24)(96) - (94)(219)| - \dfrac{433}{2}\right)^2}{(243)(190)(118)(315)} \\
&= \frac{433(18{,}065.5)^2}{1{,}716{,}138{,}900}
\end{aligned}$$

$$= \frac{433(326{,}362{,}290.25)}{1{,}716{,}138{,}900}$$

$$= \frac{141{,}314{,}871{,}677.25}{1{,}716{,}138{,}900}$$

$$= 82.34.$$

Although this value differs somewhat from the 82.26 obtained by Formula (11.2) and the 82.25 obtained by Formula (11.8), this one is correct because no figures were rounded until the final division. Actually, we do not need to use so many figures, but if a calculator is available it is sometimes easier to keep them than to decide how many to discard. If the work is done without a calculator, we could revise the right side of the last four equations so that they would read

$$_c\chi^2 = \frac{433(18{,}065)(18{,}066)}{1{,}716{,}000{,}000}$$

$$= \frac{433(326{,}362{,}000)}{1{,}716{,}000{,}000}$$

$$= \frac{141{,}315{,}000{,}000}{1{,}716{,}000{,}000}$$

$$= 82.35.$$

This difference of .01, small though it is, is greater than would usually occur when 4 to 6 significant figures are retained.

Relation Between χ^2 and ϕ If we look at Formula (11.9) and at Formula (10.10) for ϕ, we see that many of the terms are identical. In fact

$$\chi^2 = N\phi^2 \tag{11.11}$$

and

$$\phi = \sqrt{\frac{\chi^2}{N}}. \tag{11.12}$$

In using the latter formula, we should give ϕ the same sign as that of $(AD - BC)$.

If we have computed either ϕ or χ^2, it is thus a very simple matter to obtain the other. These formulas are not, however, of much use in checking for accuracy, because so many of the computations involved in Formulas (11.9) and (10.10) are identical. We may, however, illustrate their use with the pie-sex data for which we found on pages 349 and 388 that $\phi = +.31$ and $\chi^2 = 16.14$.

Substituting in Formula (11.11), we obtain

$$\begin{aligned}\chi^2 &= N\phi^2 \\ &= 168(.31)^2 \\ &= 168(.0961) \\ &= 16.14.\end{aligned}$$

Using the other formula, (11.12), we get

$$\begin{aligned}\phi &= \sqrt{\frac{\chi^2}{N}} \\ &= \sqrt{\frac{16.14}{168}} \\ &= \sqrt{.0961} \\ &= .31.\end{aligned}$$

Thus, if the correction for continuity is not used, it is a very simple matter to obtain either χ^2 or ϕ from the other in any 2×2 table.

What would happen if all the figures in this 2×2 table were multiplied by some constant? Phi would not change but the old chi square would have to be multipled by the constant to get the new one. Thus, if all our pie-sex figures had been 10 times as large as they are, ϕ would still be $+.31$, but χ^2 would be 161.4.

Relation Between Chi Square and the Contingency Coefficient Another measure of relationship sometimes used is the contingency coefficient. Its formula is

$$C_c = \sqrt{\frac{\chi^2}{N + \chi^2}}. \tag{11.13}$$

Simple algebraic manipulation enables us to write the formula for χ^2 in terms of C_c as

$$\chi^2 = \frac{NC_c^2}{1 - C_c^2}. \tag{11.14}$$

Because C_c is nearly always computed by Formula (11.13) above, the formula just given is not useful for computing χ^2. But it is useful for shedding additional light on the relationship between C_c and χ^2. Notice that, just as we found for ϕ and χ^2, multiplying all frequencies by a constant such as 10 will leave C_c unchanged but will multiply χ^2 by that constant.

In passing, it should be noted that if we are working with a 2 × 2 table

$$C_c = \sqrt{\frac{\phi^2}{1 + \phi^2}} \tag{11.15}$$

and

$$\phi = \sqrt{\frac{C_c^2}{1 - C_c^2}}. \tag{11.16}$$

Substitution of a few values of ϕ or C_c in these equations shows that ϕ is always larger in absolute value than C_c, but that for small values of either, the difference is very small. Thus, if C_c takes any value between $-.20$ and $+.20$, ϕ will differ from it by .004 or less. Or, if ϕ is assumed to be between $-.50$ and $+.50$, the difference between ϕ and C_c will not exceed .05, an amount that would not cause us to alter our interpretation of the degree of relationship in any ordinary situation. For higher degrees of correlation, the difference is, of course, greater.

CHAPTER 12 NONPARAMETRIC STATISTICS OTHER THAN CHI SQUARE

Prior to the preceding chapter on chi square, the great majority of the statistics we computed were *parametric statistics* but there was no occasion to use that expression in referring to them. These statistics frequently made assumptions (such as normality) about the nature of the population from which the samples were drawn. Because the population values are called parameters, such statistics can be called parametric, although the word is not often used.

Another class of statistical procedures makes very few or no assumptions, either about the parameters or about the shape or nature of the population from which our sample is assumed to be drawn. Such procedures are called *nonparametric statistics*. They are also called *distribution-free methods* and they are sometimes rather loosely referred to as *shortcut statistics*. Examples of nonparametric statistics are chi square (to which most of Chapter 11 was devoted), the exact probability test (also treated in Chapter 11), and Spearman's rank correlation coefficient (discussed in Supplement 10). Several others will be treated in this chapter.

THE PURPOSES OF NONPARAMETRIC STATISTICS

The primary purpose of nonparametric statistics is to provide appropriate procedures for the analysis of data without the necessity of making assumptions* about the nature of the population from which the sample was drawn. A frequently mentioned secondary purpose is that most of the nonparametric methods are quick and easy to apply. From this have come recommendations both that they be used as timesavers in ordinary studies and that they be used in making preliminary analyses of data before a research study is completed.

The chief advantages of nonparametric over other statistics are that fewer assumptions are needed, that these methods are usually easy to apply, and that they are more suitable for many analyses of dichotomous data. The chief disadvantages are that they so operate as to throw away part of the information contained in the data and that their labor-saving claims are often illusory, especially in reasonably large samples.

Those favoring nonparametric methods correctly point out that often we have little justification for assuming that our sample is drawn from a normally distributed population. Those opposed correctly point out that this and other assumptions have been shown to be essentially correct with many types of educational and psychological data. Further, many of the traditional statistical formulas have been shown to give essentially accurate results when it is known that some of the assumptions underlying them are violated. Perhaps a proper way to operate is this: If the assumptions involved in the use of some formula appear to be fairly reasonable in a particular situation, use the formula. If it appears unlikely that they hold, or if you wish to proceed with great caution, don't use the formula, but use one of the nonparametric methods if it can be justified in that situation.

Where both parametric and nonparametric methods are applicable, the former are to be preferred, primarily because they are more efficient in utilizing the refinements of the available data. Even in this situation, the research worker who likes nonparametric methods may still use them by gathering enough more data to compensate for their lower efficiency. Where both methods are not applicable, it is almost automatic that the parametric methods are inappropriate and that the nonparametric should be used.

Let us now consider some of the more common and more useful nonparametric methods, starting with the simpler ones. As usual, we shall use two-tailed rather than one-tailed tests of significance, for the reasons discussed on page 138.

---

*Other than those of an ordinarily negligible nature, such as that a score of 83 is somewhere between scores of 73 and 93, though not necessarily halfway between them.

THE SIGN TEST

When we have $2N$ observations occurring as N pairs, the sign test may be used to determine whether the difference between the two sets of scores is significant. If we note whether the individual scores in the second set are higher or lower than those in the first set, the null hypothesis would call for half the changes to be positive and half negative. The actual numbers are tested against those to be expected, as given by the binomial expansion shown in Figure 4.4. To simplify the arithmetic, Table W has been prepared.

In using Table W first determine the sign of all the differences. Omit all those showing no difference, thus reducing N. Find the row of Table W for this reduced N. Compare the number of + or − signs, whichever is smaller, with the figures in that row. These figures give the highest frequency that is significant at the indicated level. Thus, if 25 comparisons are made and 2 of them are ties, 18 are +, and 5 are −, we note that the reduced N is 23 and look for 5 in that row. It isn't there, but it is between 6 in the .05 column and 4 in the .01 column, so our difference is obviously significant at the .05, but not at the .01 or .001 levels.

An illustration will clarify the use of this simple test. One of the writers parked next to a bridge over the Nile River in Cairo and tabulated automobile headlights. At that time, cars in Cairo did not use either the high beam or the low beam, but drove with their parking lights, flashing the regular lights on from time to time if the driver wished to see where he was going. Possibly because nothing much could be seen with the parking lights, many cars continued to drive when one or even both lights burned out. Table 12.1 shows the data as tabulated where L means the left light was out, R means the right light was out, and 0 means both lights were out.

If another motorist mistakes a "one-eyed driver" for a motorcycle, the hazard is greater if the left light is out; so we shall call that negative and of course a right light out is positive. Table 12.1 may then be summarized as 21 minus signs (for L), 9 plus signs (for R), and 7 ties (for 0). There was also an uncounted number of other ties (for both lights on). Ignoring ties, as is always done in using the sign test, we have a reduced N of 30 and a smaller frequency of 9. Row 30 of Table W shows that this is significant at the .05 level but not at any more extreme level. Thus, there appears to be a

TABLE 12.1
Lights Out—Eastbound Cars

L R R L R L L L R 0 L L L L 0 L L R 0 L
0 L L R R L 0 L R L L R L 0 L L 0

TABLE 12.2
Lights Out—All Cars

	East-bound	West-bound	South-bound	North-bound	All Four Directions
Left light out; −	21	18	16	20	75
Right light out; +	9	12	14	14	49
Sum	30	30	30	34	124

tendency for cars with one headlight out to have that light out on the left side (near the center of the street) rather than on the right side (next to the curb).

We hesitate to draw firm conclusions when results are significant at the .05 level. One way to be more definite is to get more data. In this case, enough westbound cars were tabulated the same evening to get another reduced N of 30. For the westbound cars, there were 18 minus signs, 12 plus signs, and 3 ties (for 0). This result, although in the same direction, is not significant at any level. Let us get still more data. A few days later, another location in Cairo was selected and headlights were tabulated until reduced N's of 30 (southbound) and 34 (northbound) were obtained. The southbound results were 16 minus signs (for L), 14 plus signs, and 7 ties. The northbound cars had 20 minus signs, 14 plus signs, and 14 ties. Thus, the figures for all four directions are as shown in Table 12.2.

The results are exasperating. All four directions gave an excess of cars with left parking lights out; but no one of the last three sets of data was significant, even at the .05 level. If we try to combine the eastbound and westbound records, or the northbound and southbound figures, or all four directions, our reduced N is beyond the range of Table W. However, as we saw in Chapter 4, when N is large, the binomial expansion approximates the normal probability curve. For our present purposes, $p = q = ½$; the theoretical mean of the number of plus or minus signs is $\frac{N}{2}$, while the corresponding standard deviation is $\sqrt{\frac{N}{4}}$. Correcting for continuity, and using N_+ for the number of plus signs, the formula for z_+ becomes

$$z_+ = \frac{|2N_+ - N| - 1}{\sqrt{N}}, \tag{12.1}$$

where z_+ is interpreted in the usual manner (1.96 for significance at the .05

level, 2.58 for the .01, and 3.29 for the .001 level). Analyzing the data for all four directions, as given in the last column of Table 12.2, we obtain

$$
\begin{aligned}
z_{+} &= \frac{|(2)(49) - 124| - 1}{\sqrt{124}} \\
&= \frac{25.}{11.1355} \\
&= 2.25.
\end{aligned}
$$

Since this value of z_{+} is between 1.96 and 2.58, it follows that $.05 > P > .01$, so the figures on lights out are still significantly different from the expected 50–50 distribution at the .05 level, but not at any more extreme level.

THE RUNS TEST

The Wald–Wolfowitz runs test is designed to aid in determining whether two samples differ with respect to any aspect of their distribution. In using the runs test, the data of the two samples must be placed in a single rank order. (It is desirable that there be no ties.) Consider the following three possible sets of scores of 10 boys and 10 girls arranged in order with low scores at the left:

B B B B B B B B B B G G G G G G G G G G

B B B B B G G G G G G G G G G B B B B B

B G B G B G B G B G B G B G B G B G B G

In the first row, the boys have all the low scores and the girls all the high ones. The underlining shows only 2 runs. In the next row, no difference in central tendency is apparent, but the boys are much more variable, having all the extreme scores. There are only 3 runs. The last row, with 20 runs, would make anybody suspicious that chance alone was not operating. We have seen such distributions of men and women around a dinner table; nobody pretends that they are random. Either an unduly small or an unduly large number of runs is cause for casting suspicion on the null hypothesis. So, contrary to the advice given in a small number of statistics texts, a two-tailed test both seems and is far more appropriate than a one-tailed test.

The mechanics of using the runs test differ for large and for small samples. For small samples, Table Y has been prepared. N_1 is the number of persons in the first group, N_2 is the number in the second group, and n_R is the number of runs. If neither N_1 nor N_2 is greater than 20, the table gives four figures, two for the .01 level of significance and two for the .05 level. Thus, at the intersection of column (10) and row (8), we find the four figures

$\begin{matrix} 4- & 16+ \\ 5- & 15+ \end{matrix}$. This means that if $N_1 = 10$ and $N_2 = 8$, our results will be significant at the .01 level if n_R is 2 (the smallest possible value), 3, 4, 16, or 17 (the largest possible value). Further, if n_R takes any of the values 2, 3, 4, or 5 (corresponding to 5−) or 15, 16, or 17 (corresponding to 15+), our results are significant at the .05 level. The table is symmetrical, so this same information can be obtained from the identical entries in column (8) and row (10). Note that in each block of 4 figures, those in the **first** row are significant at the **.01** level, those in the other row being significant at the .05 level. NS, meaning "not significant," means that there is no number of runs that is significant at the indicated level in the direction that would be tabulated in that position.

When either N_1 or N_2 is greater than 20, Table Y obviously cannot be used. We then compute the mean number of runs in the theoretical sampling distribution for our N_1 and N_2, compare it with the obtained number of runs, and divide by the standard deviation of the sampling distribution. This theoretical mean is given between the large parentheses in the numerator of Formula (12.2), the theoretical standard deviation is the denominator, and the entire formula, incorporating a correction for continuity, is

$$z_R = \frac{\left| n_R - \left(\frac{2N_1N_2}{N_1 + N_2} + 1 \right) \right| - .50}{\sqrt{\frac{2N_1 N_2 (2N_1N_2 - N_1 - N_2)}{(N_1 + N_2)^2 (N_1 + N_2 - 1)}}}. \tag{12.2}$$

Despite its appearance, this formula contains only three variables, n_R, N_1, and N_2, and it can be computed (including the square root) in far less than five minutes. The computed value of z_R is a standard score. The two-tailed probability may be obtained by doubling the area obtained from Table E or, if preferred, we may simply note that when the absolute value of z_R reaches 1.96, n_R is significant at the .05 level; when it reaches 2.58, n_R is significant at the .01 level; and when it reaches 3.29, n_R is significant at the .001 level.

The Wald–Wolfowitz runs test is often used when each member of two groups (such as Boys and Girls or Experimental and Control rats) has a score on a test or on some other continuous measure of ability. The scores are then arranged in order and a group designation (such as B or G) is inserted, giving a series of letters somewhat like the three sample series shown on page 396. But the runs test is more general and we will now illustrate its use with a different type of data.

Let us see whether or not successive names in the phone book tend to be of about the same length. We don't want to count *every* name, because if we

TABLE 12.3
Length of 100 Consecutive Different Names of People in Daytona Beach Phone Book

4	8	5	7	8	6	6	7	6	4	11	9	12	13	11	7	4	5	9	9
5	7	5	8	10	8	6	8	7	7	7	7	4	7	6	5	6	6	5	5
5	11	10	8	6	6	6	7	8	7	9	8	7	7	7	8	7	6	8	6
7	8	9	6	10	6	7	6	9	8	6	13	8	6	8	7	6	6	6	10
8	7	8	7	7	8	8	8	14	12	10	11	11	8	9	11	8	6	10	10

did, when we got to the Smiths we would have hundreds of consecutive five-letter names. Let us confine ourselves to *different* names. We selected a Daytona Beach Telephone Directory to represent a small but fairly well-known city. Starting with the four-letter name Sapp, we worked through the 14-letter Schiffermuller and wound up with Schleicher. All names of businesses were omitted, giving 100 consecutive different names of people. Table 12.3 shows the basic data, with the runs of short names (4 through 7 letters) connected by lines below them, and the runs of long names (8 through 14 letters) connected by lines above them. When a line continues beyond the end of a row, that run continues on the next row—or is continued from the preceding row. There are 54 short names, 46 long ones, 20 runs of short names, and 20 runs of long names. Thus, $N_1 = 54$, $N_2 = 46$, and $n_R = 40$.

Substituting in Formula (12.2), we obtain

$$z_R = \frac{\left| n_R - \left(\frac{2N_1 N_2}{N_1 + N_2} + 1 \right) \right| - .50}{\sqrt{\frac{2N_1N_2\,(2N_1N_2 - N_1 - N_2)}{(N_1 + N_2)^2\,(N_1 + N_2 - 1)}}}$$

$$= \frac{\left| 40 - \left(\frac{(2)(54)(46)}{54 + 46} + 1 \right) \right| - .50}{\sqrt{\frac{(2)(54)(46)\,[(2)(54)(46) - 54 - 46]}{(54 + 46)^2\,(54 + 46 - 1)}}}$$

$$= \frac{|40 - 50.68| - .50}{\sqrt{24.43}}$$

$$= \frac{10.18}{4.94}$$

$$= 2.06.$$

From Table E, we see that when $z = 2.06$, the area in the two tails of the normal probability curve will be .0394. Hence, our 40 runs are significantly different from the expected 50.68 at the .05 but not at the .01 level. We thus have some support for our notion that big names tended to be next to other big names and that little names also tended to be next to little names. If this conclusion sounds too wishy-washy, the remedy is simple: gather more data and settle the issue. If there really is such a tendency of any appreciable size, we'll find it. If we don't find it with our larger sample, then either there is no difference or, if it does exist, it is too small to be of much importance in most practical problems.

A doubtful advantage of the Wald–Wolfowitz runs test is that it is a very nice test if you don't know what you are looking for. If it is used with test scores of two groups of people, it will detect differences between the groups in means, variability, skewness, or even kurtosis. But the "Jack of all trades, and master of none" proverb sometimes applies in such situations. For instance, if we are interested in differences between medians, both the median test and the Mann–Whitney U test are more appropriate because they are more efficient for this specific purpose. The median test and its extension to several groups will be considered next.

THE MEDIAN TEST

If we wish to know whether or not two groups of scores differ in their medians, the median test is appropriate. In using it, we are testing the null hypothesis that our two samples have been drawn from the same population or from populations with the same medians.

In using the median test, we must first determine the median for the two groups combined. We then dichotomize scores at this median and plot a 2 × 2 table. If the null hypothesis is true, there should be about the same proportion of cases above the median in each group. This is tested by the use of chi square corrected for continuity in reasonably large samples and by the use of the exact test of significance for smaller samples.

Let us illustrate this by using the 100 Daytona Beach names from Table 12.3 together with 50 additional names starting at the name Merry in the same phone book. The lengths of these 50 names are:

5, 8, 7, 6, 5, 9, 8, 8, 7, 7, 8, 8, 7, 5, 7, 5, 4, 7,
7, 5, 6, 8, 7, 7, 7, 7, 10, 7, 6, 12, 9, 6, 7, 6,
9, 8, 6, 4, 5, 5, 7, 5, 7, 7, 10, 5, 5, 6, 8, 9.

When we combine the two distributions, we find that there are 52 names with 4 through 6 letters, 37 with 7 letters, and 61 with 8 through 14 letters. The median is 7 but where should the 37 people with 7-letter names be tabulated? The writers prefer to make the split as nearly even as possible;

TABLE 12.4
Long and Short Names in Different Sections of Phone Book

	M	*S*	Total
Long (8–14)	15	46	61
Short (4–7)	35	54	89
Total	50	100	150

nearly all writers in nonparametric statistics make the division into those above the median versus those at or below the median. In this example, the two procedures give the same result, which is shown in Table 12.4.

We readily obtain the four expected frequencies, which are

20.3	40.7
29.7	59.3

and compute ${}_c\chi^2$ by Formula (11.2), obtaining ${}_c\chi^2 = 2.86$.

This is not significant, even at the .05 level. Do not conclude that we have proved that *M* names have the same median length as *S* names. All we have done is fail to prove that the two medians are different.

Notice that the median test is a straight application of chi square, with the continuous variable dichotomized at the median of the two groups combined.

The Median Test for More than Two Samples The median test may readily be extended to several independent groups. As before, we classify the data in the various groups as above or below the median of the combined samples. Then we use χ^2 to determine whether or not we are justified in rejecting the null hypothesis of merely chance deviations from the theoretical frequencies. An example will make this clear.

It isn't often that we get an opportunity to comment on research from another discipline. However, in searching for data that would provide topics that are timeless, generalizable across disciplines, while still serving as appropriate examples of various statistical procedures, we came upon a most unusual monograph. It was written by five scientists and is entitled "The Making of the Broads."* Bear in mind that this is a serious work concerned with the origin of certain small lakes in England. In fact, one of the five authors concluded, "The historical evidence confirms most strongly the theory that the broads are man-made." So much for the interdisciplinary approach; now on with the data.

*Lambert, J. M. *et al. The making of the broads*. London: Royal Geographical Society, 1960.

TABLE 12.5
Dates of Articles and Kind of Evidence

	Stratigraphical Evidence	Historical Evidence	Archeological Evidence	Total
1939–1957	10	14	20	44
1588–1938	7	22	16	45
Total	17	36	36	89

The monograph is divided into three parts considering the stratigraphical, historical, and archeological evidence, with different authors writing the different parts. Will the dates on the cited literature differ in the three parts of such a book? Before tabulating the dates, the following ground rules were adopted. References were classified by the senior author. All items written by any of the five authors of the monograph were eliminated. In each part, only one article was counted for any one author. If an author had an odd number of citations, the median date was tabulated. If he had an even number, the toss of a half-dollar determined which of the two middle ones to use. The oldest articles cited in the three parts were 1834, 1588, and 1698; the most recent were 1956, 1957, and 1957. There were 43 articles published from 1588 through 1937, there were 2 in 1938, and there were 44 from 1939 through 1957. The median is thus 1938. In using the median test, it is customary to categorize the data as scores above the median vs. scores at the median or below (though it would be just as logical to say scores at or above the median vs. scores below the median). Doing this, we obtain Table 12.5.

Because there are only 17 stratigraphical references in the table, we shall have two f_e's below 10 but above 5. There seems to be no reasonable basis for combining this column with either of the others, and it is obviously impossible to obtain additional comparable citations from the authors of that part, so we compute χ^2 with two small (but not with too small) f_e's.

The expected frequencies are obtained by getting the six products of marginal totals and dividing them by 89. This gives Table 12.6.

TABLE 12.6
Expected Frequencies for Dates of Articles

	Stratigraphical Evidence	Historical Evidence	Archeological Evidence	Total
1939–1957	8.4	17.8	17.8	44.0
1588–1938	8.6	18.2	18.2	45.0
Total	17.0	36.0	36.0	89.0

The computation of χ^2 by Formula (11.1) is straightforward.

$$\begin{aligned}\chi^2 &= \sum \frac{(f_o - f_e)^2}{f_e} \\ &= \frac{(10 - 8.4)^2}{8.4} + \frac{(14 - 17.8)^2}{17.8} + \cdots \\ &= .30 + .81 + \cdots \\ &= 2.74.\end{aligned}$$

By Formula (11.2), we see that $df = (n_c - 1)(n_r - 1) = (3 - 1)(2 - 1) = 2$. Since our value of χ^2 does not even reach the value of 5.991 required for significance at the .05 level, we have no reason to reject the null hypothesis that there are no significant differences in the dates of literature cited in the three different parts of the book.

In closing, let us note that the median test extended to several groups is nothing but χ^2 with all the groups dichotomized at the overall median.

THE MANN–WHITNEY U TEST

This test is sometimes referred to as the rank test or the sum-of-ranks test; and if the two groups are equal it is the same as the Wilcoxon T test (not to be confused with the t test discussed in Chapter 5). This test is applicable when we have scores for two separate and uncorrelated groups. The scores of both groups together are ranked. The ranks are added separately for each group and the test is applied.

The rationale behind the Mann–Whitney U test is that if the two groups really are samples from the same population, then the ratio of the sum of the ranks in the two groups will be roughly proportional to the ratio of the number of cases in the two groups. If, however, the sums of the ranks are not so proportioned, we reject the null hypothesis and conclude that the two groups differ. An example will make this clear.

Bertin* used word beginnings such as EFF, GN, and SH and developed a scoring system to differentiate between 54 male seniors in electrical engineering and 109 elementary school teachers (71 females and 38 males). Thus, the response SHut was given by 1 engineer and 17 teachers and is scored positively for teaching, while SHot was given by 7 engineers and 1 teacher and is scored negatively for teaching (or positively for engineering). The scoring system was cross-validated by trying it out later on 34 senior electrical engineers and a group of 27 elementary school teachers. (It

---

*Bertin, Morton A. Differentiating between two academic groups on the basis of a new associative test. Unpublished master's thesis, University of Florida, 1956.

differentiated extremely well.) Students in another education course were also to have been used, but it was discovered that only 9 of them were elementary school teachers, the other 5 being engaged in other occupations. Do the 14 people in this mixed group differ from the group of 27 elementary school teachers? Let us use the Mann–Whitney U test to decide. (See Table 12.7.)

We list the scores of the people in the two groups and then rank them. It does not matter whether rank 1 is assigned to the highest or to the lowest score. We add the ranks, designating the total of the N_1 ranks in the smaller group as T_1 and the other total as T_2. Here T_1 = 188 and T_2 = 673. We

TABLE 12.7
Teaching-Minus-Engineering Scores in Cross-Validation Groups

Mixed Group		Elementary Teachers	
Score	*Rank*	*Score*	*Rank*
91	36.5	114	41
67	27.5	105	40
66	26	104	39
59	20.5	97	38
50	15.5	91	36.5
49	13.5	84	34.5
42	12	84	34.5
41	10.5	78	33
36	7	74	32
27	6	73	30.5
24	5	73	30.5
17	4	69	29
15	3	67	27.5
6	1	65	24
		65	24
		65	24
		60	22
		59	20.5
		57	19
		54	18
		51	17
		50	15.5
		49	13.5
		41	10.5
		40	9
		38	8
		11	2
Total	188		673

check this by noting that $T_1 + T_2 = \dfrac{(N_1 + N_2)(N_1 + N_2 + 1)}{2}$. In this case,

$$188 + 673 = \frac{(41)(42)}{2} = 861.$$

When both N_1 and N_2 are 8 or more, the statistic

$$z_{U_1} = \frac{2T_1 - N_1(N_1 + N_2 + 1)}{\sqrt{\dfrac{N_1N_2(N_1 + N_2 + 1)}{3}}} \tag{12.3}$$

is distributed essentially normally with a mean of 0 and a standard deviation of 1.

So is the statistic

$$z_{U_2} = \frac{2T_2 - N_2(N_1 + N_2 + 1)}{\sqrt{\dfrac{N_1N_2(N_1 + N_2 + 1)}{3}}} \tag{12.4}$$

which will have the same absolute value but opposite sign from z_{U_1}.

Let us compute these.

$$\begin{aligned} z_{U_1} &= \frac{(2)(188) - 14(14 + 27 + 1)}{\sqrt{\dfrac{(14)(27)(14 + 27 + 1)}{3}}} \\ &= \frac{376 - 588}{\sqrt{5{,}292}} \\ &= \frac{-212}{72.75} \\ &= -2.91, \\ z_{U_2} &= \frac{(2)(673) - 27(14 + 27 + 1)}{\sqrt{\dfrac{(14)(27)(14 + 27 + 1)}{3}}} \\ &= \frac{1{,}346 - 1{,}134}{\sqrt{5{,}292}} \\ &= \frac{+212}{72.75} \\ &= +2.91. \end{aligned}$$

Looking up the area in Table E, we find that the area beyond 2.91 is .0018 in each tail or .0036 for the usual two-tailed test. Hence, the elementary school teachers and the mixed group differ in their test responses at the .01 but not at the .001 level. This result may appear surprising when one realizes that 9 of the 14 in the mixed group were supposedly the same kind of people as the 27 in the other group. Nevertheless, the two groups did differ significantly.

The reader may wonder about the U in the name of this test. Originally, people computed U as a function of N_1, N_2, and T_1 or T_2, but since identical results are obtained with less arithmetic by using Formula (12.3) or (12.4), there seems no need to continue computing U itself, even though we are still using the Mann–Whitney U test.

This test is somewhat similar both to the runs test and to the median test. Partly because the runs test was more general, Mann and Whitney in their original article expressed the intuitive feeling that their test was more efficient in testing for differences in central tendency. This feeling is borne out by the subsequent experience of others. The Mann–Whitney U test is more directly competitive with the median test and is more efficient than the median test. It usually takes more computational time than the median test, however. Some writers point out that the Mann–Whitney U test should not be used if there are "too many" ties in rank. While theoretically true, it has been shown that reasonably large numbers of ties have very little effect on the test.

WHICH NONPARAMETRIC STATISTIC SHOULD BE USED?

First of all, we must decide whether we wish to use a parametric or a nonparametric statistic after considering the various conditions discussed on page 393. If the decision is in favor of a nonparametric statistic, there are often several choices available. Which one the research worker uses will then be determined partly by the purpose of the study, partly by which method was used earlier in somewhat similar studies, and often to a considerable extent by the experimenter's previous familiarity with or preference for one particular procedure. (For example, some people just love to compute χ^2.) In the absence of any such bases for preference, the following guide should be useful.

Assuming that we wish a nonparametric statistic:

Use ρ when two scores or ranks are available for each case, and we wish to know the relationship existing between the two sets.

Use χ^2 when the available information consists of the frequencies in each of a relatively small number of categories or groups, and our interest is in whether or not any nonchance relation or association exists.

Also, use χ^2 when the purpose is to determine whether or not a set of frequencies is in accord with any hypothesis.

Use the *exact probability test,* but with caution, when χ^2 would be used if more records were available.

Use the *sign test* when the available records consist of differences between paired scores and one wishes to determine whether the proportion of differences in one direction departs appreciably from .50.

Also, use the *sign test* with any data that may be dichotomized (more or less, boy or girl, left or right) to determine whether the proportion in one category differs from .50.

Use the *runs test* when there are two kinds of individuals that can be placed in one overall ranking or arrangement, and one wishes to determine whether the actual arrangement might reasonably be due to chance. This tests simultaneously for differences in central tendency, variability, skewness, and kurtosis.

Use the *median test* to determine whether the medians of two groups differ significantly.

Use the *median test for more than two samples* to determine whether any of the several group medians differ significantly from the median of all the groups taken together.

Use the *Mann–Whitney U test* when there are scores for two groups such that all scores may be placed in one overall ranking, and one wishes to test for differences in central tendency between the two groups. This test is better than the runs test for this purpose and is also superior to the median test, but it requires more computational time.

OTHER NONPARAMETRIC STATISTICS

There are many other nonparametric statistics. They vary greatly both in the extent to which they make use of the refinements in the data and in their computational time. If you are interested in them, here are three sources. Two books devoted solely to them are Sidney Siegel: Nonparametric statistics for the behavioral sciences (McGraw–Hill, 1956) and Merle W. Tate and Richard C. Clelland: Nonparametric and shortcut statistics (Interstate Printers and Publishers, Danville, Ill., 1957). A comprehensive bibliography in this field is the second of two prepared by the same man. It contains about 2800 entries. It is I. Richard Savage: Bibliography of nonparametric statistics (Harvard University Press, 1962).

SUMMARY

Nonparametric statistics make few or no assumptions about the nature of the populations from which our samples are drawn. Three nonparametric statistics have been treated earlier in this book and five more are treated in this chapter. One or more of these eight will probably take care of the great majority of situations in which a nonparametric test is needed for educational or psychological research studies. Which one to use? One section of this chapter is devoted to answering that very question. One or two paragraphs tell under exactly what circumstances each of these eight nonparametric methods should be used.

"Five trillion, four hundred eighty billion, five hundred twenty-three million, two hundred ninety-seven thousand, one hundred and sixty-two . . ."

CHAPTER 13 SIMPLE ANALYSIS OF VARIANCE

Back in Chapter 5, we learned to use the *t* test to determine whether or not two means differ significantly from each other. In this chapter, we will learn to use analysis of variance to determine whether or not several means differ significantly from each other.

WHY NOT THE T TEST

When there are only two means to be compared, there is nothing wrong with the *t* test. It takes care of the situation admirably. But what happens when the number of groups is 3, 4, 6, or even a dozen or more? Can't we still use the *t* test? Let us see what happens by considering an extreme case. Jaspen* actually produced 17 different versions of an instructional film showing how to assemble the breech block of a 40 millimeter antiaircraft gun. Let us suppose that we had a few hundred subjects available, that we assigned them at random to 17 groups, and that we then measured the proficiency of each subject in actually assembling the breech block. If we

*Jaspen, Nathan. The contribution of especially designed sound motion pictures to the learning of skills: assembly of the breech block of the 40mm antiaircraft gun. Unpublished doctoral dissertation, Pennsylvania State University, 1949.

followed through on our plan to use the t test, we would run into several difficulties. First, since there are $\frac{(17)(16)}{2} = 136$ combinations of 17 methods taken two at a time, we would have to make 136 separate computations of t. This is an enormous amount of work. Second, if a few of our results turned out to be significant, we would have trouble interpreting them. Why? Because when we make 136 comparisons, it is not rare, but perfectly natural, to get one or two results significant at the .01 level, even if the null hypothesis is correct. After all, an event so rare that it is likely to happen only once in a hundred trials should, by definition, be expected to happen roughly 1.36 times (or about once or twice) in 136 trials. Similarly, since 5% of 136 is 6.80, we should not be surprised if around 5 or 10 of our differences turn out to be significant at the .05 level.

Thus, the most serious objections to the use of t when there are many comparisons are the large number of computations and the difficulty or even impossibility of determining significance levels. There are other objections. For instance, when we are testing our results against the null hypothesis, we would do better to base our estimates of the variance or standard deviation of the difference between means upon all the data, rather than just on the scores for the two groups being compared. Another difficulty is that since the 136 comparisons are not independent of each other, it is not strictly true that we should expect 6.80 differences significant at the .05 level. In some such experiments, there might not be any; but in other experiments, many more might be expected strictly on a chance basis. That is why we said *roughly* 1.36 and *around* 5 or 10 in the preceding paragraph. At times the approximation may be *very* rough. So much for the t test. Let us now consider the fundamental basis of analysis of variance.

THE BASIS OF ANALYSIS OF VARIANCE

We shall start, as we often do, by assuming the null hypothesis, H_0, even though it may subsequently be disproved. Our basic data will consist of the scores of N individuals, n of whom are randomly assigned to each of k groups. (Actually, there may be varying numbers of cases in the k groups, but this initial explanation is facilitated by having equal groups, in which case $nk = N$.)

Under H_0, should the means vary widely? Of course not. Should they all be the same? Let's consider that a moment. We shall ask someone to toss some presumably unbiased coins and report the results to us. First, we give him a penny, ask him to toss it 10 times and to report back. He returns and tells us he got 5 heads and 5 tails. We notice that this is more likely than any other combination, but say nothing about it, merely asking him to toss a nickel 10 times. He leaves, returns and reports 5 heads and 5 tails, a result we accept without comment, asking him to use a dime now. He does and—you guessed it—again says he got 5 heads and 5 tails. We ask him to use a quarter for his last 10 tosses. He returns with a final report of 5 heads and 5 tails. In such coin-tossing, it really is more likely that the number of heads should be 5, 5, 5, and 5 than that it should be 4, 6, 5, 4 or any other set of 4 figures; yet most people will be suspicious. (Those who are not may continue the experiment with a half-dollar, dollar, and foreign coins until they run out of time, money, or patience—the results always being reported back as 5 heads and 5 tails.) Why should we be suspicious of the most probable of all possible results? While it *may* be that unbiased coins were fairly tossed, correctly recorded, and honestly reported, such a run of results is most unlikely if chance alone is operating. Although the probability of getting 5 heads and 5 tails in the first 10 tosses is $\frac{252}{1{,}024} = .246$, the probability of this happening 4 times is only $(.246)^4 = .004$. (For the hard to convince, who continued through 8 or 12 trials, the probabilities become .000,013 and .000,000,05 respectively.) Thus, even under the null hypothesis, we definitely do not expect all our means to be equal.

The means of our k groups, then, should not vary widely nor should they all be equal. How much should they vary? Going back to Chapter 5 again, we can get the answer. With N cases, Formula (5.2) gave the standard error of the mean as

$$s_{\overline{X}} = \frac{s}{\sqrt{N}}.$$

Since each of our k groups contains n cases, the standard error of the mean of such a group would be

$$s_{\overline{X}} = \frac{s}{\sqrt{n}}.$$

In the past, we have always known s and n, and we have used this formula to estimate $s_{\overline{X}}$. But if, instead, we know s_X and n, we can use the same formula to estimate s. Doing this, and squaring in order to work with variances rather than with standard deviations and standard errors, we may write

$$\text{Among variance} = s^2 = ns_{\overline{X}}^2. \tag{13.1}$$

If we know n and $s_{\overline{X}}^2$, this formula for s^2 gives us a method of estimating σ^2, the population variance. We certainly know n, the number of cases in each of our groups, and $s_{\overline{X}}^2$, the variance of the means, is obtained simply by computing the variance of the means of the k groups. Because we are using this variance among the means in order to estimate the population variance, this estimate is referred to as the *among variance*.

The fundamental principle upon which the analysis of variance is based is that of obtaining two *independent* estimates of the population variance, σ^2, and then determining whether or not their difference in size is such that it might reasonably be attributed to the operation of chance. If it can, we do not reject the null hypothesis. If it cannot be attributed to the operation of chance, we reject H_0 and conclude that there is a real difference in the effectiveness of the treatments or other variables that are being investigated.

How do we get a second independent estimate of the population variance? This is easy. It is just a minor variant of what we do whenever we compute a standard deviation. We want an estimate based on the variation within the k groups. We could compute s^2 within any one group, but we shall obtain a more reliable estimate of σ^2 if we pool the results for all groups. In doing this, we measure each deviation from the mean of its own group. Thus, if we have k groups of n cases each, we will have k sums of squares, each based on n cases and hence having $n - 1$ degrees of freedom. Since each deviation, x, is measured from the mean of its own group rather than from the overall mean, we shall use subscripts 1, 2, 3, $\cdots$, k to designate the various groups. Thus, a deviation in the second group will be called not x but $(X_2 - \overline{X}_2)$. So our formula for the *within variance*, based on deviations *within* the k groups, will not be simply $\frac{\Sigma x^2}{n-1}$ but

$$\text{within variance} = s^2$$

$$= \frac{\Sigma(X_1 - \overline{X}_1)^2 + \Sigma(X_2 - \overline{X}_2)^2 + \cdots + \Sigma(X_k - \overline{X}_k)^2}{(n_1 - 1) + (n_2 - 1) + \cdots + (n_k - 1)}. \tag{13.2}$$

The denominator could, if we wished, be written $N - k$. But since Formulas (13.1) and (13.2) are not usually used for computing, the above arrangement, with each sum of squares appearing above its *df*, seems preferable.

The final step in the analysis of variance is to compare the two variances, determine whether or not they differ significantly, and draw an appropriate conclusion. This is done by dividing the among variance by the within variance, looking up the quotient in the F table, noting that the difference is or is not significant at the .05, .01, or .001 level, and concluding that the differences among the k groups either are not due to the operation of chance or that they may very well be only chance fluctuations. That is, we reject or we do not reject the null hypothesis.

The preceding discussion has given an informal presentation of the theoretical base underlying the technique known as analysis of variance. We will now illustrate the use of analysis of variance in the solution of a practical problem.

A FIRST EXAMPLE

We will use a portion of the data secured by Shaw* in a study of the effects of cooperation and competition in using a hand wheel to make a rotating cursor follow the erratic movements of a rotating target. In the cooperation group, each subject was led to believe that another subject (actually a confederate of the experimenter) was using his hand wheel to help the subject keep the cursor on the target. In the competition group, each subject was led to believe that another subject (the confederate) was using his hand wheel to keep the cursor on a different target. In the control group, the subject worked alone. Sometimes the subject and sometimes the confederate controlled the movements of the cursor, but the only records Shaw analyzed were for fifteen 15-second periods when the subject was in complete control of the situation. Because there was evidence of learning, we will confine our analysis to the raw data, kindly supplied by the author, for the last 4 of the 15 periods. Thus, our scores, as shown near the top of Table 13.1, give the number of seconds on the target for the last minute of each subject's record. Since there are 10 subjects in each group, we can readily see from the ΣX row that the means for the three groups are 41.0, 29.3, and 36.6. Are the differences among these means significant; or could such differences readily be explained as nothing more than chance fluctuations? It is this question that our analysis of variance will answer.

Although we could use Formulas (13.1) and (13.2) for computational purposes, it is usually easier and faster to avoid the decimals and negative signs which they entail. We will first define three simple functions, J, K, and L, of the raw scores and their squares. Then we shall note that the three positive differences, $J - K$, $K - L$, and $J - L$, are the numerators of the within, among, and total variances, respectively.

As is done near the top of Table 13.1, the scores in each group are listed in a column and the columns are numbered consecutively from 1 to k, where k is the number of groups. There are n_1 scores, designated as X_1, in the first column, and, in general, there are n_k scores, called X_k, in the kth column. Often, as in Table 13.1, there are the same number of cases (here 10) in each group, but this is not necessary; the formulas are perfectly general. Let us give them:

$$J = \Sigma X_1^2 + \Sigma X_2^2 + \cdots + \Sigma X_k^2. \tag{13.3}$$

Just below the 10 scores listed in the X_1 column of Table 13.1 we see that $\Sigma X = 410$ and that $\Sigma X^2 = 17{,}004$. These sums are checked by the Charlier

*Shaw, Marvin E. Some motivational factors in cooperation and competition. *Journal of Personality*, 1962, **18**, 155–169. Shaw kindly supplied us with the raw data which we used in Table 13.1.

check (thus, 17,004 + (2)(410) + 10 = 17,834) and when all k columns are correct, a check is placed in the parentheses at the right. Then, to obtain J, we simply add the k values of ΣX^2, as shown in the row for J in the table. This total of 40,011 is simply the sum of all N values of X^2.

To get K, we add the scores in each column, square each column sum and divide it by its n, and then add the k quotients. In algebraic terms,

$$K = \frac{(\Sigma X_1)^2}{n_1} + \frac{(\Sigma X_2)^2}{n_2} + \cdots + \frac{(\Sigma X_k)^2}{n_k}. \tag{13.4}$$

In the row for K in the table, the first of these three terms is $\frac{(410)^2}{10}$ and the value of K is shown to be 38,790.50.

L is one Nth of the square of the total of all the values of X. To obtain L, we first add each column, then we add the k column sums, then we square this total, and finally we divide by N. The formula is

$$L = \frac{(\Sigma X_1 + \Sigma X_2 + \cdots + \Sigma X_k)^2}{N}. \tag{13.5}$$

In the table, the row for L shows that $L = 38{,}092.03$

Now that we know J, K, and L, the formulas for the three variances are extremely simple. In each case, the numerator is equal to a sum of squares and the denominator is the appropriate number of degrees of freedom. The three formulas are

$$\text{Within variance} = \frac{J - K}{N - k}, \tag{13.6}$$

$$\text{Among variance} = \frac{K - L}{k - 1}, \tag{13.7}$$

$$\text{Total variance} = \frac{J - L}{N - 1}. \tag{13.8}$$

A mnemonic device for remembering these sums of squares is that a

hi**J**ac**K**ing plus a **K**il**L**ing often adds up to a **J**ai**L**ing
or, more briefly, **J**ac**K** plus **K**il**L** equals **J**ai**L**
or, symbolically, $(J - K) + (K - L) = (J - L)$
which means that Within + Among = Total

This shows the use of the components, J, K, and L. The equations are literally true for the numerators. In fact, Within + Among = Total is true not only for the sums of squares in the numerators but also for the degrees of freedom in the denominators. Notice also how the numerators and denominators are related; capital K is above small k; capital L is above

TABLE 13.1
Total Number of Seconds on Target in Last 4 Trials of 15 Seconds Each

	Cooperation	Competition	Control	
	X_1	X_2	X_3	
	45	18	32	
	46	26	39	
	43	40	45	
	46	32	49	
	37	29	38	
	40	36	25	
	44	18	38	
	40	38	32	
	32	31	28	
	37	25	40	
ΣX	410.	293.	366.	
ΣX^2	17,004.	9,115.	13,892.	
$\Sigma(X+1)$	420.	303.	376.	
$\Sigma(X+1)^2$	17,834.	9,711.	14,634.	(√)

$$J = \Sigma X_1^2 + \Sigma X_2^2 + \cdots + \Sigma X_k^2$$
$$= 17{,}004 + 9{,}115 + 13{,}892$$
$$= 40{,}011$$

$$K = \frac{(\Sigma X_1)^2}{n_1} + \frac{(\Sigma X_2)^2}{n_2} + \cdots + \frac{(\Sigma X_k)^2}{n_k}$$
$$= \frac{(410)^2}{10} + \frac{(293)^2}{10} + \frac{(366)^2}{10}$$
$$= 16{,}810.00 + 8{,}584.90 + 13{,}395.60$$
$$= 38{,}790.50$$

$$L = \frac{(\Sigma X_1 + \Sigma X_2 + \cdots + \Sigma X_k)^2}{N}$$
$$= \frac{(410 + 293 + 366)^2}{30} = \frac{(1{,}069)^2}{30}$$
$$= 38{,}092.03$$

$$\text{Within variance} = \frac{J - K}{N - k} = \frac{40{,}011 - 38{,}790.50}{30 - 3}$$
$$= \frac{1{,}220.50}{27} = 45.20$$

$$\text{Among variance} = \frac{K - L}{k - 1} = \frac{38{,}790.50 - 38{,}092.03}{3 - 1}$$
$$= \frac{698.47}{2} = 349.24$$

TABLE 13.1
(continued)

Cooperation	Competition	Control

$$\text{Total variance} = \frac{J - L}{N - 1} = \frac{40{,}011 - 38{,}092.03}{30 - 1}$$

$$= \frac{1{,}918.97}{29} = 66.17$$

$$F = \frac{\text{Among variance}}{\text{Within variance}} = \frac{349.24}{45.20} = 7.73$$

$df = 2/27$

$.01 > P > .001$

small 1 (actually one, not small L); and J and N go together just as they do in Jane, Jean, Joan, June, and dear John.

The substitutions in the formulas and the solutions of the three variances are shown in Table 13.1. Notice that J, K, L, all three numerators, all three denominators, and all three quotients are *always* positive. J is always greater than K, which, in turn, is always greater than L. The total variance is always intermediate in value between the other two, with the among variance usually, but not quite always, larger than the within variance.

The formula for F is

$$F = \frac{\text{Among variance}}{\text{Within variance}}. \tag{13.9}$$

We compute F to get a number that we can look up in a table, appropriately called the F Table, in order to find out whether or not the difference between the two variances is significant. In using the F Table, we need to know the ratio of the two variances, as just given in Formula (13.9), and the number of degrees of freedom upon which each is based. The latter are the denominators that were used in computing the two variances. In the example, the two *df*'s are 2 and 27, which is written $df = 2/27$, the first being for the numerator and the second for the denominator of Formula (13.9) for F. Let us now look at the F Table. Because $df = 2$ for the among variance, we want column (2) and because $df = 27$ for the within variance, we want row (27). At the intersection of column (2) and row (27), we find the numbers 3.35 **5.49** *9.02*. This means that for $df = 2/27$, an F of 3.35 is significant at the .05 level, an F of **5.49** is significant at the .01 level, and an F of *9.02* is

TABLE 13.2
Analysis of Variance Summary Table for Time on Target

Source of Variation	Sum of Squares	*df*	Estimate of Variance	*F*
Among groups	698.47	2	349.24	7.73**
Within groups	1,220.50	27	45.20	
Total	1,918.97	29		

**$.01 > P > .001$

significant at the .001 level. Since our F is 7.73, it is significant at either or both of the first two levels, and so the last line of Table 13.1 reads $.01 > P > .001$.

We have finished the computations. Because P is so small, we reject the null hypothesis of no difference between methods, and we conclude that there is a real difference or there are real differences in the effectiveness of the three methods. In Shaw's problem of following a moving target, it makes a difference whether the subject feels that he is cooperating, competing, or working alone. Using the notation introduced on page 124, we may write F = 7.73** and describe the differences among the methods as highly significant.

Before leaving this example, it seems desirable to point out that there is a fairly standard way of presenting the results of an analysis of variance. We have done this in Table 13.2. Sometimes the first column is called *Components* or simply *Source*. The second is sometimes abbreviated SS and the fourth may be labeled *Variance, Mean Square,* or *MS*. The last is sometimes omitted, often with a line below the table or in the text giving the essential information:

$$F = \frac{349.24}{45.20} = 7.73; \quad df = 2/27; \quad .01 > P > .001.$$

These differences, however, are trivial. Both the information given and the manner of presenting it are quite well agreed upon in both the educational and the psychological literature.

ASSUMPTIONS

Four assumptions are made in using analysis of variance for the solution of problems of the type discussed in this chapter. Unfortunately, we sometimes cannot tell whether or not some of the assumptions hold but, fortunately, it sometimes does not matter much. Let us consider these assumptions.

1. It is assumed that the distribution of the variable in the population from which our k samples are drawn is normal. Ordinarily, we do not know whether or not this assumption holds, but empirical studies have shown that even when the distribution departs appreciably from normality, it has very little effect on the results. Hence, if the distribution of the variable, X, in our sample is markedly skewed or very leptokurtic or platykurtic, see Chapter 7 for a method of transforming the scores and then do an analysis of variance on the transformed scores. Otherwise, don't worry; it's probably all right to go ahead with the analysis of variance of the raw data.
2. It is assumed that all k groups have the same variance. Again, we may interpret this requirement loosely. In practice, if one group has a standard deviation even two or three times as great as another group, this is sufficiently close to equality to have very little effect on the appropriateness of using analysis of variance.
3. Since analysis of variance attempts to divide the total variance into components (among groups and within groups), it is assumed that the factors which account for deviations in an individual's score are additive. This assumption probably holds sufficiently well for practically all educational and psychological measurements that give continuous rather than categorical scores.
4. It is assumed that the subjects are assigned *at random* to the k groups. If an experiment is so set up that as each subject is considered, the toss of a coin, the roll of a die, or an entry in a table of random numbers determines to which group the subject is assigned, this assumption will be met. If, however, intact classes are used, with a different teacher teaching by a different method in each class, any differences that may be found cannot be interpreted. They may be due to the different methods, but they also may be due to different initial abilities of the various classes, to variations in teaching abilities of the various teachers, or to combinations of any two or more of these or of still other unknown and uncontrolled variables. Thus, the requirements of this assumption of randomness may be either (a) extremely easy or (b) virtually impossible to fulfill.

CHECKING

Table 13.1 and the accompanying text regarding Shaw's experiment were used primarily to illustrate the method of solving an analysis of variance problem, and almost nothing was said about checking. Because computational errors—some large, some small—can be found in a majority of theses and dissertations based on psychological or educational data, and because similar mistakes, which frequently have affected the conclusions, have been found in articles published in educational and psychological journals, it is imperative that statisticians do what they can to diminish the number of such blunders. We may never achieve perfection, but let us see what we can do to try to eliminate mistakes in analysis of variance calculations.

There are seven specific steps we can (and should) take to see to it that our analysis of variance calculations are correct:

1. See that the original scores which constitute the basic data for the study are correctly obtained and accurately recorded.
2. Check to make sure the N scores are correctly copied onto the page, such as Table 13.1, to be used for the computations.
3. Compute ΣX, ΣX^2, $\Sigma(X+1)$, and $\Sigma(X + 1)^2$ in any convenient way, but check these four sums by the Charlier check, which was referred to on page 413 and which is explained more fully on pages 57–58.
4. After computing J, K, and L, compute J', K', and L' by substituting $(X + 1)$ for X and $(X + 1)^2$ for X^2 in Formulas (13.3), (13.4), and (13.5). In our example, these values turn out to be $J' = 42{,}179$; $K' = 40{,}958.50$; and $L' = 40{,}260.03$.
5. While it would be possible to use these values of J', K', and L' to compute the within, among, and total variances, a much better check is available. We compute F by the formula

$$F = \frac{(K' - L')(N - k)}{(J' - K')(k - 1)}. \tag{13.10}$$

Substituting the values just obtained in step (4) and rechecking the degrees of freedom, we obtain

$$\begin{aligned} F &= \frac{(40{,}958.50 - 40{,}260.03)(30 - 3)}{(42{,}179 - 40{,}958.50)(3 - 1)} \\ &= \frac{18{,}858.69}{2{,}441.00} \\ &= 7.73. \end{aligned}$$

In this case, the check is exact. If the difference is .02 or more, the cause of the discrepancy should be determined. If there is no difference or a difference of only .01 (which may well be due to rounding errors), this process checks not only F but also J, K, L, the within variance, and the among variance. Not bad, is it?

6. The total variance is checked by computing it with $(J' - L')$ in the numerator instead of $(J - L)$. But since most research workers do not include it in their analysis of variance summary tables, this may not be necessary. Incidentally, as shown in Table 13.2, Among + Within = Total, both for the sum of squares and for *df*. But do not use this as a check because most mistakes in J, K, or L (such as substituting any

wrong value for J, K, L, or even for all three of them) will go undetected, with among plus within still equalling total.

7. Next, it is necessary to make sure that the figures already recorded for *df* actually are $(k - 1)$ and $(N - k)$. Finally, we look in the F Table to find out whether our obtained F is significant and, if so, at what level.

Notice that, with less work than it took to do it the first time, we have now checked the analysis of variance, and that nearly all the checking has been done by an entirely different method, thus avoiding the embarrassment of making the same mistake twice and then thinking that erroneous results are correct. The use of all these checks will be illustrated in the next example, which begins on page 421.

TECHNICALITIES

Aside from the assumptions and checking procedures just discussed, a number of technical questions arise concerning the use of analysis of variance. We will now briefly consider some of them.

It is not necessary that all groups have the same number of cases. As we pointed out, the various groups are often of the same size, but all the computational formulas in this chapter are so written as to work perfectly whether or not $n_1 = n_2 = \cdots = n_k$.

What happens if $F < 1.00$? Note that every entry in the F Table (except for $df = \infty/\infty$) is greater than 1.00. Note also that in F Tables in some other statistics texts, what we have called *df* for among variance is referred to as *df* for greater variance (or *df* for greater mean square). Ordinarily, it doesn't make much difference because the among variance usually is the greater variance. What happens when it isn't? Nothing much. The F ratio, then, is clearly smaller than any entry in the F Table, so there is no evidence that the null hypothesis is false, and (without even needing to look in the table) we state that our results are not significant at any level.

When we have several methods and F is significant, which methods are significantly better than which other ones? It may seem peculiar, but actually, we cannot answer this question. Of course, if F is not significant, no further analysis of the data needs to be made or should be made. But if we naively start making t tests between pairs of extreme means, we unknowingly run into exactly the same difficulty described on page 409 of this chapter. Mathematical statisticians have worked on this problem but seem not yet to have come up with a simple solution. We offer two suggestions, both of which have their difficulties.

1. If you have adopted the .05 level as your standard of significance, and if F is significant at this level, then make t tests, but count as significant

only those that reach the .01 level. If you have adopted .01 and F meets this, insist on .001 for the t tests. If you have adopted .001 and F meets this, you will have trouble finding appropriate tables, so insist on some value clearly beyond the .001 level for your t tests.

2. Repeat the experiment, using only a few of the methods—either the most extreme ones or those which, for any reason, now seem to be of greatest interest to you. This is much more work, but it is definitely the best procedure, especially in view of the widespread feeling that there is a great need for more replication of experimental studies in Education and in Psychology.

This chapter considers but by no means covers the topic of analysis of variance. More complex designs are appropriate for handling several variables simultaneously, for considering the interaction between variables, for matched rather than randomly assigned individuals, for attempting to adjust for the varying abilities present when intact groups are used, etc. These topics are treated in some more advanced books on statistical methods and in books on the design of experiments.

The use of F may be regarded as an extension of the use of t. Or, the use of the t test is a special case of the use of the F test in which there are two groups and hence df for among groups is 1. With two groups $\sqrt{F} = t$ or $t^2 = F$. We can readily see this by comparing the entries for any row of the t Table, Table G, with those in the same row and the $df = 1$ column of the F Table, Table F. Thus for row 60, $(2.00)^2 = 4.00$; $(2.66)^2 = 7.08$; and $(3.46)^2 = 11.97$. From this, it would appear that if you wished to do so, you could use analysis of variance not just for 3 or more groups, but also for two groups. This is correct. With two groups, both procedures give exactly the same results.

In the preceding paragraph, we saw that $F = t^2$. Supplement 13 shows that a few values of F are related to those of χ^2 and that one small set of values of F is related to z of the normal probability curve.

Earlier in this chapter, we referred to the need for getting two independent estimates of the population variance, and then we developed formulas for three variances—within, among, and total. All three are estimates of the population variance, so how do we know that at least two, but not all three, really are independent of each other? Let us begin by writing the identity implied, but not specifically stated, on page 417:

$$(X_i - \overline{X}) = (X_i - \overline{X}_i) + (\overline{X}_i - \overline{X}). \tag{13.11}$$

This equation states that the deviation of any raw score, X_i, from the overall mean, $\overline{X}$, may be divided into two parts, its deviation from the mean of its group plus the deviation of the mean of its group from the overall mean. To many persons, it rightly appears intuitively obvious that any (small or large) variation within groups does not necessitate nor even imply that there should be any particular (small or large) variation between groups. To others, the mathematical demonstration of this truth is not very difficult. Let us first square such a deviation for a person in the first group,

$$(X_1 - \overline{X})^2 = (X_1 - \overline{X}_1)^2 + (\overline{X}_1 - \overline{X})^2 + 2(X_1 - \overline{X}_1)(\overline{X}_1 - \overline{X}).$$

Let us now add such figures for all n_1 persons in the first group, noting that when we add a constant, such as $(\overline{X}_1 - \overline{X})^2$, we simply get n_1 times the constant, and that when we add a constant times a variable, we may rewrite it as the constant times the sum of the variable. We obtain

$$\Sigma(X_1 - \overline{X})^2 = \Sigma(X_1 - \overline{X}_1)^2 + n_1(\overline{X}_1 - \overline{X})^2 + 2(\overline{X}_1 - \overline{X})\,\Sigma(X_1 - \overline{X}_1).$$

The last half of the last term is the sum of the deviations, x_1, from their mean, $\overline{X}_1$, and, as we know, $\Sigma x = 0$. Hence, that whole term is equal to zero. If we now write similar equations for all the other groups and combine them by adding corresponding terms, the left term is clearly the numerator for the total variance based on all N cases, the first term on the right is the numerator given in Formula (13.2) for the within variance, the second term on the right is clearly related to the among variance and is actually the numerator of one variant of Formula (13.1) for the among variance, and finally, the third term is ignored because it is the sum of k zeros. Thus, we have shown that the total sum of squares can be divided into two parts which, if they had been correlated, would have caused the third term above not to be zero. The fact that it is zero thus confirms the analysis into two components and at the same time establishes their independence. The proof that we do not have three independent estimates of the population variance is simple: since the total sum of squares is equal to the sum of the within sum of squares and the among sum of squares, the total sum of squares is obviously related to both of its uncorrelated components.

A SECOND EXAMPLE

Merrill* used several teaching methods to study the learning and retention of a hierarchical task, one in which knowledge of each successive part is a prerequisite to knowledge of the next part. His 62 Education students were randomly assigned to five groups and were presented a complex imaginary science by means of a computer-based teaching system. For our example, we shall use data on the errors made by three of the groups on a retention test given three weeks after the learning had been completed.

---

*Merrill, M. David. Correction and review on successive parts in learning a hierarchical task. *Journal of Educational Psychology*, 1965, **56**, 225–234. Merrill also provided us with the data used in Table 13.4.

In each of the five lessons, each subject in Group A was presented a short passage of material and a question. If he answered correctly he was so informed. If he made an error, he was given a general review and asked again. If the subject made more errors, he was given specific reviews until he did answer correctly. Each lesson was followed by a quiz which also provided general and specific reviews.

Group B received no correction review on either the lessons or the quizzes. Group C was a special control group included to obtain a measure of the effectiveness of the review material. This group was presented only a set of summary statements. All three groups were then given a test that also provided both general and specific reviews, as described above. Three weeks later, all groups were given a retention test. The average time and errors made by the groups on the lessons, quizzes, test, and retention test are shown in Table 13.3.

Without having seen any of the lessons or quizzes, Group C, as expected, took more time (3.42 hours) and made more errors (64.91) on the test than did the other groups. But the question of the effectiveness of the various teaching methods cannot be answered without data on the retention test. Our problem, then, is that of determining whether or not there are significant differences in the mean number of errors made on the retention test by the three groups.

The figures that constitute the basis for our analysis of variance are the 37 error scores given at the top of Table 13.4. The first four rows below the error scores are the sums of the scores, of their squares, of the numbers when each score is increased by one, and of the squares of these increased scores. Add the first three of these sums for any group and you should get

TABLE 13.3
Average Time (in Hours and Hundredths) and Average Errors in Learning and Testing

	Group A *Correction Review*	Group B *No Correction Review*	Group C *Saw Summary Only*
Time on Lessons and Quizzes	5.97	3.49	—
Time on Test	2.32	1.59	3.42
Total Time through Test	8.29	5.08	3.42
Errors on Test	30.07	20.45	64.91
Time on Retention Test	1.30	.75	.95
Errors on Retention Test	16.43	13.82	23.67

the fourth one. If this check holds for all (three) groups, put a $\surd$ between the parentheses at the right of the $\Sigma(X+1)^2$ row. This is the Charlier check. Because $\Sigma(X+1) = \Sigma X + n$, it may be written as

$$\Sigma(X+1)^2 = \Sigma X + \Sigma X^2 + \Sigma(X+1). \qquad (13.12)$$

The next two rows give n and $\overline{X}$ for the three groups. These means are, of course, identical with those given in the last line of Table 13.3.

Using Formulas (13.3), (13.4), and (13.5), we obtain J, K, and L. Then by using $(X + 1)$ instead of X in these same formulas, we obtain J', K', and L'. It is not necessary to check any of these calculations.

We first note that $k = 3$ and that $N = n_1 + n_2 + n_3 = 14 + 11 + 12 = 37$. However simple this may be, be sure that the n's, k, and N are written down correctly because if a mistake should be made in any of them, the checks would not reveal it. Now substitute in Formulas (13.6), (13.7), and (13.8) to obtain the three variances. Note that the other two numerators add to the numerator of the total variance, that this is also true of the denominators, and that the total variance is always intermediate in size between the other two. (If you are going to use the total variance for anything or if you are going to quote it, check it by using $J' - L'$ instead of $J - L$ in the numerator of Formula (13.8).)

We now compute F from Formula (13.9) by dividing the among by the within variance. We note that the *df*'s used in computing these two variances were 2 and 34, respectively, so *df* for use in the F Table is 2/34. The F Table has a column (2) for the among variance, but no row (34) for the within variance. If necessary, we would interpolate between the entries for row (30) and row (40), but since our F of 1.65 is far smaller than 3.32, 3.23, and any value interpolated between them, it is not necessary this time. Our F does not reach the .05 level, so we write "$P > .05$ Not Significant."

Formula (13.10) is used to check F. If the two values agree perfectly or differ by only .01, we have checked F and, as mentioned on page 418, most of the other values as well. Finally, we redetermine *df* and again look in the F Table to find out whether our F is significant.

We are now in a position to present our results in what we have seen is the standard fashion after an analysis of variance has been carried out. This is done in Table 13.5, the abbreviation NS standing for Not Significant.

In this case, F is not significant, even at the .05 level. What does this mean? It means that we have no evidence that these three teaching methods differ in their effectiveness, as measured by errors made on the retention test. If we ran a large number of such experiments in which three methods were

TABLE 13.4
Errors on Retention Test

	Correction Review	No Correction Review	Summary Only
	X_1	X_2	X_3
	29	17	9
	11	27	15
	14	35	34
	8	6	48
	27	3	18
	1	25	7
	30	3	8
	11	3	52
	14	20	21
	28	2	22
	19	11	48
	4		2
	3		
	31		
ΣX	230	152	284
ΣX^2	5,300	3,456	10,140
$\Sigma(X+1)$	244	163	296
$\Sigma(X+1)^2$	5,774	3,771	10,720 (√)
n	14	11	12
$\overline{X}$	16.43	13.82	23.67

$$J = \Sigma X_1^2 + \Sigma X_2^2 + \Sigma X_3^2$$
$$= 5{,}300 + 3{,}456 + 10{,}140 = 18{,}896$$

$$K = \frac{(\Sigma X_1)^2}{n_1} + \frac{(\Sigma X_2)^2}{n_2} + \frac{(\Sigma X_3)^2}{n_3}$$
$$= \frac{(230)^2}{14} + \frac{(152)^2}{11} + \frac{(284)^2}{12}$$
$$= 3{,}778.57 + 2{,}100.36 + 6{,}721.33$$
$$= 12{,}600.26$$

$$L = \frac{(\Sigma X_1 + \Sigma X_2 + \Sigma X_3)^2}{N}$$
$$= \frac{(230 + 152 + 284)^2}{37} = \frac{(666)^2}{37}$$
$$= 11{,}988.00$$

$$J' = \Sigma(X_1+1)^2 + \Sigma(X_2+1)^2 + \Sigma(X_3+1)^2$$
$$= 5{,}774 + 3{,}771 + 10{,}720$$
$$= 20{,}265$$

TABLE 13.4
(continued)

Correction Review	No Correction Review	Summary Only

$$K' = \frac{[\Sigma(X_1+1)]^2}{n_1} + \frac{[\Sigma(X_2+1)]^2}{n_2} + \frac{[\Sigma(X_3+1)]^2}{n_3}$$

$$= \frac{(244)^2}{14} + \frac{(163)^2}{11} + \frac{(296)^2}{12}$$

$$= 4{,}252.57 + 2{,}415.36 + 7{,}301.33$$

$$= 13{,}969.26$$

$$L' = \frac{[\Sigma(X_1+1) + \Sigma(X_2+1) + \Sigma(X_3+1)]^2}{N}$$

$$= \frac{(244 + 163 + 296)^2}{37}$$

$$= \frac{(703)^2}{37} = 13{,}357.00$$

$$\text{Within variance} = \frac{J - K}{N - k} = \frac{18{,}896 - 12{,}600.26}{37 - 3}$$

$$= \frac{6{,}295.74}{34} = 185.17$$

$$\text{Among variance} = \frac{K - L}{k - 1} = \frac{12{,}600.26 - 11{,}988.00}{3 - 1}$$

$$= \frac{612.26}{2} = 306.13$$

$$\text{Total variance} = \frac{J - L}{N - 1} = \frac{18{,}896 - 11{,}988.00}{37 - 1}$$

$$= \frac{6908.00}{36} = 191.89$$

$$F = \frac{\text{Among variance}}{\text{Within variance}} = \frac{306.13}{185.17} = 1.65$$

$df = 2/34$

$P > .05$ Not Significant

$$F = \frac{(K' - L')(N - k)}{(J' - K')(k - 1)}$$

$$= \frac{(13{,}969.26 - 13{,}357.00)(37 - 3)}{(20{,}265 - 13{,}969.26)(3 - 1)}$$

$$= \frac{(612.26)(34)}{(6{,}295.74)(2)} = 1.65 \qquad (\checkmark)$$

TABLE 13.5
Analysis of Variance Summary Table for Errors on Retention Test

Source of Variation	Sum of Squares	*df*	Estimate of Variance	*F*
Among groups	612.26	2	306.13	1.65 NS
Within groups	6,295.74	34	185.17	
Total	6,908.00	36		

P > .05

exactly equally good (or identical, if you prefer), the F Table has just shown us that 5% of the time we would get a difference greater than some value interpolated between 3.32 and 3.23. Consequently, more than 5% of the time we would expect to get an F larger than the one of only 1.65 which we obtained. Thus, the differences in the three means (in the last row of Table 13.3) may very well be due to the operation of chance alone. If you would have expected the correction review procedure of Group A to give the best results and the summary only procedure of Group C to be the worst, you are not alone. We were surprised at the results too. But regardless of any great expectations, there just isn't any evidence that there are any differences among these methods, as measured by errors on the retention test. This holds not merely for these three, but also for all five methods analyzed by Merrill in the article cited on page 421. We end this second analysis of variance problem by stating the now obvious conclusion: There are no significant differences among the three groups in the number of errors made on the retention test.

Summary of Formulas In order to facilitate the solution of simple problems in analysis of variance, Table 13.6 has been prepared. It gives every formula needed in order to compute and check such a problem; and it presents these formulas in the exact order in which they should be used.

SUMMARY

Sometimes subjects are randomly assigned to several groups and each group is treated differently in an experiment. Analysis of variance is then appropriate for finding out whether the obtained differences among the group means are significantly different from those to be expected on the basis of chance alone. The F test is used to determine whether or not such a null hypothesis is to be rejected.

Two examples were given to illustrate the computation, checking of results, and the drawing of the two types of conclusions to which such analyses may lead. Table 13.6 presents all the formulas in a form

TABLE 13.6
Summary of all Computational and Checking Formulas

The Charlier check
$\Sigma(X+1)^2 = \Sigma X + \Sigma X^2 + \Sigma(X+1)$

The functions J, K, and L

$$J = \Sigma X_1^2 + \Sigma X_2^2 + \cdots + \Sigma X_k^2$$
$$K = \frac{(\Sigma X_1)^2}{n_1} + \frac{(\Sigma X_2)^2}{n_2} + \cdots + \frac{(\Sigma X_k)^2}{n_k}$$
$$L = \frac{(\Sigma X_1 + \Sigma X_2 + \cdots + \Sigma X_k)^2}{N}$$

Within, among, and total sums, df's, and variances

	Within	Among	Total
Sum of squares = Numerator	$J - K$	$K - L$	$J - L$
Degrees of freedom = Denominator	$N - k$	$k - 1$	$N - 1$
Variance = Quotient	$\frac{J - K}{N - k}$	$\frac{K - L}{k - 1}$	$\frac{J - L}{N - 1}$

F

$$F = \frac{\text{Among variance}}{\text{Within variance}}$$

df for $F = (k - 1)/(N - k)$

See F Table for P and significance

The functions J', K', and L'

$$J' = \Sigma(X_1 + 1)^2 + \Sigma(X_2 + 1)^2 + \cdots + \Sigma(X_k + 1)^2$$
$$K' = \frac{[\Sigma(X_1 + 1)]^2}{n_1} + \frac{[\Sigma(X_2 + 1)]^2}{n_2} + \cdots + \frac{[\Sigma(X_k + 1)]^2}{n_k}$$
$$L' = \frac{[\Sigma(X_1 + 1) + \Sigma(X_2 + 1) + \cdots + \Sigma(X_k + 1)]^2}{N}$$

Total variance (if wanted)

$$\text{Total variance} = \frac{J' - L'}{N - 1}$$

F

$$F = \frac{(K' - L')(N - k)}{(J' - K')(k - 1)}$$

df for $F = (k - 1)/(N - k)$

See F Table again

designed to make it easy to know what to do, how to do it, and when to do it.

SUPPLEMENT 13

Three topics will be considered in this supplement. They are

Naming of F and the F Table
Proof that size of total variance is always between sizes of within and among variances
Relation of F to z and to χ^2

Naming of F and the F Table Perhaps the greatest statistician who every lived was Sir Ronald A. Fisher (1890–1962). Among his many accomplishments, he discovered the distribution of a certain ratio of two variances. Another statistician, George W. Snedecor, named this ratio F in Fisher's honor. The F Table, of course, is similarly named in honor of this famous British statistician.

Proof that size of total variance is always between the sizes of within and among variances On page 413, we saw that Within + Among = Total was true not only for the sums of squares in the numerators, but also for the degrees of freedom in the denominators. If the within and the among variance are designated by $\frac{a}{c}$ and $\frac{b}{d}$, though not necessarily in that order, our problem is to prove that the numerical value of the total variance, $\frac{a+b}{c+d}$, is between that of the other two variances, $\frac{a}{c}$ and $\frac{b}{d}$ (unless all three variances are equal). Without loss of generality, we shall use $\frac{a}{c}$ to represent the smaller and $\frac{b}{d}$ the larger of these two variances.

Since a, b, c, and d are all positive, and we assumed above that $\frac{a}{c} < \frac{b}{d}$, we may multiply both sides of the inequality by cd without altering its correctness. This gives

$$ad < bc$$

which, when we add ac to both sides, becomes

$$(ac + ad) < (ac + bc).$$

Dividing both sides by the inherently positive quantity, $c(c + d)$, gives

$$\frac{a(c + d)}{c(c + d)} < \frac{c(a + b)}{c(c + d)}$$

which results in

$$\frac{a}{c} < \frac{a + b}{c + d}.$$

By similar reasoning, we again start with

$$\frac{a}{c} < \frac{b}{d}$$

and

$$ad < bc.$$

Adding bd to both sides, we get

$$(ad + bd) < (bc + bd).$$

Dividing now by $d(c + d)$ gives

$$\frac{d(a + b)}{d(c + d)} < \frac{b(c + d)}{d(c + d)}$$

from which it follows that

$$\frac{a + b}{c + d} < \frac{b}{d}.$$

Putting our two results together, we get

$$\frac{a}{c} < \frac{a + b}{c + d} < \frac{b}{d},$$

which proves that the total variance, $\frac{a + b}{c + d}$, is greater than one variance, $\frac{a}{c}$, and smaller than the other, $\frac{b}{d}$. This proof was given by the mathematician Robert G. Blake.

Relation of F to z and to χ^2 On page 420 we saw that if we consider only the column of the F Table for which *df* for among variance = 1, the values given there were the squares of those in the t Table so that, for the first column of the F Table,

$$F = t^2. \tag{13.13}$$

There is another interesting relation pertaining to the last row of the F Table (when df for within variance = ∞). Several significance levels for the χ^2 distribution were given in Table V. If we divide any tabled value of χ^2 by its df, we get exactly the same numerical value (except for a few minor rounding errors) as that given in the ∞ row of Table F. Let us cite three examples, one for each of the significance levels given in the ∞ row of the F Table. First, from Table V for χ^2, we obtain

$$\text{When } df = 4,\ \chi^2 \text{ at .05 level} = 9.488.\ \frac{\chi^2}{df} = \frac{9.488}{4.} = 2.372.$$

$$\text{When } df = 5,\ \chi^2 \text{ at .01 level} = 15.086.\ \frac{\chi^2}{df} = \frac{15.086}{5.} = 3.017.$$

$$\text{When } df = 6,\ \chi^2 \text{ at .001 level} = 22.457.\ \frac{\chi^2}{df} = \frac{22.457}{6.} = 3.743.$$

But if we now look at the last or ∞ row of Table F, we get

When $df = 4/\infty$, F at .05 level = 2.37.
When $df = 5/\infty$, F at .01 level = 3.02.
When $df = 6/\infty$, F at .001 level = 3.74.

Since these values of F agree with those just previously obtained for $\frac{\chi^2}{df}$, this illustrates the fact that $\frac{\chi^2}{df}$ is distributed in the same manner as F with varying df for the numerator but with a constant df of ∞ for its denominator. More briefly, for the ∞ row of the F Table,

$$F = \frac{\chi^2}{df}. \tag{13.14}$$

We have now considered column (1) and row (∞) of the F Table. What happens at their intersection where df is $1/\infty$? Does the F Table agree with t^2 or with $\frac{\chi^2}{df}$? This is a very good question, but a tricky one. It might seem that the answer is both. In a sense, it is, but, more precisely, that answer is true but incomplete. Certainly Equations (13.13) and (13.14) both hold here, but this particular spot in the F Table is also related to the normal probability curve. This is to be expected. On page 123 we noted that as df increases, each column of t values approaches and eventually equals the z score or σ deviate corresponding to the same probability in the normal probability curve as the significance level. Hence, it is not at all surprising that when *df* for F is $1/\infty$, we find the distribution of F identical to that of the squared standard score, or

$$F = z^2. \tag{13.15}$$

For instance, it is well known that only 5% of the area under a normal probability curve is beyond the limits of $z = \pm 1.96$. Squaring 1.96, we get 3.8416, which agrees with the top value (3.84) in the $1/\infty$ cell of the F Table.

Anyone with work experience in a bureaucracy has not only heard of, but actually been a victim of, Murphy's Law. Below are the results of a cross-cultural, cross-validated search for cross-references to Murphy's Law.

Murphy's Law	If anything can go wrong, it will.
Cahn's Axiom	When all else fails, read the instructions.
Gumperson's Law	The probability of a given event occurring is inversely proportional to its desirability.
The Transcription Square Law	The number of errors made is equal to the sum of the "squares" employed.
The Theory of the International Society of Philosophic Engineering	In any calculation, any error which can creep in, will.
Rule of Accuracy	When working toward the solution of a problem, it always helps if you know the answer.

From *The Industrial-Organizational Psychologist,* Feb., 1978, **15**, No. 2, 13.

CHAPTER 14 STANDARD ERRORS OF DIFFERENCES

Because of our feeling that standard errors should be closely associated with the statistics to which they pertain, this book was so arranged as to make it practicable to place the chapter on *Statistical Inference* near the beginning of the text—after only a few statistics ($\overline{X}$, *Mdn, s,* and *AD*) had been presented. Subsequently, as new statistical terms were introduced they were usually accompanied by their standard errors. From these standard errors, it was a very simple matter to determine confidence limits whenever they were needed. And from these limits we could find out whether or not our sample statistic differed significantly from any hypothesized population parameter. But such parameters were always fixed points, not figures subject to variation, as the sample statistics were.

Sometimes we may wish to know whether one sample statistic differs significantly from another sample statistic. For example, do boys or girls do better on fourth-grade objective arithmetic tests? Is there a mean difference in neurotic tendency scores of college men and women? Which of two learning procedures is more effective? Will viewing an instructional film alter the mean attitude toward a nationality group? Will such a film make the viewers more or less homogeneous in their attitude scores? Are men more or less variable than women in intelligence? Can people perform better on a simulated driving test before or after drinking three cups of coffee? Such questions involve the comparison of two means or of two standard deviations, both of which are subject to error. This chapter will show how questions such as these can be answered.

STANDARD ERROR OF ANY DIFFERENCE

In all earlier chapters when we dealt with standard errors, we were dealing with the sampling distribution of statistics about the population parameter or with the significance of differences from a fixed point. The very important problem just discussed is whether two sample statistics differ, not from a fixed point, but from each other. Let us begin by deriving a formula for the standard error of the difference between any two statistics, A and B. By way of introduction, we note that

$$\begin{aligned} A - B &= (A - \overline{X}_A) - (B - \overline{X}_B) + (\overline{X}_A - \overline{X}_B) \\ &= a - b + (\overline{X}_A - \overline{X}_B), \end{aligned}$$

where a and b are deviations from the respective means of A and B. In Chapter 3 we found that adding or subtracting a constant from all the scores had no effect on the size of the standard deviation. Hence, $s_{(A-B)}$ will equal $s_{(a-b)}$, whether or not we add or subtract the constant, $\overline{X}_A - \overline{X}_B$, from all the scores. Therefore, since $s_X = \sqrt{\dfrac{\Sigma x^2}{N-1}}$ by Formula (3.1), it follows that

$$\begin{aligned} s_{(A-B)} = s_{(a-b)} &= \sqrt{\frac{\Sigma(a-b)^2}{N-1}} \\ &= \sqrt{\frac{\Sigma(a^2 - 2ab + b^2)}{N-1}} \\ &= \sqrt{\frac{\Sigma a^2}{N-1} - 2\left(\frac{\Sigma ab}{N-1}\right) + \frac{\Sigma b^2}{N-1}}. \end{aligned}$$

Now, from Formula (3.1), the meaning of the first and last terms is obvious. Formula (8.1) tells us that $r_{XY} = \dfrac{\Sigma z_x z_y}{N-1}$, from which we readily obtain

$$r_{XY} = \frac{\Sigma\left(\dfrac{x}{s_X}\right)\left(\dfrac{y}{s_Y}\right)}{N-1} = \frac{\Sigma xy}{(N-1)\, s_X s_Y}.$$

Hence it follows that $r_{XY} s_X s_Y = \dfrac{\Sigma xy}{N-1}$. Making the indicated substitutions and using A and B in place of X and Y, our formula becomes

$$s_{(A-B)} = \sqrt{s_A^2 - 2r_{AB} s_A s_B + s_B^2}. \tag{14.1}$$

This formula is perfectly general. We defined A and B as "any two statistics." They may be raw scores, they may be means, they may be correlation coefficients, . . . , in short, they may be any statistics. They may even differ so that A is a mean and B is the corresponding standard deviation, though very seldom do such comparisons make any sense. We have implied that they varied, but even this is not necessary. Suppose $A = \overline{X}_3$ and $B = 100$. Since the standard deviation of a constant is 0, because constants do not vary, s_B must be 0, so the formula reduces to

$$s_{(A-B)} = s_{(\overline{X}_3-100)} = \sqrt{s^2_{\overline{X}_3} - 0 + 0}$$
$$= s_{\overline{X}_3},$$

which is as it should be. (If both A and B are constants, $s_{(A-B)} = 0$, which is sensible, but not very enlightening, since most of us already know, for instance, that $\pi - 3$, to four decimals, is .1416 with a standard error of zero.)

STANDARD ERROR OF THE DIFFERENCE BETWEEN MEANS

The most important special case of Formula (14.1) is that obtained by substituting $\overline{X}$ and $\overline{Y}$ for A and B. Most of the substitution is straightforward, but the middle term would involve the correlation between means, and we do not yet know what this is. If we were to draw random samples of 100 couples from the data shown in Figure 9.2 until we had a large number (maybe 269 because N was 26,907) of such groups, then compute the mean age of bride and the mean age of groom for each sample and correlate these means, what would we get? These means would be far less variable than the original ages (actually about one-tenth as variable), but they would correlate with each other just as highly, on the average, as the original ages did. That is

$$r_{\overline{X}\overline{Y}} = r_{XY}. \tag{14.2}$$

Taking account of this relationship, the formula for the standard error of the difference between two means becomes

$$s_{(\overline{X}-\overline{Y})} = \sqrt{s^2_{\overline{X}} - 2r_{XY}s_{\overline{X}}s_{\overline{Y}} + s^2_{\overline{Y}}}. \tag{14.3}$$

In Chapter 9, we found that the Florida grooms averaged 3.95 years older than their brides. How confident can we be that there really is a difference here? Before we can answer this, we shall need the standard errors of the two means. Substituting the data from Table 9.3 in Formula (5.2), we obtain

Check:

$$s_{\overline{X}} = \frac{s_X}{\sqrt{N}} \qquad s^2_{\bar{x}} = \frac{s^2_X}{N}$$

$$= \frac{13.45}{\sqrt{26{,}907}} \qquad = \frac{(13.45)^2}{26{,}907}$$

$$= \frac{13.45}{164.03} \qquad = \frac{180.9025}{26{,}907}$$

$$= .0820 \qquad = .006723$$

$$s_{\overline{X}} = .0820$$

Check:

$$s_{\overline{Y}} = \frac{s_Y}{\sqrt{N}} \qquad s_{\overline{Y}}^2 = \frac{s_Y^2}{N}$$

$$= \frac{11.85}{\sqrt{26{,}907}} \qquad = \frac{(11.85)^2}{26{,}907}$$

$$= \frac{11.85}{164.03} \qquad = \frac{140.4225}{26{,}907}$$

$$= .0722 \qquad = .005219$$

$$s_{\overline{Y}} = .0722$$

Notice that our checking system not only gives us two separate determinations of the standard error of each mean, but also gives us the squared standard deviations of each, which are also needed in the formula. In this case, both standard errors checked exactly. Sometimes there will be a discrepancy of 1 in the last retained digit; any greater difference indicates that an error has been made. Notice also that it was necessary to keep more than our usual two decimals in computing these standard errors. It is also desirable to keep more than two decimals for r_{XY}.

Let us now substitute in Formula (14.3):

$$\begin{aligned} s_{(\overline{X}-\overline{Y})} &= \sqrt{s_{\overline{X}}^2 - 2r_{XY}s_{\overline{X}}s_{\overline{Y}} + s_{\overline{Y}}^2} \\ &= \sqrt{.006723 - (2)(.8756)(.0820)(.0722) + .005219} \\ &= \sqrt{.006723 - .010368 + .005219} \\ &= \sqrt{.001574} \\ &= .0397. \end{aligned}$$

The difference divided by its standard error will give an extremely large value of t:

$$t = \frac{3.95}{.0397} = 99.50.$$

Since with an N of this size, t needs to be only 3.29 to be significant at the .001 level, there is no doubt about the significance of this difference.

Despite the fact that there are several thousand couples in the cells of Figure 9.2 in which the brides are older than the grooms, there are far more instances of the opposite situation. In such marriages as these, the grooms most definitely average older than their brides.

The mean difference was 32.15 − 28.20 = 3.95 years. Let us set up confidence limits for this difference:

$$t = \frac{D - D_\infty}{s_D}, \tag{14.4}$$

where D refers to our obtained difference, s_D to its standard error, and D_∞ to the true difference. N is so large that we can well afford to operate at the .001 level of significance, where t for our N will be 3.29. Substituting, we obtain

$$\pm 3.29 = \frac{+3.95 - D_\infty}{.0397},$$

$$\pm .1306 = 3.95 - D_\infty,$$

$$D_\infty = 3.95 \pm .13,$$

$$D_\infty = 3.82 \text{ or } 4.08.$$

Finally, we conclude at the .001 level of significance that, in marriages such as these, the average difference in age is between 3.82 and 4.08 years in favor of the groom. In more common terminology, we may be very sure that the true difference in ages in the universe from which our group is regarded as a random sample is between 3 years, 10 months and 4 years, 1 month. These are very precise limits, especially considering the fact that the ages were grouped in 5-year age groups. They are quite accurate, however, because of both our large N (26,907) and our high r (+.8756). Even so, if the preceding statements seem unduly precise, just say the difference is 4 years—you'll not miss the truth by more than a month or two.

A Second Example of the Standard Error of the Difference between Means We use data from a long established (1921) American tradition—the Miss America Pageant. The criteria for winning this contest certainly have changed. Today's contestants are not only extremely attractive and talented but also bright. Many of the participants are college students and each of them is awarded a cash scholarship. We might note that college textbooks have changed also. A very short time ago you would not have expected to find this example cited in a text.

In order to get an N of sufficient size, we are using records for two consecutive years of the contest. The correlation between hip (X) and bust

(Y) measurements is shown in Figure 14.1. Straightforward calculation yields the following values:

$$\begin{array}{lll} \overline{X} = 35.43 & \overline{Y} = 35.32 & r_{XY} = +.79 \\ s_X = .92 & s_Y = 1.00 & \\ s_{\overline{X}}^2 = .0084 & s_{\overline{Y}}^2 = .0099 & \\ s_{\overline{X}} = .092 & s_{\overline{Y}} = .099 & \end{array}$$

The difference between the two means is very small, only .11 inch or 2.8 mm. The correlation between the two measurements is very high, however, so even small differences *may* be significant. Is that true in this case? Let us find out if this small difference between these means is a dependable one. Is the difference statistically significant?

From Formula (14.3), we readily obtain the standard error of the difference between the two means.

$$\begin{aligned} s_{(\overline{X}-\overline{Y})} &= \sqrt{s_{\overline{X}}^2 - 2r_{XY}s_{\overline{X}}s_{\overline{Y}} + s_{\overline{Y}}^2} \\ &= \sqrt{.0084 - (2)(.79)(.092)(.099) + .0099} \\ &= \sqrt{.0084 - .0144 + .0099} \\ &= \sqrt{.0039} \\ &= .062. \end{aligned}$$

This is a very small standard error, only about one sixteenth of an inch, but nearly two-thirds of the paired measurements were identical, so the difference between the two means is necessarily small. Is it significant? The difference between the two means divided by its standard error is

$$t = \frac{35.43 - 35.32}{.062} = \frac{+.11}{.062} = 1.77.$$

Are we interested in making a one-tailed or a two-tailed test of the significance of the difference between these two measurements? Remember that the difference might have turned out to be in either direction. Therefore, it is appropriate (as it nearly always is) to include the areas in both tails of the t distribution (or of the normal distribution) in determining the significance of this (positive) X minus Y difference. This possible problem is taken care of automatically by the way Table G of the t distribution is set up, but if we had a very large N and had used Table E of the normal probability curve to determine the exact area more than 1.77σ from the mean, it would have been necessary to double the obtained .0384 and realize that the desired area is .0768.

Going back to Table G, we readily see that for our N of 101, it would be necessary for our t to be 1.98 or higher to be significant at the .05 level.

FIGURE 14.1
Measurements of 101 contestants

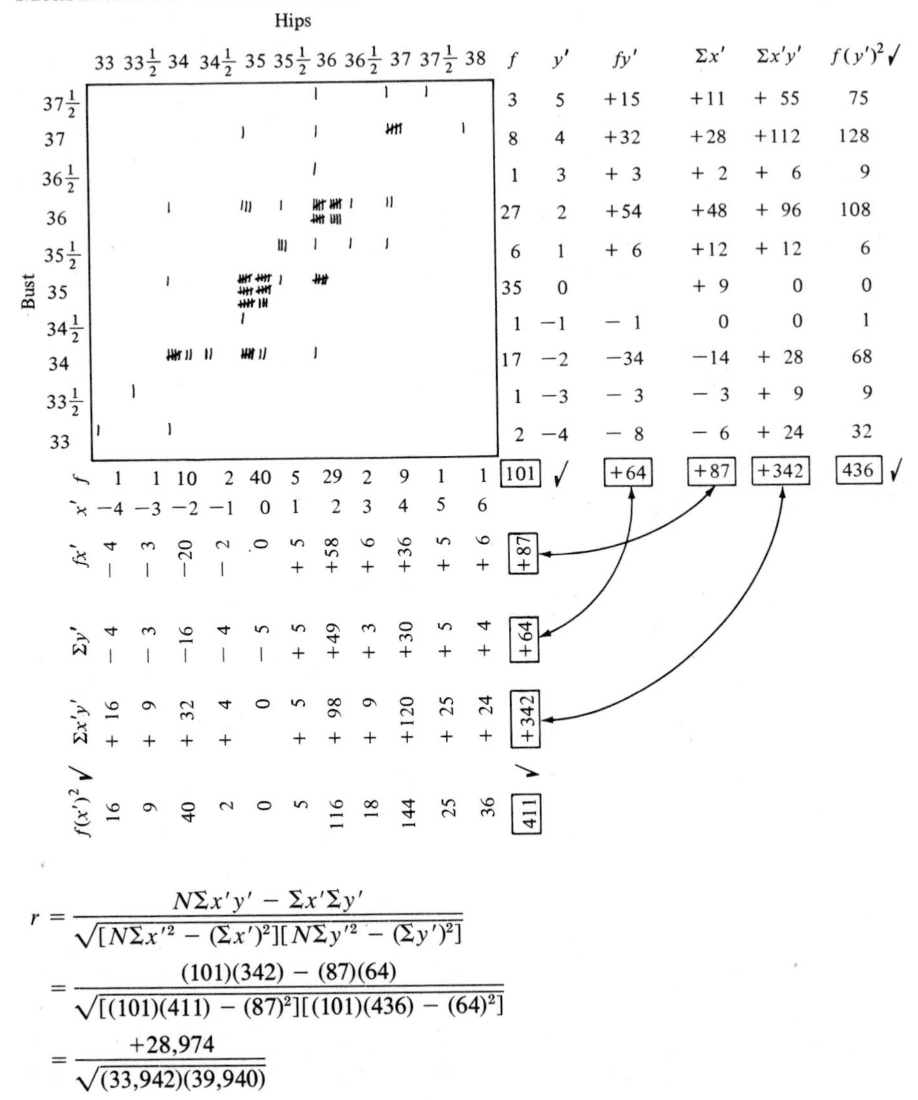

$$r = \frac{N\Sigma x'y' - \Sigma x'\Sigma y'}{\sqrt{[N\Sigma x'^2 - (\Sigma x')^2][N\Sigma y'^2 - (\Sigma y')^2]}}$$

$$= \frac{(101)(342) - (87)(64)}{\sqrt{[(101)(411) - (87)^2][(101)(436) - (64)^2]}}$$

$$= \frac{+28{,}974}{\sqrt{(33{,}942)(39{,}940)}}$$

$$= +.79$$

Since it is only 1.77, we must conclude that we have not demonstrated that there is a statistically dependable difference in these two body measurements in the universe of which these 101 contestants may be regarded as a sample. Notice two things.

First, we have not proved that there is no difference; no one can ever prove that any difference is zero. (Besides, we found a difference of +.11 and if desired, we could determine confidence limits for it. At the .01 level, these limits turn out to be −.05 and +.27, or about −1/16 and +1/4 inch.)

Second, any conclusion we draw does not apply to women in general nor even to all those of about this age (17 to 25). It must be limited to the population composed of individuals essentially similar to these participants.

Our conclusion, then, is that we have failed to demonstrate a statistically dependable difference, but that it is between −.05 and +.27 inches. These are the .99 confidence limits.

An Alternative Method for Computing the Standard Error of the Difference between Means Where paired measurements are made on the same individuals, it can be shown that the difference between means is the same as the mean of the differences. Further, since these are identical, it must also follow that the standard error of the difference between means is also equal to the standard error of the mean of the differences. In the problem under discussion the amount of work is very greatly reduced by the use of this alternative method based on the mean of the differences. In using this method, it is first necessary to obtain a difference score for every subject and to compute the mean and standard deviation of these differences.

In our present problem, we subtract and obtain a hips minus bust score for each of the 101 contestants, paying careful attention to signs. We then plot the data and obtain the frequency distribution shown in the first two columns of Table 14.2.

The mean difference, the standard deviation, and the standard error of the mean difference are computed by the usual formulas (2.2), (3.3), and (5.2). Each difference is simply treated as though it were a raw score, X, in the appropriate formula. Finally, we divide the mean difference, +.11, by its standard error, .063, and obtain the quotient, +1.75. This value of t differs by only .02 (due to rounding errors) from that obtained previously by the use of Formula (14.3).

In some circumstances, rounding errors might lead to slightly different results in comparing Formula (14.3) with the direct computation of differences. Ordinarily such differences are entirely negligible and they may, of course, be reduced simply by retaining more decimal places than might otherwise appear to be needed.

The interpretation and conclusions to be drawn from the obtained t are, of course, exactly the same, no matter which way it happens to be computed. The computation should be done in the easiest manner; this will ordinarily be by the direct computation of differences, as just illustrated in Table 14.2.

TABLE 14.2
Standard Error of Hips Minus Bust Difference

Difference	f	x'	fx'	$f(x')^2$
+2	1	+4	+4	16
+1½	1	+3	+3	9
+1	17	+2	+34	68
+ ½	6	+1	+6	6
0	65	0	0	0
− ½	3	−1	−3	3
−1	5	−2	−10	20
−1½	1	−3	−3	9
−2	2	−4	−8	32
	101		+23	163

$$\overline{X} = i\frac{\Sigma fx'}{N} + M' = ½\left(\frac{+23}{101}\right) + 0 = +.11$$

$$s = i\sqrt{\frac{N\Sigma f(x')^2 - (\Sigma fx')^2}{N(N-1)}} = ½\sqrt{\frac{(101)(163) - (+23)^2}{(101)(100)}} = .63$$

$$s_{\overline{X}} = \frac{s}{\sqrt{N}} = \frac{.63}{\sqrt{101}} = .063$$

$$t = \frac{\overline{X}}{s_{\overline{X}}} = \frac{+.11}{.063} = +1.75$$

STANDARD ERROR OF THE DIFFERENCE BETWEEN STANDARD DEVIATIONS

Just as we may want to know whether an obtained difference between two means is significant, so may we wish to know whether a difference between two standard deviations is statistically significant. The procedure is somewhat time-consuming but perfectly straightforward. We find and substitute appropriate values in Formula (14.1), solve it, determine whether the obtained difference is significant, and, perhaps, determine confidence limits for the difference.

In referring to the data on 26,907 marriages, we stated that the variability of the grooms' ages is also a little larger than that of their brides. How sure are we that it is larger? And how much larger is it? Alternatively, what do we mean by "a little larger"? To answer such questions, we first need a formula for the standard error of the difference between standard deviations. This will obviously be a minor modification of the general Formula (14.1) in which A becomes s_X and B becomes s_Y. The only term that offers any difficulty is the correlation coefficient. Mathematical statisticians have found that this correlation is usually lower than that between the original raw scores. Specifically,

$$r_{s_X s_Y} = r^2_{XY}. \tag{14.5}$$

Performing the indicated substitutions in Formula (14.1), we readily obtain

$$s_{(s_X - s_Y)} = \sqrt{s^2_{s_X} - 2r^2_{XY} s_{s_X} s_{s_Y} + s^2_{s_Y}}. \tag{14.6}$$

If we had not already computed and checked the standard error and the variance of each of these means, we should compute and check the standard errors of the standard deviations in this manner:

$$\begin{aligned} s_{s_X} &= \frac{s_X}{\sqrt{2N}} \\ &= \frac{13.45}{\sqrt{(2)(26{,}907)}} \\ &= \frac{13.45}{231.98} \\ &= .0580 \end{aligned}$$

Check:

$$\begin{aligned} s^2_{s_X} &= \frac{s^2_X}{2N} \\ &= \frac{(13.45)^2}{(2)(26{,}907)} \\ &= \frac{180.9025}{53{,}814} \\ &= .003362 \\ s_{s_X} &= .0580 \end{aligned}$$

We know, however, that the formula for s_s is $\frac{1}{\sqrt{2}}$ times that of $s_{\bar{X}}$ and that the formula for s^2_s is ½ that of $s^2_{\bar{X}}$. Hence, we can simply multiply the previously obtained values of $s_{\bar{X}}$ and $s_{\bar{Y}}$ by .7071 to obtain s_{s_X} and s_{s_Y}. Similarly, we can divide the variances of the means by two to obtain the variances of the standard deviations. Thus,

$$\begin{aligned} s_{s_X} &= .7071 s_{\bar{X}} & s^1_{s_X} &= (.5)(s^2_{\bar{X}}) \\ &= (.7071)(.0820) & &= (.5)(.006723) \\ &= .0580 & &= .003362 \\ s_{s_Y} &= (.7071)(.0722) & s^2_{s_Y} &= (.5)(.005219) \\ &= .0511 & &= .002610 \end{aligned}$$

To check all four of these values, we simply check to see that each of the values of s is the correct square root of the corresponding value of s^2. As was the case earlier, a difference of 1 in the last retained digit is permissible, any larger discrepancy indicating an error. In this case, the figures check perfectly.

We now find $s_{(s_X - s_Y)}$ by substituting the four figures just obtained and the correlation, +.8756, in Formula (14.6):

$$s_{(s_X - s_Y)} = \sqrt{s_{s_X}^2 - 2r_{XY}^2 s_{s_X} s_{s_Y} + s_{s_Y}^2}$$
$$= \sqrt{.003362 - 2(.8756)^2(.0580)(.0511) + .002610}$$
$$= \sqrt{.003362 - .004545 + .002610}$$
$$= \sqrt{.001427}$$
$$= .0378.$$

The difference, 13.45 – 11.85, divided by its standard error gives

$$t = \frac{1.60}{.0378} = 42.33.$$

This difference is significant far beyond the .001 level. Thus, there is no doubt that the grooms' ages are more variable than the brides'. How much more variable? Let us set up confidence limits for this difference. The formula is the same one used earlier in testing the difference between the mean ages, the only change being that we must keep in mind that D now refers to the difference between the two standard deviations. Using Formula (14.4), and the .999 confidence limits, we obtain

$$t = \frac{D - D_\infty}{s_D},$$
$$\pm 3.29 = \frac{1.60 - D_\infty}{.0378},$$
$$\pm .1244 = 1.60 - D_\infty,$$
$$D_\infty = 1.60 \pm .12,$$
$$D_\infty = 1.48 \text{ or } 1.72.$$

As was indicated in the preceding paragraph, the standard deviation of these grooms' ages was 13.45 years, which is 1.60 years greater than the brides' standard deviation of 11.85 years. We have now established that this difference is a highly dependable one. In fact, at the widest confidence limits (.999) ordinarily used by statisticians, we may state that the standard deviation of the ages of such grooms as these may be depended upon to be between 1.48 and 1.72 years (or, very closely, 18 to 21 months) greater than that of their brides. The high precision of this result is due both to the large number of cases and the high correlation between the two variables.

What Happens When the Correlation Is Unknown? In the examples used to illustrate the standard error of a difference, the correlation coefficient had already been computed. If the correlation is not known, one way of handling the situation, relevant only for the difference between means, is to use the alternative method explained on page 439. With this single exception, if the correlation is not known, compute it. If you don't have time to compute it, take time or abandon the project. If there is no possible way to compute the correlation (perhaps because records were obtained on two unrelated groups with no possible reason for pairing off any one person with any particular individual in the other group), then you may assume that $r = .00$ and the middle term will vanish from whatever formula you are using for the standard error of the difference.

The preceding paragraph may puzzle the reader. If might seem that all that need be said is that Formula (14.1) calls for a correlation coefficient. Such a procedure has not worked in the past. For instance, some statistics texts have given our Formula (14.3), but have preceded it by a formula with the middle term omitted. Result: people learn first about a relatively easy formula, and then incorrectly use it. Note that there is nothing intrinsically wrong with such a formula *when correctly used*. Unfortunately, the incorrect usages and confusions are so great that we will not even print the formula. Whenever it is needed, the student can obtain it by simply omitting the middle term from Formula (14.3). Let us now consider some obvious and then some perhaps not obvious situations.

The correlation should obviously be included if we are interested in determining the standard error of the difference between:

1. mean standard scores of fifth graders on an arithmetic and a spelling test;
2. mean Wechsler Intelligence Scale for Children and mean Stanford–Binet IQs for the same fourth grade pupils;
3. mean strength of grip in right hand and same in left hand for college juniors;
4. any of the above with "standard deviation of" substituted for "mean";
5. any two statistics derived from two sets of measurements of the same individuals.

The correlational term should obviously be omitted if we are interested in determining the standard error of the difference between:

1. mean trials needed by college students and by rats to learn a maze;
2. mean IQ of the girls and of the boys in a school;
3. mean annual pay of steelworkers and of university faculty members in Pittsburgh;
4. any of the above with "standard deviation of" substituted for "mean";
5. any two statistics derived from measurements of two entirely different and unrelated individuals (such that there is no way of pairing them off).

The distinction may be subtle, but the correlation must be computed and included in determining the standard error of the difference between:

1. mean IQ of father and of daughter;
2. mean Morse code receiving speed of experimental and control groups, when students have been matched on Radio Code Aptitude Test scores;
3. mean IQ of third grade pupils and of fifth grade pupils when the latter are the same pupils two years later;
4. mean errors of two groups in learning a maze, when the second group of rats consists of littermates of the first group;
5. any of the above with "standard deviation of" substituted for "mean";
6. any two statistics derived from measurements such that those in the two groups can be placed in correspondence with each other.

This next paragraph was to have started, "The distinction may be subtle, but the correlational term should be omitted," but we are unable to think of a single situation in which you might expect to use r in which it would be inappropriate to do so.

To end this section, let us simply state that the correlational term in such formulas as (14.1), (14.3), and (14.6) should be included if there is any conceivable way of pairing the scores and computing the correlation coefficient.

THE STANDARD ERROR OF OTHER DIFFERENCES

It was pointed out that Formula (14.1) was perfectly general, but we have used it only in determining the significance of differences between two means or between two standard deviations. Table 14.3 in Supplement 14 lists these two and eleven other pairs of statistics together with the formula (or a citation to it) for the correlation between the two members of the pair. With this information, the obvious substitutions can be made in Formula (14.1) and the standard error of the difference may be obtained. We have carried through this process for two such differences. Probably the two others of greatest potential usefulness are the difference between s_1^2 and s_2^2 and that between r_{12} and r_{13}. A brief discussion of the importance of this latter comparison is given in Supplement 14.

SUMMARY

The standard error of the difference between any two statistics, A and B, may be computed from Formula (14.1) or some simplification of it as soon as we know s_A, s_B, and r_{AB}. It is sometimes difficult to determine r_{AB}. Formulas for it are given (in this chapter or in Supplement 14) for comparisons among a number of common statistics. For the specific case in which A and B are means, an alternative method is given which sometimes greatly reduces the amount of computing time.

When a necessary correlation coefficient is unknown, it should be computed. When it is known to be zero, the formula for the standard error of the difference will simplify considerably.

SUPPLEMENT 14

Two books by the same author present a large number of correlations between pairs of statistics. Formula (14.2) gave the correlation between $\bar{X}$ and $\bar{Y}$ and Formula (14.5) gave that between s_X and s_Y. These two head the list in Table 14.3. The others follow, together with a page reference to them in Truman L. Kelley's* *Statistical Method* or his *Fundamentals of Statistics*. Most of the formulas assume normality; several of them make additional assumptions which the prospective user will ordinarily not be in a position to know about. For exact details, consult the two references.

In Chapter 14, we have used the first two of these formulas. Many of the others will be found useful only very rarely, but one of the formulas, the last one, should probably be used far more often than it is. Thus, if one test, X_2, has been used to predict a criterion, X_1, and another test, X_3, is proposed as a substitute, A becomes r_{12} and B becomes r_{13} in Formula (14.1), while r_{AB}, the correlation between r_{12} and r_{13}, is given by the formula just referred to. The experiment, of course, necessitates that all subjects take both tests and that then (or later) they are all assigned criterion scores. After the three correlations among these variables are computed, the substitution is straightforward and the significance of the difference can be determined on a calculator.

We find it a bit difficult to conjure up a situation in which anyone would care if any mean was significantly greater or smaller than any correlation coefficient. Hence, the formula, which is a bit complicated, is not printed in the table. In the four instances in which formulas are not given, citations are given so that if the reader ever needs such a formula, he can readily find it.

*Kelley, Truman L. *Statistical method.* New York: Macmillan, 1924 (out of print) and Kelley, Truman L. *Fundamentals of statistics.* Cambridge: Harvard University Press, 1947.

TABLE 14.3 Correlations between Statistics

The Two Statistics	Correlation Coefficient	Citation	Remarks
$\overline{X}$ $\overline{Y}$	r_{XY}	SM 178, FS 548	= Formula (14.2)
s_1 s_2	r_{12}^2	SM 178, FS 552	= Formula (14.5)
s_1^2 s_2^2	r_{12}^2	FS 550	
$\overline{X}$ s_X	$\sqrt{\dfrac{\beta_1}{\beta_2 - 1}}$	SM 178, FS 552	
$\overline{X}$ s_X^2	$\sqrt{\dfrac{\beta_1}{\beta_2 - 1}}$	SM 178, FS 552	
P_p P_p'	$\sqrt{\dfrac{pq'}{qp'}}$ where $p < p'$	SM 76	
f f'	$-\sqrt{\dfrac{pp'}{qq'}}$	FS 547	These are the correlations between the frequencies (or proportions) in two nonoverlapping rows in a frequency distribution or cells in a contingency table
p p'	$-\sqrt{\dfrac{pp'}{qq'}}$	FS 547	(same as above)
$\overline{X}$ r_{12}	Complicated	SM 178	If normality assumed, correlation is zero
s_1 r_{12}	Complicated	SM 178, FS 552	If normality assumed, correlation is $\dfrac{r}{\sqrt{2}}$
s_1 r_{23}	Complicated	SM 179, FS 553	
r_{12} r_{34}	Complicated	SM 179	
r_{12} r_{13}	$r_{23} - \dfrac{r_{12}r_{13}(1 - r_{12}^2 - r_{13}^2 - r_{23}^2 + 2r_{12}r_{13}r_{23})}{2(1 - r_{12}^2)(1 - r_{13}^2)}$	SM 179	

CHAPTER 15 REORIENTATION

There are many ways to summarize a book. One excellent way is just to go through the book, a chapter at a time, and write a summary of each. This will not be done in Chapter 15 because we have already done it at the end of nearly every chapter. Those summaries should now prove useful to you in finding out where you should place the most emphasis in studying for the final examination.

Another way to condense a book, and it is even briefer, is to write a very short digest of what is in each chapter. We won't do that either at this time because we have already done it—in Chapter 1. That preview was, we hope, helpful then, but now that you've been through the course, why not take another look at it? It should be much clearer and far more meaningful now.

MOMENTS

We shall now recapitulate some of the material by disregarding chapter boundaries and noting a few interrelationships. As we saw in Supplement 7, a moment is the mean of a set of scores, each of which has been raised to some power. If instead of using deviations (as was done in Supplement 7), we use raw scores, the first moment is $\frac{\Sigma X^1}{N}$, which equals $\frac{\Sigma X}{N}$ and, of course, is the mean. Similarly, the second moment is $\frac{\Sigma X^2}{N}$ which is rarely used for anything today, though it once had a very slight use as still another measure of central tendency. We could go on to $\frac{\Sigma X^3}{N}$ and other higher moments, but so far as the writers know, they have never had any particular use.

This is not true, however, of moments about the mean. The following tabulation lists the only moments, based on either raw scores or deviations, that we need consider:

Zeroth moment	$\frac{\Sigma X^0}{N} = 1$	$\frac{\Sigma x^0}{N} = 1$	
First moment	$\frac{\Sigma X^1}{N} = \overline{X}$	$\frac{\Sigma x^1}{N} = 0$	
Second moment	Not used	$\frac{\Sigma x^2}{N} = \sigma_N^2$	
Third moment	Not used	$\frac{\Sigma x^3}{N}$	= used in skewness
Fourth moment	Not used	$\frac{\Sigma x^4}{N}$	= used in kurtosis
Higher moments	Not used	Not used	
Product moment	$\frac{\Sigma XY}{N}$	$\frac{\Sigma xy}{N}$	= used in correlation

If you remember algebra, you may remember that $\frac{X^a}{X^b} = X^{(a-b)}$. Now if $a = b$, this becomes $\frac{X^a}{X^a} = X^0$ and since we know that $\frac{X^a}{X^a} = 1$, then any number (except 0 or infinity) raised to the 0 power = 1. Hence ΣX^0 or Σx^0 indicates that we must add 1 for each individual and the sum of these N 1's will obviously be N. Therefore, $\frac{\Sigma X^0}{N} = \frac{N}{N} = 1$, as shown in the top row.

The tabulation shows the way that some of our statistical concepts are related. Notice that the successive rows reveal that:

the zeroth moment gives a constant,
the first moment is the mean,

the second moment is basic to the standard deviation,
the third moment is basic to a measure of skewness,
the fourth moment is basic to a measure of kurtosis, and
the product moment is basic to a measure of correlation.

These first and second moments often give a good description of a frequency distribution. The first three will usually take care of it even if it is lopsided. Finally, all four do such an excellent job, even with distributions that appear to be very unusual in shape, that no statisticians have yet come up with any useful measures based on fifth or higher moments. That takes care of describing a one-dimensional frequency distribution or a two-dimensional graph of it.

CORRELATION

How do we describe a two-dimensional frequency distribution or a three-dimensional model of it? By r, which was introduced by calling it the Pearson product moment coefficient of correlation. By including the product moment in the list of moments we see that it, and hence r, is very simply related to these other concepts, thus tying together the basic formulas and material from Chapters 2, 3, 7, and 8, along with several others which are based on and closely related to those fundamental chapters.

Although regression equations and other measures of correlation are certainly related to r, the two chapters devoted to them have but little relation to the preceding discussion of moments. They are, however, very closely related to the measures included in that discussion.

Educators and psychologists tend to emphasize r rather than the regression equation. In some other fields such as Economics and Mathematics, the regression equation and predictions made from it often seem more important and are sometimes presented ahead of r. Of course, the order of presentation does not matter much, but the differential emphasis that leads to the rearrangement may. Our emphasis on r has led to the derivation of a great many other measures of relationship most of which have little, if any, value. The four outstanding exceptions are featured in Chapter 10, with those of less value being treated in Supplement 10.

Similarly, although the standard deviation is the square root of the variance, its treatment and use as described in Chapter 3 has very little to do with the analysis of variance as described in Chapter 13.

Again, χ^2 and most of the other nonparametric statistics don't have much in common with moments.

POPULAR CONCEPTS

Getting back to *this* course, what are the most important, most useful, and most widely used of the concepts covered in this book? Yes, of course, there are many answers, but here are our guesses.

1. Mean
2. Standard deviation
3. Pearson r
4. Analysis of variance
5. Statistical inference
6. Normal probability curve

Now ask your instructor what his order of preference is. Not only will it differ, but it may be of much greater value to you. If you don't already see why, just guess who writes the questions for the final examination.

FUTURE COURSES

Where do we go from here? What happens after you finish this course? There is no standard answer. As scientists rather than technicians, the future training of those who take this course will vary greatly. Some of you will never take another statistics course. Others will. If you do, what will it cover? First of all, it is *not* likely to give you more advanced material on the topics covered in Chapters 2, 3, 6, 7, 8, 9, or 11. You may not know everything there is to know about those topics—neither does anybody else—but you should now know or be able to look up whatever you may need to know about them. So what *will* be in more advanced statistics courses? Such courses are not standardized with respect either to title or to content, and the offerings vary tremendously from one college to another. Also there is no sure way to know whether the course will be taught in the Department of Education, Psychology, Behavioral Science, Mathematics, or even in a Department named Statistics. Here are some of the more advanced courses:

Multiple correlation and regression In multiple correlation, the scores on two or more variables are so weighted and combined as to give the best possible prediction of a criterion. This is clearly an extension of what we have learned in Chapters 8 and 9.

Analysis of covariance While it may be true that some students will not regard Chapter 13 as the simplest in this book, it is nevertheless true that

that chapter considers only the simplest aspects of analysis of variance. For instance, if we try out two learning methods on bright children and the same two on retarded children, it is possible that method A may turn out to be better with the bright children while method B is better with the retarded children. This is a simple illustration of what is known as interaction. If it is not feasible to make random assignments of subjects to the various method groups, so that intact groups must be used, the technique known as analysis of covariance enables us to correct our obtained differences for the effect of initial differences between the groups on variables believed to be relevant.

Multivariate analysis Since the word multivariate clearly refers to several variables, a course in multivariate analysis may include both multiple correlation and analysis of covariance along with other technical methods of analyzing complex data in order to determine which relationships among the data are statistically significant. Such a course is thus an extension of Chapters 8, 9, 13, 14, and perhaps 11.

Design of experiments Many universities give courses on the design and analysis of experiments in Education or Psychology. Such courses are ideal for students who plan to write a thesis or dissertation because they place the emphasis where statisticians and many others agree it belongs—on carefully planning an experiment rather than just gathering a large amount of data and then asking a statistician what to do with it. (He knows, but is generally too polite to refer to wastebaskets or incinerators.) Courses on experimental design will normally include a number of different topics such as those described in the three preceding paragraphs, but their emphasis will be on designing experiments in an appropriate manner and analyzing the results so obtained. They will also include the testing of hypotheses, methods of ensuring randomization when it is desired, replication (which is having more than one subject in each condition), and other more advanced concepts—always emphasizing the relation between the conditions and the analysis to be made of the data.

Statistical inference Although Chapters 5 and 14 were designed to give a rather complete treatment of statistical inference at the elementary level, a university may well give one, two, or even three more advanced courses on this topic. Such courses will often, but not necessarily, be given by the Department of Statistics rather than Psychology or Education. They may also include hypothesis testing and perhaps some other of the topics referred to in the preceding paragraph.

Methods of research A course in research methodology may be less technical and comprehensive than the course you are now completing or it may contain any or all of the material so far referred to in this survey of advanced courses. Look at the catalog description or ask people who have taken the course. The title alone can mean almost anything.

Nonparametric statistics There are a great many nonparametric statistical methods. Several new ones come into existence every year, but no one of them seems unequivocally better than all the others. Chapter 12 treated some of these methods. Advanced courses in nonparametric statistics will probably shed little, if any, new light on χ^2, but they will probably give more information about some of the methods presented in Chapter 12, and they will present a number of new methods. Such courses will probably give you more information regarding the particular circumstances under which each method is most effective.

Theory of tests and measurements Good tests are not produced simply by asking a teacher to write out 10 or 100 test questions. One must now have a good understanding of the theory of measurement, must know the characteristics of various types of test items, must try out items, analyze the results, and finally come up with a test. The theoretical considerations relating to tests of abilities, aptitudes, and achievement are different from those pertaining to the measurement of personality, feelings, and attitudes. Such a course will amplify and give new meaning to known concepts such as reliability, validity, and point biserial r, while introducing new concepts related to scaling, techniques for determining norms, and other measurement concepts.

Psychophysics and psychological scaling If someone asks you to lift two similar looking weights of 7½ and of 8 ounces and tell him which is the heavier, you will probably say you don't know, but if he insists and does this a number of times you will be right more often than wrong. If he then changes to 7¾ and 8 ounces, your proportion of right answers will decline, but it will remain above .50. The same principles apply to telling him which of two attitude statements is the more favorable. Courses in psychophysics and psychological scaling start with very simple data and yield very complex and useful results by the application of statistical and mathematical techniques to the data. Such material might be and sometimes is included in the type of course described in the preceding paragraph.

Factor analysis If we give 20 tests to 400 people and then compute all of the 190 possible Pearson r's, we will often find a large number of positive rather than negative correlation coefficients. Why? Factor analysis not only provides the answer, but it also enables us to determine how many really independent factors are involved and what they are. Thus, in the intellectual realm, two important and relatively independent factors are the verbal factor and the numerical factor. That is why tests designed to select students for graduate study commonly measure V, the verbal factor, and

N, the number factor. A course in factor analysis will tell you how to make the sort of studies that have revealed the existence of several dozen different intellectual ability factors and perhaps even more different factors in the personality domain.

Other courses The preceding list gives the advanced courses which we feel are among the most important to research workers in Education and Psychology. There are others, of course. Some of you may wish to explore some that are less closely related to our primary fields of interest. A list of such courses would probably include statistical sampling theory, probability, and curve fitting. Still other statistics courses such as time series analysis and the making of index numbers have only a limited applicability to psychological or educational problems at the present time.

Thus, we see that there are a great many statistics courses that logically follow this one and that you may wish to take. There is, however, the very real possibility that you may never take another one, so let us forget the future and come back to the present. Since you, the reader, have interacted, perhaps struggled, with the 14 previous chapters, where does that put you *now?* What have you really accomplished? What does it all mean? How can you use statistics in the future?

USING STATISTICS

Nearly all students in a first course in statistics worry because they believe that they are not learning much. There are some things that they should realize. Compared with what you can know several years after the course, you probably don't know much now, but others are in the same boat. In a course situation we must go over topics so quickly that things don't have time to soak in. Technically, we all need what psychologists call "overlearning" to thoroughly learn the concepts in statistics. Hardly anyone has enough time for enough overlearning during the course. However, if we use what we've learned soon after the course, then our grasp of the concepts is greatly improved.

But how can you use the statistical methods you have learned in the course? Chances are, you'll find some need for them in future courses. If not, you may need to take the initiative and find situations to which you can apply what you've learned.

You could make the effort to read more of the technical literature in your field and learn by observing statistics at work in reports of surveys or experimental studies. You could offer to tutor someone else who is taking a course in statistics. One of the best ways to learn something better is to teach it.

What about using statistics in future courses? This depends upon several factors, such as what your major is, what degree you are working for, and how much freedom you have in choosing electives. Also, courses lend themselves to using statistics in varying degrees.

If you are in Education at the undergraduate level, it may be unusual for you to take a statistics course at all, unless you are a Mathematics major, or unless you take it as an elective. After the Bachelor's degree, an Education major may take a statistics course as a special student or as a required course in a Master's degree program. In Education, the relevant follow-up courses applying statistics would be those in research, measurement, and evaluation. If you go into a heavy quantitative or scientific area such as Educational Psychology, Tests and Measurements, Guidance, Curriculum Improvement, School Psychology, or one of various kinds of specialties—then you may need a second or third course in statistics in which you go deeper into analysis of variance as a prerequisite for studying experimental design, or into correlation.

It has been recognized by many people but perhaps never stated any more clearly than by Cronbach* that there are two statistical approaches to scientific Psychology. Theory is implied here, but the two-way distinction also spills over into application. Cronbach calls the two points of view the *experimental* and the *correlational.* One of these can be seen in the application of analysis of variance to evaluating the outcome of controlled experimentation. The other can be seen in the application of tests to the selection of personnel or in the administration of survey tests in school systems.

The major in Psychology at the undergraduate level will probably be required to take the introductory statistics course, especially as a prerequisite to course work in experimental Psychology or in measurement.

At the graduate level in Psychology departments and in some departments or schools of Education one can progress to advanced courses such as those already mentioned in this chapter on the design of experiments, multivariate analysis, analysis of covariance, or statistical inference. Alternatively, the graduate student may take courses in which the primary emphasis is on applications of statistical methods to research problems, such as a course in scaling theory and methods or one on test construction and development.

---

*Cronbach, Lee J. The two disciplines of scientific psychology. *American Psychologist,* 1957, **12**, 671–684.

In addition to the two-way breakdown of statistics into descriptive statistics and inferential statistics that we noted earlier, there is a two-way breakdown in a course in terms of outcomes. We might call these the conceptual and the computational (the two "C's"). You could, under certain conditions, learn how to compute statistical indices without much understanding of what you are doing or of what the figures mean. Or you might go to the other extreme of learning some skills in interpreting statistical figures without being very good at computing. Ideally we should balance the two and come out well-rounded by being good in both. There is no substitute in statistics for learning by doing. There is not much retention without *using* statistics. Both computation and interpretation are facilitated by use.

As statisticians sometimes do, those who wrote this book disagreed on how it should end. Accordingly, the book has two final paragraphs. Read whichever one you prefer:

Statistics does not have to be a necessary evil, a hurdle to surmount and then to forget. It can serve as a beginning, as a threshold to new vistas and fields of endeavor.

Throughout this book, the writers have tried hard, by the use of many examples and alternative explanations, to add variety to the text so that no student will ever stagger out of a classroom mumbling Hebrews 13:8*.

*The exact quotation is, "Jesus Christ, the same yesterday, and today, and forever."

Appendix

NOTE: There are no tables I, N, O, or X.

TABLE A
Recommended Size of Interval Width for All Ranges up to 1,000

Range of scores	Interval width	Number of intervals
1–22	1	1 to 22
23–42	2	12 to 22
43–44	3	15 to 16
45–55	3	15 to 19
	or 4	12 to 15
56–63	5	12 to 14
	or 4	14 to 17
	or 3	19 to 22
64–84	5	13 to 18
	or 4	16 to 22
85–106	5	17 to 22
107–110	5	22 to 23
	or 10	11 to 12
111–211	10	12 to 22
212–220	10	22 to 23
	or 20	11 to 12
221–421	20	12 to 22
422–550	20	22 to 29
	or 50	9 to 12
551–1051	50	12 to 22

TABLE B
Values of $N(N\text{-}1)$ for Samples up to 499

	0	1	2	3	4	5	6	7	8	9	
0			2	6	12	20	30	42	56	72	0
10	90	110	132	156	182	210	240	272	306	342	10
20	380	420	462	506	552	600	650	702	756	812	20
30	870	930	992	1,056	1,122	1,190	1,260	1,332	1,406	1,482	30
40	1,560	1,640	1,722	1,806	1,892	1,980	2,070	2,162	2,256	2,352	40
50	2,450	2,550	2,652	2,756	2,862	2,970	3,080	3,192	3,306	3,422	50
60	3,540	3,660	3,782	3,906	4,032	4,160	4,290	4,422	4,556	4,692	60
70	4,830	4,970	5,112	5,256	5,402	5,550	5,700	5,852	6,006	6,162	70
80	6,320	6,480	6,642	6,806	6,972	7,140	7,310	7,482	7,656	7,832	80
90	8,010	8,190	8,372	8.556	8,742	8,930	9,120	9,312	9,506	9,702	90
100	9,900	10,100	10,302	10,506	10,712	10,920	11,130	11,342	11,556	11,772	100
110	11,990	12,210	12,432	12,656	12,882	13,110	13,340	13,572	13,806	14,042	110
120	14,280	14,520	14,762	15,006	15,252	15,500	15,750	16,002	16,256	16,512	120
130	16,770	17,030	17,292	17,556	17,822	18,090	18,360	18,632	18,906	19,182	130
140	19,460	19,740	20,022	20,306	20,592	20,880	21,170	21,462	21,756	22,052	140
150	22,350	22,650	22,952	23,256	23,562	23,870	24,180	24,492	24,806	25,122	150
160	25,440	25,760	26,082	26,406	26,732	27,060	27,390	27,722	28,056	28,392	160
170	28,730	29,070	29,412	29,756	30,102	30,450	30,800	31,152	31,506	31,862	170
180	32,220	32,580	32,942	33,306	33,672	34,040	34,410	34,782	35,156	35,532	180
190	35,910	36,290	36,672	37,056	37,442	37,830	38,220	38,612	39,006	39,402	190
200	39,800	40,200	40,602	41,006	41,412	41,820	42,230	42,642	43,056	43,472	200
210	43,890	44,310	44,732	45,156	45,582	46,010	46,440	46,872	47,306	47,742	210
220	48,180	48,620	49,062	49,506	49,952	50,400	50,850	51,302	51,756	52,212	220
230	52,670	53,130	53,592	54,056	54,522	54,990	55,460	55,932	56,406	56,882	230
240	57,360	57,840	58,322	58,806	59,292	59,780	60,270	60,762	61,256	61,752	240

	0	1	2	3	4	5	6	7	8	9	
250	62,250	62,750	63,252	63,756	64,262	64,770	65,280	65,792	66,306	66,822	250
260	67,340	67,860	68,382	68,906	69,432	69,960	70,490	71,022	71,556	72,092	260
270	72,630	73,170	73,712	74,256	74,802	75,350	75,900	76,452	77,006	77,562	270
280	78,120	78,680	79,242	79,806	80,372	80,940	81,510	82,082	82,656	83,232	280
290	83,810	84,390	84,972	85,556	86,142	86,730	87,320	87,912	88,506	89,102	290
300	89,700	90,300	90,902	91,506	92,112	92,720	93,330	93,942	94,556	95,172	300
310	95,790	96,410	97,032	97,656	98,282	98,910	99,540	100,172	100,806	101,442	310
320	102,080	102,720	103,362	104,006	104,652	105,300	105,950	106,602	107,256	107,912	320
330	108,570	109,230	109,892	110,556	111,222	111,890	112,560	113,232	113,906	114,582	330
340	115,260	115,940	116,622	117,306	117,992	118,680	119,370	120,062	120,756	121,452	340
350	122,150	122,850	123,552	124,256	124,962	125,670	126,380	127,092	127,806	128,522	350
360	129,240	129,960	130,682	131,406	132,132	132,860	133,590	134,322	135,056	135,792	360
370	136,530	137,270	138,012	138,756	139,502	140,250	141,000	141,752	142,506	143,262	370
380	144,020	144,780	145,542	146,306	147,072	147,840	148,610	149,382	150,156	150,932	380
390	151,710	152,490	153,272	154,056	154,842	155,630	156,420	157,212	158,006	158,802	390
400	159,600	160,400	161,202	162,006	162,812	163,620	164,430	165,242	166,056	166,872	400
410	167,690	168,510	169,332	170,156	170,982	171,810	172,640	173,472	174,306	175,142	410
420	175,980	176,820	177,662	178,506	179,352	180,200	181,050	181,902	182,756	183,612	420
430	184,470	185,330	186,192	187,056	187,922	188,790	189,660	190,532	191,406	192,282	430
440	193,160	194,040	194,922	195,806	196,692	197,580	198,470	199,362	200,256	201,152	440
450	202,050	202,950	203,852	204,756	205,662	206,570	207,480	208,392	209,306	210,222	450
460	211,140	212,060	212,982	213,906	214,832	215,760	216,690	217,622	218,556	219,492	460
470	220,430	221,370	222,312	223,256	224,202	225,150	226,100	227,052	228,006	228,962	470
480	229,920	230,880	231,842	232,806	233,772	234,740	235,710	236,682	237,656	238,632	480
490	239,610	240,590	241,572	242,556	243,542	244,530	245,520	246,512	247,506	248,502	490

TABLE C
Friden Method for Extracting Square Root

To Use the Table
If the number whose root is desired is between 1 and 100 the table can be entered immediately. If the number is greater than 100 move the decimal point two places to the left (or multiples thereof). Thus for purposes of using the table, 6250 becomes 62.50 and 34,567.89 becomes 3.456789. If the number is less than 1 move the decimal point two places to the right (or multiples thereof). Thus .1234 becomes 12.34 and .0004567 becomes 4.567. Now from the table, find the two consecutive values in the "N" Column between which lies the number whose square root is desired. Select the Plus and Divide factors between these two numbers.

Routine Note—All numbers are set on extreme left side of keyboard.

1. Set the number whose square root is desired on the keyboard. Touch ENTER DIVD key.
2. From the table, find the Plus (+) and Divide (÷) factors. Set Plus factor on keyboard. Touch Plus bar.
3. Set Divide factor on keyboard. Touch both Divide keys. The square root appears in the lower dials accurate to five significant digits.

Location of Decimal Point in Answer
For numbers greater than one: count the number of *pairs* of digits to the left of the decimal point. This will be the number of digits to the left of the decimal point in the answer. An odd digit at the extreme left is to be counted as an extra pair.

Examples: $\sqrt{4} = 2$; $\sqrt{25} = 5$; $\sqrt{625} = 25$; $\sqrt{15{,}129} = 123$; $\sqrt{1{,}522{,}756} = 1234$.

For numbers less than one; count the number of *pairs* of zeros to the *right* of the decimal. (A lone zero would not be counted as a pair.) This will be the number of zeros following the decimal point in the answer.

Examples: $\sqrt{.25} = .5$; $\sqrt{.025} = .158$; $\sqrt{.0025} = .05$; $\sqrt{.000\,25} = .0158$; $\sqrt{.000\,025} = .005$; $\sqrt{.000\,002\,5} = .001\,58$; $\sqrt{.000\,000\,25} = .0005$.

TABLE C
(*continued*)

Example I $\sqrt{2475.812} = 49.757$. Move ADD lever down. NON ENT key down. Set Tab Stop #7.

1. Shift decimal point in radicand two places to left. Then set 24.758 12 on keyboard. Touch ENTER DIVD.
2. From the table, the Plus factor is 248; the Divide factor is 996. Set 248 on keyboard. Touch Plus bar.
3. Set 996 on keyboard. Touch both Divide keys.

Answer (49.757 14) appears in lower dials. The root correct to five significant digits then is 49.757.

Example II $\sqrt{.000\ 006\ 78} = .002\ 603\ 8$

1. Shift decimal point in radicand six places to the right. Then set 6.78 on keyboard. Touch ENTER DIVD.
2. From the table, the Plus factor is 680; the Divide factor is 521 543. Set 680 on keyboard. Touch Plus bar.
3. Set 521 543 on keyboard. Touch both Divide keys.

Answer (.002 603 812 1) appears in the lower dials. The root correct to five significant digits then is .002 603 8.

Example III $\sqrt{79{,}682.134} = 282.280\ 240\ 1$

1. Shift decimal point in radicand four places to left. Then set 7.968 213 4 on keyboard. Touch ENTER DIVD.
2. From the table, the Plus factor is 800; the Divide factor is 565 690. Set 800 on keyboard. Touch Plus bar.
3. Set 565 690 on keyboard. Touch both Divide keys.

Answer (282.278 51) appears in the lower dials. The root correct to five significant digits then is 282.28.

TABLE C (*continued*)

N	+	÷	N	+	÷	N	+	÷	N	+	÷	N	+	÷
1.00			2.55			5.79			10.5			15.5		
	102	202 000		260	322 500		588	484 980		106	651 160		156	789 940
1.04			2.65			5.97			10.7			15.7		
	106	205 920		270	328 640		606	492 350		108	657 270		158	794 990
1.08			2.75			6.15			10.9			15.9		
	110	209 770		280	334 670		624	499 606		110	663 330		160	800 010
1.12			2.85			6.33			11.1			16.2		
	114	213 550		290	340 594		642	506 760		112	669 330		164	809 946
1.16			2.95			6.51			11.3			16.6		
	118	217 260		300	346 420		660	513 817		114	675 280		168	819 763
1.20			3.05			6.7			11.5			17.0		
	122	220 910		310	352 140		680	521 543		116	681 180		172	829 465
1.24			3.15			6.9			11.7			17.4		
	126	224 505		320	357 780		700	529 160		118	687 030		176	839 054
1.29			3.25			7.1			11.9			17.8		
	132	229 790		330	363 326		720	536 660		120	692 820		180	848 535
1.35			3.36			7.3			12.1			18.2		
	138	234 954		342	369 870		740	544 065		122	698 570		184	857 910
1.41			3.48			7.5			12.3			18.6		
	144	240 010		354	376 305		760	551 370		124	704 280		188	867 186
1.47			3.60			7.7			12.5			19.0		
	150	244 956		366	382 630		780	558 575		126	709 930		192	876 360
1.53			3.72			7.9			12.7			19.4		
	156	249 806		378	388 850		800	565 690		128	715 540		196	885 440
1.59			3.84			8.1			12.9			19.8		
	162	254 560		390	394 974		820	572 720		130	721 110		200	894 430

N	+	÷	N	+	÷	N	+	÷	N	+	÷	N	+	÷
1.65			3.96			8.3			13.1			20.2		
	168	259 235		402	401 006		840	579 660		132	726 640		204	903 330
1.71			4.09			8.5			13.3			20.6		
	174	263 820		416	407 930		860	586 520		134	732 120		208	912 150
1.77			4.23			8.7			13.5			21.0		
	180	268 330		430	414 736		880	593 300		136	737 570		212	920 870
1.83			4.37			8.9			13.7			21.4		
	186	272 770		444	421 430		900	600 000		138	742 970		216	929 520
1.90			4.51			9.1			13.9			21.8		
	194	278 575		458	428 025		920	606 630		140	748 330		220	938 090
1.98			4.65			9.3			14.1			22.2		
	202	284 260		472	434 520		940	613 190		142	753 660		224	946 580
2.06			4.80			9.5			14.3			22.6		
	210	289 834		488	441 820		960	619 680		144	758 950		228	954 990
2.14			4.96			9.7			14.5			23.0		
	218	295 300		504	449 006		980	626 100		146	764 200		232	963 330
2.22			5.12			9.9			14.7			23.4		
	226	300 670		520	456 080		995	630 880		148	769 420		236	971 600
2.30			5.28			10.0			14.9			23.8		
	234	305 950		536	463 040		1005	634 040		150	774 600		240	979 800
2.38			5.44			10.1			15.1			24.2		
	242	311 130		552	469 900		102	638 750		152	779 750		244	987 930
2.46			5.61			10.3			15.3			24.6		
	250	316 236		570	477 500		104	644 980		154	784 860		248	996 000

TABLE C (*continued*)

N	+	÷	N	+	÷	N	+	÷	N	+	÷	N	+	÷
25.0			35.5			50.0			68.5			88.5		
	252	100 400		358	119 667		504	141 9865		690	166 133		890	188 680
25.4			36.1			50.8			69.5			89.5		
	256	101 193		364	120 665		512	143 109		700	167 333		900	189 737
25.8			36.7			51.6			70.5			90.5		
	260	101 981		370	121 656		520	144 223		710	168 5235		910	190 788
26.2			37.3			52.4			71.5			91.5		
	264	102 762		376	122 638		528	145 328		720	169 706		920	191 834
26.6			37.9			53.2			72.5			92.5		
	268	103 538		382	123 613		536	146 425		730	170 881		930	192 873
27.0			38.5			54.0			73.5			93.5		
	272	104 308		388	124 580		544	147 513		740	172 047		940	193 908
27.4			39.1			54.8			74.5			94.5		
	276	105 072		394	125 539		552	148 594		750	173 206		950	194 936
27.8			39.7			55.6			75.5			95.5		
	280	105 831		400	126 492		560	149 667		760	174 356		960	195 960
28.3			40.4			56.5			76.5			96.5		
	286	106 9587		408	127 7505		570	150 9974		770	175 500		970	196 977
28.9			41.2			57.5			77.5			97.5		
	292	108 075		416	128 997		580	152 316		780	176 636		980	197 9905
29.5			42.0			58.5			78.5			98.7		
	298	109 1795		424	130 2313		590	153 6236		790	177 764		994	199 4000
30.1			42.8			59.5			79.5			100.0		
	304	110 273		432	131 454		600	154 920		800	178 886			
30.7			43.6			60.5			80.5					
	310	111 356		440	132 666		610	156 2056		810	180 000			
31.3			44.4			61.5			81.5					
	316	112 4284		448	133 866		620	157 481		820	181 108			

N	+	÷	N	+	÷	N	+	÷	N	+	÷	N	+	÷
31.9			45.2			62.5			82.5					
	322	113 491		456	135 056		630	161 246		830	182 209			
32.5			46.0			63.5			83.5					
	328	114 543		464	136 236		640	162 481		840	183 303			
33.1			46.8			64.5			84.5					
	334	115 586		472	137 405		650	163 708		850	184 391			
33.7			47.6			65.5			85.5					
	340	116 620		480	138 565		660	158 746		860	185 473			
34.3			48.4			66.5			86.5					
	346	117 644		488	139 7146		670	160 001		870	186 548			
34.9			49.2			67.5			87.5					
	352	118 660		496	140 855		680	164 925		880	187 617			

TABLE D
The Normal Curve Ordinate, y, for a Given $\frac{x}{\sigma}$ Deviation

$\frac{x}{\sigma}$	.00	.01	.02	.03	.04	.05	.06	.07	.08	.09	
.0	.3989	.3989	.3989	.3988	.3986	.3984	.3982	.3980	.3977	.3973	.0
.1	.3970	.3965	.3961	.3956	.3951	.3945	.3939	.3932	.3925	.3918	.1
.2	.3910	.3902	.3894	.3885	.3876	.3867	.3857	.3847	.3836	.3825	.2
.3	.3814	.3802	.3790	.3778	.3765	.3752	.3739	.3725	.3712	.3697	.3
.4	.3683	.3668	.3653	.3637	.3621	.3605	.3589	.3572	.3555	.3538	.4
.5	.3521	.3503	.3485	.3467	.3448	.3429	.3410	.3391	.3372	.3352	.5
.6	.3332	.3312	.3292	.3271	.3251	.3230	.3209	.3187	.3166	.3144	.6
.7	.3123	.3101	.3079	.3056	.3034	.3011	.2989	.2966	.2943	.2920	.7
.8	.2897	.2874	.2850	.2827	.2803	.2780	.2756	.2732	.2709	.2685	.8
.9	.2661	.2637	.2613	.2589	.2565	.2541	.2516	.2492	.2468	.2444	.9
1.0	.2420	.2396	.2371	.2347	.2323	.2299	.2275	.2251	.2227	.2203	1.0
1.1	.2179	.2155	.2131	.2107	.2083	.2059	.2036	.2012	.1989	.1965	1.1
1.2	.1942	.1919	.1895	.1872	.1849	.1826	.1804	.1781	.1758	.1736	1.2
1.3	.1714	.1691	.1669	.1647	.1626	.1604	.1582	.1561	.1539	.1518	1.3
1.4	.1497	.1476	.1456	.1435	.1415	.1394	.1374	.1354	.1334	.1315	1.4
1.5	.1295	.1276	.1257	.1238	.1219	.1200	.1182	.1163	.1145	.1127	1.5
1.6	.1109	.1092	.1074	.1057	.1040	.1023	.1006	.0989	.0973	.0957	1.6
1.7	.0940	.0925	.0909	.0893	.0878	.0863	.0848	.0833	.0818	.0804	1.7
1.8	.0790	.0775	.0761	.0748	.0734	.0721	.0707	.0694	.0681	.0669	1.8
1.9	.0656	.0644	.0632	.0620	.0608	.0596	.0584	.0573	.0562	.0551	1.9
2.0	.0540	.0529	.0519	.0508	.0498	.0488	.0478	.0468	.0459	.0449	2.0
2.1	.0440	.0431	.0422	.0413	.0404	.0396	.0387	.0379	.0371	.0363	2.1
2.2	.0355	.0347	.0339	.0332	.0325	.0317	.0310	.0303	.0297	.0290	2.2
2.3	.0283	.0277	.0270	.0264	.0258	.0252	.0246	.0241	.0235	.0229	2.3
2.4	.0224	.0219	.0213	.0208	.0203	.0198	.0194	.0189	.0184	.0180	2.4

	.00	.01	.02	.03	.04	.05	.06	.07	.08	.09	
2.5	.0175	.0171	.0167	.0163	.0158	.0154	.0151	.0147	.0143	.0139	2.5
2.6	.0136	.0132	.0129	.0126	.0122	.0119	.0116	.0113	.0110	.0107	2.6
2.7	.0104	.0101	.0099	.0096	.0093	.0091	.0088	.0086	.0084	.0081	2.7
2.8	.0079	.0077	.0075	.0073	.0071	.0069	.0067	.0065	.0063	.0061	2.8
2.9	.0060	.0058	.0056	.0055	.0053	.0051	.0050	.0048	.0047	.0046	2.9
3.0	.0044	.0043	.0042	.0040	.0039	.0038	.0037	.0036	.0035	.0034	3.0
3.1	.0033	.0032	.0031	.0030	.0029	.0028	.0027	.0026	.0025	.0025	3.1
3.2	.0424	.0023	.0022	.0022	.0021	.0020	.0020	.0019	.0018	.0018	3.2
3.3	.0017	.0017	.0016	.0016	.0015	.0015	.0014	.0014	.0013	.0013	3.3
3.4	.0012	.0012	.0012	.0011	.0011	.0010	.0010	.0010	.0009	.0009	3.4

$\frac{x}{\sigma}$	y	$\frac{x}{\sigma}$	y	$\frac{x}{\sigma}$	y
.0	.398,942,28	3.5	.000,872,683	7.0	.000,000,000,009,134,72
.5	.352,065,33	4.0	.000,133,830	7.5	.000,000,000,000,243,432
1.0	.241,970,72	4.5	.000,015,983,7	8.0	.000,000,000,000,005,052,27
1.5	.129,517,60	5.0	.000,001,486,72	8.5	.000,000,000,000,000,081,662,4
2.0	.053,990,97	5.5	.000,000,107,698	9.0	.000,000,000,000,000,001,027,98
2.5	.017,528,30	6.0	.000,000,006,075,88	9.5	.000,000,000,000,000,000,010,077,9
3.0	.004,431,85	6.5	.000,000,000,266,956	10.0	.000,000,000,000,000,000,000,076,946,0

TABLE E
The Normal Curve Area Between the Mean and a Given $\frac{x}{\sigma}$ Deviation

$\frac{x}{\sigma}$	.00	.01	.02	.03	.04	.05	.06	.07	.08	.09	
.0	.0000	.0040	.0080	.0120	.0160	.0199	.0239	.0279	.0319	.0359	.0
.1	.0398	.0438	.0478	.0517	.0557	.0596	.0636	.0675	.0714	.0753	.1
.2	.0793	.0832	.0871	.0910	.0948	.0987	.1026	.1064	.1103	.1141	.2
.3	.1179	.1217	.1255	.1293	.1331	.1368	.1406	.1443	.1480	.1517	.3
.4	.1554	.1591	.1628	.1664	.1700	.1736	.1772	.1808	.1844	.1879	.4
.5	.1915	.1950	.1985	.2019	.2054	.2088	.2123	.2157	.2190	.2224	.5
.6	.2257	.2291	.2324	.2357	.2389	.2422	.2454	.2486	.2517	.2549	.6
.7	.2580	.2611	.2642	.2673	.2704	.2734	.2764	.2794	.2823	.2852	.7
.8	.2881	.2910	.2939	.2967	.2995	.3023	.3051	.3078	.3106	.3133	.8
.9	.3159	.3186	.3212	.3238	.3264	.3289	.3315	.3340	.3365	.3389	.9
1.0	.3413	.3438	.3461	.3485	.3508	.3531	.3554	.3577	.3599	.3621	1.0
1.1	.3643	.3665	.3686	.3708	.3729	.3749	.3770	.3790	.3810	.3830	1.1
1.2	.3849	.3869	.3888	.3907	.3925	.3944	.3962	.3980	.3997	.4015	1.2
1.3	.4032	.4049	.4066	.4082	.4099	.4115	.4131	.4147	.4162	.4177	1.3
1.4	.4192	.4207	.4222	.4236	.4251	.4265	.4279	.4292	.4306	.4319	1.4
1.5	.4332	.4345	.4357	.4370	.4382	.4394	.4406	.4418	.4429	.4441	1.5
1.6	.4452	.4463	.4474	.4484	.4495	.4505	.4515	.4525	.4535	.4545	1.6
1.7	.4554	.4564	.4573	.4582	.4591	.4599	.4608	.4616	.4625	.4633	1.7
1.8	.4641	.4649	.4656	.4664	.4671	.4678	.4686	.4693	.4699	.4706	1.8
1.9	4713	.4719	.4726	.4732	.4738	.4744	.4750	.4756	.4761	.4767	1.9
2.0	.4772	.4778	.4783	.4788	.4793	.4798	.4803	.4808	.4812	.4817	2.0
2.1	.4821	.4826	.4830	.4834	.4838	.4842	.4846	.4850	.4854	.4857	2.1
2.2	.4861	.4864	.4868	.4871	.4875	.4878	.4881	.4884	.4887	.4890	2.2
2.3	.4893	.4896	.4898	.4901	.4904	.4906	.4909	.4911	.4913	.4916	2.3
2.4	.4918	.4920	.4922	.4925	.4927	.4929	.4931	.4932	.4934	.4936	2.4

2.5	.4938	.4940	.4941	.4943	.4945	.4946	.4948	.4949	.4951	.4952	2.5
2.6	.4953	.4955	.4956	.4957	.4959	.4960	.4961	.4962	.4963	.4964	2.6
2.7	.4965	.4966	.4967	.4968	.4969	.4970	.4971	.4972	.4973	.4974	2.7
2.8	.4974	.4975	.4976	.4977	.4977	.4978	.4979	.4979	.4980	.4981	2.8
2.9	.4981	.4982	.4982	.4983	.4984	.4984	.4985	.4985	.4986	.4986	2.9
3.0	.4987	.4987	.4987	.4988	.4988	.4989	.4989	.4989	.4990	.4990	3.0
3.1	.4990	.4991	.4991	.4991	.4992	.4992	.4992	.4992	.4993	.4993	3.1
3.2	.4993	.4993	.4994	.4994	.4994	.4994	.4994	.4995	.4995	.4995	3.2
3.3	.4995	.4995	.4995	.4996	.4996	.4996	.4996	.4996	.4996	.4997	3.3
3.4	.4997	.4997	.4997	.4997	.4997	.4997	.4997	.4997	.4997	.4998	3.4
	.00	.01	.02	.03	.04	.05	.06	.07	.08	.09	

$\frac{x}{\sigma}$	Area	$\frac{x}{\sigma}$	Area	$\frac{x}{\sigma}$	Area
.0	.000,000,00	3.5	.499,767,371	7.0	.499,999,999,998,720,19
.5	.191,462,46	4.0	.499,968,328,8	7.5	.499,999,999,999,968,091,1
1.0	.341,344,75	4.5	.499,996,602,33	8.0	.499,999,999,999,999,377,904
1.5	.433,192,80	5.0	.499,999,713,348	8.5	.499,999,999,999,999,990,520,5
2.0	.477,249,87	5.5	.499,999,981,010,4	9.0	.499,999,999,999,999,999,887,141
2.5	.493,790,33	6.0	.499,999,999,013,41	9.5	.499,999,999,999,999,999,998,950,55
3.0	.498,650,10	6.5	.499,999,999,959,840,0	10.0	.499,999,999,999,999,999,999,992,380,1

TABLE F

The *F* Table Showing All Three Levels of Significance*

*Key .05 in roman type; **.01 in bold face type;** and *.001 in italic type.*

df for Within variance	*df* for Among variance 1	2	3	4	5	6	8	12	24	∞
1	161	200	216	225	230	234	239	244	249	254
	4052	**4999**	**5403**	**5625**	**5724**	**5859**	**5981**	**6106**	**6234**	**6366**
	405284	*500000*	*540379*	*562500*	*576405*	*585937*	*598144*	*610667*	*623497*	*636619*
2	18.51	19.90	19.16	19.25	19.30	19.33	19.37	19.41	19.45	19.50
	98.49	**99.01**	**99.17**	**99.25**	**99.30**	**99.33**	**99.36**	**99.42**	**99.46**	**99.50**
	998.5	*999.0*	*999.2*	*999.2*	*999.3*	*999.3*	*999.4*	*999.4*	*999.5*	*999.5*
3	10.13	9.55	9.28	9.12	9.01	8.94	8.84	8.74	8.64	8.53
	34.12	**30.81**	**29.46**	**28.71**	**28.24**	**27.91**	**27.49**	**27.05**	**26.60**	**26.12**
	167.5	*148.5*	*141.1*	*137.1*	*134.6*	*132.8*	*130.6*	*128.3*	*125.9*	*123.5*
4	7.71	6.94	6.59	6.39	6.26	6.16	6.04	5.91	5.77	5.63
	21.20	**18.00**	**16.69**	**15.98**	**15.52**	**15.21**	**14.80**	**14.37**	**13.93**	**13.46**
	74.14	*61.25*	*56.18*	*53.44*	*51.71*	*50.53*	*49.00*	*47.41*	*45.77*	*44.05*
5	6.61	5.79	5.41	5.19	5.05	4.95	4.82	4.68	4.53	4.36
	16.26	**13.27**	**12.06**	**11.39**	**10.97**	**10.67**	**10.27**	**9.89**	**9.47**	**9.02**
	47.04	*36.61*	*33.20*	*31.09*	*29.75*	*28.84*	*27.64*	*26.42*	*25.14*	*23.78*
6	5.99	5.14	4.76	4.53	4.39	4.28	4.15	4.00	3.84	3.67
	13.74	**10.92**	**9.78**	**9.15**	**8.75**	**8.47**	**8.10**	**7.72**	**7.31**	**6.88**
	35.51	*27.00*	*23.70*	*21.90*	*20.81*	*20.03*	*19.03*	*17.99*	*16.89*	*15.75*
7	5.59	4.74	4.35	4.12	3.97	3.87	3.73	3.57	3.41	3.23
	12.25	**9.55**	**8.45**	**7.85**	**7.46**	**7.19**	**6.84**	**6.47**	**6.07**	**5.65**
	29.22	*21.69*	18.77	*17.19*	*16.21*	*15.52*	*14.63*	*13.71*	*12.73*	*11.69*

df for Within variance	1	2	3	4	5	6	8	12	24	∞
	5.32	4.46	4.07	3.84	3.69	3.58	3.44	3.28	3.12	2.93
8	**11.26**	**8.65**	**7.59**	**7.01**	**6.63**	**6.37**	**6.03**	**5.67**	**5.28**	**4.86**
	25.42	*18.49*	*15.83*	*14.39*	*13.49*	*12.86*	*12.04*	*11.19*	*10.30*	*9.34*
	5.12	4.26	3.86	3.63	3.48	3.37	3.23	3.07	2.90	2.71
9	**10.56**	**8.02**	**6.99**	**6.42**	**6.06**	**5.80**	**5.47**	**5.11**	**4.73**	**4.31**
	22.86	*16.39*	*13.90*	*12.56*	*11.71*	*11.13*	*10.37*	*9.57*	*8.72*	*7.81*
	4.96	4.10	3.71	3.48	3.33	3.22	3.07	2.91	2.74	2.54
10	**10.04**	**7.56**	**6.55**	**5.99**	**5.64**	**5.39**	**5.06**	**4.71**	**4.33**	**3.91**
	21.04	*14.91*	*12.55*	*11.28*	*10.48*	*9.92*	*9.20*	*8.45*	*7.64*	*6.76*
	4.84	3.98	3.59	3.36	3.20	3.09	2.95	2.79	2.61	2.40
11	**9.65**	**7.20**	**6.22**	**5.67**	**5.32**	**5.07**	**4.74**	**4.40**	**4.02**	**3.60**
	19.69	*13.81*	*11.56*	*10.35*	*9.58*	*9.05*	*8.35*	*7.63*	*6.85*	*6.00*
	4.75	3.88	3.49	3.26	3.11	3.00	2.85	2.69	2.50	2.30
12	**9.33**	**6.93**	**5.95**	**5.41**	**5.06**	**4.82**	**4.50**	**4.16**	**3.78**	**3.36**
	18.64	*12.97*	*10.80*	*9.63*	*8.89*	*8.38*	*7.71*	*7.00*	*6.25*	*5.42*
	4.67	3.80	3.41	3.18	3.02	2.92	2.77	2.60	2.42	2.21
13	**9.07**	**6.70**	**5.74**	**5.20**	**4.86**	**4.62**	**4.30**	**3.96**	**3.59**	**3.16**
	17.81	*12.31*	*10.21*	*9.07*	*8.35*	*7.86*	*7.21*	*6.52*	*5.78*	*4.97*
	4.60	3.74	3.34	3.11	2.96	2.85	2.70	2.53	2.35	2.13
14	**8.86**	**6.51**	**5.56**	**5.03**	**4.69**	**4.46**	**4.14**	**3.80**	**3.43**	**3.00**
	17.14	*11.78*	*9.73*	*8.62*	*7.92*	*7.43*	*6.80*	*6.13*	*5.41*	*4.60*
	4.54	3.68	3.29	3.06	2.90	2.79	2.64	2.48	2.29	2.07
15	**8.68**	**6.36**	**5.42**	**4.89**	**4.56**	**4.32**	**4.00**	**3.67**	**3.29**	**2.87**
	16.59	*11.34*	*9.34*	*8.25*	*7.57*	*7.09*	*6.47*	*5.81*	*5.10*	*4.31*

df for Among variance

TABLE F (*continued*)

*Key .05 in roman type; **.01 in bold face type;** and *.001 in italic type.*

df for Within variance	*df* for Among variance: 1	2	3	4	5	6	8	12	24	∞
16	4.49	3.63	3.24	3.01	2.85	2.74	2.59	2.42	2.24	2.01
	8.53	**6.23**	**5.29**	**4.77**	**4.44**	**4.20**	**3.89**	**3.55**	**3.18**	**2.75**
	16.12	*10.97*	*9.00*	*7.94*	*7.27*	*6.81*	*6.19*	*5.55*	*4.85*	*4.06*
17	4.45	3.59	3.20	2.96	2.81	2.70	2.55	2.38	2.19	1.96
	8.40	**6.11**	**5.18**	**4.67**	**4.34**	**4.10**	**3.79**	**3.45**	**3.08**	**2.65**
	15.72	*10.66*	*8.73*	*7.68*	*7.02*	*6.56*	*5.96*	*5.32*	*4.63*	*3.85*
18	4.41	3.55	3.16	2.93	2.77	2.66	2.51	2.34	2.15	1.92
	8.28	**6.01**	**5.09**	**4.58**	**4.25**	**4.01**	**3.71**	**3.37**	**3.00**	**2.57**
	15.38	*10.39*	*8.49*	*7.46*	*6.81*	*6.35*	*5.76*	*5.13*	*4.45*	*3.67*
19	4.38	3.52	3.13	2.90	2.74	2.63	2.48	2.31	2.11	1.88
	8.18	**5.93**	**5.01**	**4.50**	**4.17**	**3.94**	**3.63**	**3.30**	**2.92**	**2.49**
	15.08	*10.16*	*8.28*	*7.26*	*6.61*	*6.18*	*5.59*	*4.97*	*4.29*	*3.52*
20	4.35	3.49	3.10	2.87	2.71	2.60	2.45	2.28	2.08	1.84
	8.10	**5.85**	**4.94**	**4.43**	**4.10**	**3.87**	**3.56**	**3.23**	**2.86**	**2.42**
	14.82	*9.95*	*8.10*	*7.10*	*6.46*	*6.02*	*5.44*	*4.82*	*4.15*	*3.38*
21	4.32	3.47	3.07	2.84	2.68	2.57	2.42	2.25	2.05	1.81
	8.02	**5.78**	**4.87**	**4.37**	**4.04**	**3.81**	**3.51**	**3.17**	**2.80**	**2.36**
	14.59	*9.77*	*7.94*	*6.95*	*6.32*	*5.88*	*5.31*	*4.70*	*4.03*	*3.26*
22	4.30	3.44	3.05	2.82	2.66	2.55	2.40	2.23	2.03	1.78
	7.94	**5.72**	**4.82**	**4.31**	**3.99**	**3.76**	**3.45**	**3.12**	**2.75**	**2.31**
	14.38	*9.61*	*7.80*	*6.81*	*6.19*	*5.76*	*5.19*	*4.58*	*3.92*	*3.15*
23	4.28	3.42	3.03	2.80	2.64	2.53	2.38	2.20	2.00	1.76
	7.88	**5.66**	**4.76**	**4.26**	**3.94**	**3.71**	**3.41**	**3.07**	**2.70**	**2.26**
	14.19	*9.47*	*7.67*	*6.69*	*6.08*	*5.65*	*5.09*	*4.48*	*3.82*	*3.05*

df for Within variance	1	2	3	4	5	6	8	12	24	∞
24	4.26	3.40	3.01	2.78	2.62	2.51	2.36	2.18	1.98	1.73
	7.82	**5.61**	**4.72**	**4.22**	**3.90**	**3.67**	**3.36**	**3.03**	**2.66**	**2.21**
	14.03	*9.34*	*7.55*	*6.59*	*5.98*	*5.55*	*4.99*	*4.39*	*3.74*	*2.97*
25	4.24	3.38	2.99	2.76	2.60	2.49	2.34	2.16	1.96	1.71
	7.77	**5.57**	**4.68**	**4.18**	**3.86**	**3.63**	**3.32**	**2.99**	**2.62**	**2.17**
	13.88	*9.22*	*7.45*	*6.49*	*5.88*	*5.46*	*4.91*	*4.31*	*3.66*	*2.89*
26	4.22	3.37	2.98	2.74	2.59	2.47	2.32	2.15	1.95	1.69
	7.72	**5.53**	**4.64**	**4.14**	**3.82**	**3.59**	**3.29**	**2.96**	**2.58**	**2.13**
	13.74	*9.12*	*7.36*	*6.41*	*5.80*	*5.38*	*4.83*	*4.24*	*3.59*	*2.82*
27	4.21	3.35	2.96	2.73	2.57	2.46	2.30	2.13	1.93	1.67
	7.68	**5.49**	**4.60**	**4.11**	**3.78**	**3.56**	**3.26**	**2.93**	**2.55**	**2.10**
	13.61	*9.02*	*7.27*	*6.33*	*5.73*	*5.31*	*4.76*	*4.17*	*3.52*	*2.75*
28	4.20	3.34	2.95	2.71	2.56	2.44	2.29	2.12	1.91	1.65
	7.64	**5.45**	**4.57**	**4.07**	**3.75**	**3.53**	**3.23**	**2.90**	**2.52**	**2.06**
	13.50	*8.93*	*7.19*	*6.25*	*5.66*	*5.24*	*4.69*	*4.11*	*3.46*	*2.70*
29	4.18	3.33	2.93	2.70	2.54	2.43	2.28	2.10	1.90	1.64
	7.60	**5.42**	**4.54**	**4.04**	**3.73**	**3.50**	**3.20**	**2.87**	**2.49**	**2.03**
	13.39	*8.85*	*7.12*	*6.19*	*5.59*	*5.18*	*4.64*	*4.05*	*3.41*	*2.64*
30	4.17	3.32	2.92	2.69	2.53	2.42	2.27	2.09	1.89	1.62
	7.56	**5.39**	**4.51**	**4.02**	**3.70**	**3.47**	**3.17**	**2.84**	**2.47**	**2.01**
	13.29	*8.77*	*7.05*	*6.12*	*5.53*	*5.12*	*4.58*	*4.00*	*3.36*	*2.59*
40	4.08	3.23	2.84	2.61	2.45	2.34	2.18	2.00	1.79	1.51
	7.31	**5.18**	**4.31**	**3.83**	**3.51**	**3.29**	**2.99**	**2.66**	**2.29**	**1.80**
	12.61	*8.25*	*6.60*	*5.70*	*5.13*	*4.73*	*4.21*	*3.64*	*3.01*	*2.23*
	df for Among variance									
	1	2	3	4	5	6	8	12	24	∞

TABLE F (*continued*)

*Key .05 in roman type; **.01 in bold face type;** and *.001 in italic type.*

df for Within variance	*df* for Among variance									
	1	2	3	4	5	6	8	12	24	∞
60	4.00	3.15	2.76	2.52	2.37	2.25	2.10	1.92	1.70	1.39
	7.08	**4.98**	**4.13**	**3.65**	**3.34**	**3.12**	**2.82**	**2.50**	**2.12**	**1.60**
	11.97	*7.76*	*6.17*	*5.31*	*4.76*	*4.37*	*3.87*	*3.31*	*2.69*	*1.90*
120	3.92	3.07	2.68	2.45	2.29	2.17	2.02	1.83	1.61	1.25
	6.85	**4.79**	**3.95**	**3.48**	**3.17**	**2.96**	**2.66**	**2.34**	**1.95**	**1.38**
	11.38	*7.31*	*5.79*	*4.95*	*4.42*	*4.04*	*3.55*	*3.02*	*2.40*	*1.56*
∞	3.84	2.99	2.60	2.37	2.21	2.09	1.94	1.75	1.52	1.00
	6.64	**4.60**	**3.78**	**3.32**	**3.02**	**2.80**	**2.51**	**2.18**	**1.79**	**1.00**
	10.83	*6.91*	*5.42*	*4.62*	*4.10*	*3.74*	*3.27*	*2.74*	*2.13*	*1.00*
	1	2	3	4	5	6	8	12	24	∞

Table F is reprinted, in rearranged form, from Table V of Fisher and Yates: *Statistical tables for biological, agricultural and medical research,* Oliver and Boyd, Ltd., Edinburgh, by permission of the authors and publishers. Thanks are also due to Quinn McNemar: *Psychological Statistics,* John Wiley and Sons, Inc., New York, for the useful rearrangement.

TABLE G
The *t* Table

Degrees of freedom (1)	.05 level of significance (2)	.01 level of significance (3)	.001 level of significance (4)
1	12.71	63.66	636.62
5	2.57	4.03	6.86
10	2.23	3.17	4.59
15	2.13	2.95	4.07
20	2.09	2.85	3.85
25	2.06	2.79	3.72
30	2.04	2.75	3.65
40	2.02	2.70	3.55
50	2.01	2.68	3.50
60	2.00	2.66	3.46
70	1.99	2.65	3.44
80	1.99	2.64	3.42
90	1.99	2.63	3.40
100	1.98	2.63	3.39
150	1.98	2.61	3.36
200	1.97	2.60	3.34
300	1.97	2.59	3.32
400	1.97	2.59	3.32
500	1.96	2.59	3.31
1000	1.96	2.58	3.30
∞	1.96	2.58	3.29

TABLE H
Interpretation of *P* Levels

(1) Precise Designation	(2) Usual Designation	(3) Symbol	(4) Interpretation
$P > .05$	$P > .05$	NS	If P is greater than .05, the difference has not been shown to be significant
$.05 > P > .01$	$P < .05$	*	If P is between .05 and .01, the difference is barely significant
$.01 > P > .001$	$P < .01$	**	If P is between .01 and .001, the difference is highly significant
$P < .001$	$P < .001$	***	If P is less than .001, the difference is very highly significant

TABLE J
Multiplication Table

1	2	3	4	5	6	7	8	9	10	11	12	13	14	15	16	17	18	19	20	1
2	4	6	8	10	12	14	16	18	20	22	24	26	28	30	32	34	36	38	40	2
3	6	9	12	15	18	21	24	27	30	33	36	39	42	45	48	51	54	57	60	3
4	8	12	16	20	24	28	32	36	40	44	48	52	56	60	64	68	72	76	80	4
5	10	15	20	25	30	35	40	45	50	55	60	65	70	75	80	85	90	95	100	5
6	12	18	24	30	36	42	48	54	60	66	72	78	84	90	96	102	108	114	120	6
7	14	21	28	35	42	49	56	63	70	77	84	91	98	105	112	119	126	133	140	7
8	16	24	32	40	48	56	64	72	80	88	96	104	112	120	128	136	144	152	160	8
9	18	27	36	45	54	63	72	81	90	99	108	117	126	135	144	153	162	171	180	9
10	20	30	40	50	60	70	80	90	100	110	120	130	140	150	160	170	180	190	200	10
11	22	33	44	55	66	77	88	99	110	121	132	143	154	165	176	187	198	209	220	11
12	24	36	48	60	72	84	96	108	120	132	144	156	168	180	192	204	216	228	240	12
13	26	39	52	65	78	91	104	117	130	143	156	169	182	195	208	221	234	247	260	13
14	28	42	56	70	84	98	112	126	140	154	168	182	196	210	224	238	252	266	280	14
15	30	45	60	75	90	105	120	135	150	165	180	195	210	225	240	255	270	285	300	15
16	32	48	64	80	96	112	128	144	160	176	192	208	224	240	256	272	288	304	320	16
17	34	51	68	85	102	119	136	153	170	187	204	221	238	255	272	289	306	323	340	17
18	36	54	72	90	108	126	144	162	180	198	216	234	252	270	288	306	324	342	360	18
19	38	57	76	95	114	133	152	171	190	209	228	247	266	285	304	323	342	361	380	19
20	40	60	80	100	120	140	160	180	200	220	240	260	280	300	320	340	360	380	400	20
21	42	63	84	105	126	147	168	189	210	231	252	273	294	315	336	357	378	399	420	21
22	44	66	88	110	132	154	176	198	220	242	264	286	308	330	352	374	396	418	440	22
23	46	69	92	115	138	161	184	207	230	253	276	299	322	345	368	391	414	437	460	23
24	48	72	96	120	144	168	192	216	240	264	288	312	336	360	384	408	432	456	480	24

25	50	75	100	125	150	175	200	225	250	275	300	325	350	375	400	425	450	475	500	25
26	52	78	104	130	156	182	208	234	260	286	312	338	364	390	416	442	468	494	520	26
27	54	81	108	135	162	189	216	243	270	297	324	351	378	405	432	459	486	513	540	27
28	56	84	112	140	168	196	224	252	280	308	336	364	392	420	448	476	504	532	560	28
29	58	87	116	145	174	203	232	261	290	319	348	377	406	435	464	493	522	551	580	29
30	60	90	120	150	180	210	240	270	300	330	360	390	420	450	480	510	540	570	600	30
31	62	93	124	155	186	217	248	279	310	341	372	403	434	465	496	527	558	589	620	31
32	64	96	128	160	192	224	256	288	320	352	384	416	448	480	512	544	576	608	640	32
33	66	99	132	165	198	231	264	297	330	363	396	429	462	495	528	561	594	627	660	33
34	68	102	136	170	204	238	272	306	340	374	408	442	476	510	544	578	612	646	680	34
35	70	105	140	175	210	245	280	315	350	385	420	455	490	525	560	595	630	665	700	35
36	72	108	144	180	216	252	288	324	360	396	432	468	504	540	576	612	648	684	720	36
37	74	111	148	185	222	259	296	333	370	407	444	481	518	555	592	629	666	703	740	37
38	76	114	152	190	228	266	304	342	380	418	456	494	532	570	608	646	684	722	760	38
39	78	117	156	195	234	273	312	351	390	429	468	507	546	585	624	663	702	741	780	39
40	80	120	160	200	240	280	320	360	400	440	480	520	560	600	640	680	720	760	800	40
41	82	123	164	205	246	287	328	369	410	451	492	533	574	615	656	697	738	779	820	41
42	84	126	168	210	252	294	336	378	420	462	504	546	588	630	672	714	756	798	840	42
43	86	129	172	215	258	301	344	387	430	473	516	559	602	645	688	731	774	817	860	43
44	88	132	176	220	264	308	352	396	440	484	528	572	616	660	704	748	792	836	880	44
45	90	135	180	225	270	315	360	405	450	495	540	585	630	675	720	765	810	855	900	45
46	92	138	184	230	276	322	368	414	460	506	552	598	644	690	736	782	828	874	920	46
47	94	141	188	235	282	329	376	423	470	517	564	611	658	705	752	799	846	893	940	47
48	96	144	192	240	288	336	384	432	480	528	576	624	672	720	768	816	864	912	960	48
49	98	147	196	245	294	343	392	441	490	539	588	637	686	735	784	833	882	931	980	49
1	2	3	4	5	6	7	8	9	10	11	12	13	14	15	16	17	18	19	20	1

TABLE J (*continued*)

1	21	22	23	24	25	26	27	28	29	30	31	32	33	34	1
2	42	44	46	48	50	52	54	56	58	60	62	64	66	68	2
3	63	66	69	72	75	78	81	84	87	90	93	96	99	102	3
4	84	88	92	96	100	104	108	112	116	120	124	128	132	136	4
5	105	110	115	120	125	130	135	140	145	150	155	160	165	170	5
6	126	132	138	144	150	156	162	168	174	180	186	192	198	204	6
7	147	154	161	168	175	182	189	196	203	210	217	224	231	238	7
8	168	176	184	192	200	208	216	224	232	240	248	256	264	272	8
9	189	198	207	216	225	234	243	252	261	270	279	288	297	306	9
10	210	220	230	240	250	260	270	280	290	300	310	320	330	340	10
11	231	242	253	264	275	286	297	308	319	330	341	352	363	374	11
12	252	264	276	288	300	312	324	336	348	360	372	384	396	408	12
13	273	286	299	312	325	338	351	364	377	390	403	416	429	442	13
14	294	308	322	336	350	364	378	392	406	420	434	448	462	476	14
15	315	330	345	360	375	390	405	420	435	450	465	480	495	510	15
16	336	352	368	384	400	416	431	448	464	480	496	512	528	544	16
17	357	374	391	408	425	422	459	476	493	510	527	544	561	578	17
18	378	396	414	432	450	468	486	504	522	540	558	576	594	612	18
19	399	418	437	456	475	494	513	532	551	570	589	608	627	646	19
20	420	440	460	480	500	520	540	560	580	600	620	640	660	680	20
21	441	462	483	504	525	546	567	588	609	630	651	672	693	714	21
22	462	484	506	528	550	572	594	616	638	660	682	704	726	748	22
23	483	506	529	552	575	598	621	644	667	690	713	736	759	782	23
24	504	528	552	576	600	624	648	672	696	720	744	768	792	816	24

1	21	22	23	24	25	26	27	28	29	30	31	32	33	34	1
25	525	550	575	600	625	650	675	700	725	750	775	800	825	850	25
26	546	572	598	624	650	676	702	728	754	780	806	832	858	884	26
27	567	594	621	648	675	702	729	756	783	810	837	864	891	918	27
28	588	616	644	672	700	728	756	784	812	840	868	896	924	952	28
29	609	638	667	696	725	754	783	812	841	870	899	928	957	986	29
30	630	660	690	720	750	780	810	840	870	900	930	960	990	1020	30
31	651	682	713	744	775	806	837	868	899	930	961	992	1023	1054	31
32	672	704	736	768	800	832	864	896	928	960	992	1024	1056	1088	32
33	693	726	759	792	825	858	891	924	957	990	1023	1056	1089	1122	33
34	714	748	782	816	850	884	918	952	986	1020	1054	1088	1122	1156	34
35	735	770	805	840	875	910	945	980	1015	1050	1085	1120	1155	1190	35
36	756	792	828	864	900	936	972	1008	1044	1080	1116	1152	1188	1224	36
37	777	814	851	888	925	962	999	1036	1073	1110	1147	1184	1221	1258	37
38	798	836	874	912	950	988	1026	1064	1102	1140	1178	1216	1254	1292	38
39	819	858	897	936	975	1014	1053	1092	1131	1170	1209	1248	1287	1326	39
40	840	880	920	960	1000	1040	1080	1120	1160	1200	1240	1280	1320	1360	40
41	861	902	943	984	1025	1066	1107	1148	1189	1230	1271	1312	1353	1394	41
42	882	924	966	1008	1050	1092	1134	1176	1218	1260	1302	1344	1386	1428	42
43	903	946	989	1032	1075	1118	1161	1204	1247	1290	1333	1376	1419	1462	43
44	924	968	1012	1056	1100	1144	1188	1232	1276	1320	1364	1408	1452	1496	44
45	945	990	1035	1080	1125	1170	1215	1260	1305	1350	1395	1440	1485	1530	45
46	966	1012	1058	1104	1150	1196	1242	1288	1334	1380	1426	1472	1518	1564	46
47	987	1034	1081	1128	1175	1222	1269	1316	1363	1410	1457	1504	1551	1598	47
48	1008	1056	1104	1152	1200	1248	1296	1344	1392	1440	1488	1536	1584	1632	48
49	1029	1078	1127	1176	1225	1274	1323	1372	1421	1470	1519	1568	1617	1666	49

TABLE K
Transformation of r into z'

	.000	.001	.002	.003	.004	.005	.006	.007	.008	.009	
.00											.00
.01	To four decimal places, $r = z'$ for all values of r or z' from .0000 to .0530										.01
.02											.02
.03											.03
.04											.04
.05	.0500	.0510	.0520	.0530	.0541	.0551	.0561	.0571	.0581	.0591	.05
.06	.0601	.0611	.0621	.0631	.0641	.0651	.0661	.0671	.0681	.0691	.06
.07	.0701	.0711	.0721	.0731	.0741	.0751	.0761	.0772	.0782	.0792	.07
.08	.0802	.0812	.0822	.0832	.0842	.0852	.0862	.0872	.0882	.0892	.08
.09	.0902	.0913	.0923	.0933	.0943	.0953	.0963	.0973	.0983	.0993	.09
.10	.1003	.1013	.1024	.1034	.1044	.1054	.1064	.1074	.1084	.1094	.10
.11	.1104	.1115	.1125	.1135	.1145	.1155	.1165	.1175	.1186	.1196	.11
.12	.1206	.1216	.1226	.1236	.1246	.1257	.1267	.1277	.1287	.1297	.12
.13	.1307	.1318	.1328	.1338	.1348	.1358	.1368	.1379	.1389	.1399	.13
.14	.1409	.1419	.1430	.1440	.1450	.1460	.1471	.1481	.1491	.1501	.14
.15	.1511	.1522	.1532	.1542	.1552	.1563	.1573	.1583	.1593	.1604	.15
.16	.1614	.1624	.1634	.1645	.1655	.1665	.1676	.1686	.1696	.1706	.16
.17	.1717	.1727	.1737	.1748	.1758	.1768	.1779	.1789	.1799	.1809	.17
.18	.1820	.1830	.1841	.1851	.1861	.1872	.1882	.1892	.1903	.1913	.18
.19	.1923	.1934	.1944	.1955	.1965	.1975	.1986	.1996	.2007	.2017	.19
.20	.2027	.2038	.2048	.2059	.2069	.2079	.2090	.2100	.2111	.2121	.20
.21	.2132	.2142	.2153	.2163	.2174	.2184	.2195	.2205	.2216	.2226	.21
.22	.2237	.2247	.2258	.2268	.2279	.2289	.2300	.2310	.2321	.2331	.22
.23	.2342	.2352	.2363	.2374	.2384	.2395	.2405	.2416	.2427	.2437	.23
.24	.2448	.2458	.2469	.2480	.2490	.2501	.2512	.2522	.2533	.2543	.24

	.000	.001	.002	.003	.004	.005	.006	.007	.008	.009	
.25	.2554	.2565	.2575	.2586	.2597	.2608	.2618	.2629	.2640	.2650	.25
.26	.2661	.2672	.2683	.2693	.2704	.2715	.2726	.2736	.2747	.2758	.26
.27	.2769	.2779	.2790	.2801	.2812	.2823	.2833	.2844	.2855	.2866	.27
.28	.2877	.2888	.2899	.2909	.2920	.2931	.2942	.2953	.2964	.2975	.28
.29	.2986	.2997	.3008	.3018	.3029	.3040	.3051	.3062	.3073	.3084	.29
.30	.3095	.3106	.3117	.3128	.3139	.3150	.3161	.3172	.3183	.3194	.30
.31	.3205	.3217	.3228	.3239	.3250	.3261	.3272	.3283	.3294	.3305	.31
.32	.3316	.3328	.3339	.3350	.3361	.3372	.3383	.3395	.3406	.3417	.32
.33	.3428	.3440	.3451	.3462	.3473	.3484	.3496	.3507	.3518	.3530	.33
.34	.3541	.3552	.3564	.3575	.3586	.3598	.3609	.3620	.3632	.3643	.34
.35	.3654	.3666	.3677	.3689	.3700	.3712	.3723	.3734	.3746	.3757	.35
.36	.3769	.3780	.3792	.3803	.3815	.3826	.3838	.3850	.3861	.3873	.36
.37	.3884	.3896	.3907	.3919	.3931	.3942	.3954	.3966	.3977	.3989	.37
.38	.4001	.4012	.4024	.4036	.4047	.4059	.4071	.4083	.4094	.4106	.38
.39	.4118	.4130	.4142	.4153	.4165	.4177	.4189	.4201	.4213	.4225	.39
.40	.4236	.4248	.4260	.4272	.4284	.4296	.4308	.4320	.4332	.4344	.40
.41	.4356	.4368	.4380	.4392	.4404	.4416	.4428	.4441	.4453	.4465	.41
.42	.4477	.4489	.4501	.4513	.4526	.4538	.4550	.4562	.4574	.4587	.42
.43	.4599	.4611	.4624	.4636	.4648	.4660	.4673	.4685	.4698	.4710	.43
.44	.4722	.4735	.4747	.4760	.4772	.4784	.4797	.4809	.4822	.4834	.44
.45	.4847	.4860	.4872	.4885	.4897	.4910	.4922	.4935	.4948	.4960	.45
.46	.4973	.4986	.4999	.5011	.5024	.5037	.5049	.5062	.5075	.5088	.46
.47	.5101	.5114	.5126	.5139	.5152	.5165	.5178	.5191	.5204	.5217	.47
.48	.5230	.5243	.5256	.5269	.5282	.5295	.5308	.5321	.5334	.5347	.48
.49	.5361	.5374	.5387	.5400	.5413	.5427	.5440	.5453	.5466	.5480	.49

TABLE K *(continued)*

	.000	.001	.002	.003	.004	.005	.006	.007	.008	.009	
.50	.5493	.5506	.5520	.5533	.5547	.5560	.5573	.5587	.5600	.5614	.50
.51	.5627	.5641	.5654	.5668	.5682	.5695	.5709	.5722	.5736	.5750	.51
.52	.5763	.5777	.5791	.5805	.5818	.5832	.5846	.5860	.5874	.5888	.52
.53	.5901	.5915	.5929	.5943	.5957	.5971	.5985	.5999	.6013	.6027	.53
.54	.6042	.6056	.6070	.6084	.6098	.6112	.6127	.6141	.6155	.6169	.54
.55	.6184	.6198	.6213	.6227	.6241	.6256	.6270	.6285	.6299	.6314	.55
.56	.6328	.6343	.6358	.6372	.6387	.6401	.6416	.6431	.6446	.6460	.56
.57	.6475	.6490	.6505	.6520	.6535	.6550	.6565	.6580	.6595	.6610	.57
.58	.6625	.6640	.6655	.6670	.6685	.6700	.6716	.6731	.6746	.6761	.58
.59	.6777	.6792	.6807	.6823	.6838	.6854	.6869	.6885	.6900	.6916	.59
.60	.6931	.6947	.6963	.6978	.6994	.7010	.7026	.7042	.7057	.7073	.60
.61	.7089	.7105	.7121	.7137	.7153	.7169	.7185	.7201	.7218	.7234	.61
.62	.7250	.7266	.7283	.7299	.7315	.7332	.7348	.7365	.7381	.7398	.62
.63	.7414	.7431	.7447	.7464	.7481	.7498	.7514	.7531	.7548	.7565	.63
.64	.7582	.7599	.7616	.7633	.7650	.7667	.7684	.7701	.7718	.7736	.64
.65	.7753	.7770	.7788	.7805	.7823	.7840	.7858	.7875	.7893	.7910	.65
.66	.7928	.7946	.7964	.7981	.7999	.8017	.8035	.8053	.8071	.8089	.66
.67	.8107	.8126	.8144	.8162	.8180	.8199	.8217	.8236	.8254	.8273	.67
.68	.8291	.8310	.8328	.8347	.8366	.8385	.8404	.8423	.8441	.8460	.68
.69	.8480	.8499	.8518	.8537	.8556	.8576	.8595	.8614	.8634	.8653	.69
.70	.8673	.8693	.8712	.8732	.8752	.8772	.8792	.8812	.8832	.8852	.70
.71	.8872	.8892	.8912	.8933	.8953	.8973	.8994	.9014	.9035	.9056	.71
.72	.9076	.9097	.9118	.9139	.9160	.9181	.9202	.9223	.9245	.9266	.72
.73	.9287	.9309	.9330	.9352	.9373	.9395	.9417	.9439	.9461	.9483	.73
.74	.9505	.9527	.9549	.9571	.9594	.9616	.9639	.9661	.9684	.9707	.74

	.000	.001	.002	.003	.004	.005	.006	.007	.008	.009	
.75	.9730	.9752	.9775	.9798	.9822	.9845	.9868	.9892	.9915	.9939	.75
.76	.9962	.9986	1.0010	1.0034	1.0058	1.0082	1.0106	1.0130	1.0154	1.0179	.76
.77	1.0203	1.0228	1.0253	1.0277	1.0302	1.0327	1.0352	1.0378	1.0403	1.0428	.77
.78	1.0454	1.0479	1.0505	1.0531	1.0557	1.0583	1.0609	1.0635	1.0661	1.0688	.78
.79	1.0714	1.0741	1.0768	1.0795	1.0822	1.0849	1.0876	1.0903	1.0931	1.0958	.79
.80	1.0986	1.1014	1.1042	1.1070	1.1098	1.1127	1.1155	1.1184	1.1212	1.1241	.80
.81	1.1270	1.1299	1.1329	1.1358	1.1388	1.1417	1.1447	1.1477	1.1507	1.1538	.81
.82	1.1568	1.1599	1.1630	1.1660	1.1692	1.1723	1.1754	1.1786	1.1817	1.1849	.82
.83	1.1881	1.1914	1.1946	1.1979	1.2011	1.2044	1.2077	1.2111	1.2144	1.2178	.83
.84	1.2212	1.2246	1.2280	1.2315	1.2349	1.2384	1.2419	1.2454	1.2490	1.2526	.84
.85	1.2562	1.2598	1.2634	1.2671	1.2707	1.2745	1.2782	1.2819	1.2857	1.2895	.85
.86	1.2933	1.2972	1.3011	1.3050	1.3089	1.3129	1.3169	1.3209	1.3249	1.3290	.86
.87	1.3331	1.3372	1.3414	1.3456	1.3498	1.3540	1.3583	1.3626	1.3670	1.3714	.87
.88	1.3758	1.3802	1.3847	1.3892	1.3938	1.3984	1.4030	1.4077	1.4124	1.4171	.88
.89	1.4219	1.4268	1.4316	1.4365	1.4415	1.4465	1.4516	1.4566	1.4618	1.4670	.89
.90	1.4722	1.4775	1.4828	1.4882	1.4937	1.4992	1.5047	1.5103	1.5160	1.5217	.90
.91	1.5275	1.5334	1.5393	1.5453	1.5513	1.5574	1.5636	1.5698	1.5762	1.5826	.91
.92	1.5890	1.5956	1.6022	1.6089	1.6157	1.6226	1.6296	1.6366	1.6438	1.6510	.92
.93	1.6584	1.6658	1.6734	1.6811	1.6888	1.6967	1.7047	1.7129	1.7211	1.7295	.93
.94	1.7380	1.7467	1.7555	1.7645	1.7736	1.7828	1.7923	1.8019	1.8117	1.8216	.94
.95	1.8318	1.8421	1.8527	1.8635	1.8745	1.8857	1.8972	1.9090	1.9210	1.9333	.95
.96	1.9459	1.9588	1.9721	1.9857	1.9996	2.0139	2.0287	2.0439	2.0595	2.0756	.96
.97	2.0923	2.1095	2.1273	2.1457	2.1649	2.1847	2.2054	2.2269	2.2494	2.2729	.97
.98	2.2976	2.3235	2.3507	2.3796	2.4101	2.4427	2.4774	2.5147	2.5550	2.5987	.98
.99	2.6467	2.6996	2.7587	2.8257	2.9031	2.9945	3.1063	3.2504	3.4534	3.8002	.99

TABLE L
Standard Error of z' and Multiples of $s_{z'}$ with Their Confidence Limits

N	$s_{z'}$	$1.96s_{z'}$ (.95)	$2.58s_{z'}$ (.99)	$3.29s_{z'}$ (.999)
50	.1459	.2859	.3757	.4800
51	.1443	.2829	.3718	.4749
52	.1429	.2800	.3680	.4701
53	.1414	.2772	.3643	.4654
54	.1400	.2744	.3607	.4608
55	.1387	.2718	.3572	.4563
56	.1374	.2692	.3538	.4520
57	.1361	.2667	.3505	.4478
58	.1348	.2643	.3473	.4437
59	.1336	.2619	.3442	.4397
60	.1325	.2596	.3412	.4358
61	.1313	.2574	.3382	.4321
62	.1302	.2552	.3353	.4284
63	.1291	.2530	.3325	.4248
64	.1280	.2509	.3298	.4213
65	.1270	.2489	.3271	.4179
66	.1260	.2469	.3245	.4146
67	.1250	.2450	.3220	.4113
68	.1240	.2431	.3195	.4081
69	.1231	.2413	.3171	.4050
70	.1222	.2394	.3147	.4020
71	.1213	.2377	.3124	.3990
72	.1204	.2360	.3101	.3961
73	.1195	.2343	.3079	.3933
74	.1187	.2326	.3057	.3905
75	.1179	.2310	.3036	.3878
76	.1170	.2294	.3015	.3851
77	.1162	.2278	.2994	.3825
78	.1155	.2263	.2974	.3800
79	.1147	.2248	.2955	.3774
80	.1140	.2234	.2935	.3750
81	.1132	.2219	.2917	.3726
82	.1125	.2205	.2898	.3702
83	.1118	.2191	.2880	.3679
84	.1111	.2178	.2862	.3656
85	.1104	.2164	.2845	.3634
86	.1098	.2151	.2827	.3612
87	.1091	.2138	.2810	.3590
88	.1085	.2126	.2794	.3569
89	.1078	.2113	.2778	.3548

TABLE L
(continued)

N	$s_{z'}$	$1.96s_{z'}$ (.95)	$2.58s_{z'}$ (.99)	$3.29s_{z'}$ (.999)
90	.1072	.2101	.2762	.3528
91	.1066	.2089	.2746	.3508
92	.1060	.2078	.2730	.3488
93	.1054	.2066	.2715	.3469
94	.1048	.2055	.2700	.3449
95	.1043	.2043	.2685	.3431
96	.1037	.2032	.2671	.3412
97	.1031	.2022	.2657	.3394
98	.1026	.2011	.2643	.3376
99	.1021	.2000	.2629	.3358
100	.1015	.1990	.2615	.3341
101	.1010	.1980	.2602	.3324
102	.1005	.1970	.2589	.3307
103	.1000	.1960	.2576	.3291
104	.0995	.1950	.2563	.3274
105	.0990	.1941	.2550	.3258
106	.0985	.1931	.2538	.3242
107	.0981	.1922	.2526	.3227
108	.0976	.1913	.2514	.3211
109	.0971	.1904	.2502	.3196
110	.0967	.1895	.2490	.3181
111	.0962	.1886	.2479	.3166
112	.0958	.1877	.2467	.3152
113	.0953	.1869	.2456	.3137
114	.0949	.1860	.2445	.3123
115	.0945	.1852	.2434	.3109
116	.0941	.1844	.2423	.3095
117	.0937	.1836	.2412	.3082
118	.0933	.1828	.2402	.3068
119	.0928	.1820	.2392	.3055
120	.0925	.1812	.2381	.3042
121	.0921	.1804	.2371	.3029
122	.0917	.1797	.2361	.3016
123	.0913	.1789	.2351	.3004
124	.0909	.1782	.2342	.2991
125	.0905	.1774	.2332	.2979
126	.0902	.1767	.2323	.2967
127	.0898	.1760	.2313	.2955
128	.0894	.1753	.2304	.2943
129	.0891	.1746	.2295	.2931

TABLE L
(*continued*)

N	$s_{z'}$	$1.96s_{z'}$ (.95)	$2.58s_{z'}$ (.99)	$3.29s_{z'}$ (.999)
130	.0887	.1739	.2286	.2920
131	.0884	.1732	.2277	.2908
132	.0880	.1726	.2268	.2897
133	.0877	.1719	.2259	.2886
134	.0874	.1712	.2251	.2875
135	.0870	.1706	.2242	.2864
136	.0867	.1700	.2234	.2853
137	.0864	.1693	.2225	.2843
138	.0861	.1687	.2217	.2832
139	.0857	.1681	.2209	.2822
140	.0854	.1675	.2201	.2811
141	.0851	.1668	.2193	.2801
142	.0848	.1662	.2185	.2791
143	.0845	.1656	.2177	.2781
144	.0842	.1651	.2169	.2771
145	.0839	.1645	.2162	.2761
146	.0836	.1639	.2154	.2752
147	.0833	.1633	.2147	.2742
148	.0830	.1628	.2139	.2733
149	.0828	.1622	.2132	.2723
150	.0825	.1617	.2125	.2714
155	.0811	.1590	.2089	.2669
160	.0798	.1564	.2056	.2626
165	.0786	.1540	.2024	.2585
170	.0774	.1517	.1993	.2546
175	.0762	.1494	.1964	.2509
180	.0752	.1473	.1936	.2473
185	.0741	.1453	.1909	.2439
190	.0731	.1433	.1884	.2406
195	.0722	.1414	.1859	.2375
200	.0712	.1396	.1835	.2344
205	.0704	.1379	.1812	.2315
210	.0695	.1362	.1790	.2287
215	.0687	.1346	.1769	.2260
220	.0679	.1331	.1749	.2234
225	.0671	.1315	.1729	.2208
230	.0664	.1301	.1710	.2184
235	.0657	.1287	.1691	.2160
240	.0650	.1273	.1673	.2137
245	.0643	.1260	.1656	.2115

TABLE L
(*continued*)

N	$s_{z'}$	$1.96s_{z'}$ (.95)	$2.58s_{z'}$ (.99)	$3.29s_{z'}$ (.999)
250	.0636	.1247	.1639	.2094
255	.0630	.1235	.1623	.2073
260	.0624	.1223	.1607	.2053
265	.0618	.1211	.1591	.2033
270	.0612	.1199	.1576	.2013
275	.0606	.1188	.1562	.1995
280	.0601	.1178	.1548	.1977
285	.0595	.1167	.1534	.1959
290	.0590	.1157	.1520	.1942
295	.0585	.1147	.1507	.1926
300	.0580	.1137	.1495	.1909
310	.0571	.1119	.1470	.1878
320	.0562	.1101	.1447	.1848
330	.0553	.1084	.1424	.1820
340	.0545	.1068	.1403	.1792
350	.0537	.1052	.1383	.1766
360	.0529	.1037	.1363	.1742
370	.0522	.1023	.1345	.1718
380	.0515	.1009	.1327	.1695
390	.0508	.0996	.1309	.1673
400	.0502	.0984	.1293	.1651
410	.0496	.0972	.1277	.1631
420	.0490	.0960	.1261	.1611
430	.0484	.0948	.1247	.1592
440	.0478	.0938	.1232	.1574
450	.0473	.0927	.1218	.1556
460	.0468	.0917	.1205	.1539
470	.0463	.0907	.1192	.1523
480	.0458	.0897	.1179	.1507
490	.0453	.0888	.1167	.1491
500	.0449	.0879	.1155	.1476
520	.0440	.0862	.1133	.1447
540	.0432	.0846	.1112	.1420
560	.0424	.0830	.1091	.1394
580	.0416	.0816	.1072	.1370
600	.0409	.0802	.1054	.1347
620	.0403	.0789	.1037	.1325
640	.0396	.0777	.1021	.1304
660	.0390	.0765	.1005	.1284
680	.0384	.0753	.0990	.1265

TABLE L
(*continued*)

N	$s_{z'}$	$1.96s_{z'}$ (.95)	$2.58s_{z'}$ (.99)	$3.29s_{z'}$ (.999)
700	.0379	.0742	.0976	.1246
720	.0373	.0732	,0962	.1229
740	.0368	.0722	.0949	.1212
760	.0363	.0712	.0936	.1196
780	.0359	.0703	.0924	.1180
800	.0354	.0694	.0912	.1166
820	.0350	.0686	.0901	.1151
840	.0346	.0677	.0890	.1137
860	.0342	.0670	.0880	.1124
880	.0338	.0662	.0870	.1111
900	.0334	.0654	.0860	.1099
920	.0330	.0647	.0851	.1087
940	.0327	.0640	.0841	.1075
960	.0323	.0634	.0833	.1064
980	.0320	.0627	.0824	.1053
1000	.0317	.0621	.0816	.1042
1100	.0302	.0592	.0778	.0993
1200	.0289	.0567	.0745	.0951
1300	.0278	.0544	.0715	.0914
1400	.0268	.0524	.0689	.0880
1500	.0258	.0507	.0666	.0850
1600	.0250	.0490	.0645	.0823
1700	.0243	.0476	.0625	.0799
1800	.0236	.0462	.0608	.0776
1900	.0230	.0450	.0591	.0755
2000	.0224	.0439		.0736
2100	.0218	.0428	.0562	.0719
2200	.0213	.0418	.0550	.0702
2300	.0209	.0409	.0537	.0687
2400	.0204	.0400	.0526	.0672
2500	.0200	.0392	.0515	.0659
3000	.0183	.0358	.0471	.0601
3500	.0169	.0331	.0436	.0556
4000	.0158	.0310	.0407	.0520
4500	.0149	.0292	.0384	.0491
5000	.0141	.0277	.0364	.0465
6000	.0129	.0253	.0333	.0425
7000	.0120	.0234	.0308	.0393
8000	.0112	.0219	.0288	.0368
9000	.0105	.0207	.0272	.0347

TABLE M
The Smallest Values of Pearson *r* that Are Significant at the .05 and .01 Levels

N	*r* at .05 Level	*r* at .01 Level
10	.632	.765
15	.514	.641
20	.444	.561
25	.396	.505
30	.361	.463
32	.349	.449
37	.325	.418
42	.304	.393
47	.288	.372
52	.273	.354
62	.250	.325
72	.232	.302
82	.217	.283
92	.205	.267
102	.195	.254
127	.174	.228
152	.159	.208
202	.138	.181
302	.113	.148
402	.098	.128
502	.088	.115
1002	.062	.081

Note: *N* here refers to sample size, not *df* as in similar tables.

TABLE P

The Normal Curve Ordinate, y, for a Given p or q

p	.000	.001	.002	.003	.004	.005	.006	.007	.008	.009	
.000	.0000	.0034	.0063	.0091	.0118	.0145	.0170	.0195	.0219	.0243	**.000**
.010	.0267	.0290	.0312	.0335	.0357	.0379	.0400	.0422	.0443	.0464	**.010**
.020	.0484	.0505	.0525	.0545	.0565	.0584	.0604	.0623	.0643	.0662	**.020**
.030	.0680	.0699	.0718	.0736	.0755	.0773	.0791	.0809	.0826	.0844	**.030**
.040	.0862	.0879	.0897	.0914	.0931	.0948	.0965	.0982	.0998	.1015	**.040**
.050	.1031	.1048	.1064	.1080	.1096	.1112	.1128	.1144	.1160	.1176	**.050**
.060	.1191	.1207	.1222	.1237	.1253	.1268	.1283	.1298	.1313	.1328	**.060**
.070	.1343	.1357	.1372	.1387	.1401	.1416	.1430	.1444	.1458	.1473	**.070**
.080	.1487	.1501	.1515	.1529	.1542	.1556	.1570	.1583	.1597	.1610	**.080**
.090	.1624	.1637	.1651	.1664	.1677	.1690	.1703	.1716	.1729	.1742	**.090**
.100	.1755	.1768	.1781	.1793	.1806	.1818	.1831	.1843	.1856	.1868	**.100**
.110	.1880	.1893	.1905	.1917	.1929	.1941	.1953	.1965	.1977	.1989	**.110**
.120	.2000	.2012	.2024	.2035	.2047	.2059	.2070	.2081	.2093	.2104	**.120**
.130	.2115	.2127	.2138	.2149	.2160	.2171	.2182	.2193	.2204	.2215	**.130**
.140	.2226	.2237	.2247	.2258	.2269	.2279	.2290	.2300	.2311	.2321	**.140**
.150	.2332	.2342	.2352	.2362	.2373	.2383	.2393	.2403	.2413	.2423	**.150**
.160	.2433	.2443	.2453	.2463	.2473	.2482	.2492	.2502	.2511	.2521	**.160**
.170	.2531	.2540	.2550	.2559	.2568	.2578	.2587	.2596	.2606	.2615	**.170**
.180	.2624	.2633	.2642	.2651	.2660	.2669	.2678	.2687	.2696	.2705	**.180**
.190	.2714	.2722	.2731	.2740	.2748	.2757	.2766	.2774	.2783	.2791	**.190**
.200	.2800	.2808	.2816	.2825	.2833	.2841	.2849	.2858	.2866	.2874	**.200**
.210	.2882	.2890	.2898	.2906	.2914	.2922	.2930	.2938	.2945	.2953	**.210**
.220	.2961	.2969	.2976	.2984	.2992	.2999	.3007	.3014	.3022	.3029	**.220**
.230	.3036	.3044	.3051	.3058	.3066	.3073	.3080	.3087	.3095	.3102	**.230**
.240	.3109	.3116	.3123	.3130	.3137	.3144	.3151	.3157	.3164	.3171	**.240**

.250	.3178	.3184	.3191	.3198	.3204	.3211	.3218	.3224	.3231	.3237	**.250**
.260	.3244	.3250	.3256	.3263	.3269	.3275	.3282	.3288	.3294	.3300	**.260**
.270	.3306	.3313	.3319	.3325	.3331	.3337	.3343	.3349	.3355	.3360	**.270**
.280	.3366	.3372	.3378	.3384	.3389	.3395	.3401	.3406	.3412	.3417	**.280**
.290	.3423	.3429	.3434	.3440	.3445	.3450	.3456	.3461	.3466	.3472	**.290**
.300	.3477	.3482	.3487	.3493	.3498	.3503	.3508	.3513	.3518	.3523	**.300**
.310	.3528	.3533	.3538	.3543	.3548	.3552	.3557	.3562	.3567	.3571	**.310**
.320	.3576	.3581	.3585	.3590	.3595	.3599	.3604	.3608	.3613	.3617	**.320**
.330	.3621	.3626	.3630	.3635	.3639	.3643	.3647	.3652	.3656	.3660	**.330**
.340	.3664	.3668	.3672	.3676	.3680	.3684	.3688	.3692	.3696	.3700	**.340**
.350	.3704	.3708	.3712	.3715	.3719	.3723	.3727	.3730	.3734	.3738	**.350**
.360	.3741	.3745	.3748	.3752	.3755	.3759	.3762	.3766	.3769	.3772	**.360**
.370	.3776	.3779	.3782	.3786	.3789	.3792	.3795	.3798	.3801	.3804	**.370**
.380	.3808	.3811	.3814	.3817	.3820	.3823	.3825	.3828	.3831	.3834	**.380**
.390	.3837	.3840	.3842	.3845	.3848	.3850	.3853	.3856	.3858	.3861	**.390**
.400	.3863	.3866	.3868	.3871	.3873	.3876	.3878	.3881	.3883	.3885	**.400**
.410	.3887	.3890	.3892	.3894	.3896	.3899	.3901	.3903	.3905	.3907	**.410**
.420	.3909	.3911	.3913	.3915	.3917	.3919	.3921	.3922	.3924	.2926	**.420**
.430	.3928	.3930	.3931	.3933	.3935	.3936	.3938	.3940	.3941	.3943	**.430**
.440	.3944	.3946	.3947	.3949	.3950	.3951	.3953	.3954	.3955	.3957	**.440**
.450	.3958	.3959	.3961	.3962	.3963	.3964	.3965	.3966	.3967	.3968	**.450**
.460	.3969	.3970	.3971	.3972	.3973	.3974	.3975	.3976	.3977	.3977	**.460**
.470	.3978	.3979	.3980	.3980	.3981	.3982	.3982	.3983	.3983	.3984	**.470**
.480	.3984	.3985	.3985	.3986	.3986	.3987	.3987	.3987	.3988	.3988	**.480**
.490	.3988	.3988	.3989	.3989	.3989	.3989	.3989	.3989	.3989	.3989	**.490**
.500	.3989										**.500**
	.000	**.001**	**.002**	**.003**	**.004**	**.005**	**.006**	**.007**	**.008**	**.009**	

TABLE Q

Table of $\frac{\sqrt{pq}}{y}$

p	.000	.001	.002	.003	.004	.005	.006	.007	.008	.009	
.500	1.2533	1.2533	1.2533	1.2533	1.2533	1.2533	1.2534	1.2534	1.2534	1.2534	**.500**
.510	1.2535	1.2535	1.2535	1.2536	1.2536	1.2536	1.2537	1.2537	1.2538	1.2538	**.510**
.520	1.2539	1.2539	1.2540	1.2541	1.2541	1.2542	1.2543	1.2544	1.2544	1.2545	**.520**
.530	1.2546	1.2547	1.2548	1.2549	1.2550	1.2551	1.2552	1.2553	1.2554	1.2555	**.530**
.540	1.2556	1.2557	1.2559	1.2560	1.2561	1.2562	1.2564	1.2565	1.2566	1.2568	**.540**
.550	1.2569	1.2571	1.2572	1.2574	1.2575	1.2577	1.2578	1.2580	1.2582	1.2583	**.550**
.560	1.2585	1.2587	1.2589	1.2591	1.2592	1.2594	1.2596	1.2598	1.2600	1.2602	**.560**
.570	1.2604	1.2606	1.2608	1.2611	1.2613	1.2615	1.2617	1.2619	1.2622	1.2624	**.570**
.580	1.2626	1.2629	1.2631	1.2634	1.2636	1.2639	1.2641	1.2644	1.2646	1.2649	**.580**
.590	1.2652	1.2654	1.2657	1.2660	1.2663	1.2666	1.2669	1.2671	1.2674	1.2677	**.590**
.600	1.2680	1.2683	1.2687	1.2690	1.2693	1.2696	1.2699	1.2702	1.2706	1.2709	**.600**
.610	1.2712	1.2716	1.2719	1.2723	1.2726	1.2730	1.2733	1.2737	1.2741	1.2744	**.610**
.620	1.2748	1.2752	1.2756	1.2759	1.2763	1.2767	1.2771	1.2775	1.2779	1.2783	**.620**
.630	1.2787	1.2791	1.2795	1.2800	1.2804	1.2808	1.2813	1.2817	1.2821	1.2826	**.630**
.640	1.2830	1.2835	1.2839	1.2844	1.2848	1.2853	1.2858	1.2863	1.2867	1.2872	**.640**
.650	1.2877	1.2882	1.2887	1.2892	1.2897	1.2902	1.2907	1.2913	1.2918	1.2923	**.650**
.660	1.2928	1.2934	1.2939	1.2945	1.2950	1.2956	1.2961	1.2967	1.2972	1.2978	**.660**
.670	1.2984	1.2990	1.2996	1.3002	1.3007	1.3014	1.3020	1.3026	1.3032	1.3038	**.670**
.680	1.3044	1.3051	1.3057	1.3063	1.3070	1.3076	1.3083	1.3089	1.3096	1.3103	**.680**
.690	1.3109	1.3116	1.3123	1.3130	1.3137	1.3144	1.3151	1.3158	1.3165	1.3173	**.690**
.700	1.3180	1.3187	1.3195	1.3202	1.3210	1.3217	1.3225	1.3233	1.3240	1.3248	**.700**
.710	1.3256	1.3264	1.3272	1.3280	1.3288	1.3296	1.3305	1.3313	1.3321	1.3330	**.710**
.720	1.3338	1.3347	1.3356	1.3364	1.3373	1.3382	1.3391	1.3400	1.3409	1.3418	**.720**
.730	1.3427	1.3436	1.3446	1.3455	1.3464	1.3474	1.3484	1.3493	1.3503	1.3513	**.730**
.740	1.3523	1.3533	1.3543	1.3553	1.3563	1.3574	1.3584	1.3594	1.3605	1.3616	**.740**

	.000	.001	.002	.003	.004	.005	.006	.007	.008	.009	
.750	1.3626	1.3637	1.3648	1.3659	1.3670	1.3681	1.3692	1.3704	1.3715	1.3727	**.750**
.760	1.3738	1.3750	1.3762	1.3773	1.3785	1.3797	1.3810	1.3822	1.3834	1.3847	**.760**
.770	1.3859	1.3872	1.3885	1.3897	1.3910	1.3923	1.3937	1.3950	1.3963	1.3977	**.770**
.780	1.3990	1.4004	1.4018	1.4032	1.4046	1.4060	1.4074	1.4089	1.4103	1.4118	**.780**
.790	1.4133	1.4148	1.4163	1.4178	1.4193	1.4209	1.4224	1.4240	1.4256	1.4272	**.790**
.800	1.4288	1.4304	1.4320	1.4337	1.4353	1.4370	1.4387	1.4404	1.4422	1.4439	**.800**
.810	1.4457	1.4474	1.4492	1.4510	1.4528	1.4547	1.4565	1.4584	1.4603	1.4622	**.810**
.820	1.4641	1.4661	1.4680	1.4700	1.4720	1.4740	1.4761	1.4781	1.4801	1.4823	**.820**
.830	1.4844	1.4865	1.4887	1.4909	1.4931	1.4953	1.4975	1.4998	1.5021	1.5044	**.830**
.840	1.5067	1.5091	1.5115	1.5139	1.5163	1.5188	1.5213	1.5238	1.5263	1.5289	**.840**
.850	1.5315	1.5341	1.5367	1.5394	1.5421	1.5448	1.5476	1.5504	1.5532	1.5561	**.850**
.860	1.5590	1.5619	1.5648	1.5678	1.5708	1.5739	1.5770	1.5801	1.5833	1.5865	**.860**
.870	1.5897	1.5930	1.5964	1.5997	1.6031	1.6066	1.6101	1.6136	1.6172	1.6208	**.870**
.880	1.6245	1.6282	1.6320	1.6358	1.6397	1.6436	1.6475	1.6516	1.6557	1.6598	**.880**
.890	1.6640	1.6682	1.6726	1.6769	1.6814	1.6859	1.6905	1.6951	1.6998	1.7046	**.890**
.900	1.7094	1.7143	1.7193	1.7244	1.7296	1.7348	1.7401	1.7455	1.7510	1.7566	**.900**
.910	1.7623	1.7681	1.7740	1.7799	1.7860	1.7922	1.7985	1.8049	1.8114	1.8181	**.910**
.920	1.8248	1.8317	1.8388	1.8459	1.8532	1.8607	1.8683	1.8760	1.8840	1.8920	**.920**
.930	1.9003	*1.9087*	*1.9173*	*1.9261*	*1.9351*	*1.9443*	*1.9537*	*1.9633*	*1.9732*	*1.9833*	**.930**
.940	*1.9936*	*2.0042*	*2.0151*	*2.0262*	*2.0377*	*2.0494*	*2.0615*	*2.0739*	*2.0866*	*2.0997*	**.940**
.950	*2.1132*	*2.1271*	*2.1414*	*2.1561*	*2.1713*	*2.1871*	*2.2033*	*2.2201*	*2.2374*	*2.2554*	**.950**
.960	*2.2740*	*2.2933*	*2.3134*	*2.3342*	*2.3558*	*2.3784*	*2.4019*	*2.4265*	*2.4521*	*2.4790*	**.960**
.970	*2.5071*	*2.5366*	*2.5677*	*2.6003*	*2.6348*	*2.6713*	*2.7100*	*2.7510*	*2.7947*	*2.8414*	**.970**
.980	*2.8915*	*2.9453*	*3.0033*	*3.0662*	*3.1347*	*3.2097*	*3.2923*	*3.3838*	*3.4861*	*3.6016*	**.980**
.990	*3.7332*	*3.8854*	*4.0641*	*4.2784*	*4.5420*	*4.8779*	*5.3278*	*5.9776*	*7.0466*	*9.3870*	**.990**
	.000	**.001**	**.002**	**.003**	**.004**	**.005**	**.006**	**.007**	**.008**	**.009**	

WARNING: Formulas (6.3) and (10.4) are not applicable when p or q is small unless N is unusually large. Certain values of $\frac{\sqrt{pq}}{y}$ are italicized in order to emphasize that they should be used only in exceptional circumstances.

TABLE R

.99 and .95 Confidence Limits for r_{pb} in Terms of N and Obtained r_{pb}

	.00	.10	.20	.30	.40	.50	.60	.70	.80	.90	
15	−.55 .55	−.50 .60	−.43 .65	−.36 .70	−.28 .74	−.17 .78	−.05 .82	.10 .86	.31 .91	.60 .95	15
	−.45 .45	−.38 .52	−.31 .57	−.22 .63	−.12 .68	−.01 .74	.12 .79	.28 .84	48 .89	.72 .94	
16	−.54 .54	−.49 .59	−.42 .64	−.34 .69	−.26 .74	−.15 .78	−.03 .82	.13 .86	.34 .91	.62 .95	16
	−.44 .44	−.37 .51	−.29 .56	−.20 .62	−.10 .68	.01 .74	.14 .78	.30 .84	.50 .89	.73 .94	
17	−.53 .53	−.47 .58	−.40 .63	−.32 .68	−.24 .73	−.13 .77	.00 .81	.16 .86	.37 .90	.64 .95	17
	−.43 .43	−.36 .50	−.28 .56	−.19 .62	−.09 .67	.03 .73	.16 .78	.32 .83	.51 .89	.74 .94	
18	−.52 .52	−.46 .58	−.39 .63	−.31 .68	−.22 .72	−.11 .77	.02 .81	.18 .86	.39 .90	.66 .95	18
	−.42 .42	−.35 .49	−.26 .55	−.17 .61	−.07 .67	.04 .72	.18 .78	.34 .83	.53 .88	.75 .94	
19	−.51 .51	−.44 .57	−.37 .62	−.29 .67	−.20 .71	−.09 .76	.04 .81	.20 .85	.41 .90	.67 .95	19
	−.41 .41	−.34 .48	−.26 .54	−.16 .60	−.06 .66	.06 .72	.20 .77	.36 .83	.54 .88	.76 .94	
20	−.50 .50	−.43 .56	−.36 .61	−.28 .67	−.18 .71	−.07 .76	.06 .81	.22 .85	.43 .90	.68 .95	20
	−.40 .40	−.33 .47	−.24 .54	−.14 .60	−.04 .66	.07 .72	.21 .77	.37 .82	.55 .88	.76 .94	
21	−.49 .49	−.42 .55	−.35 .61	−.27 .66	−.16 .71	−.05 .75	.08 .80	.24 .85	.45 .90	.70 .95	21
	−.39 .39	−.32 .46	−.23 .53	−.13 .59	−.03 .65	.08 .71	.22 .77	.38 .82	.56 .88	.77 .94	
22	−.48 .48	−.41 .54	−.34 .60	−.26 .65	−.15 .70	−.04 .75	.09 .80	.26 .85	.46 .90	.71 .95	22
	−.38 .38	−.31 .46	−.22 .52	−.12 .58	−.02 .65	.10 .71	.24 .76	.39 .82	.57 .88	.78 .94	
23	−.47 .47	−.40 .53	−.33 .60	−.24 .65	−.14 .70	−.03 .75	.11 .80	.28 .84	.47 .89	.72 .95	23
	−.38 .38	−.30 .45	−.21 .52	−.12 .58	−.02 .65	.11 .71	.25 .76	.40 .82	.58 .88	.78 .94	
24	−.47 .47	−.40 .53	−.32 .59	−.23 .64	−.13 .70	−.02 .75	.12 .80	.29 .84	.48 .89	.72 .95	24
	−.38 .38	−.29 .44	−.21 .52	−.11 .58	.00 .64	.12 .70	.26 .76	.41 .82	.59 .88	.78 .94	
25	−.46 .46	−.39 .52	−.31 .58	−.22 .64	−.12 .69	.00 .74	.13 .79	.30 .84	.50 .89	.73 .94	25
	−.37 .37	−.28 .44	−.20 .51	−.10 .58	.01 .64	.13 .70	.27 .76	.42 .82	.60 .88	.79 .94	
26	−.45 .45	−.38 .52	−.30 .58	−.21 .63	−.11 .69	.01 .74	.15 .79	.31 .84	.51 .89	.74 .94	26
	−.36 .36	−.28 .44	−.19 .50	−.09 .58	.02 .64	.14 .70	.28 .76	.43 .82	.60 .88	.79 .94	
27	−.44 .44	−.37 .51	−.29 .57	−.20 .63	−.10 .68	.02 .73	.16 .79	.32 .84	.52 .89	.75 .94	27
	−.35 .35	−.27 .43	−.18 .50	−.08 .57	.03 .63	.14 .69	.28 .75	.44 .81	.61 .87	.79 .94	

	.00	.10	.20	.30	.40	.50	.60	.70	.80	.90	
28	−.44 .44	−.37 .51	−.28 .57	−.19 .63	−.09 .68	.03 .73	.17 .79	.33 .84	.52 .89	.75 .94	28
	−.34 .34	−.26 .42	−.18 .50	−.08 .56	.04 .63	.15 .69	.29 .75	.44 .81	.62 .87	.80 .94	
29	−.44 .44	−.36 .50	−.27 .56	−.18 .62	−.08 .67	.04 .73	.18 .78	.34 .84	.53 .89	.75 .94	29
	−.34 .34	−.26 .42	−.17 .49	−.07 .56	.04 .63	.16 .69	.30 .75	.45 .81	.62 .87	.80 .94	
30	−.43 .43	−.35 .49	−.27 .56	−.18 .62	−.07 .67	.05 .73	.19 .78	.35 .84	.54 .89	.76 .94	30
	−.34 .34	−.26 .42	−.16 .48	−.06 .56	.04 .62	.17 .68	.30 .75	.46 .81	.62 .87	.80 .94	
31	−.42 .42	−.34 .49	−.26 .55	−.17 .62	−.06 .67	.06 .73	.20 .78	.36 .83	.54 .89	.77 .94	31
	−.33 .33	−.25 .41	−.16 .48	−.06 .55	.05 .62	.18 .68	.31 .75	.46 .81	.63 .87	.81 .93	
32	−.41 .41	−.34 .48	−.25 .55	−.16 .61	−.05 .67	.07 .73	.21 .78	.37 .83	.55 .89	.77 .94	32
	−.32 .32	−.24 .40	−.16 .48	−.06 .54	.06 .62	.19 .68	.32 .75	.46 .81	.64 .87	.82 .93	
37	−.39 .39	−.31 .46	−.22 .53	−.13 .59	−.02 .65	.10 .72	.24 .77	.40 .83	.58 .88	.79 .94	37
	−.31 .31	−.22 .39	−.13 .46	−.02 .54	.09 .61	.21 .67	.34 .74	.49 .80	.65 .87	.82 .93	
42	−.37 .37	−.29 .44	−.20 .52	−.10 .58	.01 .64	.13 .70	.27 .77	.43 .82	.60 .88	.80 .94	42
	−.29 .29	−.20 .37	−.11 .45	.00 .52	.11 .60	.23 .66	.36 .74	.51 .80	.66 .86	.83 .93	
47	−.35 .35	−.27 .40	−.17 .50	−.07 .57	.04 .63	.16 .69	.30 .76	.45 .82	.62 .87	.81 .93	47
	−.28 .28	−.18 .36	−.09 .44	.01 .52	.13 .59	.25 .66	.38 .73	.52 .80	.67 .86	.84 .93	
52	−.34 .34	−.25 .42	−.15 .49	−.04 .57	.06 .62	.18 .68	.32 .76	.47 .82	.64 .87	.81 .93	52
	−.26 .26	−.17 .35	−.07 .43	.03 .51	.14 .58	.26 .65	.39 .72	.53 .79	.68 .86	.84 .93	
62	−.32 .32	−.23 .40	−.13 .47	−.02 .55	.08 .61	.20 .68	.34 .75	.49 .81	.65 .87	.82 .93	62
	−.24 .24	−.15 .33	−.05 .41	.05 .49	.16 .57	.28 .64	.41 .72	.55 .78	.70 .86	.84 .93	
72	−.30 .30	−.21 .38	−.11 .45	.00 .53	.11 .60	.23 .67	.36 .74	.51 .81	.67 .87	.83 .93	72
	−.23 .23	−.13 .31	−.03 .40	.07 .48	.19 .56	.30 .63	.43 .71	.57 .78	.71 .85	.85 .93	
82	−.28 .28	−.19 .36	−.09 .44	.02 .52	.13 .59	.25 .67	.38 .73	.53 .80	.68 .86	.84 .93	82
	−.21 .21	−.11 .30	−.01 .39	.09 .47	.20 .55	.32 .62	.44 .70	.58 .78	.72 .85	.86 .92	
92	−.27 .27	−.17 .35	−.07 .43	.04 .51	.15 .58	.27 .66	.40 .72	.54 .79	.69 .86	.85 .93	92
	−.20 .20	−.10 .29	.00 .38	.10 .46	.21 .54	.33 .62	.45 .70	.58 .78	.72 .85	.86 .92	
102	−.25 .25	−.16 .34	−.05 .42	.05 .50	.16 .57	.28 .65	.41 .72	.55 .79	.69 .86	.85 .93	102
	−.19 .19	−.09 .28	.01 .37	.11 .45	.22 .53	.34 .62	.46 .70	.59 .77	.72 .85	.86 .92	

TABLE R *(continued)*

	.00	.10	.20	.30	.40	.50	.60	.70	.80	.90	
127	−.23 .23	−.13 .32	−.03 .40	.07 .48	.19 .56	.31 .64	.43 .71	.57 .79	.71 .86	.86 .93	127
	−.17 .17	−.07 .26	.03 .35	.13 .44	.24 .52	.36 .61	.48 .69	.60 .76	.73 .84	.87 .92	
152	−.21 .21	−.11 .30	.00 .38	.09 .47	.21 .55	.33 .63	.45 .70	.58 .78	.72 .85	.86 .92	152
	−.16 .16	−.06 .25	.04 .34	.15 .43	.26 .52	.37 .60	.49 .68	.62 .76	.74 .84	.87 .92	
202	−.18 .18	−.08 .27	.01 .36	.13 .45	.24 .53	.35 .62	.48 .69	.60 .77	.73 .85	.87 .92	202
	−.14 .14	−.04 .23	.06 .32	.17 .41	.28 .50	.39 .59	.51 .67	.63 .75	.75 .84	.88 .92	
302	−.15 .15	−.05 .24	.05 .33	.16 .42	.27 .51	.38 .59	.50 .68	.62 .76	.75 .84	.87 .92	302
	−.12 .12	−.02 .21	.08 .30	.19 .39	.30 .48	.41 .57	.53 .66	.64 .74	.76 .84	.88 .92	
402	−.13 .13	−.03 .22	.07 .32	.18 .41	.29 .50	.40 .58	.52 .67	.63 .75	.76 .84	.88 .92	402
	−.10 .10	.00 .19	.10 .29	.21 .38	.32 .47	.43 .56	.54 .65	.65 .74	.77 .83	.88 .91	
502	−.11 .11	−.01 .21	.09 .31	.19 .40	.30 .49	.41 .57	.53 .66	.64 .74	.76 .83	.88 .91	502
	−.09 .09	.01 .18	.11 .28	.22 .37	.33 .46	.44 .56	.54 .64	.66 .74	.77 .83	.88 .91	
1002	−.08 .08	.02 .18	.12 .28	.23 .37	.33 .46	.44 .55	.55 .64	.66 .73	.77 .82	.89 .91	1002
	−.06 .06	.04 .16	.14 .26	.24 .35	.35 .45	.45 .54	.56 .63	.67 .73	.78 .82	.89 .91	
	.00	.10	.20	.30	.40	.50	.60	.70	.80	.90	

TABLE S

F_{r_t}, a Function of Tetrachoric r

	.00	.01	.02	.03	.04	.05	.06	.07	.08	.09	
.00	1.0000	.9999	.9997	.9994	.9989	.9982	.9975	.9966	.9955	.9943	.00
.10	.9930	.9915	.9899	.9881	.9862	.9841	.9819	.9796	.9771	.9745	.10
.20	.9717	.9688	.9657	.9625	.9591	.9556	.9520	.9482	.9442	.9401	.20
.30	.9358	.9314	.9268	.9221	.9172	.9122	.9070	.9016	.8961	.8904	.30
.40	.8845	.8785	.8723	.8659	.8594	.8527	.8458	.8388	.8315	.8241	.40
.50	.8165	.8087	.8007	.7926	.7842	.7756	.7669	.7579	.7488	.7394	.50
.60	.7298	.7200	.7099	.6997	.6892	.6785	.6675	.6563	.6448	.6331	.60
.70	.6211	.6088	.5962	.5834	.5702	.5568	.5430	.5288	.5144	.4995	.70
.80	.4843	.4687	.4526	.4362	.4192	.4018	.3838	.3652	.3461	.3262	.80
.90	.3057	.2843	.2620	.2387	.2142	.1882	.1605	.1305	.0972	.0585	.90
	.00	.01	.02	.03	.04	.05	.06	.07	.08	.09	

This table of $\sqrt{(1 - r_t^2)\left[1 - \left(\frac{\sin^{-1} r_t}{90°}\right)^2\right]}$ was calculated by Julia Bell and first published in Pearson, Karl. On the probable error of a coefficient of correlation as found from a fourfold table. *Biometrika*, 1913, **9**, 22–27. It is reproduced by special permission of the Biometrika Trust.

TABLE T
The Two Functions of p and of p' Used in Computing the Standard Error of ϕ

p	F_{p_4}	F_{p_5}	p	F_{p_4}	F_{p_5}	p	F_{p_4}	F_{p_5}
.99	+9.8494	+97.0101	.64	+.5833	+.3403	.29	−.9256	+.8567
.98	6.8571	47.0204	.63	.5385	.2900	.28	−.9800	.9603
.97	5.5104	30.3643	.62	.4945	.2445	.27	−1.0361	1.0736
.96	4.6949	22.0417	.61	.4511	.2034	.26	−1.0943	1.1975
.95	4.1295	17.0526	.60	.4082	.1667	.25	−1.1547	1.3333
.94	+3.7055	+13.7305	.59	+.3660	+.1339	.24	−1.2176	+1.4825
.93	3.3706	11.3610	.58	.3242	.1051	.23	−1.2832	1.6465
.92	3.0963	9.5870	.57	.2828	.0800	.22	−1.3519	1.8275
.91	2.8653	8.2100	.56	.2417	.0584	.21	−1.4240	2.0277
.90	2.6667	7.1111	.55	.2010	.0404	.20	−1.5000	2.2500
.89	+2.4929	+6.2145	.54	+.1605	+.0258	.19	−1.5804	+2.4977
.88	2.3387	5.4697	.53	.1202	.0145	.18	−1.6659	2.7751
.87	2.2004	4.8417	.52	.0801	.0064	.17	−1.7570	3.0872
.86	2.0750	4.3056	.51	.0400	.0016	.16	−1.8549	3.4405
.85	1.9604	3.8431	.50	.0000	.0000	.15	−1.9604	3.8431
.84	+1.8549	+3.4405	.49	−.0400	+.0016	.14	−2.0750	+4.3056
.83	1.7570	3.0872	.48	−.0801	.0064	.13	−2.2004	4.8417
.82	1.6659	2.7751	.47	−.1202	.0145	.12	−2.3387	5.4697
.81	1.5804	2.4977	.46	−.1605	.0258	.11	−2.4929	6.2145
.80	1.5000	2.2500	.45	−.2010	.0404	.10	−2.6667	7.1111
.79	+1.4240	+2.0277	.44	−.2417	+.0584	.09	−2.8653	+8.2100
.78	1.3519	1.8275	.43	−.2828	.0800	.08	−3.0963	9.5870
.77	1.2832	1.6465	.42	−.3242	.1051	.07	−3.3706	11.3610
.76	1.2176	1.4825	.41	−.3660	.1339	.06	−3.7055	13.7305
.75	1.1547	1.3333	.40	−.4082	.1667	.05	−4.1295	17.0526

p	F_{p_4}	F_{p_5}	p	F_{p_4}	F_{p_5}	p	F_{p_4}	F_{p_5}
.74	+1.0943	+1.1975	.39	−.4511	+.2034	.04	−4.6949	+22.0417
.73	1.0361	1.0736	.38	−.4945	.2445	.03	−5.5104	30.3643
.72	.9800	.9603	.37	−.5385	.2900	.02	−6.8571	47.0204
.71	.9256	.8567	.36	−.5833	.3403	.01	−9.8494	97.0101
.70	.8729	.7619	.35	−.6290	.3956			
.69	+.8216	+.6751	.34	−.6755	+.4563			
.68	.7717	.5956	.33	−.7231	.5228			
.67	.7231	.5228	.32	−.7717	.5956			
.66	.6755	.4563	.31	−.8216	.6751			
.65	.6290	.3956	.30	−.8729	.7619			

TABLE U
The Three Functions of ϕ Used in Computing the Standard Error of ϕ

ϕ	F_{ϕ_1}	F_{ϕ_2}	F_{ϕ_3}	ϕ	F_{ϕ_1}	F_{ϕ_2}	F_{ϕ_3}
+.99	+.0199	+1.4751	+.7351	+.49	+.7599	+.5488	+.1801
+.98	.0396	1.4506	.7203	+.48	.7696	.5353	.1728
+.97	.0591	1.4263	.7057	+.47	.7791	.5219	.1657
+.96	.0784	1.4024	.6912	+.46	.7884	.5087	.1587
+.95	.0975	1.3787	.6769	+.45	.7975	.4956	.1519
+.94	+.1164	+1.3553	+.6627	+.44	+.8064	+.4826	+.1452
+.93	.1351	1.3322	.6487	+.43	.8151	.4698	.1387
+.92	.1536	1.3093	.6348	+.42	.8236	.4570	.1323
+.91	.1719	1.2868	.6211	+.41	.8319	.4445	.1261
+.90	.1900	1.2645	.6075	+.40	.8400	.4320	.1200
+.89	+.2079	+1.2425	+.5941	+.39	+.8479	+.4197	+.1141
+.88	.2256	1.2207	.5808	+.38	.8556	.4074	.1083
+.87	.2431	1.1993	.5677	+.37	.8631	.3953	.1027
+.86	.2604	1.1780	.5547	+.36	.8704	.3833	.0972
+.85	.2775	1.1571	.5419	+.35	.8775	.3714	.0919
+.84	+.2944	+1.1364	+.5292	+.34	+.8844	+.3597	+.0867
+.83	.3111	1.1159	.5167	+.33	.8911	.3480	.0817
+.82	.3276	1.0957	.5043	+.32	.8976	.3364	.0768
+.81	.3439	1.0757	.4921	+.31	.9039	.3249	.0721
+.80	.3600	1.0560	.4800	+.30	.9100	.3135	.0675
+.79	+.3759	+1.0365	+.4681	+.29	+.9159	+.3022	+.0631
+.78	.3916	1.0173	.4563	+.28	.9216	.2910	.0588
+.77	.4071	.9983	.4447	+.27	.9271	.2798	.0547
+.76	.4224	.9795	.4332	+.26	.9324	.2688	.0507
+.75	.4375	.9609	.4219	+.25	.9375	.2578	.0469

ϕ	F_{ϕ_1}	F_{ϕ_2}	F_{ϕ_3}	ϕ	F_{ϕ_1}	F_{ϕ_2}	F_{ϕ_3}
+.74	+.4524	+.9426	+.4107	+.24	+.9424	+.2469	+.0432
+.73	.4671	.9245	.3997	+.23	.9471	.2361	.0397
+.72	.4816	.9066	.3888	+.22	.9516	.2253	.0363
+.71	.4959	.8890	.3781	+.21	.9559	.2146	.0331
+.70	.5100	.8715	.3675	+.20	.9600	.2040	.0300
+.69	+.5239	+.8543	+.3571	+.19	+.9639	+.1934	+.0271
+.68	.5376	.8372	.3468	+.18	.9676	.1829	.0243
+.67	.5511	.8204	.3367	+.17	.9711	.1725	.0217
+.66	.5644	.8037	.3267	+.16	.9744	.1620	.0192
+.65	.5775	.7873	.3169	+.15	.9775	.1517	.0169
+.64	+.5904	+.7711	+.3072	+.14	+.9804	+.1414	+.0147
+.63	.6031	.7550	.2977	+.13	.9831	.1311	.0127
+.62	.6156	.7392	.2883	+.12	.9856	.1209	.0108
+.61	.6279	.7235	.2791	+.11	.9879	.1107	.0091
+.60	.6400	.7080	.2700	+.10	.9900	.1005	.0075
+.59	+.6519	+.6927	+.2611	+.09	+.9919	+.0904	+.0061
+.58	.6636	.6776	.2523	+.08	.9936	.0803	.0048
+.57	.6751	.6626	.2437	+.07	.9951	.0702	.0037
+.56	.6864	.6478	.2352	+.06	.9964	.0601	.0027
+.55	.6975	.6332	.2269	+.05	.9975	.0501	.0019
+.54	+.7084	+.6187	+.2187	+.04	+.9984	+.0400	+.0012
+.53	.7191	.6044	.2107	+.03	.9991	.0300	.0007
+.52	.7296	.5903	.2028	+.02	.9996	.0200	.0003
+.51	.7399	.5763	.1951	+.01	.9999	.0100	.0001
+.50	.7500	.5625	.1875	+.00	1.0000	.0000	.0000

TABLE U (*continued*)

ϕ	F_{ϕ_1}	F_{ϕ_2}	F_{ϕ_3}	ϕ	F_{ϕ_1}	F_{ϕ_2}	F_{ϕ_3}
−.00	+1.0000	−.0000	+.0000	−.50	+.7500	−.5625	+.1875
−.01	.9999	−.0100	.0001	−.51	.7399	−.5763	.1951
−.02	.9996	−.0200	.0003	−.52	.7296	−.5903	.2028
−.03	.9991	−.0300	.0007	−.53	.7191	−.6044	.2107
−.04	.9984	−.0400	.0012	−.54	.7084	−.6187	.2187
−.05	+.9975	−.0501	+.0019	−.55	+.6975	−.6332	+.2269
−.06	.9964	−.0601	.0027	−.56	.6864	−.6478	.2352
−.07	.9951	−.0702	.0037	−.57	.6751	−.6626	.2437
−.08	.9936	−.0803	.0048	−.58	.6636	−.6776	.2523
−.09	.9919	−.0904	.0061	−.59	.6519	−.6927	.2611
−.10	+.9900	−.1005	+.0075	−.60	+.6400	−.7080	+.2700
−.11	.9879	−.1107	.0091	−.61	.6279	−.7235	.2791
−.12	.9856	−.1209	.0108	−.62	.6156	−.7392	.2883
−.13	.9831	−.1311	.0127	−.63	.6031	−.7550	.2977
−.14	.9804	−.1414	.0147	−.64	.5904	−.7711	.3072
−.15	+.9775	−.1517	+.0169	−.65	+.5775	−.7873	+.3169
−.16	.9744	−.1620	.0192	−.66	.5644	−.8037	.3267
−.17	.9711	−.1725	.0217	−.67	.5511	−.8204	.3367
−.18	.9676	−.1829	.0243	−.68	.5376	−.8372	.3468
−.19	.9639	−.1934	.0271	−.69	.5239	−.8543	.3571
−.20	+.9600	−.2040	+.0300	−.70	+.5100	−.8715	+.3675
−.21	.9559	−.2146	.0331	−.71	.4959	−.8890	.3781
−.22	.9516	−.2253	.0363	−.72	.4816	−.9066	.3888
−.23	.9471	−.2361	.0397	−.73	.4671	−.9245	.3997
−.24	.9424	−.2469	.0432	−.74	.4524	−.9426	.4107

ϕ	F_{ϕ_1}	F_{ϕ_2}	F_{ϕ_3}	ϕ	F_{ϕ_1}	F_{ϕ_2}	F_{ϕ_3}
−.25	+.9375	−.2578	+.0469	−.75	+.4375	−.9609	+.4219
−.26	.9324	−.2688	.0507	−.76	.4224	−.9795	.4332
−.27	.9271	−.2798	.0547	−.77	.4071	−.9983	.4447
−.28	.9216	−.2910	.0588	−.78	.3916	−1.0173	.4563
−.29	.9159	−.3022	.0631	−.79	.3759	−1.0365	.4681
−.30	+.9100	−.3135	+.0675	−.80	+.3600	−1.0560	+.4800
−.31	.9039	−.3249	.0721	−.81	.3439	−1.0757	.4921
−.32	.8976	−.3364	.0768	−.82	.3276	−1.0957	.5043
−.33	.8911	−.3480	.0817	−.83	.3111	−1.1159	.5167
−.34	.8844	−.3597	.0867	−.84	.2944	−1.1364	.5292
−.35	+.8775	−.3714	+.0919	−.85	+.2775	−1.1571	+.5419
−.36	.8704	−.3833	.0972	−.86	.2604	−1.1780	.5547
−.37	.8631	−.3953	.1027	−.87	.2431	−1.1993	.5677
−.38	.8556	−.4074	.1083	−.88	.2256	−1.2207	.5808
−.39	.8479	−.4197	.1141	−.89	.2079	−1.2425	.5941
−.40	+.8400	−.4320	+.1200	−.90	+.1900	−1.2645	+.6075
−.41	.8319	−.4445	.1261	−.91	.1719	−1.2868	.6211
−.42	.8236	−.4570	.1323	−.92	.1536	−1.3093	.6348
−.43	.8151	−.4698	.1387	−.93	.1351	−1.3322	.6487
−.44	.8064	−.4826	.1452	−.94	.1164	−1.3553	.6627
−.45	+.7975	−.4956	+.1519	−.95	+.0975	−1.3787	+.6769
−.46	.7884	−.5087	.1587	−.96	.0784	−1.4024	.6912
−.47	.7791	−.5219	.1657	−.97	.0591	−1.4263	.7057
−.48	.7696	−.5353	.1728	−.98	.0396	−1.4506	.7203
−.49	.7599	−.5488	.1801	−.99	.0199	−1.4751	.7351

TABLE V

Critical Values of Chi Square

For each degree of freedom, shown at the left, the table entry shows the value of χ^2 which is just barely significant at the level shown at the top of the table. Thus, for $df = 6$, $\chi^2 = 18$ is significant at the .01 level, or $.01 > P > .001$.

df	*P* = .99	.95	.05	.01	.001	*df*
1	.000157	.00393	3.841	6.635	10.827	1
2	.0201	.103	5.991	9.210	13.815	2
3	.115	.352	7.815	11.345	16.266	3
4	.297	.711	9.488	13.277	18.467	4
5	.554	1.145	11.070	15.086	20.515	5
6	.872	1.635	12.592	16.812	22.457	6
7	1.239	2.167	14.067	18.475	24.322	7
8	1.646	2.733	15.507	20.090	26.125	8
9	2.088	3.325	16.919	21.666	27.877	9
10	2.558	3.940	18.307	23.209	29.588	10
11	3.053	4.575	19.675	24.725	31.264	11
12	3.571	5.226	21.026	26.217	32.909	12
13	4.107	5.892	22.362	27.688	34.528	13
14	4.660	6.571	23.685	29.141	36.123	14
15	5.229	7.261	24.996	30.578	37.697	15
16	5.812	7.962	26.296	32.000	39.252	16
17	6.408	8.672	27.587	33.409	40.790	17
18	7.015	9.390	28.869	34.805	42.312	18
19	7.633	10.117	30.144	36.191	43.820	19
20	8.260	10.851	31.410	37.566	45.315	20
21	8.897	11.591	32.671	38.932	46.797	21
22	9.542	12.338	33.924	40.289	48.268	22
23	10.196	13.091	35.172	41.638	49.728	23
24	10.856	13.848	36.415	42.980	51.179	24
25	11.524	14.611	37.652	44.314	52.620	25
26	12.198	15.379	38.885	45.642	54.052	26
27	12.879	16.151	40.113	46.963	55.476	27
28	13.565	16.928	41.337	48.278	56.893	28
29	14.256	17.708	42.557	49.588	58.302	29
30	14.953	18.493	43.773	50.892	59.703	30
40	22.164	26.509	55.759	63.691	73.402	40
50	29.707	34.764	67.505	76.154	86.661	50
60	37.485	43.188	79.082	88.379	99.607	60
70	45.442	51.739	90.531	100.425	112.317	70
80	53.540	60.391	101.879	112.329	124.839	80
90	61.754	69.126	113.145	124.116	137.208	90
100	70.065	77.929	124.342	135.807	149.449	100
	P = .99	.95	.05	.01	.001	

TABLE W

The Number of + or − Signs, Whichever is Smaller, Needed for Various Significance Levels

Reduced N	.05 Level	.01 Level	.001 Level	Reduced N	.05 Level	.01 Level	.001 Level
5	—	—	—	30	9	7	5
6	0	—	—	31	9	7	6
7	0	—	—	32	9	8	6
8	0	0	—	33	10	8	6
9	1	0	—	34	10	9	7
10	1	0	—	35	11	9	7
11	1	0	0	36	11	9	7
12	2	1	0	37	12	10	8
13	2	1	0	38	12	10	8
14	2	1	0	39	12	11	8
15	3	2	1	40	13	11	9
16	3	2	1	41	13	11	9
17	4	2	1	42	14	12	10
18	4	3	1	43	14	12	10
19	4	3	2	44	15	13	10
20	5	3	2	45	15	13	11
21	5	4	2	46	15	13	11
22	5	4	3	47	16	14	11
23	6	4	3	48	16	14	12
24	6	5	3	49	17	15	12
25	7	5	4	50	17	15	13
26	7	6	4	51	18	15	13
27	7	6	4	52	18	16	13
28	8	6	5	53	18	16	14
29	8	7	5	54	19	17	14
				55	19	17	14

TABLE Y

.01 and .05 Significance Levels for Number of Runs, n_R, for All Values of N_1 and N_2 from 1 to 20

	N_1										
	2	3	4	5	6	7	8	9	10	11	
2	NS NS NS NS	NS NS NS NS	NS NS NS NS	NS NS NS NS	NS NS NS NS	NS NS NS NS	NS NS NS NS	NS NS NS NS	NS NS NS NS	NS NS NS NS	2
3	NS NS NS NS	NS NS NS NS	NS NS NS NS	NS NS NS NS	NS NS 2 NS	NS NS 2 NS	NS NS 2 NS	NS NS 2 NS	NS NS 2 NS	NS NS 2 NS	3
4	NS NS NS NS	NS NS NS NS	NS NS NS NS	NS NS 2 9	NS NS 2 9	NS NS 2 NS	2 NS 3−NS	2 NS 3−NS	2 NS 3−NS	2 NS 3−NS	4
5	NS NS NS NS	NS NS NS NS	NS NS 2 9	NS NS 2 10	2 11 3− 10+	2 NS 3− 11	2 NS 3− 11	2 NS 3−NS	3−NS 3−NS	3−NS 4−NS	5
6	NS NS NS NS	NS NS 2 NS	NS NS 2 9	2 11 3− 10+	2 12 3− 11+	2 13 3− 12+	3− 13 3− 12+	3−NS 4− 13	3−NS 4− 13	3−NS 4− 13	6
N_2 7	NS NS NS NS	NS NS 2 NS	NS NS 2 NS	2 NS 3− 11	2 13 3− 12+	3− 13+ 3− 13+	3− 14+ 4− 13+	3− 15 4− 14+	3− 15 5− 14+	4− 15 5− 14+	7
8	NS NS NS NS	NS NS 2 NS	2 NS 3−NS	2 NS 3− 11	3− 13 3− 12+	3− 14+ 4− 13+	3− 15+ 4− 14+	3− 15+ 5− 14+	4− 16+ 5− 15+	4− 16+ 5− 15+	8
9	NS NS NS NS	NS NS 2 NS	2 NS 3−NS	2 NS 3−NS	3−NS 4− 13	3− 15 4− 14+	3− 15+ 5− 14+	4− 16+ 5− 15+	4− 17+ 5− 16+	5− 17+ 6− 16+	9
10	NS NS NS NS	NS NS 2 NS	2 NS 3−NS	3−NS 3−NS	3−NS 4− 13	3− 15 5− 14+	4− 16+ 5− 15+	4− 17+ 5− 16+	5− 17+ 6− 16+	5− 18+ 6− 17+	10
11	NS NS NS NS	NS NS 2 NS	2 NS 3−NS	3−NS 4−NS	3−NS 4− 13	4− 15 5− 14+	4− 16+ 5− 15+	5− 17+ 6− 16+	5− 18+ 6− 17+	5− 19+ 7− 17+	11
N_2 12	NS NS 2 NS	2 NS 2 NS	2 NS 3−NS	3−NS 4−NS	3−NS 4− 13	4−NS 5− 14+	4− 17 6− 16+	5− 18+ 6− 16+	5− 19+ 7− 17+	6− 19+ 7− 18+	12

	13	NS NS 2 NS	2 NS 2 NS	2 NS 3−NS	3−NS 4−NS	3− NS 5− NS	4− NS 5− 15	5− 17 6− 16+	5− 18+ 6− 17+	5− 19+ 7− 18+	6− 20+ 7− 19+ 13
	14	NS NS 2 NS	2 NS 2 NS	2 NS 3−NS	3−NS 4−NS	4−NS 5−NS	4−NS 5− 15	5− 17 6− 16+	5− 18+ 7− 17+	6− 19+ 7− 18+	6− 20+ 8− 19+ 14
	15	NS NS 2 NS	2 NS 3−NS	3−NS 3−NS	3−NS 4−NS	4−NS 5−NS	4−NS 6− 15	5−NS 6− 16+	6− 19 7− 18+	6− 20+ 7− 18+	7− 21+ 8− 19+ 15
	16	NS NS 2 NS	2 NS 3−NS	3−NS 4−NS	3−NS 4−NS	4−NS 5−NS	5−NS 6−NS	5−NS 6− 17	6− 19 7− 18+	6− 20+ 8− 19+	7− 21+ 8− 20+ 16
N_2	17	NS NS 2 NS	2 NS 3−NS	3−NS 4−NS	3−NS 4−NS	4−NS 5−NS	5−NS 6−NS	5−NS 7− 17	6− 19 7− 18+	7− 20+ 8− 19+	7− 22+ 9− 20+ 17
	18	NS NS 2 NS	2 NS 3−NS	3−NS 4−NS	4−NS 5−NS	4−NS 5−NS	5−NS 6−NS	6−NS 7− 17	6−NS 8− 18+	7− 21 8− 19+	7− 22+ 9− 20+ 18
	19	NS NS 2 NS	2 NS 3−NS	3−NS 4−NS	4−NS 5−NS	4−NS 6−NS	5−NS 6−NS	6−NS 7− 17	6−NS 8− 18+	7− 21 8− 20+	8− 22+ 9− 21+ 19
	20	NS NS 2 NS	2 NS 3−NS	3−NS 4−NS	4−NS 5−NS	4−NS 6−NS	5−NS 6−NS	6−NS 7− 17	7−NS 8− 18+	7− 21 9− 20+	8− 22+ 9− 21+ 20
		2	3	4	5	6	7	8	9	10	11

TABLE Y (*continued*)

	N_1 12	13	14	15	16	17	18	19	20	
2	NS NS 2 NS	NS NS 2 NS	NS NS 2 NS	NS NS 2 NS	NS NS 2 NS	NS NS 2 NS	NS NS 2 NS	NS NS 2 NS	NS NS 2 NS	2
3	2 NS 2 NS	2 NS 2 NS	2 NS 2 NS	2 NS 3− NS	2 NS 3− NS	2 NS 3− NS	2 NS 3− NS	2 NS 3− NS	2 NS 3− NS	3
4	2 NS 3− NS	2 NS 3− NS	2 NS 3− NS	3− NS 3− NS	3− NS 4− NS	3− NS 4− NS	3− NS 4− NS	3− NS 4− NS	3− NS 4− NS	4
N_2 5	3− NS 4− NS	3− NS 4− NS	3− NS 4− NS	3− NS 4− NS	3− NS 4− NS	3− NS 4− NS	4− NS 5− NS	4− NS 5− NS	4− NS 5− NS	5
6	3− NS 4− 13	3− NS 5− NS	4− NS 5− NS	4− NS 5− NS	4− NS 5− NS	4− NS 5− NS	4− NS 5− NS	4− NS 6− NS	4− NS 6− NS	6
7	4− NS 5− 14+	4− NS 5− 15	4− NS 5− 15	4− NS 6− 15	5− NS 6− NS	5− NS 6− NS	5− NS 6− NS	5− NS 6− NS	5− NS 6− NS	7
8	4− 17 6− 16+	5− 17 6− 16+	5− 17 6− 16+	5− NS 6− 16+	5− NS 6− 17	5− NS 7− 17	6− NS 7− 17	6− NS 7− 17	6− NS 7− 17	8
9	5− 18+ 6− 16+	5− 18+ 6− 17+	5− 18+ 7− 17+	6− 19 7− 18+	6− 19 7− 18+	6− 19 7− 18+	6− NS 8− 18+	6− NS 8− 18+	7− NS 8− 18+	9
10	5− 19+ 7− 17+	5− 19+ 7− 18+	6− 19+ 7− 18+	6− 20+ 7− 18+	6− 20+ 8− 19+	7− 20+ 8− 19+	7− 21 8− 19+	7− 21 8− 20+	7− 21 9− 20+	10
11	6− 19+ 7− 18+	6− 20+ 7− 19+	6− 20+ 8− 19+	7− 21+ 8− 19+	7− 21+ 8− 20+	7− 22+ 9− 20+	7− 22+ 9− 20+	8− 22+ 9− 21+	8− 22+ 9− 21+	11
N_2 12	6− 20+ 7− 19+	6− 21+ 8− 19+	7− 21+ 8− 20+	7− 22+ 8− 20+	7− 22+ 9− 21+	8− 22+ 9− 21+	8− 23+ 9− 21+	8− 23+ 10− 22+	8− 23+ 10− 22+	12

	12	13	14	15	16	17	18	19	20	
13	6− 21+ 8− 19+	7− 21+ 8− 20+	7− 22+ 9− 20+	7− 22+ 9− 21+	8− 23+ 9− 21+	8− 23+ 10− 22+	8− 24+ 10− 22+	9− 24+ 10− 23+	9− 24+ 10− 23+	13
14	7− 21+ 8− 20	7− 22+ 9− 20+	7− 23+ 9−	8− 23+ 9− 22+	8− 24+ 10− 22+	8− 24+ 10− 23+	9− 25+ 10− 23+	9− 25+ 11− 23+	9− 25+ 11− 24+	14
15	7− 22+ 8− 20+	7− 22+ 9− 21+	8− 23+ 9− 22+	8− 24+ 10− 22+	9− 24+ 10− 23+	9− 25+ 11− 23+	9− 25+ 11− 24+	10− 26+ 11− 24+	10− 26+ 12− 25+	15
16	7− 22+ 9− 21+	8− 23+ 9− 21+	8− 24+ 10− 22+	9− 24+ 10− 23+	9− 25+ 11− 23+	9− 26+ 11− 24+	10− 26+ 11− 25+	10− 27+ 12− 25+	10− 27+ 12− 25+	16
N_2 17	8− 22+ 9− 21+	8− 23+ 10− 22+	8− 24+ 10− 23+	9− 25+ 11− 23+	9− 26+ 11− 24+	10− 26+ 11− 25+	10− 27+ 12− 25+	10− 27+ 12− 26+	11− 28+ 13− 26+	17
18	8− 23+ 9− 21+	8− 24+ 10− 22+	9− 25+ 10− 23+	9− 25+ 11− 24+	10− 26+ 11− 25+	10− 27+ 12− 25+	11− 27+ 12− 26+	11− 28+ 13− 26+	11− 29+ 13− 27+	18
19	8− 23+ 10− 22+	9− 24+ 10− 23+	9− 25+ 11− 23+	10− 26+ 11− 24+	10− 27+ 12− 25+	10− 27+ 12− 26+	11− 28+ 13− 26+	11− 29+ 13− 27+	12− 29+ 13− 27+	19
20	8− 23+ 10− 22+	9− 24+ 10− 23+	9− 25+ 11− 24+	10− 26+ 12− 25+	10− 27+ 12− 25+	11− 28+ 13− 26+	11− 29+ 13− 27+	12− 29+ 13− 27+	12− 30+ 14− 28+	20
	12	13	14	15	16	17	18	19	20	

TABLE Z
Squares and Square Roots

N	N^2	$\sqrt{N}$	$\sqrt{10\,N}$	N	N^2	$\sqrt{N}$	$\sqrt{10\,N}$
				50	2500	7.0711	22.3607
1	1	1.0000	3.1623	51	2601	7.1414	22.5832
2	4	1.4142	4.4721	52	2704	7.2111	22.8035
3	9	1.7321	5.4772	53	2809	7.2801	23.0217
4	16	2.0000	6.3246	54	2916	7.3485	23.2379
5	25	2.2361	7.0711	55	3025	7.4162	23.4521
6	36	2.4495	7.7460	56	3136	7.4833	23.6643
7	49	2.6458	8.3666	57	3249	7.5498	23.8747
8	64	2.8284	8.9443	58	3364	7.6158	24.0832
9	81	3.0000	9.4868	59	3481	7.6811	24.2899
10	100	3.1623	10.0000	60	3600	7.7460	24.4949
11	121	3.3166	10.4881	61	3721	7.8102	24.6982
12	144	3.4641	10.9545	62	3844	7.8740	24.8998
13	169	3.6056	11.4018	63	3969	7.9373	25.0998
14	196	3.7417	11.8322	64	4096	8.0000	25.2982
15	225	3.8730	12.2474	65	4225	8.0623	25.4951
16	256	4.0000	12.6491	66	4356	8.1240	25.6905
17	289	4.1231	13.0384	67	4489	8.1854	25.8844
18	324	4.2426	13.4164	68	4624	8.2462	26.0768
19	361	4.3589	13.7840	69	4761	8.3066	26.2679
20	400	4.4721	14.1421	70	4900	8.3666	26.4575
21	441	4.5826	14.4914	71	5041	8.4261	26.6458
22	484	4.6904	14.8324	72	5184	8.4853	26.8328
23	529	4.7958	15.1658	73	5329	8.5440	27.0185
24	576	4.8990	15.4919	74	5476	8.6023	27.2029

N	N^2	$\sqrt{N}$	$\sqrt{10N}$	N	N^2	$\sqrt{N}$	$\sqrt{10N}$
25	625	5.0000	15.8114	75	5625	8.6603	[illegible]
26	676	5.0990	16.1245	76	5776	8.717[illegible]	[illegible]
27	729	5.1962	16.4317	77	5929	8.775[illegible]	[illegible]
28	784	5.2915	16.7332	78	6084	8.8318	[illegible]
29	841	5.3852	17.0294	79	6241	8.8882	[illegible]
30	900	5.4772	17.3205	80	6400	8.9443	[illegible]
31	961	5.5678	17.6068	81	6561	9.0000	[illegible]605
32	1024	5.6569	17.8885	82	6724	9.0554	28.6356
33	1089	5.7446	18.1659	83	6889	9.1104	28.8097
34	1156	5.8310	18.4391	84	7056	9.1652	28.9828
35	1225	5.9161	18.7083	85	7225	9.2195	29.1548
36	1296	6.0000	18.9737	86	7396	9.2736	29.3258
37	1369	6.0828	19.2354	87	7569	9.3274	29.4958
38	1444	6.1644	19.4936	88	7744	9.3808	29.6648
39	1521	6.2450	19.7484	89	7921	9.4340	29.8329
40	1600	6.3246	20.0000	90	8100	9.4868	30.0000
41	1681	6.4031	20.2485	91	8281	9.5394	30.1662
42	1764	6.4807	20.4939	92	8464	9.5917	30.3315
43	1849	6.5574	20.7364	93	8649	9.6437	30.4959
44	1936	6.6332	20.9762	94	8836	9.6954	30.6594
45	2025	6.7082	21.2132	95	9025	9.7468	30.8221
46	2116	6.7823	21.4476	96	9216	9.7980	30.9839
47	2209	6.8557	21.6795	97	9409	9.8489	31.1448
48	2304	6.9282	21.9089	98	9604	9.8995	31.3050
49	2401	7.0000	22.1359	99	9801	9.9499	31.4643

TABLE Z *(continued)*

N	N^2	$\sqrt{N}$	$\sqrt{10\,N}$	N	N^2	$\sqrt{N}$	$\sqrt{10\,N}$
100	10000	10.0000	31.6228	150	22500	12.2474	38.7298
101	10201	10.0499	31.7805	151	22801	12.2882	38.8587
102	10404	10.0995	31.9374	152	23104	12.3288	38.9872
103	10609	10.1489	32.0936	153	23409	12.3693	39.1152
104	10816	10.1980	32.2490	154	23716	12.4097	39.2428
105	11025	10.2470	32.4037	155	24025	12.4499	39.3700
106	11236	10.2956	32.5576	156	24336	12.4900	39.4968
107	11449	10.3441	32.7109	157	24649	12.5300	39.6232
108	11664	10.3923	32.8634	158	24964	12.5698	39.7492
109	11881	10.4403	33.0151	159	25281	12.6095	39.8748
110	12100	10.4881	33.1662	160	25600	12.6491	40.0000
111	12321	10.5357	33.3167	161	25921	12.6886	40.1248
112	12544	10.5830	33.4664	162	26244	12.7279	40.2492
113	12769	10.6301	33.6155	163	26569	12.7671	40.3733
114	12996	10.6771	33.7639	164	26896	12.8062	40.4969
115	13225	10.7238	33.9116	165	27225	12.8452	40.6202
116	13456	10.7703	34.0588	166	27556	12.8841	40.7431
117	13689	10.8167	34.2053	167	27889	12.9228	40.8656
118	13924	10.8628	34.3511	168	28224	12.9615	40.9878
119	14161	10.9087	34.4964	169	28561	13.0000	41.1096
120	14400	10.9545	34.6410	170	28900	13.0384	[illegible]
121	14641	11.0000	34.7851	171	29241	13.0767	[illegible]
122	14884	11.0454	34.9285	172	29584	13.1149	[illegible]
123	15129	11.0905	35.0714	173	29929	13.1529	[illegible]
124	15376	11.1355	35.2136	174	30276	13.1909	[illegible]

N	N^2	$\sqrt{N}$	$\sqrt{10\,N}$	N	N^2	$\sqrt{N}$	$\sqrt{10\,N}$
125	15625	11.1803	35.3553	175	30625	13.2288	41.8330
126	15876	11.2250	35.4965	176	30976	13.2665	41.9524
127	16129	11.2694	35.6371	177	31329	13.3041	42.0714
128	16384	11.3137	35.7771	178	31684	13.3417	42.1900
129	16641	11.3578	35.9166	179	32041	13.3791	42.3084
130	16900	11.4018	36.0555	180	32400	13.4164	42.4264
131	17161	11.4455	36.1939	181	32761	13.4536	42.5441
132	17424	11.4891	36.3318	182	33124	13.4907	42.6615
133	17689	11.5326	36.4692	183	33489	13.5277	42.7785
134	17956	11.5758	36.6060	184	33856	13.5647	42.8952
135	18225	11.6190	36.7423	185	34225	13.6015	43.0116
136	18496	11.6619	36.8782	186	34596	13.6382	43.1277
137	18769	11.7047	37.0135	187	34969	13.6748	43.2435
138	19044	11.7473	37.1484	188	35344	13.7113	43.3590
139	19321	11.7898	37.2827	189	35721	13.7477	43.4741
140	19600	11.8322	37.4166	190	36100	13.7840	43.5890
141	19881	11.8743	37.5500	191	36481	13.8203	43.7035
142	20164	11.9164	37.6829	192	36864	13.8564	43.8178
143	20449	11.9583	37.8153	193	37249	13.8924	43.9318
144	20736	12.0000	37.9473	194	37636	13.9284	44.0454
145	21025	12.0416	38.0789	195	38025	13.9642	44.1588
146	21316	12.0830	38.2099	196	38416	14.0000	44.2719
147	21609	12.1244	38.3406	197	38809	14.0357	44.3847
148	21904	12.1655	38.4708	198	39204	14.0712	44.4972
149	22201	12.2066	38.6005	199	39601	14.1067	44.6094

TABLE Z (*continued*)

N	N^2	$\sqrt{N}$	$\sqrt{10\,N}$	N	N^2	$\sqrt{N}$	$\sqrt{10\,N}$
200	40000	14.1421	44.7214	250	62500	15.8114	50.0000
201	40401	14.1774	44.8330	251	63001	15.8430	50.0999
202	40804	14.2127	44.9444	252	63504	15.8745	50.1996
203	41209	14.2478	45.0555	253	64009	15.9060	50.2991
204	41616	14.2829	45.1664	254	64516	15.9374	50.3984
205	42025	14.3178	45.2769	255	65025	15.9687	50.4975
206	42436	14.3527	45.3872	256	65536	16.0000	50.5964
207	42849	14.3875	45.4973	257	66049	16.0312	50.6952
208	43264	14.4222	45.6070	258	66564	16.0624	50.7937
209	43681	14.4568	45.7165	259	67081	16.0935	50.8920
210	44100	14.4914	45.8258	260	67600	16.1245	50.9902
211	44521	14.5258	45.9347	261	68121	16.1555	51.0882
212	44944	14.5602	46.0435	262	68644	16.1864	51.1859
213	45369	14.5945	46.1519	263	69169	16.2173	51.2835
214	45796	14.6287	46.2601	264	69696	16.2481	51.3809
215	46225	14.6629	46.3681	265	70225	16.2788	51.4782
216	46656	14.6969	46.4758	266	70756	16.3095	51.5752
217	47089	14.7309	46.5833	267	71289	16.3401	51.6720
218	47524	14.7648	46.6905	268	71824	16.3707	51.7687
219	47961	14.7986	46.7974	269	72361	16.4012	51.8652
220	48400	14.8324	46.9042	270	72900	16.4317	51.9615
221	48841	14.8661	47.0106	271	73441	16.4621	52.0577
222	49284	14.8997	47.1169	272	73984	16.4924	52.1536
223	49729	14.9332	47.2229	273	74529	16.5227	52.2494
224	50176	14.9666	47.3286	274	75076	16.5529	52.3450

N	N^2	$\sqrt{N}$	$\sqrt{10N}$	N	N^2	$\sqrt{N}$	$\sqrt{10N}$
225	50625	15.0000	47.4342	275	75625	16.5831	52.4404
226	51076	15.0333	47.5395	276	76176	16.6132	52.5357
227	51529	15.0665	47.6445	277	76729	16.6433	52.6308
228	51984	15.0997	47.7493	278	77284	16.6733	52.7257
229	52441	15.1327	47.8539	279	77841	16.7033	52.8205
230	52900	15.1658	47.9583	280	78400	16.7332	52.9150
231	53361	15.1987	48.0625	281	78961	16.7631	53.0094
232	53824	15.2315	48.1664	282	79524	16.7929	53.1037
233	54289	15.2643	48.2701	283	80089	16.8226	53.1977
234	54756	15.2971	48.3735	284	80656	16.8523	53.2917
235	55225	15.3297	48.4768	285	81225	16.8819	53.3854
236	55696	15.3623	48.5798	286	81796	16.9115	53.4790
237	56169	15.3948	48.6826	287	82369	16.9411	53.5724
238	56644	15.4272	48.7852	288	82944	16.9706	53.6656
239	57121	15.4596	48.8876	289	83521	17.0000	53.7587
240	57600	15.4919	48.9898	290	84100	17.0294	53.8516
241	58081	15.5242	49.0918	291	84681	17.0587	53.9444
242	58564	15.5563	49.1935	292	85264	17.0880	54.0370
243	59049	15.5885	49.2950	293	85849	17.1172	54.1295
244	59536	15.6205	49.3964	294	86436	17.1464	54.2218
245	60025	15.6525	49.4975	295	87025	17.1756	54.3139
246	60516	15.6844	49.5984	296	87616	17.2047	54.4059
247	61009	15.7162	49.6991	297	88209	17.2337	54.4977
248	61504	15.7480	49.7996	298	88804	17.2627	54.5894
249	62001	15.7797	49.8999	299	89401	17.2916	54.6809

TABLE Z (*continued*)

N	N^2	$\sqrt{N}$	$\sqrt{10\,N}$	N	N^2	$\sqrt{N}$	$\sqrt{10\,N}$
300	90000	17.3205	54.7723	350	122500	18.7083	59.1608
301	90601	17.3494	54.8635	351	123201	18.7350	59.2453
302	91204	17.3781	54.9545	352	123904	18.7617	59.3296
303	91809	17.4069	55.0454	353	124609	18.7883	59.4138
304	92416	17.4356	55.1362	354	125316	18.8149	59.4979
305	93025	17.4642	55.2268	355	126025	18.8414	59.5819
306	93636	17.4929	55.3173	356	126736	18.8680	59.6657
307	94249	17.5214	55.4076	357	127449	18.8944	59.7495
308	94864	17.5499	55.4977	358	128164	18.9209	59.8331
309	95481	17.5784	55.5878	359	128881	18.9473	59.9166
310	96100	17.6068	55.6776	360	129600	18.9737	60.0000
311	96721	17.6352	55.7674	361	130321	19.0000	60.0833
312	97344	17.6635	55.8570	362	131044	19.0263	60.1664
313	97969	17.6918	55.9464	363	131769	19.0526	60.2495
314	98596	17.7200	56.0357	364	132496	19.0788	60.3324
315	99225	17.7482	56.1249	365	133225	19.1050	60.4152
316	99856	17.7764	56.2139	366	133956	19.1311	60.4979
317	100489	17.8045	56.3028	367	134689	19.1572	60.5805
318	101124	17.8326	56.3915	368	135424	19.1833	60.6630
319	101761	17.8606	56.4801	369	136161	19.2094	60.7454
320	102400	17.8885	56.5685	370	136900	19.2354	60.8276
321	103041	17.9165	56.6569	371	137641	19.2614	60.9098
322	103684	17.9444	56.7450	372	138384	19.2873	60.9918
323	104329	17.9722	56.8331	373	139129	19.3132	61.0737
324	104976	18.0000	56.9210	374	139876	19.3391	61.1555

N	N^2	$\sqrt{N}$	$\sqrt{10N}$	N	N^2	$\sqrt{N}$	$\sqrt{10N}$
325	105625	18.0278	57.0088	375	140625	19.3649	61.2372
326	106276	18.0555	57.0964	376	141376	19.3907	61.3188
327	106929	18.0831	57.1839	377	142129	19.4165	61.4003
328	107584	18.1108	57.2713	378	142884	19.4422	61.4817
329	108241	18.1384	57.3585	379	143641	19.4679	61.5630
330	108900	18.1659	57.4456	380	144400	19.4936	61.6441
331	109561	18.1934	57.5326	381	145161	19.5192	61.7252
332	110224	18.2209	57.6194	382	145924	19.5448	61.8061
333	110889	18.2483	57.7062	383	146689	19.5704	61.8870
334	111556	18.2757	57.7927	384	147456	19.5959	61.9677
335	112225	18.3030	57.8792	385	148225	19.6214	62.0484
336	112896	18.3303	57.9655	386	148996	19.6469	62.1289
337	113569	18.3576	58.0517	387	149769	19.6723	62.2093
338	114244	18.3848	58.1378	388	150544	19.6977	62.2896
339	114921	18.4120	58.2237	389	151321	19.7231	62.3699
340	115600	18.4391	58.3095	390	152100	19.7484	62.4500
341	116281	18.4662	58.3952	391	152881	19.7737	62.5300
342	116964	18.4932	58.4808	392	153664	19.7990	62.6099
343	117649	18.5203	58.5662	393	154449	19.8242	62.6897
344	118336	18.5472	58.6515	394	155236	19.8494	62.7694
345	119025	18.5742	58.7367	395	156025	19.8746	62.8490
346	119716	18.6011	58.8218	396	156816	19.8997	62.9285
347	120409	18.6279	58.9067	397	157609	19.9249	63.0079
348	121104	18.6548	58.9915	398	158404	19.9499	63.0872
349	121801	18.6815	59.0762	399	159201	19.9750	63.1664
N	N^2	$\sqrt{N}$	$\sqrt{10N}$	N	N^2	$\sqrt{N}$	$\sqrt{10N}$

TABLE Z (*continued*)

N	N^2	$\sqrt{N}$	$\sqrt{10\,N}$	N	N^2	$\sqrt{N}$	$\sqrt{10\,N}$
400	160000	20.0000	63.2456	450	202500	21.2132	67.0820
401	160801	20.0250	63.3246	451	203401	21.2368	67.1565
402	161604	20.0499	63.4035	452	204304	21.2603	67.2309
403	162409	20.0749	63.4823	453	205209	21.2838	67.3053
404	163216	20.0998	63.5610	454	206116	21.3073	67.3795
405	164025	20.1246	63.6396	455	207025	21.3307	67.4537
406	164836	20.1494	63.7181	456	207936	21.3542	67.5278
407	165649	20.1742	63.7966	457	208849	21.3776	67.6018
408	166464	20.1990	63.8749	458	209764	21.4009	67.6757
409	167281	20.2237	63.9531	459	210681	21.4243	67.7495
410	168100	20.2485	64.0312	460	211600	21.4476	67.8233
411	168921	20.2731	64.1093	461	212521	21.4709	67.8970
412	169744	20.2978	64.1872	462	213444	21.4942	67.9706
413	170569	20.3224	64.2651	463	214369	21.5174	68.0441
414	171396	20.3470	64.3428	464	215296	21.5407	68.1175
415	172225	20.3715	64.4205	465	216225	21.5639	68.1909
416	173056	20.3961	64.4981	466	217156	21.5870	68.2642
417	173889	20.4206	64.5755	467	218089	21.6102	68.3374
418	174724	20.4450	64.6529	468	219024	21.6333	68.4105
419	175561	20.4695	64.7302	469	219961	21.6564	68.4836
420	176400	20.4939	64.8074	470	220900	21.6795	68.5565
421	177241	20.5183	64.8845	471	221841	21.7025	68.6294
422	178084	20.5426	64.9615	472	222784	21.7256	68.7023
423	178929	20.5670	65.0385	473	223729	21.7486	68.7750
424	179776	20.5913	65.1153	474	224676	21.7715	68.8477

N	N^2	$\sqrt{N}$	$\sqrt{10\ N}$	N	N^2	$\sqrt{N}$	$\sqrt{10\ N}$
425	180625	20.6155	65.1920	475	225625	21.7945	68.9202
426	181476	20.6398	65.2687	476	226576	21.8174	68.9928
427	182329	20.6640	65.3452	477	227529	21.8403	69.0652
428	183184	20.6882	65.4217	478	228484	21.8632	69.1375
429	184041	20.7123	65.4981	479	229441	21.8861	69.2098
430	184900	20.7364	65.5744	480	230400	21.9089	69.2820
431	185761	20.7605	65.6506	481	231361	21.9317	69.3542
432	186624	20.7846	65.7267	482	232324	21.9545	69.4262
433	187489	20.8087	65.8027	483	233289	21.9773	69.4982
434	188356	20.8327	65.8787	484	234256	22.0000	69.5701
435	189225	20.8567	65.9545	485	235225	22.0227	69.6419
436	190096	20.8806	66.0303	486	236196	22.0454	69.7137
437	190969	20.9045	66.1060	487	237169	22.0681	69.7854
438	191844	20.9284	66.1816	488	238144	22.0907	69.8570
439	192721	20.9523	66.2571	489	239121	22.1133	69.9285
440	193600	20.9762	66.3325	490	240100	22.1359	70.0000
441	194481	21.0000	66.4078	491	241081	22.1585	70.0714
442	195364	21.0238	66.4831	492	242064	22.1811	70.1427
443	196249	21.0476	66.5582	493	243049	22.2036	70.2140
444	197136	21.0713	66.6333	494	244036	22.2261	70.2851
445	198025	21.0950	66.7083	495	245025	22.2486	70.3562
446	198916	21.1187	66.7832	496	246016	22.2711	70.4273
447	199809	21.1424	66.8581	497	247009	22.2935	70.4982
448	200704	21.1660	66.9328	498	248004	22.3159	70.5691
449	201601	21.1896	67.0075	499	249001	22.3383	70.6399

TABLE Z (*continued*)

N	N^2	$\sqrt{N}$	$\sqrt{10\ N}$	N	N^2	$\sqrt{N}$	$\sqrt{10\ N}$
500	250000	22.3607	70.7107	550	302500	23.4521	74.1620
501	251001	22.3830	70.7814	551	303601	23.4734	74.2294
502	252004	22.4054	70.8520	552	304704	23.4947	74.2967
503	253009	22.4277	70.9225	553	305809	23.5160	74.3640
504	254016	22.4499	70.9930	554	306916	23.5372	74.4312
505	255025	22.4722	71.0634	555	308025	23.5584	74.4983
506	256036	22.4944	71.1337	556	309136	23.5797	74.5654
507	257049	22.5167	71.2039	557	310249	23.6008	74.6324
508	258064	22.5389	71.2741	558	311364	23.6220	74.6994
509	259081	22.5610	71.3442	559	312481	23.6432	74.7663
510	260100	22.5832	71.4143	560	313600	23.6643	74.8331
511	261121	22.6053	71.4843	561	314721	23.6854	74.8999
512	262144	22.6274	71.5542	562	315844	23.7065	74.9667
513	263169	22.6495	71.6240	563	316969	23.7276	75.0333
514	264196	22.6716	71.6938	564	318096	23.7487	75.0999
515	265225	22.6936	71.7635	565	319225	23.7697	75.1665
516	266256	22.7156	71.8331	566	320356	23.7908	75.2330
517	267289	22.7376	71.9027	567	321489	23.8118	75.2994
518	268324	22.7596	71.9722	568	322624	23.8328	75.3658
519	269361	22.7816	72.0417	569	323761	23.8537	75.4321
520	270400	22.8035	72.1110	570	324900	23.8747	75.4983
521	271441	22.8254	72.1803	571	326041	23.8956	75.5645
522	272484	22.8473	72.2496	572	327184	23.9165	75.6307
523	273529	22.8692	72.3187	573	328329	23.9374	75.6968
524	274576	22.8910	72.3878	574	329476	23.9583	75.7628

N	N^2	$\sqrt{N}$	$\sqrt{10N}$	N	N^2	$\sqrt{N}$	$\sqrt{10N}$
525	275625	22.9129	72.4569	575	330625	23.9792	75.8288
526	276676	22.9347	72.5259	576	331776	24.0000	75.8947
527	277729	22.9565	72.5948	577	332929	24.0208	75.9605
528	278784	22.9783	72.6636	578	334084	24.0416	76.0263
529	279841	23.0000	72.7324	579	335241	24.0624	76.0920
530	280900	23.0217	72.8011	580	336400	24.0832	76.1577
531	281961	23.0434	72.8697	581	337561	24.1039	76.2234
532	283024	23.0651	72.9383	582	338724	24.1247	76.2889
533	284089	23.0868	73.0068	583	339889	24.1454	76.3544
534	285156	23.1084	73.0753	584	341056	24.1661	76.4199
535	286225	23.1301	73.1437	585	342225	24.1868	76.4853
536	287296	23.1517	73.2120	586	343396	24.2074	76.5506
537	288369	23.1733	73.2803	587	344569	24.2281	76.6159
538	289444	23.1948	73.3485	588	345744	24.2487	76.6812
539	290521	23.2164	73.4166	589	346921	24.2693	76.7463
540	291600	23.2379	73.4847	590	348100	24.2899	76.8115
541	292681	23.2594	73.5527	591	349281	24.3105	76.8765
542	293764	23.2809	73.6206	592	350464	24.3311	76.9415
543	294849	23.3024	73.6885	593	351649	24.3516	77.0065
544	295936	23.3238	73.7564	594	352836	24.3721	77.0714
545	297025	23.3452	73.8241	595	354025	24.3926	77.1362
546	298116	23.3666	73.8918	596	355216	24.4131	77.2010
547	299209	23.3880	73.9594	597	356409	24.4336	77.2658
548	300304	23.4094	74.0270	598	357604	24.4540	77.3305
549	301401	23.4307	74.0945	599	358801	24.4745	77.3951

TABLE Z (*continued*)

N	N^2	$\sqrt{N}$	$\sqrt{10\,N}$	N	N^2	$\sqrt{N}$	$\sqrt{10\,N}$
600	360000	24.4949	77.4597	650	422500	25.4951	80.6226
601	361201	24.5153	77.5242	651	423801	25.5147	80.6846
602	362404	24.5357	77.5887	652	425104	25.5343	80.7465
603	363609	24.5561	77.6531	653	426409	25.5539	80.8084
604	364816	24.5764	77.7174	654	427716	25.5734	80.8703
605	366025	24.5967	77.7817	655	429025	25.5930	80.9321
606	367236	24.6171	77.8460	656	430336	25.6125	80.9938
607	368449	24.6374	77.9102	657	431649	25.6320	81.0555
608	369664	24.6577	77.9744	658	432964	25.6515	81.1172
609	370881	24.6779	78.0385	659	434281	25.6710	81.1788
610	372100	24.6982	78.1025	660	435600	25.6905	81.2404
611	373321	24.7184	78.1665	661	436921	25.7099	81.3019
612	374544	24.7386	78.2304	662	438244	25.7294	81.3634
613	375769	24.7588	78.2943	663	439569	25.7488	81.4248
614	376996	24.7790	78.3582	664	440896	25.7682	81.4862
615	378225	24.7992	78.4219	665	442225	25.7876	81.5475
616	379456	24.8193	78.4857	666	443556	25.8070	81.6088
617	380689	24.8395	78.5493	667	444889	25.8263	81.6701
618	381924	24.8596	78.6130	668	446224	25.8457	81.7313
619	383161	24.8797	78.6766	669	447561	25.8650	81.7924
620	384400	24.8998	78.7401	670	448900	25.8844	81.8535
621	385641	24.9199	78.8036	671	450241	25.9037	81.9146
622	386884	24.9399	78.8670	672	451584	25.9230	81.9756
623	388129	24.9600	78.9303	673	452929	25.9422	82.0366
624	389376	24.9800	78.9937	674	454276	25.9615	82.0975

N	N^2	$\sqrt{N}$	$\sqrt{10N}$	N	N^2	$\sqrt{N}$	$\sqrt{10N}$
625	390625	25.0000	79.0569	675	455625	25.9808	82.1584
626	391876	25.0200	79.1202	676	456976	26.0000	82.2192
627	393129	25.0400	79.1833	677	458329	26.0192	82.2800
628	394384	25.0599	79.2465	678	459684	26.0384	82.3408
629	395641	25.0799	79.3095	679	461041	26.0576	82.4015
630	396900	25.0998	79.3725	680	462400	26.0768	82.4621
631	398161	25.1197	79.4355	681	463761	26.0960	82.5227
632	399424	25.1396	79.4984	682	465124	26.1151	82.5833
633	400689	25.1595	79.5613	683	466489	26.1343	82.6438
634	401956	25.1794	79.6241	684	467856	26.1534	82.7043
635	403225	25.1992	79.6869	685	469225	26.1725	82.7647
636	404496	25.2190	79.7496	686	470596	26.1916	82.8251
637	405769	25.2389	79.8123	687	471969	26.2107	82.8855
638	407044	25.2587	79.8749	688	473344	26.2298	82.9458
639	408321	25.2784	79.9375	689	474721	26.2488	83.0060
640	409600	25.2982	80.0000	690	476100	26.2679	83.0662
641	410881	25.3180	80.0625	691	477481	26.2869	83.1264
642	412164	25.3377	80.1249	692	478864	26.3059	83.1865
643	413449	25.3574	80.1873	693	480249	26.3249	83.2466
644	414736	25.3772	80.2496	694	481636	26.3439	83.3067
645	416025	25.3969	80.3119	695	483025	26.3629	83.3667
646	417316	25.4165	80.3741	696	484416	26.3818	83.4266
647	418609	25.4362	80.4363	697	485809	26.4008	83.4865
648	419904	25.4558	80.4984	698	487204	26.4197	83.5464
649	421201	25.4755	80.5605	699	488601	26.4386	83.6062
N	N^2	$\sqrt{N}$	$\sqrt{10N}$	N	N^2	$\sqrt{N}$	$\sqrt{10N}$

TABLE Z (*continued*)

N	N^2	$\sqrt{N}$	$\sqrt{10\,N}$	N	N^2	$\sqrt{N}$	$\sqrt{10\,N}$
700	490000	26.4575	83.6660	750	562500	27.3861	86.6025
701	491401	26.4764	83.7257	751	564001	27.4044	86.6603
702	492804	26.4953	83.7854	752	565504	27.4226	86.7179
703	494209	26.5141	83.8451	753	567009	27.4408	86.7756
704	495616	26.5330	83.9047	754	568516	27.4591	86.8332
705	497025	26.5518	83.9643	755	570025	27.4773	86.8907
706	498436	26.5707	84.0238	756	571536	27.4955	86.9483
707	499849	26.5895	84.0833	757	573049	27.5136	87.0057
708	501264	26.6083	84.1427	758	574564	27.5318	87.0632
709	502681	26.6271	84.2021	759	576081	27.5500	87.1206
710	504100	26.6458	84.2615	760	577600	27.5681	87.1780
711	505521	26.6646	84.3208	761	579121	27.5862	87.2353
712	506944	26.6833	84.3801	762	580644	27.6043	87.2926
713	508369	26.7021	84.4393	763	582169	27.6225	87.3499
714	509796	26.7208	84.4985	764	583696	27.6405	87.4071
715	511225	26.7395	84.5577	765	585225	27.6586	87.4643
716	512656	26.7582	84.6168	766	586756	27.6767	87.5214
717	514089	26.7769	84.6759	767	588289	27.6948	87.5785
718	515524	26.7955	84.7349	768	589824	27.7128	87.6356
719	516961	26.8142	84.7939	769	591361	27.7308	87.6926
720	518400	26.8328	84.8528	770	592900	27.7489	87.7496
721	519841	26.8514	84.9117	771	594441	27.7669	87.8066
722	521284	26.8701	84.9706	772	595984	27.7849	87.8635
723	522729	26.8887	85.0294	773	597529	27.8029	87.9204
724	524176	26.9072	85.0882	774	599076	27.8209	87.9773

N	N^2	$\sqrt{N}$	$\sqrt{10N}$	N	N^2	$\sqrt{N}$	$\sqrt{10N}$
725	525625	26.9258	85.1469	775	600625	27.8388	88.0341
726	527076	26.9444	85.2056	776	602176	27.8568	88.0909
727	528529	26.9629	85.2643	777	603729	27.8747	88.1476
728	529984	26.9815	85.3229	778	605284	27.8927	88.2043
729	531441	27.0000	85.3815	779	606841	27.9106	88.2610
730	532900	27.0185	85.4400	780	608400	27.9285	88.3176
731	534361	27.0370	85.4985	781	609961	27.9464	88.3742
732	535824	27.0555	85.5570	782	611524	27.9643	88.4308
733	537289	27.0740	85.6154	783	613089	27.9821	88.4873
734	538756	27.0924	85.6738	784	614656	28.0000	88.5438
735	540225	27.1109	85.7321	785	616225	28.0179	88.6002
736	541696	27.1293	85.7904	786	617796	28.0357	88.6566
737	543169	27.1477	85.8487	787	619369	28.0535	88.7130
738	544644	27.1662	85.9069	788	620944	28.0713	88.7694
739	546121	27.1846	85.9651	789	622521	28.0891	88.8257
740	547600	27.2029	86.0233	790	624100	28.1069	88.8819
741	549081	27.2213	86.0814	791	625681	28.1247	88.9382
742	550564	27.2397	86.1394	792	627264	28.1425	88.9944
743	552049	27.2580	86.1974	793	628849	28.1603	89.0505
744	553536	27.2764	86.2554	794	630436	28.1780	89.1067
745	555025	27.2947	86.3134	795	632025	28.1957	89.1628
746	556516	27.3130	86.3713	796	633616	28.2135	89.2188
747	558009	27.3313	86.4292	797	635209	28.2312	89.2749
748	559504	27.3496	86.4870	798	636804	28.2489	89.3308
749	561001	27.3679	86.5448	799	638401	28.2666	89.3868
N	N^2	$\sqrt{N}$	$\sqrt{10N}$	N	N^2	$\sqrt{N}$	$\sqrt{10N}$

TABLE Z (*continued*)

N	N^2	$\sqrt{N}$	$\sqrt{10\,N}$	N	N^2	$\sqrt{N}$	$\sqrt{10\,N}$
800	640000	28.2843	89.4427	850	722500	29.1548	92.1954
801	641601	28.3019	89.4986	851	724201	29.1719	92.2497
802	643204	28.3196	89.5545	852	725904	29.1890	92.3038
803	644809	28.3373	89.6103	853	727609	29.2062	92.3580
804	646416	28.3549	89.6660	854	729316	29.2233	92.4121
805	648025	28.3725	89.7218	855	731025	29.2404	92.4662
806	649636	28.3901	89.7775	856	732736	29.2575	92.5203
807	651249	28.4077	89.8332	857	734449	29.2746	92.5743
808	652864	28.4253	89.8888	858	736164	29.2916	92.6283
809	654481	28.4429	89.9444	859	737881	29.3087	92.6823
810	656100	28.4605	90.0000	860	739600	29.3258	92.7362
811	657721	28.4781	90.0555	861	741321	29.3428	92.7901
812	659344	28.4956	90.1110	862	743044	29.3598	92.8440
813	660969	28.5132	90.1665	863	744769	29.3769	92.8978
814	662596	28.5307	90.2219	864	746496	29.3939	92.9516
815	664225	28.5482	90.2774	865	748225	29.4109	93.0054
816	665856	28.5657	90.3327	866	749956	29.4279	93.0591
817	667489	28.5832	90.3881	867	751689	29.4449	93.1128
818	669124	28.6007	90.4434	868	753424	29.4618	93.1665
819	670761	28.6182	90.4986	869	755161	29.4788	93.2202
820	672400	28.6356	90.5539	870	756900	29.4958	93.2738
821	674041	28.6531	90.6091	871	758641	29.5127	93.3274
822	675684	28.6705	90.6642	872	760384	29.5296	93.3809
823	677329	28.6880	90.7193	873	762129	29.5466	93.4345
824	678976	28.7054	90.7744	874	763876	29.5635	93.4880

N	N^2	$\sqrt{N}$	$\sqrt{10N}$	N	N^2	$\sqrt{N}$	$\sqrt{10N}$
825	680625	28.7228	90.8295	875	765625	29.5804	93.5414
826	682276	28.7402	90.8845	876	767376	29.5973	93.5949
827	683929	28.7576	90.9395	877	769129	29.6142	93.6483
828	685584	28.7750	90.9945	878	770884	29.6311	93.7017
829	687241	28.7924	91.0494	879	772641	29.6479	93.7550
830	688900	28.8097	91.1043	880	774400	29.6648	93.8083
831	690561	28.8271	91.1592	881	776161	29.6816	93.8616
832	692224	28.8444	91.2140	882	777924	29.6985	93.9149
833	693889	28.8617	91.2688	883	779689	29.7153	93.9681
834	695556	28.8791	91.3236	884	781456	29.7321	94.0213
835	697225	28.8964	91.3783	885	783225	29.7489	94.0744
836	698896	28.9137	91.4330	886	784996	29.7658	94.1276
837	700569	28.9310	91.4877	887	786769	29.7825	94.1807
838	702244	28.9482	91.5423	888	788544	29.7993	94.2338
839	703921	28.9655	91.5969	889	790321	29.8161	94.2868
840	705600	28.9828	91.6515	890	792100	29.8329	94.3398
841	707281	29.0000	91.7061	891	793881	29.8496	94.3928
842	708964	29.0172	91.7606	892	795664	29.8664	94.4458
843	710649	29.0345	91.8150	893	797449	29.8831	94.4987
844	712336	29.0517	91.8695	894	799236	29.8998	94.5516
845	714025	29.0689	91.9239	895	801025	29.9166	94.6044
846	715716	29.0861	91.9783	896	802816	29.9333	94.6573
847	717409	29.1033	92.0326	897	804609	29.9500	94.7101
848	719104	29.1204	92.0869	898	806404	29.9666	94.7629
849	720801	29.1376	92.1412	899	808201	29.9833	94.8156

TABLE Z (*continued*)

N	N^2	$\sqrt{N}$	$\sqrt{10\,N}$	N	N^2	$\sqrt{N}$	$\sqrt{10\,N}$
900	810000	30.0000	94.8683	950	902500	30.8221	97.4679
901	811801	30.0167	94.9210	951	904401	30.8383	97.5192
902	813604	30.0333	94.9737	952	906304	30.8545	97.5705
903	815409	30.0500	95.0263	953	908209	30.8707	97.6217
904	817216	30.0666	95.0789	954	910116	30.8869	97.6729
905	819025	30.0832	95.1315	955	912025	30.9031	97.7241
906	820836	30.0998	95.1840	956	913936	30.9192	97.7753
907	822649	30.1164	95.2365	957	915849	30.9354	97.8264
908	824464	30.1330	95.2890	958	917764	30.9516	97.8775
909	826281	30.1496	95.3415	959	919681	30.9677	97.9285
910	828100	30.1662	95.3939	960	921600	30.9839	97.9796
911	829921	30.1828	95.4463	961	923521	31.0000	98.0306
912	831744	30.1993	95.4987	962	925444	31.0161	98.0816
913	833569	30.2159	95.5510	963	927369	31.0322	98.1326
914	835396	30.2324	95.6033	964	929296	31.0483	98.1835
915	837225	30.2490	95.6556	965	931225	31.0644	98.2344
916	839056	30.2655	95.7079	966	933156	31.0805	98.2853
917	840889	30.2820	95.7601	967	935089	31.0966	98.3362
918	842724	30.2985	95.8123	968	937024	31.1127	98.3870
919	844561	30.3150	95.8645	969	938961	31.1288	98.4378
920	846400	30.3315	95.9166	970	940900	31.1448	98.4886
921	848241	30.3480	95.9687	971	942841	31.1609	98.5393
922	850084	30.3645	96.0208	972	944784	31.1769	98.5901
923	851929	30.3809	96.0729	973	946729	31.1929	98.6408
924	853776	30.3974	96.1249	974	948676	31.2090	98.6914

N	N^2	$\sqrt{N}$	$\sqrt{10N}$	N	N^2	$\sqrt{N}$	$\sqrt{10N}$
925	855625	30.4138	96.1769	975	950625	31.2250	98.7421
926	857476	30.4302	96.2289	976	952576	31.2410	98.7927
927	859329	30.4467	96.2808	977	954529	31.2570	98.8433
928	861184	30.4631	96.3328	978	956484	31.2730	98.8939
929	863041	30.4795	96.3846	979	958441	31.2890	98.9444
930	864900	30.4959	96.4365	980	960400	31.3050	98.9949
931	866761	30.5123	96.4883	981	962361	31.3209	99.0454
932	868624	30.5287	96.5401	982	964324	31.3369	99.0959
933	870489	30.5450	96.5919	983	966289	31.3528	99.1464
934	872356	30.5614	96.6437	984	968256	31.3688	99.1968
935	874225	30.5778	96.6954	985	970225	31.3847	99.2472
936	876096	30.5941	96.7471	986	972196	31.4006	99.2975
937	877969	30.6105	96.7988	987	974169	31.4166	99.3479
938	879844	30.6268	96.8504	988	976144	31.4325	99.3982
939	881721	30.6431	96.9020	989	978121	31.4484	99.4485
940	883600	30.6594	96.9536	990	980100	31.4643	99.4987
941	885481	30.6757	97.0052	991	982081	31.4802	99.5490
942	887364	30.6920	97.0567	992	984064	31.4960	99.5992
943	889249	30.7083	97.1082	993	986049	31.5119	99.6494
944	891136	30.7246	97.1597	994	988036	31.5278	99.6995
945	893025	30.7409	97.2111	995	990025	31.5436	99.7497
946	894916	30.7571	97.2625	996	992016	31.5595	99.7998
947	896809	30.7734	97.3139	997	994009	31.5753	99.8499
948	898704	30.7896	97.3653	998	996004	31.5911	99.8999
949	900601	30.8058	97.4166	999	998001	31.6070	99.9500

INDEX

Finite Markov Chains
By **J. G. Kemeny** and **J. L. Snell**
1976. ix, 210p. cloth
(Undergraduate Texts in Mathematics)

Here is a comprehensive treatment of the theory of finite Markov chains, suitable as a text for an undergraduate probability course as well as a reference book for those in fields outside mathematics. Two unique features of this work are its complete avoidance of the theory of eigen-values, and the fact that the elementary matrix operations discussed can easily be programmed for a high-speed computer.

Elementary Probability Theory with Stochastic Processes
By **K. L. Chung**
1975. x, 325p. 36 illus. 1 table. cloth
(Undergraduate Texts in Mathematics)

"Every student of probability and statistics should have this book in his library for reference. Advanced juniors and seniors may find [it] so stimulating and motivating that they might wish to pursue graduate studies in the subject area."

The Mathematics Teacher

" . . . an extremely well-conceived and executed text. The material chosen would provide the reader enough background to read most post-calculus probability texts . . . many exercises and examples . . ."

The American Mathematical Monthly

Encounter with Mathematics
By **L. Garding**
1977. ix, 270p. 82 illus. cloth

This text provides a historical, scientific and cultural framework for the study of the fundamental mathematics encountered in college. Chapters discuss: number theory; geometry and linear algebra; limiting processes of analysis and topology; differentiation and integration; series and probability; applications; models and reality; the sociology, psychology and teaching of mathematics; and the mathematics of the seventeenth century (for a fuller historical background of infinitesimal calculus.)

1989